Classification of Musculoskeletal Trauma

Classification of Musculoskeletal Trauma

Edited by

P.B. Pynsent PhD
Director, Research and Teaching Centre, Royal Orthopaedic Hospital NHS Trust, Birmingham, UK

J.C.T. Fairbank MD FRCS
Consulting Orthopaedic Surgeon, Nuffield Orthopaedic Centre, Oxford, UK

A.J. Carr ChM FRCS
Consulting Orthopaedic Surgeon, Nuffield Orthopaedic Centre, Oxford, UK

OXFORD AUKLAND BOSTON JOHANNESBURG MELBOURNE NEW DELHI

Butterworth-Heinemann
Linacre House, Jordan Hill, Oxford OX2 8DP
225 Wildwood Avenue, Woburn, MA 01801-2041
A division of Reed Educational and Professional Publishing Ltd

 A member of the Reed Elsevier plc group

First published 1999

British Library Cataloguing in Publication Data
Classification of musculoskeletal trauma
1. Musculoskeletal system – Wounds and injuries
I. Pynsent, P. B. (Paul B.), 1945– II. Fairbank, J. C. T. (Jeremy Charles Thomas), 1948– III. Carr, A. J. (Andrew Jonathan) 1958–
616.7

Library of Congress Cataloguing in Publication Data
Classification of musculoskeletal trauma / edited by P. B. Pynsent, J. C. T. Fairbank, A. J. Carr
p. cm.
Includes bibliographical references and index.
ISBN 0 7506 2722 0
1. Musculoskeletal system – Wounds and injuries – Classification
2. Musculoskeletal system – Diseases – Classification. I. Pynsent, P. B., 1945–. II. Fairbank, J. C. T., 1948–. III. Carr, A., 1958–
[DNLM: 1. Musculoskeletal Diseases – classification.
2. Musculoskeletal System – injuries. WE 15C614]
RD680.C53 98–30092
617.4′7044–dc21 CIP

ISBN 0 7506 2722 0

Printed and bound in Great Britain by the Alden Group, Oxford

Contents

Preface

Classification is an essential aid in the management of injuries, the assessment of outcomes and in clinical research. It provides a mirror and a historical record of our clinical thinking (or lack of it!). If a classification is widely used it usually means that management is clear-cut. When many systems are available for a particular injury, it is likely that there is confusion and controversy over management. This book brings together all the musculoskeletal trauma classification systems that we have been able to find. We have, as far as possible, sought out the original version, often submerged in chapters of unavailable textbooks or obscure journals. Many have suffered at the hands of subsequent interpreters and adjusters. Some of these modifications have been helpful, and we have documented these. If you know of a system which we have missed out, please tell us. We would like to have citations as well as publications to substantiate the value of the classification.

We recommend systems which are appropriate for clinical practice, audit and research. Often a single system cannot satisfy all three of these facets of clinical practice adequately. We found, time and time again, that even widely used systems are unreliable. Much more work needs to be done in establishing the value of all these systems in prognosis.

The AO group have evolved the only systematic classification of fractures. Inevitably their systems occur widely throughout the book. Their approach has been to encompass a totality of fracture patterns into triads. This seems to work better in long bone fractures than in those involving joints. Problems with reliability are reported and in some anatomical areas have been found wanting. Soft-tissue injuries are not generally addressed.

This volume appears as a sequel to our text books on outcome measures and methodology, Pynsent et al. (1993, 1994 and 1997). As the later volumes developed it became clear to us that the concept of severity was an essential component to use with instruments of outcome for auditing practice in orthopaedic surgery and trauma. The ideal classification should provide a reliable and valid description of the injury. Usually reliability is directly related to simplicity. In addition, it should provide an indication of the treatment and prognosis.

The question of the quality and value of classification systems has been raised by Bernstein et al. (1997), who have suggested a 'classification of classifications'. We think this is laudable but unworkable. They criticize the AO system on similar grounds to ours. Colton's (1997) reaction to this article is likely to be generally supported. He feels that although the latter classification is widely used, this is not necessarily analogous to being widely useful. There is still some way to go before there is clarity of thinking about this issue.

In the compilation of this book, it has been observed that, in some cases, classifications have emphasized the morphological characteristics of fractures, whilst others have put more emphasis on classification with respect to treatment. The reliability (that is the inter- and intra-rater error) of many classifications has not been measured.

Some classifications have not been able to withstand changes in investigative technology (e.g. MRI) and changing treatments (e.g. external fixation). For example, most surgeons would now use MRI to assess spinal cord injury but classifications based on this technique are sparse (see Chapter 9).

As far as possible the line drawings have been copied from the original work. The standard of artwork in the original publications varies considerably. In some cases this has contributed to the confusion attached to the use of a classified system. Where paired structures are involved in a drawing, all illustrations have been produced for a right-sided structure.

This book was compiled in a similar way to its predecessors in that the authors and their chapters were subjected to discussion at a meeting on the Dahlem principle (Dixon, 1987). These special interests groups discussed the relevant chapters. Those who helped organize and chair this meeting (along with the editors) are mentioned in the acknowledgement that follows.

References

Bernstein J., Monaghan, B., Silber, J. and DeLong, W. (1997) Taxonomy and treatment – a classification of fracture classifications. *J. Bone Joint Surg.*, **79B**, 706–707.

Colton, C. (1997) Fracture classification. *J. Bone Joint Surg.*, **79B**, 708–709.

Dixon, B. (1987) Scientifically speaking. *Br. Med. J.*, **294**, 1424.

Pynsent, P., Fairbank, J. and Carr, A. (1993) *Outcome Measures in Orthopaedics*. Oxford: Butterworth-Heinemann.

Pynsent, P., Fairbank, J. and Carr, A. (1994) *Outcome Measures in Trauma*. Oxford: Butterworth-Heinemann.

Pynsent, P., Fairbank, J. and Carr, A. (1997) *Assessment Methodologies in Orthopaedics*. Oxford: Butterworth-Heinemann.

Contributors

B.L. Atkins MBBS MA MRCP MSc DipRCPath
Department of Microbiology, John Radcliffe Hospital, Oxford, UK

Anthony R. Berendt BM BCh MRCP
Consulting Physician, Bone Infection Unit, The Nuffield Orthopaedic Centre, Oxford, UK

N. Blewitt MB ChB FRCS FRCS(Orth)
Consulting Orthopaedic Surgeon, Frenchay Hospital, Bristol, UK

Gavin B. Bowyer MA MChir FRCS(Orth)
Consulting Trauma and Orthopaedic Surgeon, The Royal Hospital Haslar, Gosport, UK

Mark F. Brown MA PhD FRCS(Ed)(Orth)
Consulting Orthopaedic Spinal Surgeon, North Staffs Hospital, Stoke-on-Trent, UK

Jai Chitnavis MA FRCS FRCS(Ed)
Orthopaedic Registrar, The Nuffield Orthopaedic Centre, Oxford, UK

Sanjiv Chugh FRCS(Orth)
Consulting Orthopaedic Surgeon, Newcross Hospital, Wolverhampton, UK

Deborah M. Eastwood FRCS
Consulting Orthopaedic Surgeon, The Royal Free Hospital and The Royal National Orthopaedic Hospital, London, UK

C.L.M.H. Gibbons MA FRCS
Consulting Orthopaedic Surgeon, The Nuffield Orthopaedic Centre, Oxford, UK

Marcus A. Green MB ChB BSCI FRCS(Tr) FRCS (Orth)
Specialist Orthopaedic Registrar, Royal Orthopaedic Hospital, Birmingham, UK

L. Hamilton
Senior House Officer, The Royal Free Hospital, London, UK

Amir W.B. Hanna MS (Orthopaedic Surgery)
Orthopaedic Research Fellow, The Royal Orthopaedic Hospital, Birmingham, UK

H. Hegde MBBS MS(Orth)
Bhailalbai General Hospital, Alembic Road, Gujurut, India

Timothy E.J. Hems BM BCh MA DM FRCS FRCS(Ed)(Orth)
Consulting Hand and Orthopaedic Surgeon, The Victoria Infirmary, Glasgow, UK

Stephen J. Krikler BSc PhD FRCS(Orth)
Consulting Orthopaedic Surgeon, Coventry and Warwickshire Hospital, Coventry, UK

Jeremy M. Latham MA FRCS(Orth)
Consulting Orthopaedic Surgeon, Southampton General Hospital, Southampton, UK

Raymond Y.L. Liow MBBS
Specialist Registrar in Orthopaedics, Freeman Hospital, Newcastle-upon-Tyne, Newcastle, UK

Martyn J. Parker FRCS MS
Orthopaedic Research Fellow, Peterborough District Hospital, Peterborough, UK

Daniel E. Porter BSc MB ChB MD FRCS(Ed) FRCS(Glas)
Clinical Lecturer, Royal Berkshire Hospital, Reading, UK

Simon L. Royston FRCS
Specialist Registrar in Orthopaedic Surgery, Northern General Hospital, Sheffield, UK

Godwin V. Scerri LRCP MRCS FRCS(Eng) FRCS(Plast) RAF
RAF Consulting Plastic Surgeon, Royal Hospital Haslar, Gosport, UK

Pradipkumar Sett MBBS MS FRCS FRCS(SN)
Clinical Director, Regional Spinal Injuries Centre, Southport Formby NHS Trust, Southport, UK

Matthew Scott-Young MBBS OID FRACS FA(Ortho)
Consulting Orthopaedic Surgeon, AHC House, Benowa, Queensland, Australia

John R. Williams MA DM FRCS(Orth)
Senior Lecturer and Honorary Consulting Orthopaedic Surgeon, Department of Trauma and Orthopaedic Surgery, The Medical School, Newcastle-upon-Tyne, UK

Acknowledgements

The editors are grateful to the following who helped chair the discussion groups: Messrs Andrew Macey FRCS, Jeremy Plewes FRCS and Professor David Marsh FRCS. We gratefully acknowledge Biomet Ltd and Smith and Nephew Surgical Products Ltd for their generous sponsorship of this meeting. Thanks are also due to Melanie Tait and Myriam Brearley of Butterworth-Heinemann for their support and enthusiasm for the production of this volume with over 1000 re-drawn figures. Once again we are immensely grateful to Ann Weaver for her help in organizing the meeting and collating and editing the many versions of the typescript.

1

General classifications

G. W. Bowyer

Introduction

'General' Classifications in Orthopaedic Trauma

Most classification systems in common use in musculoskeletal trauma concern themselves with specific tissues (e.g. the bones or soft tissues) or a localized anatomical region (e.g. the spine, the pelvis etc.) and are the subject of the remainder of this book. There are, however, some other classifications which deal with injury from a different perspective, which may be of importance to those involved in the care of musculoskeletal trauma.

There are three broad areas which might be considered under the heading of 'general classifications':

- Classifications dealing with the assessment of injury severity and the generalized physiological effects of trauma. These include the anatomical classifications such as the *Abbreviated Injury Scale (AIS)* and *Anatomic Profile (AP)*, and derived measures such as the *Injury Severity Score (ISS)*. Physiological measures include the *Trauma Score (TS)*, and *Revised Trauma Score (RTS)*. The combination of anatomical and physiological scores is involved in the *TRISS* methodology, and similar methods for assessing the severity of trauma, such as *A Severity Characterization of Trauma (ASCOT)*, and the *Pediatric Trauma Score (PTS)*. There is also a recognition that pre-existing medical conditions may have a bearing on prognosis, and this is accounted for in combination with physiological factors in the *Acute Physiology and Chronic Health Evaluation (APACHE)*, used in the intensive care setting.
- Scaling of injury to the entire limb. This is exemplified by the *Mangled Extremity Severity Score (MESS)*.
- Classifications of trauma from a specific mechanism of injury. The study and classification of vehicular injury gave rise to the *Abbreviated Injury Scale (AIS)*. Experience of war surgery, particularly involving penetrating ballistic wounds, has been the stimulus for the development of the *Red Cross Wound Score*.

Each of these classifications, and related scoring systems, will be considered in detail in this chapter.

The Potential Role of Classification Systems

When considering the utility and value of any classification system, it is essential to keep in mind the purposes for which that system was designed. There are three major functions which a classification or scaling system might fulfil:

- To facilitate communication. To allow valid comparison between treatments or outcomes it is essential that the case-mix is identified and characterized; a classification system might be intended to permit this.
- To allow prognostication. If an outcome probability for a case is to be predicted then it is important to know the results from similar such cases, and to match the case to a known and characterized series, using a consistent classification.
- To guide treatment. Arising from the previous two functions, it may be clear that a particular pattern of injury carries particular risks, or responds to certain treatments in a predictable manner. A properly defined classification system is essential if such cases are to be identified.

It will quickly be recognized that the classification systems in common usage in orthopaedic trauma have not all been designed to meet every potential rôle. If a classification system is to be fairly assessed and appropriately used then it is important to avoid what might be termed 'usage-creep'.

Assessing the Validity of a Classification System

In assessing the validity and value of a classification system there are three key factors which should be borne in mind:

- Is the system *usable*?
- Is it *useful*?
- Is the system actually *used* by workers other than those who originally defined it?

Considering each of these factors in turn:

Is the system usable?

In order to fulfil this criterion the classification must be clearly and consistently defined. Definitions should be set out in a publication which is readily available to workers, with no ambiguity as to how the classification is to be applied. The need for any special investigations should be set out in the defining publication. Furthermore it should be clear on the circumstances in which it is appropriate to use the classification.

Is it useful?

Having defined the circumstances in which it is appropriate to use a given classification, it should be apparent what the proposed role or purpose of the classification is. The success of the system in meeting this aim should be tested, and the results published in the medical literature, as testament to its applicability and utility.

Use by other workers

A further test of a classification's worth is whether it is taken up and used by other workers with similar interests to those of the original authors of the classification system. It is the adoption of the system by these investigators, and the subsequent assessment of the usability and usefulness of the classification scheme, which stands to tell if a classification is truly valid and of value.

Classifications of Trauma Severity

Classifications of trauma severity tend to take account of the physiological consequences of the injury, or the anatomy of injury. Numerous scales have been described for use in the quantification of individual injury severity and the characterization of case-mix for a particular institution. Those with an impact on the work of orthopaedic surgeons will be considered here. Many classifications in this category have been concerned with life/death in the early post-injury period as the outcome measure which the scales were designed to predict. Alternatively, they have arisen out of an attempt to define criteria for triage of casualties, i.e. deciding whether or not they should be taken to a specific trauma centre. Few papers on trauma scoring have examined the effects of trauma over time, or the impact on survivors.

The Glasgow Coma Scale

The Glasgow Coma Scale (GCS) (Teasdale and Jennett, 1974) is based on three variables: eye opening (4 point scale), verbal response (5 point scale) and motor response (6 point scale). The sum of these scores is the GCS, an index used to assess brain damage and to follow patient progress. This scale is summarized in Table 1.1.

Table 1.1 Glasgow Coma Scale (Teasdale and Jennett, 1974)
Scores for best response

E. Eye Opening	
Spontaneous	4
To voice	3
To pain	2
None	1
V. Verbal Response	
Oriented	5
Confused	4
Inappropriate words	3
Incomprehensible words/ sounds	2
None	1
M. Motor Response	
Obeys command	6
Purposeful movement (localizes pain)	5
Withdrawal from pain	4
Flexion response to pain	3
Extension response to pain	2
None	1
Total GCS = E + V + M	

Table 1.2 Revised Trauma Score (Champion et al., 1989)

Glasgow Coma Score	*Systolic BP (mmHg)*	*Respiratory Rate minute^{-1}*	*Coded Value*
13–15	> 89	10–29	4
9–12	76–89	> 29	3
6–8	50–75	6–9	2
4–5	1–49	1–5	1
3	0	0	0

Weighting factors are applied to Coded Values
RTS = GCSc × 0.9368 + SBPc × 0.7326 + RRc × 0.2908
Where GCSc, SBPc, RRc represent the coded values

The scale has been in use for more than 20 years, and has been adopted throughout the world. The GCS is simple to apply and well defined (Teasdale and Jennett, 1974); problems may arise in applying the system to small children, and modifications have been proposed to allow for the altered verbal responses (Yates, 1990). The scale has been correlated with outcome in terms of survival and neurological function using the Glasgow Outcome Scale (Jennett, Teasdale and Braakman, 1976). The GCS is an important component of the Trauma Score (TS) and several other scaling systems, including the Revised Trauma Score (RTS) described below.

In the assessment of the spectrum of head injuries presenting to a trauma centre, GCS remains a valued tool (Trunkey et al., 1983), but it is recognized that further factors, beyond GCS, are important in establishing the prognosis of the more severe head injuries, when the casualty is in a coma (GCS < 8) (Levati et al., 1982). Of the three responses scored in the GCS the motor pattern is the best prognostic indicator, and motor responses have been shown to correlate well with mortality in the more severely head injured casualty (Gennarelli, Champion and Sacco, 1989).

Trauma Score

The Trauma Score (TS) (Champion et al., 1981) is a physiological index of injury severity. It is based on the Glasgow Coma Scale (GCS) and indicators of cardiovascular status (systolic blood pressure and capillary return) and respiratory status (rate and effort). These five variables are weighted and added to produce the TS. Scores range from 1 (worst prognosis) through to 16 (best prognosis).

The variables in the TS are well defined (Champion, Sacco and Carnazzo, 1981), and familiar to staff working in the pre-hospital as well as hospital setting. It has been shown to have a high degree of inter-rater reliability (Moreau et al., 1985), and has been extensively used in characterizing the physiological status of traumatized patients. The TS has been used extensively for triage purposes in the field (Champion, Sacco and Hunt, 1983), and its results have been shown to correlate well with survival in both blunt (Champion, Sacco and Carnazzo, 1981) and penetrating (Champion and Sacco, 1984) trauma.

Revised Trauma Score

The Revised Trauma Score (RTS) (Champion et al., 1989) arose from an attempt to improve the useability of the TS, particularly in the field and at night. In addition, it had been recognized that TS underestimated the severity of some head injured patients. The capillary refill and respiratory effort variables were dropped from the TS; the remaining variables are GCS, respiratory rate and systolic blood pressure. Two versions of the revision have been described, for triage (T-RTS), and for outcome studies and case-mix control (RTS). The variables in RTS are divided into intervals and a coded value assigned, as shown in Table 1.2.

Coded values less than the norm (< 4) are considered an indication for care in a trauma centre, or may be used to indicate that a nominated 'trauma team' be called to treat the patient (Gilpin and Nelson, 1991). In order to evaluate in-patient outcomes, the coded values are multiplied by coefficients to weight the variables, improving the predictive accuracy of the RTS over the TS, particularly where there is a serious head injury (Champion et al., 1989).

The RTS has been incorporated in mathematical models designed to predict outcome, most notably TRISS, described below, and as such the RTS has been applied across a large trauma database.

Table 1.3 Circulation, Respiration, Abdomen, Motor, Speech Scale (Gormican, 1982)

Circulation	*Respiration*	*Abdomen*	*Motor*	*Speech*	*Score*
Normal capillary refill and BP > 100	Normal	Abdomen and thorax non-tender	Normal	Normal	2
Delayed capillary refill or 85 < BP < 100	Laboured or slow	Abdomen or thorax tender	Responds only to pain (but not decerebrate)	Confused	1
No capillary refill or BP≪85	Absent	Abdomen rigid or chest flail	No response, or decerebrate	No intelligible words	0

CRAMS score = sum of codes values
(BP in mmHg)

The Circulation, Respiration, Abdomen, Motor, Speech Scale

The Circulation, Respiration, Abdomen, Motor, Speech (CRAMS) scale (Gormican, 1982) also arose from an attempt to simplify the TS. It omits eye opening, but retains the other two elements of the GCS whilst it simplifies their scoring. An attempt is made to indicate the presence of thoraco-abdominal injury. Hence this scale incorporates indicators of injury or derangement in the major systems affecting mortality. This is outlined in Table 1.3.
The CRAMS scale was intended for use in field triage, indicating whether a casualty ought to be transported to a centre capable of dealing with major trauma (CRAMS > 8).

The rather loosely defined thoraco-abdominal element has not been subjected to investigation as to its reproducibility or validity, but several North American paramedic and emergency services have taken up the scale for use in field triage.

Abbreviated Injury Scale

The Abbreviated Injury Scale (AIS) comprises a list of injuries which are scored according to their severity from 1 (minor) to 6 (usually fatal). Originally drawn up to characterize blunt injuries in vehicular accidents, by the American Association for Automotive Medicine (AAAM), it has undergone several modifications and now includes coding for penetrating trauma (American Association for Automotive Medicine, 1990). The scale provides an indication of injury severity based solely on anatomical location, without consideration of physiological consequences, and on the basis of individual injuries, without taking account of multiple trauma. The scale is ordinal, but not interval — an injury with a value 2 is not half as serious as one with an assigned value of 4.

Injuries are well defined, but need to be looked-up in the manual. The AIS is widely used, and is employed to derive the Injury Severity Score (ISS) and used in TRISS methodology described below. As a component of these systems it has been extensively evaluated (Trunkey et al., 1983).

Injury Severity Score

The Injury Severity Score (ISS) provides a means of scoring multiple trauma (Baker et al., 1974). The ISS considers the injuries in six regions: head and neck; face; thorax; abdomen and pelvic contents; extremities; external. The score is calculated by summing the squares of the highest AIS score in the three most severely injured body regions. The maximum score is 75 ($5^2 + 5^2 + 5^2$); any casualty with an AIS score of 6 in any region (indicating a very high chance of mortality) is automatically given an ISS of 75.

Example of ISS scoring:

A 25-year-old front seat passenger is struck from the side in his car, sustaining mainly left-sided injuries. These include a head injury (CT demonstrates parietal lobe swelling), fractures of three ribs and pulmonary contusion, with liver laceration and ipsilateral femoral and tibial fractures, both closed.
The AIS values, looked up in the manual are as follows, with the scores which count towards ISS underlined:

Head	<u>Parietal lobe swelling</u>	<u>AIS = 3</u>
Chest	<u>Multiple fractured ribs + pneumothorax</u>	<u>AIS = 4</u>
	Pulmonary contusion	AIS = 3
Abdomen	<u>Minor liver laceration, no tissue loss</u>	<u>AIS = 3</u>
Extremities	Femoral fracture (closed, displaced)	AIS = 3
	Tibial fracture (closed displaced)	AIS = 3

$$ISS = 3^2 + 4^2 + 3^2 = 34$$

The quality of definition of ISS matches that of the AIS protocol on which it is based, and is regarded as being sound. The ISS has been shown to correlate with mortality (Baker and O'Neill, 1976), and has been extensively adopted.

There are, however, a number of limitations and pitfalls. Only the worst injury in each region is represented in the ISS: hence a patient with fractures of major long bones in all four limbs has the same score as the patient with a similar fracture pattern affecting just one limb. Furthermore, a given AIS score does not represent the same injury severity in every body region; hence similar scores may actually represent different prognoses for survival (Champion et al., 1990). The summation across regions in ISS may actually hide, rather than reveal, differences in prognosis: a score of ISS = 25 may represent AIS scores of 3 in one region plus 4 in another, or may arise from an AIS score of 5 in just one region.

A further consequence of the way in which the score is calculated is that the scale is discontinuous; some values cannot be obtained, and others tend to occur with a frequency which is not represented by a normal distribution. Hence it is essential to use distribution-free, non-parametric statistics to compare ISS scores between different populations (Yates, 1990); the 'mean ISS score' is an entirely inappropriate term (a fact which many papers in the orthopaedic literature fail to recognize).

The problem of applying ISS to penetrating trauma has been reported (Bellamy and Vayer, 1988), and the failure to account for different prognoses in different age-groups has also been recognized. This may lead to inappropriate conclusions when populations with differing case-mix are compared.

The ISS is of particular importance to surgeons dealing with musculoskeletal trauma, since some papers have described treatment protocols for limb trauma, based on the severity of general trauma as measured by this scale (Seibel et al., 1985). Others have tried to relate complications of femoral nailing in multiple trauma to the AIS score reflecting thoracic trauma (Pape et al., 1993) although some of the pitfalls of doing so were brought out by Burgess in the discussion following that paper.

Despite its short-comings, ISS has become established as the most commonly used summary measure of injury severity in multiple trauma. There is currently activity to improve the deficiencies (Champion, Gennarelli and Moore, 1995). This severity scale is a key component in the TRISS methodology described below, and as such has been thoroughly validated across a number of very large trauma databases.

International Classification of Diseases

The International Classification of Diseases, 9th revision, clinical modification (ICD-9-CM) provided a nomenclature of injury as well as other diagnoses (Commission of Professional Hospital Activities, 1977). The classification has no scalar or ordinal properties, it is purely nominal. However, a mapping of ICD-9-CM rubrics to AIS-85 severity scores has been produced, effective for the majority of trauma diagnoses (MacKenzie, Steinwachs and Shankkar, 1989). This allows scaling of injuries from the format recorded in many ICD-9 based trauma or clinical registries.

The 9th edition of ICD is in the process of being replaced by ICD-10; ICD-9 achieved widespread usage throughout the United States and Europe, but data processed using ICD-10 has yet to be evaluated and reported. A mapping of ICD-10-CM classifications to AIS-90 rubrics to produce severity scores would be a worthwhile undertaking but has not yet been reported.

Anatomic Profile

The Anatomic Profile (AP) was designed in order to overcome some of the perceived limitations of ISS (Champion, Gennarelli and Moore, 1995). The AP forms a summary of all injuries with an AIS > 2, grouping these in three categories: A = head, brain and spinal cord; B = thorax and front of neck; C = remaining serious injuries. The score in each of these categories is the square root of the sum of the squares of each AIS > 2. The 4th category, D, comprises the remaining minor injuries (AIS = 1 or 2) in all regions, and facial injuries. This is summarized in Table 1.4.

The AP is built on the AIS and ICD-9-CM, and shares the quality of definition inherent in those classifications. It has been combined with RTS to produce a measure of injury severity, A Severity Classification of Trauma (ASCOT); as such it has been the subject to recent investigation and validation, described below.

The anatomic profile has been drawn up by a committee of experienced trauma surgeons in the United States. The grouping of injuries in the four categories makes allowance for the greater relative contribution to mortality attributable to head and CNS injuries (Gennarelli, Champion and Sacco, 1989), followed by thoracic trauma and then other major injuries. This gives the scale considerable face validity. There is the potential for adjustment of severity, perhaps to prognosticate for outcomes other than death, by adjusting the weighting of each category by a coefficient.

Table 1.4 Anatomic Profile Components (Champion, Copes and Sacco et al., 1990)

Component	*Injury*	*AIS Severity*	*ISS Region*
A	Head/Brain	3–5	1
	Spinal cord	3–5	1/3/4
B	Thoracic	3–5	3
	Front of neck	3–5	1
C	Abdomen/pelvis	3–5	4
	Spine, not cord	3	1/3/4
	Pelvic fracture	4–5	5
	Femoral artery	4–5	5
	Crush above knee	4–5	5
	Amputation above knee	4–5	5
	Popliteal artery	4	5
D	Face	1–4	2
	All others	1–2	1–6

Severity Scales Derived from Physiological and Anatomical Factors

The severity scales described above have been combined and incorporated in models designed to predict outcome from trauma. These do not themselves represent classifications, but are included here as they are an important way in which the classifications are utilized, particularly in comparison of trauma care between institutions or highlighting cases with an 'unexpected' outcome for audit.

TRISS Methodology

The probability of patient survival may be calculated using the TRISS method (Boyd, Tolson and Copes, 1987). According to this methodology the patient injury severity may be characterized by a physiological measure (RTS), an anatomical severity indicator (ISS), with the patient's age also factored into the equation:

$$b = b_0 + b_1 \cdot RTS + b_2 \cdot ISS + b_3 \cdot Age$$

Probability of survival, P_s, can then be calculated:

$$P_s = \frac{1}{1 + e^{-b}}$$

Example of TRISS Methodology:

Consider the driver struck from the side, as in the example of ISS scaling detailed above. Along with the anatomical injuries described, he arrives in the emergency department with the following vital signs: pulse 145, respiratory rate 35, blood pressure 85/40, Glasgow Coma Scale 7 (E1, V2, M4)

The casualty's RTS is therefore: GCSc = 2 SBPc = 3 RRc = 3 $RTS = (2 \times 0.9368) + (3 \times 0.7326) + (3 \times 0.2908) = 4.9438$

The ISS, previously calculated, is 34

Following TRISS methodology, age (A) = 1 if casualty > 54 years old, 0 if < 54.

Hence RTS = 4.9438, ISS = 34, A = 0

TRISS coefficients depend on whether the trauma is blunt or penetrating, and from which database the figures are derived, using here the current UK figures:

$b = 0.945 + (0.642 \times 4.9438) + (-0.122 \times 34) + (-1.886 \times 0) = -0.0291$

Probability of survival, $P_s = \frac{1}{1+e^{-b}} = 0.493$

It must be borne in mind that this statistical method provides a probability of outcome which should be interpreted on a population, rather than individual basis: a $P_s = 0.5$ means that 50% of patients with such an injury or set of injuries may be expected to survive.

The Major Trauma Outcome Study (MTOS) was originally set up in the United States to refine methods for trauma scoring as well as to provide an objective evaluation of outcome and quality assurance (Sacco et al., 1994). The coefficients b_0, b_1, b_2 and b_3 were derived retrospectively from a large patient database (Champion, Sacco and Copes, 1995), which has subsequently been used to validate the methodology (Smith, Ward and Smith, 1990). The TRISS methodology has also been assessed using data from a non-MTOS database (Hannan et al., 1995) and the UK has also developed an MTOS database (Yates, Woodford and Hollis, 1993). It is not yet clear to what extent coefficients should be altered to allow for a true difference in case-mix across populations or between institutions (Hannan et al., 1995; Hollis et al., 1995);

Table 1.5 TRISS coefficients based on AIS-90 in MTOS controlled sites (USA) (Champion et al., 1995)

	Constant	*RTS*	*ISS*	*Age*
Blunt	−0.4499	0.8085	−0.0835	−1.7430
Penetrating	−2.5330	0.9934	−0.0651	−1.1360

certainly there is a demonstrated need for different coefficients for blunt and penetrating injuries (Champion, Sacco and Copes, 1995). The most recent calibration of the coefficients, from the United States MTOS, produced the values in Table 1.5.

A Severity Classification of Trauma

The authors of the TRISS methodology (Boyd, Tolson and Copes, 1987) have gone on to recommend a new method for calculating probability of survival, A Severity Classification Of Trauma (ASCOT) (Champion et al., 1990). This combines the three factors from RTS, coded as in the RTS: GCS (G), respiratory rate (R) and systolic blood pressure (S). The anatomical severity scoring from AP is included in ASCOT, utilising the A, B and C categories, but the minor and facial injuries (D category) are excluded from ASCOT. Age is encoded as in Table 1.6, which represents a refinement of the age coding of TRISS.

A set of coefficients (k_1 to k_8) has been produced to weight the various ASCOT components:

$$k = k_1 + k_2{\cdot}G + k_3{\cdot}S + k_4{\cdot}R + k_5{\cdot}A + k_6{\cdot}B + k_7{\cdot}C + k_8{\cdot}Age$$

These coefficients are given in Table 1.7. The probability of survival is then calculated using the logistic model, as for TRISS. A summary of ASCOT calculation is given in the example below:

Example of ASCOT calculation:

Consider the same patient as used in the examples above. Coded values for the physiological components of ASCOT are:

GCSc = 2 SBPc = 3 RRc = 3

The anatomical components from AP are:

A = 3 (Swollen parietal lobe)

B = 5 (Ribs/pneumothorax + pulmonary contusion, $\sqrt{4^2 + 3^2}$)

C = 3 (Liver)

D injuries are not used in ASCOT

ASCOT may then be calculated, using the coefficients from Table 1.6:

$-1.1570 + (0.7705 \times 2) + (0.9583 \times 3) + (0.281 \times 3) + (-0.1961 \times 5) + (-0.2086 \times 3) = 0.695$

ASCOT $P_s = 0.667$

Table 1.6 ASCOT patient age characterization (Champion et al., 1990)

Age (years)	*Age Code*
0–54	0
55–54	1
65–74	2
75–84	3
>84	4

As AIS-90 becomes fully implemented, so the ASCOT coefficients will need to be adjusted and the methodology fully validated and compared with TRISS (Markle et al., 1992). It is possible that further adjustment of the coefficients may allow ASCOT to be used in prognostication concerning outcomes other than life/death or ICU stay — for example delay to return to work or functional outcomes, particularly if the D category of ASCOT undergoes further refinement (Champion, Sacco and Copes, 1995).

Pediatric Trauma Score

Until the mid-1980s trauma scores for children tended to be modifications of the corresponding adult score. The Pediatric Trauma Score (PTS) was designed as a means of rapid assessment of the injured child, and as a triage guide (Tepas et al., 1987). It was derived from paediatric trauma data on the University of Florida Pediatric Trauma

Table 1.7 ASCOT model weights (Champion et al., 1994)

Variable	*Blunt*	*Penetrating*
Constant	–1.1570	–1.1350
GCS	0.7705	1.0626
Systolic BP	0.6583	0.3638
Respiratory rate	0.2810	0.3332
Anatomic Profile A	–0.3002	–0.3702
Anatomic Profile B	–0.1961	–0.2053
Anatomic Profile C	–0.2086	–0.3188
Age	–0.6355	–0.8365

Table 1.8 The Pediatric Trauma Score (Tepas et al., 1987)

Component	*Category*		
	+2	*+1*	*–1*
Weight	> 20 kg	10–20 kg	< 10 kg
Airway	Normal	Maintainable	Unmaintainable
Systolic BP	> 90 mm Hg	90–50 mm Hg	< 50 mm Hg
CNS	Awake	Obtunded/LoC	Coma/decerebrate
Open wound	None	Minor	Major/penetrating
Skeletal	None	Closed fracture	Open fracture

BP = Blood pressure, CNS = central nervous system, LoC = loss of consciousness
If proper sized BP cuff not available assign: +2 for pulse palpable at wrist
+1 for palpable at groin
– 1 no pulse palpable
Pediatric Trauma Score is the sum of grades for each of the six components

Registry and the National Pediatric Trauma Registry (NPTR).

Each of six components is graded for life-threatening, moderate or minimal derangement. The components include physiological variables: systolic blood pressure, airway and CNS assessment. Cutaneous and skeletal scores comprize anatomical components, and the size component is intended to identify the very small infant as being at potentially increased risk. The scoring system is set out in Table 1.8.

The triage function of PTS has been evaluated using data input into the NPTR. This showed an inverse relationship between mortality and PTS (Tepas et al., 1988). The score has been validated as sensitive and specific for mortality (Ramenofsky et al., 1988), and as having a very close correlation with ISS (Tepas et al., 1988). The critical point for mortality is a score of 8, and this has been used as an indicator of the need for transfer or referral to a specialized paediatric trauma facility (Aprahamian et al., 1990; Jubelirer et al., 1990).

There are, however, doubts as to whether the PTS offers advantages over the established adult trauma classifications. As a triage tool, the PTS has been shown to offer no advantage over the Revised Trauma Score (RTS) (Eichelberger et al., 1989; Kaufmann et al., 1990) or the Trauma Score (TS) (Nayduch et al., 1991). Use of the adult system, possibly with adjustment for rapid respirations in the pre-school child, obviates the need to learn a separate scoring system. As a predictor of outcome, in terms of mortality, the PTS has been shown to offer no advantage over TRISS (Kaufmann et al., 1991).

Acute Physiology and Chronic Health Evaluation

The Acute Physiology and Chronic Health Evaluation (APACHE) classification comprises physiological variables measured in the first 24 hours of admission to the intensive care unit (ICU) as well as pre-admission conditions, intended to give an indication of severity of illness in the critically ill patient. The system has evolved, starting in the early 1980s with APACHE (Knaus et al., 1981), through the more widely accepted APACHE II (Knaus, Draper and Wagner, 1985), to the more recent version III which includes 16 physiological variables (Knaus et al., 1991). Variables encoded include mean blood pressure and oxygenation, as well as haematological and biochemical factors such as haematocrit, creatinine, albumin, bilirubin and glucose. The normative database for the latest version is much larger than previously (> 17 000 patients) and trauma cases make up a greater percentage than in the earlier versions, although these still comprise a minority of the cases. A study comparing APACHE II and III has suggested that the latter may correct the underestimation of mortality which was sometimes encountered with the previous system, but in this study fewer than 10% of the patients were trauma casualties (Barie, Hydo and Fischer, 1995).

There have been no studies evaluating this system for orthopaedic trauma, and the instrument was not designed for such disease-specific applications. The system takes no account of anatomical injury severity, and this compounds the problem of applying the system to a trauma population.

There is considerable face validity in allowing for pre-existing illness in assessing the likelihood of survival after trauma. There are, however, problems in applying these factors within existing injury classification systems. Pre-injury illness conditions (PICs) are defined in APACHE II and III, and the effect of these factors on survival in an MTOS study group has been evaluated (Sacco et al., 1993). Five PICs were found to have a profound effect on indi-

Table 1.9 The Mangled Extremity Severity Score (MESS) (Johansen et al., 1990)

	Score
Skeletal/soft-tissue injury	
Low energy (stab; simple fracture; civilian gunshot)	1
Medium energy (open or multiple fractures, dislocation)	2
High energy (shotgun; military gunshot; crush)	3
Very high energy (as above plus gross contamination, soft-tissue avulsion)	4
Limb ischaemia	
Pulse reduced or absent, but normal perfusion	1
Pulseless; paraesthesiae; reduced capillary refill	2
Cool, paralysed, insensate, numb	3
(score doubled for ischaemia > 6 hours)	
Shock	
Systolic BP maintained >90 mmHg	0
Transient hypotension	1
Persistent hypotension	2
Age	
< 30 years	0
30–50 years	1
> 50 years	2

vidual patient outcome: diabetes, hepatic, cardiovascular, respiratory, and renal diseases. However, the relatively low incidence of these conditions in trauma patients meant that PICs did not greatly influence the institutional outcome performance in this study.

Scaling of Limb Injury Severity — Mangled Extremity Severity Scores

The ability to provide soft tissue coverage for the severely injured limb led, in the 1980s, to an increase in the number of severely injured limbs which could, technically, be salvaged. This was particularly true for tibial fractures with soft tissue loss or vascular injury (Gustilo grade IIIb and c) (Gustilo and Anderson, 1976; Gustilo, Mendoza and Williams, 1984). However, by the latter half of the decade the costs to the health care system and the patient, in terms of finance, function and social outcome, had become apparent (Bondurant et al., 1988; Sanders et al., 1992). There was a recognition on both sides of the Atlantic that some of these cases of limb salvage represented 'triumphs of technique over reason' (Hansen, 1987) and that amputation should, in selected cases, be considered not as a surgical failure but an appropriate and courageous treatment (Heatley, 1988). At this time a point scoring system was developed to provide an ordinal severity scale for the mangled extremity severity (MES) (Gregory et al., 1985). The initial paper describing the index was limited to few cases, and lacked definition of the fractures or specification of the number of primary amputations (Sanders et al., 1993). However, Lange et al. were working on indications for amputations and identified prognostic factors in lower extremity trauma (Lange et al., 1985). The variables they described were later formalized in a paper describing decision-making in massive lower extremity trauma (Lange, 1989); these factors were then incorporated in the Mangled Extremity Severity Score (MESS) (Johansen et al., 1990). The MESS scale is set out in Table 1.9.

This system has been used to predict amputation in both retrospective and prospective series by the authors of the system as well as others dealing with orthopaedic trauma. A score of 7 is predictive of amputation in 100% of cases (Helfet et al., 1990), indicating a high specificity, but the scale lacks sensitivity, as several cases with a score of <7 go on to amputation (Robertson, 1991).

Recent work has confirmed that the scoring system is not predictive of success of salvage (as opposed to indication for not attempting salvage) (Bonanni, Rhodes and Lucke, 1993). Lange has recently stated (Lange, 1994) that he feels the strength of the system is in indicating when amputation is appropriate, but not indicating when salvage will succeed; he claims that the major advance has been in changing the mind-set regarding amputation as a primary proce-

Table 1.10 Red Cross wound score (Coupland, 1991)

Feature	*Score*	*Note*
E = Entry	Estimate maximum diameter of entry wound (cm)	
X = eXit	Estimate maximum diameter of exit (cm)	Score X = 0 if no exit
C = Cavity	Assess if wound cavity will admit 2 fingers before surgery: C = 0, no C = 1, yes	An indicator of permanent cavitation in the wound For chest and abdominal wounds this refers to the wound of the wall
F = Fracture	F = 0, no fracture F = 1, fracture but no clinically significant comminution F = 2, fracture with clinically significant comminution	Some wounds will fall between these categories; guidance is given in the ICRC Booklet (Coupland, 1991)
V = Vital structure	Assess for breach of dura, pleura, peritoneum, or major vessel: V = 0, no V = 1, yes	Vascular damage as far distal as brachial or popliteal vessels is considered major in this classification
M = Metallic body	Bullets/fragments on radiograph M = 0, none M = 1, one M = 2, more than one	An indicator of bullet break-up

dure, as advocated by Hansen (1989) and Heatley (1988).

Classification of Injuries from a Specific Wounding Mechanism

Some interested groups have sought to classify trauma arising from a specific cause; the aim, generally, being to improve communication concerning the injuries.

Motor Vehicle Accidents

The AIS severity scoring system was originally designed to classify injuries arising from automobile accidents. Its uses have been extended, and it is described in the sections above.

Ballistic Wounds

Ballistic wounds from gunshot and fragmenting weapons are commonly seen in war. Unfortunately there is an increase in the incidence of these wounds in the civilian 'peacetime' setting, in the United States' 'Un-Civil War' (Schwab, 1993), as well as increasingly in the UK and Europe. If appropriate lessons concerning management of wounds from each setting are to be carried across into the other, then a nomenclature for classifying these wounds is necessary. It was in an attempt to meet this need for communication and to provide a scientific basis for war surgery that the Red Cross Wound Classification was devised (Coupland, 1991).

The Red Cross wound classification

The wound scoring and classification system developed by the International Committee of the Red Cross (ICRC) is used on all casualties in their war surgery hospitals (Coupland, 1991, 1992, 1994). Wounds are scored by taking account of six features: entry wound (E), exit wound (X), cavity (C), fracture (F), involvement of vital structure (V) and presence of metallic body in the wound (M). The scores for these elements are shown in Table 1.10. The scores for the two most severe wounds are recorded in casualties with multiple injuries.

The analysis of the score comprises two phases: the clinical decision based on the findings, and a more formal analysis at a later stage. The clinical decision determines the need for surgery and its priority, and is reached by coupling the assessment of the wound with the patient's clinical condition. More formal analysis of the wound score involves classification according to the amount of tissue damage (*grade*, as in Table 1.11) and the structures involved (*type*, as in Table 1.11). The nominal category of the wound, according to the Red Cross classification, combines the grade and type: e.g. large wound (Grade 2) involving fracture (type F) = 2F.

This scoring system provides a means of focusing attention on the injury, and may be a useful guide for those with limited experience of ballistic trauma. The system requires no prior knowledge of weapon details, such as bullet velocity or mass and is not based on descriptions from terminal ballistics such as high or low energy transfer. Rather, the wound is classified on the basis of the surgical lesion: its size and type of tissue involved. The system is nominal, and ordinal at the extremes (hence a Grade 1 type ST wound is certainly more benign than a Grade 3 type VF wound), but its ranking of wound severity for the wounds between these extremes has not been established.

This wound classification system is quick and simple to apply; thus it tends to be willingly adopted by surgeons in the field so that the majority of casualties have their wounds encoded. The ICRC treats many thousands of war wounded each year and now has a database of more than 15 000 encoded wounds (Coupland, 1993).

There are, however, areas of weakness in the classification system, even as a nominal system, or taxonomy. The definition of severity of fracture by 'clinically significant comminution' as an indicator of energy transfer is an area of incomplete definition, and has not been subjected to validation or evaluation of inter- or intra-rater agreement. This factor is one which needs considerable attention if the scoring system is to adopted as a nomenclature for ballistic wounds in a civilian setting, such as a trauma centre (Bowyer et al., 1995).

The presence of metallic bodies in the wound is scored (M) but not subsequently used in analysis. The presence of metallic fragments from break-up of a bullet is a useful indicator of wound severity, but the scoring system makes no distinction between fragmentation which has occurred in the tissues through break-up of a projectile, and injury caused by small fragments — such as those from grenades, artillery shells or bomblets. This is an important distinction, as the former suggests a very high energy transfer, whereas the latter group may be associated with a lesser wound. The presence of metal from bullet fragmentation may be of particular interest to the ICRC in their role in upholding international humanitarian law (Coupland et al., 1992).

Neurological damage is not encoded in the Red Cross system, and this could be a criticism (Bowyer,

Table 1.11 Red Cross wound classification—grade and type of wound (Coupland, 1992)

Grade *(The extent of tissue damage)*	*Definition*
1	E plus X scores < 10, with C = 0 and F = 0 or 1 (Low Energy Transfer inferred)
2	E plus X < 10, but with C = 1 or F = 2 (High Energy Transfer inferred)
3	E plus X > 10, with C = 1 or F = 2 (High Energy Transfer inferred) Wounds from explosive blast, such as those from mines, with massive tissue injury are graded 3, despite the absence of a true entry or exit wound
Type *(The structures involved)*	
ST (Soft Tissue Only)	Wounds with F = 0 and V = 0
F (Fracture)	Wounds with F = 1 or 2 and V = 0
V (Vital Structure)	Wounds with F = 0 and V = 1
VF (Fracture + Vital Structure)	Wounds with F = 1 or 2 and V = 1

Stewart and Ryan, 1993), particularly if the system is to be used in an ordinal manner. In many situations damage to a major nerve would significantly alter management: amputation would be strongly considered in the case of an injured lower limb with severe soft tissue disruption and a complete sciatic nerve lesion, whereas without the neurological lesion a more conservative approach might have been adopted.

To build up a realistic picture of multiple wounds it is necessary to record the number of wounds sustained and their distribution. This is not part of the Red Cross wound scoring protocol, but would add significantly to the value of data.

The database of wound scores using this classification has been used for examination of the resources required by various types of injury in terms of surgical interventions, blood usage and hospital stay (Eshaya-Chauvin and Coupland, 1992). It may also be used as an audit filter, flagging cases for review of clinical activity and quality assurance — for example there should be few deaths among those casualties with type ST or type F wounds (Coupland, 1992). The scoring system has been extensively used in casualties coming through the War Surgery Hospitals of the ICRC (Coupland, 1993). It was also utilized in the Gulf War (Bowyer, Stewart and Ryan, 1993), and in a civilian hospital dealing with shotgun wounds (Stewart and Kinninmonth, 1993). However, its use in a North American Level I Trauma Center (sic) has highlighted several deficiencies in the scoring system which could be adjusted better to meet the requirements of that setting (Bowyer, Kaplan and Burgess, 1996).

Summary

General classifications of trauma have an influence and importance for musculoskeletal trauma in a variety of ways. They may be indicators of trauma severity, based on physiological parameters, anatomical descriptors and pre-existing patient factors, influencing the urgency of management or place in which the casualty is managed. From these scoring systems some prognosis may be inferred regarding outcome in terms of survival, although most systems are not robust enough to permit prognostication regarding functional outcome; their role in guiding treatment methods or predicting complications has not yet been fully validated.

The classification systems designed to aid decision making regarding salvage or amputation have been shown to be useful to a limited extent, proving specific for those in which amputation is likely to be required, but lacking sensitivity.

The classification for ballistic wounds has provided a nomenclature for war injuries, but probably requires further refinement to improve its usefulness across both a wartime and civilian setting.

References

American Association for Automotive Medicine (1990) The Abbreviated Injury Scale (AIS) – 1990 Revision. Des Plaines, Illinois: AAAM.

Aprahamian, C., Cattey, R.P., Walker, A.P. et al. (1990) Pediatric Trauma Score. Predictor of hospital resource use? *Arch. Surg.*, **125**, 1128–1131.

Baker, S.P., O'Neill, B., Haddon, W. and Long, W.B. (1974) The Injury Severity Score: a method for describing patients with multiple injuries and evaluating emergency care. *J. Trauma*, **14**, 187–196.

Baker, S.P., and O'Neill, B. (1976) The injury severity score: an update. *J. Trauma*, **16**, 882–885.

Barie, P.S., Hydo, L.J. and Fischer, E. (1995) Comparison of APACHE II and III scoring systems for mortality prediction in critical surgical illness. *Arch. Surg.*, **130**, 77–82.

Bellamy, R.F. and Vayer, J.S. (1988) Assessment of penetrating injury severity. In *Advances in Trauma* (K.I. Maull, ed.), Vol. 3. Chicago: Year Book Medical Publishers, pp. 163–181.

Bonanni, F., Rhodes, M. and Lucke, J.F. (1993) The futility of predictive scoring of mangled lower extremities. *J. Trauma*, **34**, 99–102.

Bondurant, F.J., Cotler, H.B., Buckle, R. et al. (1988) The medical and economic impact of severely injured lower extremities. *J. Trauma*, **28**, 1270–1272.

Bowyer, G.W., Brown, M., Marsicano, J. and Burgess, A.R. (1995) Civilian gunshot wounds to the limbs — patterns of injury. *Injury*, **26**, 135.

Bowyer, G.W., Kaplan, T. and Burgess, A.R. (1996) The application of a war wound scoring system in a civilian trauma center: findings and suggested modifications to the dataset. (Unpublished results).

Bowyer, G.W., Stewart, M.P.M. and Ryan, J.M. (1993) Gulf War wounds; application of the Red Cross wound classification. *Injury*, **24**, 597–600.

Boyd, C.R., Tolson, M.A. and Copes, W.S. (1987) Evaluating trauma care: the TRISS method. *J. Trauma*, **27**, 370–378.

Champion, H.R., Copes, W.S., Sacco, W.J. et al. (1990) A new characterisation of injury severity. *J. Trauma*, **30**, 539–546.

Champion, H.R., Gennarelli, T.A. and Moore, E. (Convenors) (1995) The Future of Injury Severity Scaling. Conference, Annapolis, Maryland, July.

Champion, H.R. and Sacco, W.J. (1984) The trauma score as applied to penetrating injury. *Ann. Emerg. Med.*, **13**, 6–10.

Champion, H.R., Sacco, W.J. and Carnazzo, A.J. (1981) Trauma Score. *Crit. Care Med.*, **9**, 672–676.

Champion, H.R., Sacco, W.J. and Copes, W.S. (1995) Injury severity scoring again. *J. Trauma*, **38**, 94–95.

Champion, H.R., Sacco, W.J., Copes, W.S. et al. (1989) A revision of the Trauma Score. *J. Trauma*, **29**, 623–629.

Champion, H.R., Sacco, W.J. and Hunt, T. (1983) Trauma severity scoring to predict mortality. *World J. Surg.*, **7**, 4–11.

Commission of Professional Hospital Activities (1977) *International Classification of Diseases*, 9th Revision. Clinical Modification. Ann Arbor, Michigan: Edwards Brothers.

Coupland, R.M. (1991) *The Red Cross Wound Classification*. Geneva: International Committee of the Red Cross.

Coupland R.M. (1992) The Red Cross classification of war wounds: the E.X.C.F.V.M. scoring system. *World J. Surg.*, **16**, 910–917.

Coupland, R.M. (1993) *War Wounds of Limbs: Surgical Management*. Oxford: Butterworth-Heinemann.

Coupland, R.M. (1994) Classification and Management of War Wounds. In *Recent Advances in Surgery* (Taylor, I. and Johnson, C.D., eds). Edinburgh: Churchill, pp. 121–134.

Coupland, R.M., Hoikka, V., Sjoeklint, O.G. et al. (1992) Assessment of bullet disruption in armed conflicts. *Lancet*, **339**, 35–37.

Eichelberger, M.R., Gotshall, C.S., Sacco, W.J. et al. (1989) A comparison of the trauma score, the revised trauma score, and the pediatric trauma score. *Ann. Emerg. Med.*, **18**, 1053–1058.

Eshaya-Chauvin, B. and Coupland, R.M. (1992) Transfusion requirements for the management of war injured: the experience of the International Committee of the Red Cross. *Br. J. Anaesth.*, **68**, 221–223.

Gennarelli, T.A., Champion, H.R. et al. (1989) Mortality of patients with head injury and extracranial injury treated in trauma centers. *J. Trauma*, **29**, 1193–1195.

Gilpin, D.A. and Nelson, P.G. (1991) Revised trauma score: a triage tool in the accident and emergency department. *Injury*, **22**, 35–37.

Gormican, S.P. (1982) CRAMS scale: field triage of trauma victims. *Ann. Emerg. Med.*, **11**, 132–135.

Gregory, R.T., Gould, R.J., Peclet, M. et al. (1985) The Mangled Extremity Syndrome (M.E.S.): a severity grading system for multisystem injury of the extremity. *J. Trauma*, **25**, 1147–1150.

Gustilo, R.B. and Anderson, J.T. (1976) Prevention of infection in the management of one thousand and twenty-five open fractures of long bones. *J. Bone Joint Surg.*, **58A**, 453–458.

Gustilo, R.B., Mendoza, R.M. and Williams, D.N. (1984) Problems in the management of Type III (severe) open fractures: a new classification of Type III open fractures. *J. Trauma*, **24**, 742–746.

Hannan, E.L., Mendelof, J., Szypulski Farrell, L. et al. (1995) Validation of TRISS and ASCOT using a non-MTOS trauma registry. *J. Trauma*, **38**, 83–88.

Hansen, S.T. (1987) The Type III-C tibial fracture. Salvage or amputation. *J. Bone Joint Surg.*, **69A**, 799–800.

Hansen, S.T. (1989) Overview of the severely traumatized lower limb. Reconstruction versus amputation. *Clin. Orthop.*, **243**, 17–19.

Heatley, F. W. (1988) Severe open fractures of the tibia: the courage to amputate. *Br. Med. J.*, **296**, 229.

Helfet, D.L., Howey, T., Sanders, R. and Johansen, K. (1990) Limb salvage versus amputation. Preliminary results of the mangled extremity severity score. *Clin. Orthop.*, **256**, 80–86.

Hollis, S., Yates, D.W., Woodford, M. and Foster, P. (1995) Standardized comparison of performance indicators in trauma: a new approach to case-mix variation. *J. Trauma*, **38**, 763–766.

Jennett, B., Teasdale, G. and Braakman, R. (1976) Predicting outcome in individual patients after severe head injury. *Lancet*, 1031–1034.

Johansen, K., Daines, M., Howey, T. et al. (1990) Objective criteria accurately predict amputation following lower extremity trauma. *J. Trauma*, **30**, 568–573.

Jubelirer, R.A., Agarwal, N.N., Beyer, F.C. et al. (1990) Pediatric trauma triage: review of 1307 cases. *J. Trauma*, **30**, 1544–1547.

Kaufmann, C.R., Maier, R.V., Rivara, F.P. et al. (1990) Evaluation of the Pediatric Trauma Score. *JAMA*, **263**, 69–72.

Kaufmann, C.R., Maier, R.V., Kaufmann, E.J. et al. (1991) Validity of applying adult TRISS analysis to injured children. *J. Trauma*, **31**, 691–698.

Knaus, W.A., Draper, E.A. and Wagner, D.P. (1985) APACHE II: a severity of disease classification system. *Crit. Care Med.*, **13**, 818–824.

Knaus, W.A., Zimmerman, J.E., Wagner, D.P. et al. (1981) APACHE acute physiology and chronic health evaluation: a physiologically based classification system. *Crit. Care, Med.*, **9**, 951–956.

Knaus, W.A., Wagner, D.P., Draper, E.A. et al. (1991) The APACHE III prognostic system: risk prediction of hospital mortality for critically ill hospitalized adults. *Chest*, **100**, 1619–1636.

Lange, R.H. (1989) Limb reconstruction versus amputation decision making in massive lower extremity trauma. *Clin. Orthop.*, **243**, 92–99.

Lange, R.H. (1994) MESS and its significance. Lecture at AO/ASIF Advanced Course, Phoenix, Arizona, April.

Lange, R.H., Bach, A.W., Hansen, S.T. and Johansen, K.H. (1985) Open tibial fractures with associated vascular injuries. Prognosis for limb salvage. *J. Trauma*, **25**, 203–208.

Levati, A., Farina, M.L., Vecchi, G. et al. (1982) Prognosis of severe head injuries. *J. Neurosurg.*, **57**, 779–783.

MacKenzie, E.J., Steinwachs, D.M. and Shankkar, B. (1989) Classifying trauma severity based on hospital discharge diagnoses. *Med. Care*, **27**, 412–422.

Markle, J., Cayten, C.G., Byrne, D.W. et al. (1992)

Comparison between TRISS and ASCOT methods in controlling for injury severity. *J. Trauma*, **33**, 326–332.

Moreau, M., Gainer, P., Champion, H.R. and Sacco, W.J. (1985) Application of the trauma score in the pre-hospital setting. *Ann. Emerg. Med.*, **14**, 1049–1054.

Nayduch, D.A., Moylan, J., Rutledge, R. et al. (1991) Comparison of the ability of adult and pediatric trauma scores to predict pediatric outcome following major trauma. *J. Trauma*, **31**, 457–458.

Pape, H., Auf'm'Kolk, M., Paffrath, T. et al. (1993) Primary intramedullary femur fixation in multiple trauma patients with associated lung contusion — a cause of posttraumatic ARDS? *J. Trauma*, **34**, 540–548.

Ramenofsky, M.L., Ramenofsky, M.B., Jurkovich, G.J. et al. (1988) The predictive validity of the Pediatric Trauma Score. *J. Trauma*, **28**, 1038–1042.

Robertson, P.A. (1991) Prediction of amputation after severe lower limb trauma. *J. Bone Joint Surg.*, **73B**, 816–818.

Sacco, W.J., Copes, W.S., Bain, L.W. et al. (1993) Effect of preinjury illness on trauma patient survival outcome. *J. Trauma*, **35**, 538–543.

Sacco, W.J., Forrester Staz, C., Smith, J.S. and Buckman, R.F. (1994) Status of trauma patient management as measured by survival/death outcomes: looking toward the 21st century. *J. Trauma*, **36**, 297–298.

Sanders, R., Pappas, J., Mast, J. and Helfet, D. (1992) The salvage of open grade IIIB ankle and talus fractures. *J. Orthop. Trauma*, **6**, 201–208.

Sanders, R., Swiontkowski, M., Nunley, J. and Spiegel, P. (1993) The management of fractures with soft-tissue disruptions. *J. Bone Joint Surg.*, **75A**, 778–789.

Schwab, C.W. (1993) Violence: America's uncivil war — Presidential Address, sixth scientific assembly of the Eastern Association for the Surgery of Trauma. *J. Trauma*, **35**, 657–665.

Seibel, R., Laduca, J., Hassett, J.M. et al. (1985) Blunt multiple trauma (ISS 36), femur traction, and the pulmonary failure-septic state. *Ann. Surg.*, **202**, 283–295.

Smith, E.J., Ward, A.J. and Smith, D. (1990) Trauma scoring methods. *Br. J. Hosp. Med.*, **44**, 114–118.

Stewart, M.P.M. and Kinninmonth, A. (1993) Shotgun wounds of the limbs. *Injury*, **24**, 667–670.

Teasdale, G. and Jennett, B. (1974) Assessment of coma and impaired consciousness. A practical scale. *Lancet*, **ii**, 81–84.

Tepas, J.J., Mollitt, D.L., Talbert, J.L. and Bryant, M. (1987) The pediatric trauma score as a predictor of injury severity in the injured child. *J. Pediatric Surg.*, **22**, 14–18.

Tepas, J.J., Ramenofsky, M.L., Mollitt, D.L. et al. (1988) The Pediatric Trauma Score as a predictor of injury severity: an objective assessment. *J. Trauma*, **28**, 425–429.

Trunkey, D.D., Siegel, J., Baker, S.P. and Gennarelli, T.A. (1983) Panel: current status of trauma severity indices. *J. Trauma*, **23**, 185–201.

Yates, D.W. (1990) Scoring systems for trauma. *Br. Med. J.*, **301**, 1090–1094.

Yates, D.W., Woodford, M. and Hollis, S. (1993) Trauma audit: clinical judgement or statistical analysis? *Ann. R. Coll. Surg. Engl.*, **75**, 321–324.

2

General classifications of fractures

S. J. Krikler

Classification in General

To classify is to arrange or distribute knowledge in classes, according to a system. Classifiers, that is people who classify, have themselves traditionally been classified into 'lumpers' and 'splitters', i.e. those who tend to lump everything into a few, broad groups, or those who tend to split any group into smaller and more detailed groups. The most sophisticated are the 'hierarchists', who can occupy both camps; however, the question still remains as to why there is any advantage to classifying in the first place.

In everyday clinical practice, the surgeon would like to believe that, if an injury can be recognized as being of a specific type, a prognosis may be given and appropriate management suggested. For this to be practical, the means by which this recognition can be made must be reproducible, simple and memorable. More complex classifications are likely to be useful only in a research or clinical audit context. Sadly, many authors seem to suggest a classification more for their own immortality than for a clinical use. Some clinicians use classification as a one-upmanship tool to be used when competing with their peers or putting their juniors down.

This discussion of classification is based on the assumption that the classification has been derived by clinicians with a view to guiding clinical management or at least giving some sort of working prognosis. This assumption fails to recognize that some classifications are for the benefit of health managers or for public health/epidemiology purposes. ICD-9 is a means of giving codes to a large number of diseases but it was not designed to aid treatment of specific injuries and it is essentially a non-clinically based system. This chapter aims to discuss only those classifications of use in clinical practice. For a more complete discussion of the classifications and coding systems available in orthopaedics, see Macey (1997).

Good clinical practice demands that the treatment given will lead to a better outcome than alternative treatment, or, indeed, no treatment. In order to predict an outcome and in order to compare different treatments, it is essential that we can compare like with like. However, no two injuries are identical, so there will be some subjective measure of how like or unlike injuries must be to qualify for their particular classification. This ought to be determined by the outcome; if an apparent difference has no relevance to the outcome, it is not a real difference in terms of the classification. Unfortunately, many, perhaps most, classifications are devised and published with very little reference to outcome.

Even if a classification is based on outcome, to be useful in clinical practice it must be reliable in its application, i.e. have an acceptable level of inter- and intra-observer error. Ideally, no classification should be published before its reliability has been demonstrated (Burstein, 1993). In the relatively small number of cases where classifications have been subjected to this type of validation, the results have been very disappointing.

Although proximal humeral fractures are commonly classified by Neer's system (1970), when this has been subjected to formal appraisal, the correlation between different observers has been disappointing (Sidor et al., 1993; Siebenrock and Gerber 1993; Brien et al. 1995). A plea was made that no new classification system be published until its reliability was assessed in a rigorously conducted study (Burstein, 1993) but this provoked a hostile reaction from clinicians used to applying an 'established' classification (Neer, 1994; Rockwood, 1994; Bigliani, Flatow and Pollock, 1994; Cuomo, 1994; Bernstein, 1994).

Lauge-Hansen's classification of ankle fractures (1950) may be a little more robust but agreement is still poor for the more detailed analysis of the more complex fractures (Nielsen, Dons-Jensen and Sorensen, 1990; Thomsen et al., 1991). Similarly, the classification of open fractures has been shown to be less straightforward than its proponents would suggest (*vide infra*).

Ideally, any classification should be subjected to confirmation by criterion validity, the differences between different classifications being confirmed by differences in outcome. In practice, most classifications in common usage have face validity, in that the definitions used for the classification appear to make sense but very few have proven criterion validity.

There is a difference between a classification and a grading. The former simply groups like with like and distinguishes unlike, whereas a grading implies a degree of severity. An open book pelvic fracture has a different set of potential complications from a lateral compression fracture but it is arbitrary which is a 'Type 1' and which a 'Type 2' injury. Is it better for the radius to displace anteriorly or posteriorly in a Monteggia fracture? Names are more descriptive and easier to remember, so it seems sensible to try to use names whenever this is practical and to use numbers only where there is a clear progression and the system really is a grading. Sadly, many classifications use numbers to masquerade as grading systems.

By implication, classification of fractures refers only to the bony injury and therefore neglects many other important factors which must be taken into account by a treating clinician. Although this chapter is concerned with the classification of fractures, as clinicians we do not treat fractures but patients with fractures. In determining the appropriate treatment, many other considerations apply. These include other details of the injury, such as soft-tissue damage, joint instability or neurovascular complications, as well as details of the patient, such as age and functional requirement and the pre-injury state of the limb and the patient. In practice, the facilities and expertise available must also influence the choice of management. This collection of parameters which should be considered and integrated before a treatment decision is made has been termed the 'personality of the fracture' (Schatzker and Tile, 1987).

Soft-tissue Classifications

The state of the soft tissues is one of the most important determining features of a bony injury. For this reason, this chapter starts with a discussion of fracture classifications associated with injuries of the soft tissues. Injuries affecting only soft tissue are discussed in the next chapter.

At the simplest level, a fracture may be either open or closed. Although this has significant implications for the potential for infection, there is clearly a difference between the infection rates in open fractures of varying severity (Gustilo and Anderson, 1976). Even in the absence of a break in the skin, associated periosteal damage may interfere with bone healing, the development of a compartment syndrome may delay union in a closed fracture (Turen, Burgess and Vanco, 1995) and neurovascular injury may compromise the final function regardless of the bony injury and its resolution.

Gustilo and Anderson

Perhaps the most widely used is that first proposed by Gustilo and Anderson (1976). Based on their experience of 1025 open fractures in long bones (tibia and fibula, femur, radius and ulna and humerus), they recommended a clinical strategy based on their system of classification. In fact, only 352 of these open fractures were assessed prospectively after being managed with the proposed strategy. Of these, 326 had at least six weeks follow-up and only 158 had complete bacteriological studies. They recognized three types of open fracture:

Type I — Wound less than 1 cm long and clean.

Type II — Laceration more than 1 cm long without extensive soft-tissue damage, flaps or avulsions.

Type III — Either an open segmental fracture, an open fracture with extensive soft-tissue damage, or a traumatic amputation, gun shot wound, farm injury or open fracture accompanying vascular injury requiring repair.

It subsequently became clear that type III fractures covered a wide variety of injuries, so this was divided into a further subtypes (Gustilo, Mendoza and Williams, 1984):

Type IIIA — Adequate soft-tissue coverage of a fractured bone, despite extensive soft-tissue laceration or flaps, or high-energy trauma irrespective of the size of the wound.

Type IIIB — Extensive soft-tissue injury with periosteal stripping and bony exposure. This is usually associated with massive contamination.

Type IIIC Open fracture associated with arterial injury requiring repair.

In a later paper (Gustilo, Gruninger and Davis, 1987) the classification is almost identical to the previously published version but with an additional inclusion in the Type III group of 'open fractures older than eight hours'.

Further detail was subsequently added to the definitions of the types and subtypes (Gustilo, Merkow and Templeman, 1990):

Type I The wound is less than 1 cm long. It is usually a moderately clean puncture, through which a spike of bone has pierced the skin. There is little soft-tissue damage and no sign of crushing injury. The fracture is usually simple, transverse or short oblique, with little comminution.

Type II Laceration more than 1 cm long and there is no extensive soft-tissue damage, flap or avulsion. There is a slight or moderate crushing injury, moderate comminution of the fracture and moderate contamination.

Type III Extensive damage to soft tissue, including muscles, skin and neurovascular structures and a high degree of contamination. The fracture is often caused by high velocity trauma, resulting in a great deal of comminution and instability.

Type IIIA Soft-tissue coverage of the fractured bone is adequate, despite extensive laceration, flaps or high energy trauma. This subtype includes segmental or severely comminuted fractures from high energy trauma, regardless of the size of the wound.

Type IIIB Extensive injury to or loss of soft tissue, with periosteal stripping and exposure of bone, massive contamination and severe comminution of the fracture from high energy trauma. After debridement and irrigation is completed, a segment of bone is exposed and a local or free flap is needed for coverage.

Type IIIC Any open fracture associated with an arterial injury that must be repaired, regardless of the degree of soft-tissue injury.

This system has been quoted (and sometimes misquoted) in many papers, where it is often referred to the 'Gustilo and Anderson' classification. No formal attempt seems to have been made to assess its reliability until 1993 (Horn and Rettig). The authors showed photographic slides of 10 open fractures to 14 residents and eight attending surgeons in one institution. Inter-observer agreement as determined by Kappa analysis was described as moderate.

A much larger study was undertaken by Brumback and Jones (1994), who attended three orthopaedic meetings and showed a 26 minute videotape to a total of 289 orthopaedic surgeons of varying seniority and expertise. The tape showed details of 13 open tibial fractures and the 'Gustilo–Anderson' classification was reviewed before the tape was shown. The first case was used to illustrate the manner in which the participants would be viewing the open fractures, the remaining 12 were then classified by each participant. Only the answers of surgeons who stated that they used this classification were analysed. The average agreement among observers for the 12 fractures shown was 60% (range 42–94%) but for 2 of the 12 fractures, the agreement was less than 50%. Even among 'experts' (academic, trauma-fellowship-trained, treating more than 12 open fractures of the tibia each year), the average agreement was only 66% (range 33–100%). It was less than 65% for six fractures and less than 50% for three.

Not surprisingly, this paper was attacked by Gustilo (1995) but his criticisms were firmly defended (Brumback and Jones, 1995).

Despite these doubts over reliability, within many published series of patients, the infection rate does seem to increase with increasing grade of soft-tissue damage, e.g. Caudle and Stern, 1987; Court-Brown et al., 1991; Ostermann, Seligson and Henry, 1995.

Oestern and Tscherne

An alternative and more detailed classification was proposed by Oestern and Tscherne (1984). They recognized that, although the presence or absence of communication between the fracture and the skin surface is a fundamental distinction, within each of these two categories, open or closed, the degree of soft-tissue damage may vary considerably, so their system attempts to classify fractures by the amount of soft-tissue damage, both in closed and open fractures. They applied four grades of closed fracture (from 0 to III) and four grades of open fracture (from I to IV). Their system is summarized as follows:

Closed fractures

Grade C 0 Soft-tissue damage is absent or negligible. Simple fracture caused by indirect violence.

Grade C I — Superficial abrasion or contusion of skin from within. The fracture has a mild to moderately severe configuration.

Grade C II — Deep, contaminated abrasion with localized skin or muscle contusion from direct trauma. Moderately severe to severe fracture configuration. Impending compartment syndrome is included in this category.

Grade C III — The skin is extensively contused or crushed and muscle damage may be severe. Also, subcutaneous avulsion, decompensated compartment syndrome and rupture of major blood vessel associated with a closed fracture.

Open fractures

The primary concerns are not the size of the skin wound but the extent of soft-tissue damage and muscle contusion. Definitive classification may not be possible until after exploration:

Grade O I — Little or no skin contusion. Bacterial contamination is negligible. Usually the skin is pierced by only one bone fragment. The fracture severity is mild.

Grade O II — Circumscribed skin and soft-tissue contusions and moderate contamination. The severity of the fracture is variable.

Grade O III — Extensive soft-tissue destruction, heavily contaminated. Often associated vascular and neurological lesions. Includes any open fracture with ischaemia and extensive comminution, gunshot wounds, contaminated farm injuries.

Grade O IV — Total or subtotal amputation. The definition of subtotal amputation is that of the Replantation Committee of the International Society for Reconstructive Microsurgery.

Evidence of any attempt to assess the inter-observer error of this classification could not be found. The inclusion of terms such as 'extensive soft-tissue destruction' and 'moderate contamination' are clearly highly subjective, so there is likely to be no better agreement than with the 'Gustilo and Anderson' system.

AO Classification

A similar but more detailed approach to that of Oestern and Tscherne was adopted by the AO group (Rüedi, Border and Allgöwer, 1990). They use three headings: integuments (i.e. skin), muscle and tendon, and neurovascular injury. Each heading is graded from 1 to 5. Grade 1 is normal (except for open fractures), Grades 2–4 imply increasing severity and Grade 5 is 'something special'. Using this classification, any fracture may be described by a 9-digit alphanumeric code.

The skin is coded as either IC (integument closed) or IO (integument open), the grades are as follows:

IC 1 — No skin lesion
IC 2 — Contusion, no laceration
IC 3 — Circumscribed degloving
IC 4 — Extensive closed degloving
IC 5 — Necrosis from contusion
IO 1 — Skin broken from inside out
IO 2 — Skin broken from outside in <5 cm, contused edges
IO 3 — Skin breakage >5 cm, increased contusion, devitalized edges
IO 4 — Considerable, full-thickness contusion, abrasion, extensive open degloving, skin loss.

The muscle and tendon injury is graded as follows:

MT 1 — No muscle injury
MT 2 — Circumscribed muscle injury, one compartment only
MT 3 — Considerable muscle injury, two compartments
MT 4 — Muscle defect, tendon laceration, extensive muscle contusion
MT 5 — Compartment syndrome/crush syndrome with wide injury zone.

The neurovascular injury is graded as follows:

NV 1 — No neurovascular injury
NV 2 — Isolated nerve injury
NV 3 — Localized vascular injury
NV 4 — Extensive segmental vascular injury
NV 5 — Combined neurovascular injury, including subtotal or even total amputation.

This system shares much with the previous two systems described but with the addition of separate consideration of the muscle/tendon and neurovascular components of the injury. No account is made of the degree of contamination, a factor which seems likely to influence the risk of infection but which is difficult to quantify.

Open Joint Injuries

It does not seem unreasonable to suppose that an open injury which involves a major joint might lead to later problems with that joint. None of the above systems make any specific reference to whether or not any joints are involved. A classification of open joint injuries has been proposed (Collins and Temple,

1988) but it has not gained widespread usage and I have been unable to find any assessment of its correlation with outcome.

In this system, open joint injuries are classified into types I–IV, according to the state of the soft tissues:

I Punctures, low-velocity missiles, or lacerations of any dimension which can be closed by primary or delayed primary suture.

II More severe, resulting from one or more missiles, punctures or lacerations. Portions of the capsule or overlying skin may be destroyed.

III Intra-articular extension of an open periarticular fracture.

IV Open dislocation or fracture dislocation, or with an associated nerve or vascular injury requiring repair.

Within types I–III, the injury is further classified according to the extent of the intra-articular injury:

A With minimal (< 2 mm stepoff or gap; area of comminution < 1 cm^2) or no injury to any articular surface.

B With significant (> 2 mm stepoff or gap; area of comminution > 1 cm^2) injury to a single articular surface.

C With significant (> 2 mm stepoff or gap; area of comminution > 1 cm^2) injury to biarticular surfaces or meniscoligamentous disruption.

Summary

The 'standard' system of classifying the soft-tissue component of a fracture remains that of 'Gustilo and Anderson', at least in the English-speaking literature (Sanders et al., 1993). Despite the doubts which have been raised over its reliability, it seems likely to remain in common usage as it is fairly simple to remember and apply.

The AO system gives more detail but, perhaps because it is more detailed and therefore more complex to apply, it does not seem to have found widespread usage in the English-speaking literature.

Collins and Temple offered a classification for open joint injuries but this has not found widespread use.

Classification of Bony Injuries

Fracture classifications may broadly be grouped into genetic and anatomic, that is according to their mechanism or according to the structures injured. This is illustrated in classifications of ankle fractures. The Lauge-Hansen system (1950) divides fractures (and associated ligamentous injuries) by the names of the presumed causative forces, while the Weber system (1972) depends purely on the level of the fibular fracture. Which is the more appropriate may depend on the context. If surgical treatment is planned, it may be more logical to concentrate on the patho-anatomy, whereas if non-operative treatment is used, understanding the forces applied at the time of injury may direct the manipulative forces applied to reduce and hold the injury.

In order for a fracture to occur, there must be a force applied which is sufficient to deform the bone beyond its yield point. To some extent this is a rate sensitive phenomenon, so the amount of force required may depend on the rate at which it is applied. Repeated application of a force insufficient to break a bone may cause fatigue and eventual fracture. As this type of fracture was seen in the metatarsals of army recruits on forced marches, it was called a march fracture. This is too narrow a title but it is more rational than the more commonly applied 'stress fracture'. All fractures are caused by stress applied to a bone, a more accurate and descriptive term would be 'fatigue fracture'.

In general, when stresses within the normal physiological range are applied to normal bones, they do not break. Fractures occur when the forces are abnormally large, or the bones are abnormally weak. If the fracture results from an underlying bony abnormality, it is termed a pathological fracture. There are many causes of pathological fractures and an exhaustive list is inappropriate here. The causes can be described in the classical surgical sieve: congenital and acquired, local and systemic. Congenital causes include all the inborn errors of metabolism resulting in, for example, osteogenesis imperfecta or osteodystrophy. The commonest acquired causes are metastatic tumours and infection. Osteoporosis in the elderly clearly results in grossly weakened bone which fractures easily but it is not usually grouped with pathological fractures.

Most classifications have been derived for specific locations but a few have subsequently been applied more generally.

Classification by Morphology

From their first encounters with musculoskeletal trauma, medical students are taught to describe the pattern of a fracture.

In children, the radiographic appearance may show a fracture through 'only one cortex'. This appearance is, of course, an artefact of projection; there is only one cortex, it lies circumferentially around the medulla but if it is fractured through only part of its circumference, it may appear to be intact on one side and disrupted on the other. If the fractured cortex shows separation, it is described as a greenstick fracture. If the cortex is impacted, it may be described as a buckle fracture. If the buckling occurs all around the cortex, it forms the shape of a doughnut or torus and the injury may be described as a toroidal fracture.

When the bone is completely broken, there may be two fragments (a simple fracture) or more fragments (a comminuted or multifragmentary fracture).

In a simple fracture, the fracture line may be transverse, oblique or it may describe what is usually but incorrectly called a spiral. The first two fractures divide the bone in a simple plane but in the third type, the surface of the fracture describes a three-dimensional curve, which is correctly termed a helix. It would be more accurate to describe this pattern as a helical rather than a spiral fracture.

If there are more than two fragments, there may be a triangular fragment at the main fracture site. This has been called a butterfly fragment. As it is produced by the two main fragments being pushed longitudinally towards each other, it is sometimes called an extrusion wedge.

Where there is an intact section of bone which lies between two fracture lines and has a complete circumferential cortex, it is termed a segmental fracture.

Classification by Mechanism of Injury

The only general classification of fractures offered in one of the standard textbooks is by mechanism of injury (Rockwood, Green and Bucholz, 1991). Fractures are said to occur either by direct or indirect trauma. The former includes tapping fractures, crush fractures and penetrating (gunshot) fractures. Indirect trauma is divided into traction fractures, angulation fractures, rotational fractures, compression fractures and fractures due to a combination of the above.

No discussion is offered as to the clinical usefulness of this classification, nor any relationship to treatment or outcome.

Shaft Comminution (Winquist)

In a paper on intramedullary nailing of femoral shaft fractures (Winquist and Hansen, 1980), a classification of shaft comminution was proposed. In this initial paper, the fracture was classified according to its potential instability, which was partly dependant on the ability of an intramedullary nail to stabilize the fracture. Thus, a very proximal or distal fracture in which the nail had poor purchase in the smaller fragment, might be classified as Grade III. This seems to have been abandoned later (Winquist, Hansen and Clawson, 1984) in favour of classifying according to the fracture pattern alone.

I Transverse, with only a small piece of bone broken off

II Larger butterfly fragment but with at least 50% of the cortex intact

III Large butterfly fragment with < 50% of the cortex intact

IV No abutment of cortices at the level of the fracture.

Although both papers dealt only with femoral fractures, this classification is sometimes used to describe comminution of the shaft of other bones

The Comprehensive Classification of Fractures (the AO classification)

Until the 1970s, although there had been many classifications proposed for fractures in specific anatomic areas and coding systems for recording some data for public health/epidemiology purposes, there had been no attempt to apply a standard set of principles in a systematic manner to the entire skeleton. Anyone interested in classifying fractures had tended to assemble a series of patients with injuries to a specific area and had then suggested a classification relevant to this area. However, the principles of damage affecting a joint or the diaphysis of a long bone may be similar in a variety of areas and for this reason the AO group set out to develop a unified scheme of classification (Müller, 1990). This scheme has evolved over almost 20 years and has now been adopted by the Orthopaedic Trauma Association of North America.

The system of classification is hierarchical and each layer in the hierarchy has been given a single alphanumeric code. This was intended to aid computerization and was not intended to be used for personal communication, although inevitably, there is a tendency for this to happen.

One of the stated aims of the system is to grade fractures by the severity of the injury. In general, a type 'B' fracture is more severe than a type 'A' but less severe than a type 'C'. Within each type, a group '2' fracture should be less severe than a group '3' and so on.

Since the system is to be applicable to any region, the code for the fracture starts with a code for location. Each limb segment is given a number and the segment is then divided into proximal, diaphyseal and distal. Thus the forearm is segment 2 and the distal forearm is region 23. There is no advantage to saying '23' over saying 'distal forearm', it is simply an aid to computer database management. The fact that this coding system is easily memorized does not make it worth memorizing; the principles of the grouping are more important.

The first fundamental distinction (and the second number in the code) describes the segment affected, as diaphyseal fractures are classified differently from those in the proximal or distal segment. Many classifications of fractures near joints use an arbitrary distance from a bony landmark but the absolute distance varies depending on the size of the individual and this rule is impractical to a general classification. This difficulty was overcome by developing a rule of squares. If a square is constructed at the end of a long bone, with the maximum width of the epiphysis being the length of one side of the square, fractures in which the centre of the fracture lies within the square are deemed to be epiphyseal, while fractures centred outside this area are diaphyseal.

While this rule does give some degree of consistency throughout the skeleton, it also illustrates some of the disadvantages of trying to develop a comprehensive classification. The anatomy of each part of each segment is heavily influenced by its functional requirements, so the proximal femur is defined by the lower margin of the lesser trochanter and fractures of the distal tibia are arbitrarily distinguished from malleolar fractures, which have an entire group to themselves.

Defining the centre of a fracture may be difficult and there is an additional rule. A fracture with a displaced articular component is always classified as an epiphyseal fracture, even if the main fracture is clearly centred outside the epiphyseal square but if there is an undisplaced fissure extending into the joint, the fracture is classified according to its centre.

The next three figures in the code are for the type of fracture, its group and subgroup; each of these has three possibilities. The final code, the qualification, may have several options. In order to apply the classification down to this level, it is essential to have the reference book to hand and in many cases the code will be changed after surgery or more detailed imaging.

Diaphyseal injuries of the humerus, femur and tibia are divided into simple, where there is a single fracture dividing the shaft into two fragments (A), wedge fractures, which are multifragmentary but where there is contact between the two main fragments (B) and complex fractures, which are multifragmentary and there is no direct contact between the two main fragments after reduction (C).

Simple fractures (A) are grouped into 'spiral' (A1), oblique (A2) and transverse (A3). Wedge fractures may have a 'spiral' wedge (B1), a bending wedge (B2) or a fragmented wedge (B3). Complex fractures which are thought to have been caused by a twisting force are classified as C1, segmental fractures are classified as C2 and irregular complex fractures are classified as C3.

A different approach is adopted for the forearm; fractures are said to be simple (A), wedge (B) or complex (C). If there is an isolated ulnar fracture, it is defined as a Group 1 fracture, an isolated radial fracture is Group 2 and both bone fractures are Group 3.

The definitions of the subgroups and qualifications for the diaphyseal fractures vary with the bone involved and the type and group of the fracture and are only likely to be used in major research studies. At this level, the classification ceases to be a general classification but is 'customized' for each region.

The anatomy and fracture patterns of the shafts of long bones is fairly consistent throughout the skeleton but the anatomy of the articular surfaces of each joint is highly specific to that joint and the mechanisms of injury show greater variation, so the fracture patterns show considerable variation between joints. For this reason, the classification of these injuries becomes more complex.

The proximal femur and humerus and the malleolar region of the ankle are each given their own specific classification rules. For the remaining proximal and distal segments, the injuries are divided into extra-articular, partial articular and complete articular, types A, B and C respectively. Type A is self explanatory. In a partial articular fracture, there is a portion of the articular surface which is in continuity with the diaphysis, while in a complete articular fracture, the fracture extends into the articular surface and there is no continuity between the shaft and any of the articular surface.

In the proximal femur, A, B and C refer to trochanteric, transcervical and head fractures respectively. In the proximal humerus, they refer to extra-articular unifocal, extra-articular bifocal and articular fractures respectively. In the malleolar segment, the three types are those of the Weber classification (1972).

Strengths and Weaknesses of the AO Classification

The classification has been developed with a set of principles which may be applied to any fracture and the definitions of the fracture types are relatively straightforward. Being a hierarchical classification, a fracture may be classified with as much or as little detail as the classifier feels is required for a particular set of circumstances. This makes it a useful research or audit tool, although its use in everyday clinical practice has been questioned (Newey, Ricketts and Roberts, 1993).

There is no note of soft-tissue involvement, although admittedly there is a separate AO classification of soft-tissue injuries. In particular, a fractured joint which has sustained ligament injuries and is therefore very unstable cannot be distinguished from a similar bony injury without ligamentous damage. The extreme example of this is a dislocation without fracture; the limb function may be severely and permanently impaired, yet the injury does not warrant any mention in the classification.

The classification was derived from study of large numbers of fractures in normal bone. It does not legislate for abnormal bone, such as in osteoporosis, malignancy, or osteomalacia or other metabolic bone disease.

There is no measure of displacement of the fragments or articular surface. A fracture which is basically extra-articular but which technically has an extension into the articular surface is given the same classification as a fracture with major articular disruption, although the outcome may be related to the degree of displacement. A fracture which has gross displacement of the fragments is more likely to have major soft-tissue stripping and so may be much more unstable and more at risk of problems with union than a similar bony injury with little displacement.

The groups and more particularly the subgroups may be very difficult to distinguish on plain radiographs. Fractures may be reclassified after better imaging or operative treatment. How reliable is the classification of those fractures in which no subsequent imaging has been performed? In 10 out of 21 tibial plateau fractures, the AO classification was changed following MRI scan (Holt, Williams and Dent, 1995).

Fractures which include both the epiphysis and diaphysis, as defined by the rule of squares, may be classified into either area, depending on which area the clinician feels is the major injury. This inevitably introduces a further subjective variability into the classification.

Unpublished, anecdotal studies have suggested that there is reasonable correlation between observers for the types but the concordance deteriorates in the more detailed layers of the hierarchy. One study has tried to assess this formally, with rather disappointing results (Johnstone, Radford and Parnell, 1993).

The system tends to treat fractures as though they occur in generic bones but the details of the anatomy of a specific area tends to give rise to features unique to that particular area. This is an inevitable consequence of any system which tries to be all-embracing. When the classification is applied at the more detailed level, it becomes a collection of separate classifications, each with their own strengths and weaknesses (Gillham, 1991).

References

Bernstein, J. (1994) *J. Bone Joint Surg.*, **76A**, 792–793 (letter).

Bigliani, L.U., Flatow, E.L., and Pollock, R.G. (1994) *J. Bone Joint Surg.*, **76A**; 790–791 (letter).

Brien, H., Noftall, F., MacMaster, S. et al. (1995) Neer's classification system: a critical appraisal. *J. Trauma*, **38**, 257–260.

Brumback, R.J. and Jones, A.L. (1994) Interobserver agreement in the classification of open fractures of the tibia. The results of a survey of two hundred and forty-five orthopaedic surgeons. *J. Bone Joint Surg.*, **76A**, 1162–1166.

Brumback, R.J. and Jones, A.L. (1995) *J. Bone Joint Surg.*, **77A**, 1292 (reply).

Burstein, A.H. (1993) Fracture classification systems: do they work and are they useful? (Editorial) *J. Bone Joint Surg.*, **75A**, 1743–1744.

Caudle, R.J. and Stern, P.J. (1987) Severe open fractures of the tibia. *J. Bone Joint Surg.*, **69A**, 801–807.

Collins, D.N. and Temple, S.D. (1988) Open joint injuries: classification and treatment. *Clin. Orthop.*, **243**, 48–56.

Court-Brown, C.M., McQueen, M.M., Qaba, A.A. and Christie, J. (1991) Locked intramedullary nailing of open tibial fractures. *J. Bone Joint Surg.*, **73B**, 959–964.

Cuomo, F. (1994) *J. Bone Joint Surg.*, **76A**, 792 (letter).

Gillham, N.R. (1991) Classifying fractures. *J. Bone Joint Surg.*, **73B**, 1009 (letter).

Gustilo, R.B. (1995) *J. Bone Joint Surg.*, **77A**, 1291–1292 (letter).

Gustilo, R.B. and Anderson, J.T. (1976) Prevention of infection in the treatment of one thousand and twenty-five open fractures of long bones: retrospective and prospective analysis. *J. Bone Joint Surg.*, **58A**, 453–458.

Gustilo, R.B., Gruninger, R.P. and Davis, T. (1987) Classification of type III (severe) open fractures relative to treatment and results. *Orthopaedics*, **10**, 1781–1788.

Gustilo, R.B., Mendoza, R.M. and Williams, D.N. (1984) Problems in the management of Type III (severe) open fractures: a new classification of Type III open fractures. *J. Trauma*, **24**, 742–746.

Gustilo, R.B., Merkow, R.L. and Templeman, D. (1990) Current Concepts Review. The Management of Open Fractures. *J. Bone Joint Surg.*, **72A**, 299–304.

Holt, M.D., Williams, L.A. and Dent, C.M. (1995) MRI in the management of tibial plateau fractures. *Injury*, **26**, 595–599.

Horn, B.D. and Rettig, M.E. (1993) Interobserver reliability in the Gustilo and Anderson classification of open fractures. *J. Orth. Trauma*, **7**, 357–360.

Johnstone, D.J., Radford, W.J.P. and Parnell, E.J. (1993) Interobserver variation using the AO/ASIF classification of long bone fractures. *Injury*, **24**, 163–165.

Lauge-Hansen, N. (1950) Fractures of the ankle II. Combined experimental-surgical and experimental-roentgenological investigations. *Arch. Surg.*, **60**, 957–985.

Macey, A.C. (1997) Nomenclature, Classification and Coding in Orthopaedics. In *Assessment Methodology in Orthopaedics* (P.B. Pynsent, J.C.T. Fairbank and A.J. Carr, eds) Oxford: Butterworth-Heinemann.

Müller, M.E., Nazarian, S., Koch, P. and Schatzker, J. (1990) *The Comprehensive Classification of Fractures of Long Bones*. Berlin: Springer-Verlag.

Neer, C.S. (1970) Displaced proximal humeral fractures. Part 1. Classification and evaluation. *J. Bone Joint Surg*. **52A**, 1077–1089.

Neer, C.S. (1994) *J. Bone Joint Surg.*, **76A**, 789 (letter).

Newey, M.L., Ricketts, D. and Roberts, L. (1993) The AO classification of long bone fractures: an early study of its use in clinical practice. *Injury*, **24**, 309–312.

Nielsen, J.O., Dons-Jensen, H. and Sorensen, H.T. (1990) Lauge-Hansen classification of malleolar fractures. An assessment of the reproducibility in 118 cases. *Acta Orthop. Scandinavica*, **61**, 385–387.

Ostermann, P.A.W., Seligson, D. and Henry, S.L. (1995) Local antibiotic therapy for severe open fractures: a review of 1085 consecutive cases. *J. Bone Surg.*, **77B**, 93–97.

Oestern, H.J. and Tscherne, H. (1984) Pathophysiology and classification of soft tissue injuries associated with fractures. In *Fractures with Soft Tissue Injuries* (H. Tscherne and L. Gotzen, eds). Berlin: Springer-Verlag, pp. 1–9.

Rockwood, C.A. (1994) *J. Bone Joint Surg.*, **76A**, 790 (letter).

Rockwood, C.A., Green, D.P. and Bucholz, R.W. (eds) (1991) In *Rockwood and Green's Fractures in Adults*, 3rd edn. Philadelphia: Lippincott, pp. 9–21.

Rüedi, T., Border, J.R. and Allgöwer, M. (1990) Classification of Soft Tissue Injuries. In *Manual of Internal Fixation*, 3rd edn (M.E. Müller, M. Allgöwer, R. Schneider and H. Willeneger, eds). Berlin: Springer-Verlag, pp. 151–158.

Sanders, R., Swiontkowski, M., Nunley, J. and Spiegel, P. (1993) The management of fractures with soft-tissue disruptions. *J. Bone Joint Surg.*, **75A**, 778–789.

Schatzker, J. and Tile, M. (1987) *The Rationale of Operative Fracture Treatment*. Berlin: Springer-Verlag.

Sidor, M.L., Zuckerman, J.D., Lyon, T. et al. (1993). The Neer classification system for proximal humeral fractures. An assessment of interobserver reliability and intraobserver reproducibility. *J. Bone Joint Surg.*, **75A**, 1745–1750.

Siebenrock, K. A. and Gerber, C. (1993) The reproducibility of classification of fractures of the proximal end of the humerus. *J. Bone Joint Surg.*, **75A**, 1751–1755.

Thomsen, N.O., Overgaard, S., Olsen, L.H. et al. (1991). Observer variation in the radiographic classification of ankle fractures. *J. Bone Joint Surg.*, **73B**, 676–678.

Turen, C.H., Burgess, A.R. and Vanco, B. (1995) Skeletal stabilization for tibial fractures associated with acute compartment syndrome. *Clin. Orthop.*, **315**, 163–168.

Weber, B.G. (1972) *Die Verletzungen des oberen Sprunggelenks*. 2 Auflage. Berne: Hans Huber-Verlag.

Winquist, R.A. and Hansen, S.T. (1980) Comminuted fractures of the femoral shaft treated by intramedullary nailing. *Orthop. Clin. North Am.*, **11**, 633–648.

Winquist, R.A., Hansen, S.T. and Clawson, D.K. (1984) Closed intramedullary nailing of femoral fractures: a report of five hundred and twenty cases. *J. Bone Joint Surg.*, **66A**, 529–539.

3

General classifications of soft-tissue injuries

G. V. Scerri

Introduction

The soft-tissue wound has a substantial effect upon the ultimate recovery of the patient. Appropriate early management of a traumatic wound reduces the degree of subsequent morbidity. Poor assessment of the extent of damage often results in delay in treatment which may have unfavourable consequences. A good system of classification should offer the clinician a quick and effective way of estimating the nature and significance of the injury, so that a suitable treatment plan can be formulated.

The factors which determine the merit of any classification of soft-tissue injuries are:

- Convenience — is it easy and accurate to apply?
- Communication — does it facilitate interchange of information between clinicians?
- Value — does it guide the clinician towards optimal treatment and outcome?

The last of these factors is the most important. The classification should lead the surgeon towards identification of the place the injury occupies on the so-called 'reconstructive ladder' (see below). With evolving surgical technology, classifications have become more complex and comprehensive. Many aim to predict the results of treatment of injured extremities but are often difficult to apply.

Although there are several accepted classifications of soft-tissue injuries in use, many are 'traditional' and therefore not attributable to any single author. They fall into the following broad groups, which will be discussed in greater detail in this chapter:

1. *'Athletic' injuries*: A classification of soft-tissue injuries in terms of aetiology and severity.
2. *Traumatic wounds*: This deals with the extent of damage to the wound edges and to underlying structures.
3. *Burns*: A widely-used system of classifying thermal injury by causative agent and extent of skin damage.

'Athletic' Injuries

There is general agreement among authors (Garrick, 1981; Muckle, 1982; Williams and Sperryn 1972) that these should be classified in two ways (Oakes, 1972, 1995):

A. Aetiology

1. *Direct or extrinsic injuries* — usually due to blunt trauma, e.g. muscle contusions, lacerations. Forces involved are usually great, hence frequent accompanying fractures, joint dislocations and ligament injuries.
2. *Indirect injuries ('sprains and tears')* — due to sudden overloading of musculotendinous units, e.g. 'torn muscles'.
3. *Chronic or overuse* — due to repeated overload and/or frictional resistance, e.g. tenosynovitis.
4. *Acute on chronic* — sudden rupture of a chronic lesion due to repetitive micro-fatigue, e.g. rupture of chronic Achilles tendonitis.

B. Severity

1. *Grade 1* — mild pain within 24 hours of injury, especially when stressed, with or without local tenderness.
2. *Grade 2* — pain during activity necessitating cessation of activity, with moderate to severe pain when stressed.
3. *Grade 3* — severe pain or loss of function due to complete or near complete rupture or avulsion of at least a portion of a ligament or tendon; a palpable defect may be present. One may define this grade more accurately as *failure of load-carrying capacity* (and therefore *loss of*

function) — a ligament may appear intact microscopically but still have no load-carrying capability (Butler, Grood and Zernicke, 1978).

The aetiological classification may not always be easy and accurate to apply but is useful for interchange of information between clinicians. It may also be of relevance to epidemiological study, but does not really help with treatment options, particularly in cases of direct trauma. In practice this does not present a problem, as in these cases the therapeutic indications are usually clear.

The severity classification is also easily applicable. The predominant attribute is its value in determining indications for treatment. Grade 1 injuries respond well to anti-inflammatory medication and support. Grade 2 injuries require rest and elevation in addition to the above, while Grade 3 injuries very often require surgical management. Its major drawback is poor accuracy, since it relies heavily upon the patient's and clinician's subjective interpretation of the severity of pain caused by the injury.

Muscular Haematomas

These have been classified by Muckle (1982):

A. *Intermuscular haematoma* — the blood may track distally from the site of the haematoma. It appears as a bruise some distance from the site of injury, often after some time.

B. *Intramuscular haematoma* — remains confined by epimysium and may take up to three times as long to heal as an intermuscular haematoma.

C. *Mixed (inter- and intramuscular) haematoma* — this is the most common presentation.

This system would seem to have little clinical and prognostic significance. It has not been validated by clinical study.

Traumatic Wounds

There is general consensus that these should be classified in the following categories, as exemplified by Jurkiewicz and Flint (1981):

A. Tidy Wounds
   1. *Lacerations without loss of skin*
   2. *Injuries with minimal loss of skin*

   Both of the above may be present in conjunction with neural, vascular and skeletal injuries.

B. Untidy Wounds
   1. *Lacerations with marginal skin necrosis only*
   2. *Avulsion injury with substantial loss of skin* — this includes degloving injuries and may be present in conjunction with neural, vascular and skeletal injuries, e.g. gunshot wounds, high energy trauma in road traffic accidents
   3. *Crush injury* — may also be associated with compartment syndrome.

This classification is easy to apply and is useful as a means of communicating the nature and magnitude of injury. It is based on the causative agent and on the amount of tissue damage, and in particular skin damage and/or loss. It has distinct value as a guide to management. Clearly the need for reconstruction is greater as one moves down the list, and this correlates well with the steps on the so-called reconstructive ladder. This starts with the simplest form of reconstruction, namely direct repair of damaged structures, and continues through skin grafting to local flaps, then distant flaps and ultimately microvascular free tissue transfer (Figure 3.1). Last on the list are crush injuries with untidy wounds: in such cases even major reconstruction is insufficient to restore useful function and early amputation is often indicated.

The major headings of *tidy* and *untidy wounds* are, however, slightly ambiguous as they may be taken either to refer to the edges of the wound or to its bacteriological environment (they refer to the former). The terms *incised* and *ragged wounds* would perhaps be more suitable. Also, the term *substantial loss of skin* is highly subjective and reduces the accuracy of the system.

This classification does not give any indication as to whether the wound is clean, contaminated or

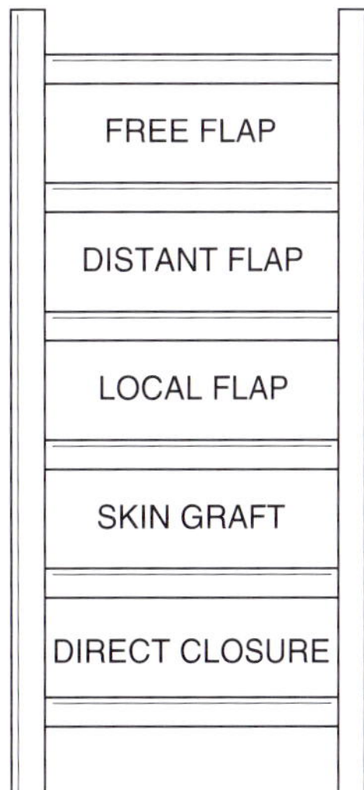

Figure 3.1. The reconstructive ladder. The complexity of reconstruction increases as one moves up the 'ladder'. If direct skin closure is not possible, a skin graft is indicated. If this is inappropriate for the site, or there is not a suitable graft bed, then a flap is indicated. If a local flap is not available, a distant or free flap may be necessary.

infected and, as these factors have considerable influence on management, this has to be considered as a major weakness. It has little value as an aid to prognostication and this is partly due to its inherent broad scope. It is imprecise in dealing with other prognostic factors, such as neural and vascular damage and whether or not bone is exposed. It has not been validated by clinical study.

Lister's Classification (1977)

This is more objective and perhaps more reliable. Although originally devised to cover all types of hand injury, it can be applied to other sites:

A. Tidy Wounds

These are inflicted by sharp instruments and have well-defined edges. They are associated with little tissue destruction. Fractures are uncommon but if present are not comminuted. Primary healing normally occurs with minimal scarring. Primary repair of tendons, arteries and nerves is easily achieved.

1. *Sliced wounds*
   - i. With tissue loss
   - ii. With flaps
2. *Puncture wounds*
3. *Finger tip injuries*
   - i. Nail bed
   - ii. Pulp
4. *Fractures*
5. *Amputations*
   - i. Partial
   - ii. Complete

B. Untidy Wounds

These result from tearing or bursting of the skin caused, for example, by explosions or machinery injuries. Skin edges are irregular with areas of doubtful viability. Primary healing is less likely than with tidy incised wounds and results in extensive scarring.

1. *Crush*
2. *Avulsion*
3. *Pressure gun*
4. *Mixed injuries*

Tissue Damage

This forms the basis of a further classification proposed by Oakes (1995):

1. Skin and deep fascia
2. Muscle-tendon unit and periosteal attachments
3. Muscle compartment
4. Joints and associated structures, e.g. dislocations, meniscal damage, tenosynovitis

The tissue damage classification seems to be easy to apply and would be useful as a precise communication tool for clinicians. It may have value as a therapeutic indicator but this has not been formally tested.

Mangled Extremity Severity Score (MESS)

Johansen et al. (1990) devised this aid to deciding which severely injured lower limbs are salvageable and which require amputation. The degree of soft-tissue damage forms part of a simple point scoring system. The authors were able to prove that the MESS score is highly accurate and readily applicable at initial assessment of the injured patient. Further proof of its reliability came when tested by Robertson (1991) and later by McNamara, Heckman and Corley (1994) who applied it specifically to Type IIIB and Type IIIC open fractures of the tibia (Gustilo and Anderson, 1976; Gustilo, Mendoza and Williams, 1984). It has been shown to be equally robust when applied to severe injuries of the upper limb (Slauterbeck et al., 1994).

The MESS scoring system is described in greater detail in Chapter 1.

Burn Injury

Burn injury represents one of the most common forms of soft-tissue trauma, with a huge range of severity and sequelae. Burn injury is often sustained in conjunction with musculoskeletal injury due to mechanical trauma. Some types of burns, notably chemical and electrical burns and frostbite, themselves frequently cause musculoskeletal injury. Thermal burns at certain sites, e.g. the hands, are more likely to cause damage to underlying muscle and bone due to the proximity of these elements to the surface.

There are two principal classifications of burn injury in widespread clinical use. They are classical and time-honoured, and I have not been able to trace either to the work of any single author. One deals with the *causative agent* of the burn and the other addresses the *depth of the burn*. In addition, there are a number of more recent subclassifications that will be discussed.

I. Causative Agent of Burn Injury

The pathological definition of a burn is an injury that causes coagulative necrosis of the tissues. Injuries

that fit this definition have been traditionally classified under the following headings:
a. Thermal
b. Chemical
c. Electrical
d. Friction
e. Radiation
f. Frostbite
Although this is a very general classification, it is of considerable value as a basis for epidemiological study. For example, Edlich et al. (1982) analysed major burn injuries under similar headings to the above and found that cause and frequency of injury differed according to age, race and sex.

The classification is of value as a guide for treatment and prognosis. Its value within the context of musculoskeletal trauma increases if the following subclassifications are added:

Thermal burns

Classically, the causes of these have been subdivided into:
a. *Dry heat*
b. *Moist heat (scalds)*
This subclassification separates thermal injuries due to dry heat or to contact with a hot surface, which often cause underlying musculoskeletal damage, from those caused by scalding, which seldom do.

Chemical burns

Most chemical burns can cause damage to muscle and bone elements, the magnitude of which depends upon the concentration of the substance involved and the length of time of application. They are usually classified as follows:
a. *Acid*
b. *Alkali*
c. *Other*
This classification is easy and accurate to apply if one obtains a good clinical history. Acid and alkali are the commonest agents causing chemical burn injuries. There is specific treatment for each category of chemical, with specific antidotes and treatment regimes for many named agents that fall under the heading of 'other'.

Electrical burns

Two classifications have been devised:
A. *Esses and Peters (1981)*
   1. Thermal
   2. Arc
   3. Direct contact
B. *Luce and Gottlieb (1984)*
   1. Flash burns
   2. True electrical burns

Of these, the former is of little descriptive, clinical or epidemiological value, as it conveys no information as to the nature and magnitude of the injury itself. The authors claim that each of the three types of burn has its own distinct appearance and prognosis, but the number of patients studied was too small ($n = 9$) for valid conclusions to be reached. Luce and Gottlieb's classification is, by comparison, very useful. It identifies those cases where the electrical current has travelled along the deep structures of the limb (a preferred route in view of the relatively high resistance of the skin). In such cases, in spite of the normal appearance of the overlying skin, there is damage ranging from compartment oedema to frank tissue destruction including burnt bone. Severe cases require major reconstruction or early amputation. Flash burn tends to be limited to the skin surface alone. It is managed conservatively and carries an excellent prognosis.

Frostbite

This type of injury fits the pathological description of a burn, i.e. a wound in which there is coagulative necrosis of tissue. Localized cold injuries are classified in two ways: one is based upon the acute physical findings observed after cold exposure and subsequent rewarming; the other notes both the physical findings and the environmental conditions under which the injury occured (Robson and Smith, 1990):
A. *Physical findings alone*
   1. First degree — whitish plaque in the area of injury (Knize, 1977), with oedema and erythema. There is no tissue necrosis.
   2. Second degree — superficial blistering containing clear or milky fluid, with surrounding oedema and erythema. These heal spontaneously, with little or no tissue loss.
   3. Third degree — deeper blisters containing reddish/purple fluid or areas of dark skin discoloration without blisters. Tissue necrosis is common.
   4. Fourth degree — deep cyanosis of injured part without blistering or oedema.
B. *Environmental conditions*
   1. Chilblains — due to chronic intermittent exposure to high humidity/low temperature environment but without tissue freezing. Characterized by discomfort in involved limbs

Table 3.1 Classifications of burn depth in common use today

British Classification[1]	*US Classification*[2]	*Depth of Injury*
Superficial epidermal	First degree	Within epidermis
Superficial partial thickness	Second degree	Loss of epidermis and part of dermis
Deep partial thickness		Deep dermal damage
Full thickness	Third degree	Complete loss of dermis
	Fourth degree	Damage to tendon, muscle, bone

[1]Adapted from Muir, Barclay and Settle (1987)
[2]After Press (1990)

which resolves spontaneously without tissue loss.

2. Trench foot — due to prolonged (hours to days) exposure of wet feet to temperatures between 1°C and 10°C but without tissue freezing. Initial anaesthesia with erythema, progressing to pallor and swelling with blistering. There is involvement of the deep soft tissues but some spontaneous resolution may occur beneath the non-viable tissue cover.
3. Frostbite — due to exposure to freezing temperatures or below; clinically this has been shown to be −6.5°C or below for a duration of 1 hour or more (Knize et al., 1969)

Both these classifications are useful, although the first is more popular due to its ease of application and accuracy, with good inter-rater reliability due to the precise definitions of the various categories. Its value is in the prediction of extent of tissue recovery. The second classification is of use in diagnosis of the exact type of cold injury. It is not as precise as the first, except in the case of frostbite where the parameters are very clearly defined. It is of limited use as a tool for estimating degree of injury and for prognostication. The two classifications cannot be combined.

II. Depth of Burn

The classification of burns according to depth can be traced back to Italy in 1483, when Jean de Vigo first differentiated between different levels of skin damage (Wallace, 1980). It was Fabricius Hildanus who first devised three categories of burn depth in his publication *De Combustionibus* in 1607, while the Frenchman Dupuytren again related the degree of injury to depth in a treatise published in 1832.

Classifications in common use today are given in Table 3.1. These classifications provide a reliable guide to management of the burn wound. They are, however, not always easy to apply in the acute clinical situation. The parameters used to estimate burn depth are general appearance, vascularity and sensation. Accurate estimation of burn depth from general appearance is difficult unless the clinician is experienced in dealing with burn injury. The apparent vascularity of the burnt surface can change markedly in the first 48 hours or so after the injury. Skin sensation varies according to the different regions of the body and cannot be adequately assessed in cases where there is no non-injured contralateral region with which to compare. All of these factors mitigate against good inter-rater reliability and lead to difficulties in communication.

Clearly the only category of significance to musculoskeletal trauma is that of full thickness burns (US Fourth degree). Unfortunately the classifications were devised to distinguish those injuries that would resolve spontaneously from those requiring skin grafting, i.e. they concentrated mainly on the skin injury. For example, the inclusion of a Fourth degree in the US classification is a recent modification. Furthermore, although the distinction between burns down to muscle but not involving bone, and burns down to and including bone, is important from the clinical and prognostic viewpoint, this is not included in either classification.

Conclusion

There are few general classifications of soft-tissue injury, and the majority of these have their origins in the early history of surgery. The fact that they have survived the test of time, albeit with slight modification in some cases, is a testimony to their value in surgical practice. The classifications in use today facilitate the appraisal of trauma severity. They are also useful as pointers to appropriate early management and reconstruction. Unfortunately their value as prognosticating tools has not been established.

References

Butler, D.L., Grood, E.S. and Zernicke, R.G. (1978) Biomechanics of ligaments and tendons. *Exerc. Sports. Sci. Rev.*, **6**, 125–181.

Edlich, R.F., Glasheen, W., Attinger, E.O. et al. (1982) Epidemiology of serious burn injuries. *Surg. Gynecol. Obstet.*, **154**, 505–509.

Esses, S.I. and Peters, W.J. (1981) Electrical burns; pathophysiology and complications. *Can. J. Surg.* **24**, 11–14.

Fabricius Hildanus, G. (1607) *De Combustionibus*. Basle: sumpt Ludovici Regis.

Garrick, J.G. (1981) The sports medicine patient. *Nurs. Clin. North Am.*, **16**, 759–766.

Gustilo, R.B. and Anderson, J.T. (1976) Prevention of infection in the management of one thousand and twenty five open fractures of long bones. *J. Bone Joint Surg.*, **58A**, 453–458.

Gustilo, R.B., Mendoza, R.M. and Williams, D.N. (1984) Problems in the management of Type III (severe) open fractures: a new classification of Type III open fractures. *J. Trauma*, **24**, 742–746.

Johansen, K., Daines, M., Howey, T. et al. (1990) Objective criteria accurately predict amputation following lower extremity trauma. *J. Trauma*, **30**, 568–573.

Jurkiewicz, M.J. and Flint, L.M. (1981) Special problems in skeletal and soft tissue trauma. In *Trauma* (D. Carter and H.C. Polk, eds). London: Butterworths, p. 47.

Knize, D.M. (1977) Cold injury. In *Plastic Surgery* (J.M. Converse, ed.), Vol. 1. Philadelphia: Saunders.

Knize, D.M., Weatherly-White, R.C. Paton, B.C. and Owens, J.C. (1969) Prognostic factors in the management of frostbite. *J. Trauma*, **9**, 749–751.

Lister, G. (1977) *The Hand: Diagnosis and Indications*. Edinburgh: Churchill Livingstone, p.11.

Luce, A. and Gottlieb, S.E. (1984) 'True' high-tension electrical injuries. *Ann. Plast. Surg.*, **12**, 321–326.

McNamara, M.G., Heckman, J.D. and Corley, F.G. (1994) Severe open fractures of the lower extremity: a retrospective evaluation of the mangled extremity severity score (MESS). *J. Orth. Trauma*, **2**, 81–87.

Muckle, D.S. (1982) Injuries in sport. *Royal Soc. Health J.*, **10**, 93–94.

Muir, I.F.K., Barclay, T.J. and Settle, J.A.D. (1987) *Burns and their Treatment*. London: Butterworths, pp. 55–59.

Oakes, B.W. (1982) Acute soft tissue injuries: nature and management. *Austr. Family Physician*, **10** (Suppl.), 3–16.

Oakes, B.W. (1995) Classification of injuries and mechanisms of injury, repair healing and soft tissue remodelling. In *Science and Medicine in Sport* (J. Bloomfield ed.), 2nd ed. Cambridge: Blackwell Science, pp. 224–245.

Press, B. (1990) Thermal and electrical injuries. In *Plastic Surgery* (W. Smith and J. Aston, eds). London: Little, Brown, pp. 675–730.

Robertson, P.A. (1991) Prediction of amputation after severe lower limb trauma. *J. Bone Joint Surg.*, **73B**, 816–818.

Robson, M.C. and Smith, D.J. Jr (1990) Cold injuries. In *Plastic Surgery* (J.G. McCarthy, ed.), Vol. 1. Philadelphia: Saunders, pp. 851–852.

Slauterbeck, J.R., Britton, C., Moneim, M.S. and Clevenger, F.W. (1994) Mangled extremity severity score: an accurate guide to treatment of the severely injured upper extremity. *J. Orth. Trauma*, **8**, 282–285.

Wallace, A.F. (1980) Burns. In *The Progress of Plastic Surgery*. London: MEEUWS, pp. 1–10.

Williams, J.G.P. and Sperryn, P. (1972) Overuse injury in sport and work. *Br. J. Sports Med.*, **6**, 50–51.

4

Infection

B. L. Atkins and A. R. Berendt

Introduction

After the initial risks of airway obstruction and haemorrhage have been successfully managed, infection has the greatest influence on subsequent mortality and morbidity in the trauma patient.

It is important to identify the factors that predispose a patient to infection in order to take measures to reduce the risk and to predict patients who have a worse prognosis following trauma. It is also important to be able to describe the patient population entering a trauma unit in terms of their risk of infection to make valid comparisons of infection rates between units. The first part of this chapter discusses the classification systems available for predicting the risk of infection following trauma. Most of the systems that attempt to predict the risk of local infection have been validated primarily on surgical wounds in general. Many of the principles however also apply to traumatic wounds and to surgical wounds following trauma. The use of trauma severity scores to predict systemic sepsis is also discussed.

The second part of the chapter discusses the classification of infection once it has occurred. This has been confined to a discussion of soft-tissue infections in general and post surgical wound and deep tissue infections including osteomyelitis. We have not included classification of sepsis at distant sites to the injury (such as ventilator associated pneumonia, urinary tract infections, etc.) although there is a brief section on scoring systems for systemic sepsis, mainly used on intensive care units.

In order to facilitate the understanding of surgical infections we include some basic microbiological definitions:

Colonization: This refers to the presence of a population of bacteria that are in a stable relationship with the host and not invading to cause infection. A postoperative wound may be healing well, yet a routine wound swab grows *Staphylococcus aureus* or *Pseudomonas aeruginosa*. If the wound is healthy this represents colonization only and needs no treatment.

Infection: This is usually a clinical diagnosis. The organism has invaded (or is producing toxins) to cause disease. Surface swabs are not helpful in differentiating infection from invasion and often lead to unnecessary antibiotic treatment. When deep infection is suspected, high quality, aseptically obtained deep samples such as an aspirate, pus, soft tissue or bone for bacteriology (and histology where possible) are crucial to the effective management of the patient. This should be done before starting antibiotics unless the patient is haemodynamically compromised or surgery is likely to be delayed. Even if antibiotics have been commenced it is still important to send specimens for culture.

Contamination: This is used in two contexts; first, a traumatic wound may be contaminated at the roadside with dirt and bacteria. With adequate cleaning and debridement the majority of these bacteria will be removed. As already discussed, the degree of likely bacterial contamination is associated with the subsequent risk of wound infection. Second, when specimens are taken for bacterial culture these can become contaminated either intra-operatively or in the laboratory. The organisms are usually skin flora. When a device-related infection (often caused by skin flora) is suspected, multiple specimens should be taken from different sites using different instruments to reduce the risk of false positive results from contamination.

The Risk of Infection

The Risk of Wound Infection

There are several important risk factors for wound infection. These include host and local wound factors as well as the degree of bacterial contamination of the wound.

NRC definitions of wounds

The National Research Council (1964) created a set of definitions to help predict the probability of surgical wound infection based on the likely degree of intra-operative bacterial contamination. They divided wounds into four categories. The categories and the associated infection rates are as follows:

a. *Clean* — elective, primarily closed, no acute inflammation encountered, no entrance of normally or frequently colonized body cavities (gastrointestinal, oropharyngeal, genitourinary, biliary or tracheobronchial tracts) and no break in sterile technique.

Clean wounds were further subdivided into refined clean and other clean. Refined clean were elective, primarily closed and undrained wounds. Other clean were not primarily closed or were drained mechanically through the incision or a separate stab wound. (Refined clean wounds had an infection rate of 3.3% and other clean wounds an incidence of 7.4%.)

b. *Clean-contaminated* — non-elective case that is otherwise a clean, controlled opening of a normally colonized body cavity, minimal spillage or break in sterile technique, re-operation through clean incision within 7 days, negative exploration through intact skin. (Infection rate 10.8%.)

c. *Contaminated* — acute non-purulent inflammation encountered, major break in technique or spillage from hollow organ, penetrating trauma less than 4 hours old, chronic open wounds for grafting. (Infection rate 16.3%.)

d. *Dirty* — purulence or abscess encountered or drained, preoperative perforation of colonized body cavity, penetrating trauma more than four hours old. (Infection rate 28.6%.)

The NRC classification of wound risk was further validated in a 10-year prospective study of surgical wounds at the Foothills Hospital, Calgary in 62 939 patients (Cruse and Foord, 1980). They found infection rates as shown in Table 4.1.

Table 4.1 Infection rates according to bacterial contamination category (Cruse and Foord, 1980)

Category of wound	*Infection rate (%)*
Clean	1.5
Clean-contaminated	7.7
Contaminated	15.2
Dirty	40
Overall	4.7

Lennard, Hargiss and Schwenknecht (1985) used the same categories to survey infection rates prospectively in 7129 surgical wounds over two years. Their rates were lower. This illustrates the problem that the nature of the host population and the method of defining a wound infection need to be taken into account when comparing infection rates between different institutions (Table 4.2).

Table 4.2 Infection rates according to bacterial contamination category (Leonard, Hargiss and Schwenknecht, 1985)

Category of wound	*Infection rate (%)*
Clean	0.8
Clean-contaminated	3.4
Contaminated	3.6
Dirty	9.9
Overall	1.7

Weigelt (1985) modified the NRC classification of surgical incisions to apply specifically to trauma operations. The original NRC classification indicated that fresh traumatic wounds are contaminated wounds and therefore have infection rates of 20–25%. This is not representative of true rates of infection after trauma. The modified system was evaluated prospectively on 1436 trauma patients; 1079 had penetrating injuries and 257 had blunt injuries. The overall infection rate was 11.3%.

The Weigelt categories were:

Clean:

a. Abdominal incision with no intra-abdominal injuries. (Infection rate 3.2%)
b. Blunt extremity wound. Less than six hour delay in patient going to theatre. (Infection rate 5.8%)

Clean contaminated:

Penetrating extremity wound, unless theatre delayed for greater than 6 hours. (Infection rate abdomen 8.1%, extremity 14.2%)

Contaminated:

Abdominal incision with colonic injury or any procedure with a six hour or more delay in going to theatre. (Infection rate abdomen 24.6%, extremity 25.9%)

None of the above classifications take any account of host factors. They base their prediction of infection purely on likely bacterial contamination of the wound. The NRC classification was extended to include host factors by the following systems.

The SENIC index

In the 1970 SENIC project (Study on the Efficacy of Nosocomial Infection Control) (Haley et al., 1985), 58 498 patients were followed postoperatively for infection. Analysing 10 risk factors with stepwise logistic regression techniques they identified four independent risk factors for wound infection and created the SENIC risk index. This index was validated on a further 59 352 surgical patients in 1975–1976. By including factors that measure patient susceptibility, as well as level of wound contamination, it was found to be more powerful in separating patients into groups with different levels of infection risk (Table 4.3). The rates of infection are shown in Table 4.4. Within each category of the index there was little variation in the rates over the categories of the traditional (NRC) system indicating that the traditional system had little predictive value after controlling for the SENIC index.

The hospital infections programme, CDC Atlanta (Culver et al., 1991) went on to modify this classification system after surveillance of 84 691 surgical procedures in 44 hospitals participating in the National Nosocomial Infections Surveillance System (Table 4.5). Their data showed surgical infection rates of 2.1, 3.3, 6.4 and 7.1% in the four traditional wound classifications. When their risk index was applied, rates of infection were 1.5, 2.9, 6.8 and 13.0% (risk index 0, 1, 2 and 3 respectively).

The risk index created provided a better predictor of surgical wound infection than the traditional wound classification. These included a broad range of surgical procedures including 1292 limb amputations, 5657 spinal fusions, 4419 open reductions of fractures, 5696 joint prostheses and 5552 'other' musculoskeletal procedures. The surgical wound infection rate increased significantly with the numbers of risk factors present. There were six procedures that were exceptions to this rule and one of these was limb amputation.

Table 4.3 SENIC Risk index (Haley et al., 1985)

Variable	*Score*
Operation involving the abdomen	1
Operation lasting more than two hours	1
Contaminated or dirty infected wound by the NRC classification system	1
Having three or more diagnoses at discharge	1
Total	4

Table 4.4 Rates of infection with SENIC index (Haley et al., 1985)

SENIC index	*Infection rate (%)*
0	1.9
1	3.6
2	8.9
3	17.2
4	27.0

None of these scoring systems have been specifically developed for trauma cases. However, trauma cases were included in the validation of these schemes and therefore a system such as the Culver classification could be used quite appropriately in a study of traumatic wounds. Other factors should perhaps also be included. Pories et al. (1991) looked at the epidemiological features of patients with nosocomial infections following trauma (229 infections in 2496 patients of whom 16 had wound or deep wound infections). They found that patients developing infection were significantly older and had a higher Injury Severity Score (Baker et al., 1974) (described in Chapter 1) than other patients. Patients who were hypotensive on admission were more likely to develop infection than those who were not. Polk (1993) emphasizes the importance of this shock time in predisposing to infection. A system for classifying

Table 4.5 Wound classification (Culver et al., 1991)

Variable	*Score*
An American Society of Anesthesiologists (ASA) preoperative physical status score of 3, 4 or 5	1
An operation classified as contaminated or dirty infected	1
An operation lasting over T hours where T depends on the operation performed	1
Total	3

the risk of wound infection following trauma should therefore include the following elements:

1. The likely degree of bacterial contamination of the wound. The Weigelt modification of the NRC categories seem to relate the closest to traumatic wounds, although it should be noted that the majority of these were gunshot wounds.
2. Host factor status. The American Society of Anesthesiologists (1982) pre-operative assessment used in the Culver score would be appropriate. The age of the patient is also important according to Pories et al. (1991).
3. The duration of hypotension on admission.
4. The delay in taking the patient to theatre. Four hours was used in the NRC definitions and six hours in the Weigelt definitions. Neither have been validated in being the appropriate cut-off but it is clear that a delay of some hours does predispose to infection.
5. The duration of the operation as used in the SENIC and Culver classifications.
6. Some assessment of the severity of trauma such as the Injury Severity Score (ISS) (see Chapter 1, p. 4).

The Risk of Major (Systemic) Infection

Several trauma assessment scores have been developed since the early 1980s for the purposes of triage of patients, epidemiology and prediction of mortality (see Chapter 1). Few scoring system evaluations have specifically looked at major infection rates in addition to mortality. Most deaths from trauma in hospital are due to infection so the major infection rate is likely to parallel mortality. Pories et al. (1991) showed a significant relationship between the development of infection and the ISS score, the Glasgow coma score and the trauma score. Hershman et al. (1988) looked at infection rates and correlated this with an Outcome Predictive Score (OPS). The score is derived from the severity of the injury overall (ISS) adjusted for age, degree of bacterial contamination and monocyte HLA-DR antigen expression on hospital admission. They evaluated this score in 61 patients and found that the OPS correlated with the development of major infection as well as with mortality. The same patients were assessed by five other trauma scores (including the ISS score) that did not include an assessment of immunological competence, no correlation with major infection rates was found. Such a score is unlikely to be widely applicable as it relies on fluorescence activated cell sorter (ALS) analysis of monocyte antigens at admission.

The Classification of Established Infection

Soft-tissue Infections

Soft-tissue infections can be described according to the tissue level involved, the organism(s) responsible or the surgical treatment required to treat the infection. These elements are variably emphasized in different classification systems. They can also be primary or secondary to trauma or surgery. In practice, most classifications of soft-tissue infections are based on simply naming the entity (e.g. necrotizing fasciitis) and the organism.

Mandell, Douglas and Bennett classification

Mandell, Douglas and Bennett (1995), in their textbook of infectious diseases, describe a classification of soft tissue infections primarily using a description of the clinical entity with a secondary category of the organism responsible (Table 4.6).

Ahrenholtz classification

Ahrenholtz (1988) uses a similar classification using both the organism and the tissue level to define the entity (Table 4.7). This is quite a useful classification although there is usually more than one distinct clinical entity within each group, for example if a different site in the body is affected. The classification is confusing because *Staphylococcus aureus* and *Streptococcus pyogenes* are represented twice, both in individual columns and together under Staph. and Strep. This seems to be a problem in classification of soft-tissue infections as the spectrum and variety of clinical entities caused by these two organisms is wide.

Baxter classification

Baxter (1972) described a very practical classification of soft-tissue infections according to the surgical management required. He describes soft-tissue infections in terms of the following categories:

1. Those requiring extensive radical surgical excision and drainage.

These principally involve a single tissue plane with advancing infection. Diagnosis is often delayed and Baxter comments that the late signs often lead to the misnaming of an infection on the basis of the extensive deep tissue involvement rather than the initial tissue plane. In this category he includes acute streptococcal haemolytic gangrene, necrotizing fasciitis,

Table 4.6 A classification of soft-tissue infection (modified from Mandell, Douglas and Bennett, 1995, with permission)

Type of lesion	*Aetiological agent*
Primary pyodermas	*S. aureus*, Group A streptococcus
Impetigo	*S. aureus, Candida, P. aeruginosa, Pityrosporum ovale*
Folliculitis	*S. aureus*
Furuncles and carbuncles	*S. aureus*, Group A streptococcus, *Candida, P. aeruginosa*
Paronychia	Group A streptococcus
Ecthyma	Group A streptococcus
Erysipelas	*T. pallidum, H. ducreyi, Sporothrix, B. anthracis,*
Chancriform lesions	*F. tularensis, M. ulcerans, M. marinum*
	Corynebacterium diphtheriae
Membranous ulcers	Group A streptococcus, *S. aureus*, rarely various other
Cellulitis	organisms
Infectious gangrene and gangrenous cellulitis	Group A streptococcus; mixed infections with
Streptococcal gangrene and necrotizing fasciitis	Enterobacteriacae and anaerobes
Progressive bacterial synergistic gangrene	Anaerobic streptococci plus a second organism
Gangrenous balanitis	(*S. aureus, Proteus*)
Gas gangrene; crepitant cellulitis	Group A streptococcus: mixed infections with enteric
Gangrenous cellulitis in immunosuppressed patients	bacteria (*E. coli, Klebsiella* etc.) and anaerobes
	Clostridium perfringens and other clostridial species;
	Bacteroides, peptostreptococci, *Klebsiella, E. coli*
	Pseudomonas, Aspergillus, agents of mucormycosis
Erythrasma	*Corynebacterium minutissimum*
Nodular lesions	*Candida, Sporothrix, S. aureus, M. marinum, Nocardia*
	brasiliensis, Leishmania
Hyperplastic and proliferative lesions	*Nocardia, Pseudoallerischia boydii, Blastomyces*
	dermatidis, Paracoccidiodes, Phialophora,
	Cladosporum
Bacilliary angiomatosis, epithelioid angiomatosis	*Bartonella (Rochlimaea) henselae, Bartonella*
	(Rochlimaea) quintana
Erythema chronicum migrans	*Borrelia burgdorferi*
Secondary bacterial infections including those of:	*P. aeruginosa, Enterobacter*, other Gram-negative
Burns	bacilli, various streptococci, *S. aureus, Candida,*
	Aspergillus
Eczematous dermatitis and exfoliative erthrodermas	*S. aureus*, Group A streptococcus
Chronic ulcers	
Traumatic lesions (abrasions, animal bites, insect bites	Coliforms, *P. aeruginosa*, peptostreptococci,
etc.)	enterococci, *Bacteroides, C. perfringens*
Pyoderma gangrenosa	*P. multicida, C. diphtheriae. S. aureus*, group A
Cutaneous involvement in systemic bacterial and mycotic	streptococcus
infections	

streptococcal myositis, clostridial cellulitis and Gram-negative anaerobic cutaneous gangrene.

2. Those requiring excision of tissue.

These are generally less fulminant (except for the clostridial infections) and tend to involve several types of tissue simultaneously. In this group he includes gas gangrene (Clostridium spp.), tetanus, progressive bacterial synergistic gangrene, non-clostridial gas producing infections, non-clostridial myonecrosis and human bite lesions.

3. Those not requiring surgical treatment.

Although in the non-surgical category, when signs of inflammation occur in closed wounds opening of such wounds is mandatory as primary therapy. In this category Baxter puts entities such as erysipelas, lymphangitis, lymphadenitis and purpura fulminans.

Freischlag

In order to see if the accurate classification of necrotizing soft-tissue infections made a difference to therapy, morbidity and mortality, Freischlag, Ajalat and

Table 4.7 A classification of soft-tissue infections (Ahrenholtz, 1988)

Tissue level	*Common surgical pathogens*				
	Streptococcus pyogenes	*Staph. aureus*	*Clostridium perfringens*	Staph. and Strep.	Mixed enteric bacteria
Epidermis	Ecthyma contagiosum	Scalded skin syndrome		Impetigo	
Dermis and subdermis	Erysipelas	Folliculitis/ abscess	Clostridial cellulitis or abscess	Synergistic gangrene	Tropical ulcer
Fascial planes	Streptococcal gangrene	Carbuncle		Necrotizing cellulitis	
Muscle	Streptococcal myositis	Muscular abscess/ pyomyositis	Clostridial myonecrosis	Non-clostridial myonecrosis	

Busuttil (1985) reviewed the notes of 21 patients treated at the UCLA Medical Center. Infections were divided into necrotizing fasciitis, clostridial gangrene, synergistic gangrene and streptococcal gangrene. Diabetes or immunosuppression was present in 71% of the patients. The cause was a perforated viscus in 43% and a traumatic injury in 43%. No single sign was diagnostic for a specific type of necrotizing soft-tissue infection. Mortality was greater if surgery was delayed more than 24 hours or if surgery was not radical and encompassing all devitalized tissue. The accurate diagnosis did not impact on management and outcome. They emphasize the role of radical surgery and broad spectrum antibiotics plus aggressive fluid resuscitation for all cases.

Dellinger

Dellinger (1988) also emphasizes that it is more important to recognize the urgent surgical nature of soft-tissue infection rather than to be too concerned with the precise label placed on a given case. He divides severe necrotizing infections into clostridial, severe non-clostridial and combined infection. Within these categories he describes several entities and comments that they are multiple disease entities which all require a common approach.

The classification of soft-tissue infections is difficult and confusing because there is a wide spectrum of clinical entities with much overlap of clinical signs, organisms and management.

There is also variation in the terminology used for each condition. If a classification system was required in order to perform an audit or an interventional trial it may be that documentation of individual signs (e.g. gas in tissues, fever, hypotension) would be more useful than attempting to name each entity.

Nosocomial Post-surgical Wound and Acute Bone Infection

This section specifically discusses post-surgical wound infections. It should be noted that there is little emphasis in these classification systems on microbiological results, particularly of surface swabs (see microbiological defintions at the beginning of the chapter). In the CDC Atlanta definitions positive cultures are only taken into account if they are from deep sites taken aseptically. This is an important concept. It is not possible to differentiate mere colonization of a wound from a true infection of the wound by surface swabs, no matter what organism is grown. There is no role for surface swabs in diagnosing the presence of postoperative infection though they are of value in detecting organisms with infection control implications (such as methicillin-resistant *Staphylococcus aureus*).

The definition that should be used for a wound infection depends on the purpose of the study being done. The presence of pus, for example, is highly specific for infection but lacks sensitivity. Scoring systems that measure the degree of erythema and discharge are more sensitive but less specific. Relatively simple definitions are required for studies involving large numbers of patients or for simple surgical audit. The definitions used by Glenister et al. (1992) are convenient for this purpose. For smaller studies a more discriminatory classification is needed which can detect lesser degrees of infection and quantify these in order to be able to detect a significant difference within a reasonable number of cases. The ASEPSIS score is an example of such a system (see Table 4.8).

Table 4.8 ASEPSIS score (Wilson et al., 1986)

Criterion	*Points*
Additional treatment	
Antibiotics	10
Drainage of pus under local anaesthesia	5
Débridement of wound (general anaesthesia)	10
Serous discharge	daily 0–5
Erythema	daily 0–5
Purulent exudate	daily 0–10
Separation of deep tissues	daily 0–10
Isolation of bacteria	10
Stay as inpatient prolonged over 14 days	5

Glenister et al. (1992) wound definitions

All wound infections must have purulent discharge in or exuding from a wound or seen on direct examination at the operative site.

Major infection is present when the wound is broken down, gaping or completely dehisced or there is evidence of septicaemia, spreading cellulitis or lymphangitis.

Minor infection is present when the wound is not broken down, gaping or completely dehisced and there is no evidence of septicaemia, spreading cellulitis or lymphangitis.

UK Surgical Infection Group definitions

Proposed definitions for the audit of postoperative infection were published by the Surgical Infection Study Group (Peel and Taylor, 1991) in the UK. They stated that the definition of wound infection should not depend on the results of bacteriological studies. Wound infection may be classified according to aetiology, time or severity. The definition of severity (major and minor) is as above. The definitions of aetiology and timing are as follows:

Primary and secondary wound infection:

The infection should be considered *primary* unless there is a predisposing complication. *Secondary* infection may follow a complication which results in the discharge of serum, haematoma, cerebrospinal fluid, urine, bile, pancreatic juice, gastric or intestinal contents from the wound, contaminated by bacteria from within the patient or from the environment.

Time:

With regard to time, wound infection may be divided into:

Early:	presenting within 30 days of operation.
Intermediate:	presenting between 1 and 3 months of operation.
Late:	presenting more than three months after operation.

CDC definitions 1992

In 1988 CDC Atlanta published their definitions of nosocomial infections (Garner et al., 1988). These were modified in 1992 by Horan et al. The term 'surgical wound' infection was changed to 'surgical site' infection. This was so that the anatomical location of the infection would be clear and also includes the term 'organ/space' as the term 'wound' signifies only the incision through soft tissues. Systemic or remote complicating infections, e.g. postoperative pneumonia after cholecystectomy are not classified as surgical site infections. Specific sites of organ/space surgical site infection (SSI) in trauma include disc space, joint or bursa, osteomyelitis and spinal abscess.

For surveillance classification purposes SSIs are divided into incisional and organ/space SSIs. Incisional SSIs are further classified as involving only the skin (superficial incisional SSIs) or involving deep soft-tissues (e.g. fascial and muscle layers) of the incision (deep incisional SSIs). Organ/space SSIs involve any part of the anatomy (organs or spaces) other than the incision opened or manipulated during the operative procedure.

1. Superficial incisional SSI

Infection within 30 days after the operative procedure involving only skin or subcutaneous tissue of the incision and at least one of the following:

a. Purulent drainage from the superficial incision.
b. Organisms isolated from an aseptically obtained culture of fluid or tissue from the superficial incision (i.e. not simply a surface swab).

Table 4.9 ASEPSIS: daily score of wound (Wilson et al., 1986)

Wound characteristic	*Proportion of wound infected (%)*					
	0	*< 20*	*20–39*	*40–59*	*60–79*	*> 80*
Serous exudate	0	1	2	3	4	5
Erythema	0	1	2	3	4	5
Purulent exudate	0	2	4	6	8	10
Separation of deep tissues	0	2	4	6	8	10

c. At least one of the following signs or symptoms of infection: pain or tenderness, localized swelling, redness or heat — and superficial incision is deliberately opened by surgeon, unless culture of incision is negative.
d. Diagnosis of superficial incisional SSI by the surgeon or attending physician.

A stitch abscess is not classified as a superficial incisional SSI.

2. Deep incisional SSI

Infection occurring within 30 days if no implant is left in place or within one year if an implant is in place and the infection appears to be related to the surgical procedure and infection involves deep soft-tissues (e.g. fascial and muscle layers) of the incision and at least one of the following is present:

a. Purulent drainage from the deep incision but not from the organ/space component of the surgical site.
b. A deep incision spontaneously dehisces or is deliberately opened by the surgeon when the patient has at least one of the following signs or symptoms: fever (> 38°C), localized pain or tenderness, unless culture of the incision is negative.
c. An abscess or other evidence of infection involving the deep incision is found on direct examination, during re-operation or by histopathological or radiological examination.
d. Diagnosis of a deep incisional SSI by a surgeon or attending physician.

3. Organ/space SSI

Infection occurring within 30 days after the operative procedure if no implant is left in place or within one year if an implant is in place and the infection appears to be related to the operative procedure and infection involves any part of the anatomy (e.g. organs or spaces) other than the incision opened or manipulated during the operative procedure and at least one of the following is present:

a. Purulent drainage from a drain that is placed through a stab wound into the organ/space.
b. Organisms isolated from an aseptically obtained culture of fluid or tissue in the organ/space.
c. An abscess or other evidence of infection involving the organ/space on direct examination, during reoperation, or by histological or radiological examination.

4. Diagnosis of an organ/space SSI by a surgeon or attending physician.

ASEPSIS score

Wilson et al. (1986) proposed a system for scoring wound infections in order to detect quantifiable differences within reasonable numbers of cases. They called this the ASEPSIS score (Table 4.8).

A points scale was developed for daily wound infection which measured the proportion of the wound affected at each point in time. Wounds are classified on the basis of the total ASEPSIS score over a length of time (Table 4.9). In the initial study this scoring method was used on 250 cardiothoracic patients with sternotomy and leg wounds. A further 51 patients were later assessed by a second observer and the score was found to be reproducible.

The ASEPSIS system was criticized by Byrne et al. (1989) who felt it was a significant advance in scoring systems in terms of objectivity and accuracy but had certain limitations particularly an assumption regarding the linearity of the index.

The ASEPSIS score has never been used for traumatic wounds and although it has face validity, it is very labour intensive and has not been subjected to rigorous validation.

The most practical and precise definitions for wound infections are those of CDC Atlanta described above. Although designed to cover surgical wounds in general, they would be appropriate for trauma cases. A caution, however, is that an infection has to occur within one year when an implant is *in situ*, to be classified as a postoperative infection. Infections in and around joint prostheses and internal fixation devices may occur later than this and still be due to organisms of low virulence introduced at the time of implantation. In any surveillance study or clinical trial, follow-up of patients in the community after discharge is crucial. A survey by Law, Mishriki and Jeffery (1990) showed that 59% of 83 patients who developed wound infection following a 'clean' procedure were diagnosed by follow-up in the community.

The Classification of Chronic Osteomyelitis

There is no universally accepted classification system for osteomyelitis although several have been proposed in order to rationalize the management of and research into this condition.

Waldvogel classification

Waldvogel, Medoff and Swartz (1970) described three categories of osteomyelitis:

1. Haematogenous
2. Contiguous focus
3. Osteomyelitis associated with vascular insufficiency.

This classification has limited use when it is applied to infection after musculoskeletal trauma as the majority of cases will fall into the contiguous focus category, although some are compounded by vascular insufficiency.

Kelly classification

Kelly (1984) published data from 268 patients with osteomyelitis seen in the Mayo Clinic between 1967 and 1980. He classified the cases as follows:

1. Haematogenous osteomyelitis
2. Osteomyelitis in a united fracture (fracture with union)
3. Osteomyelitis in a non-union (fracture with non-union)
4. Postoperative osteomyelitis without fracture. This category includes cases in which infection occurred after an operation but was not associated with a fracture, for example after removal of a bone graft.

Weiland classification

Weiland, Moore and Daniel (1984) defined chronic osteomyelitis as a wound with exposed bone, positive results on bacterial culture and drainage for more than six months. They further divided cases in order to allow meaningful comparison of treatment options at other institutions. Their categories were as follows:

Type I open exposed bone without evidence of osseous infection but with evidence of soft-tissue infection.

Type II circumferential, cortical and endosteal infection with radiographs that show a diffuse inflammatory response, increased density and spindle-shaped sclerotic thickening of the cortex in addition to areas of bone resorption and often an involucrum surrounding a sequestrum.

Type III cortical and endosteal infection associated with segmental bone defect.

Gordon and Chiu classification

Gordon and Chiu (1988) described 14 patients with tibial infections treated with debridement and microvascular transplantation of muscle. In order to look at prognosis they classified patients according to the severity of the underlying osseous disease. They categorized patients into types A, B and C.

Type A a tibial defect and non-union without significant segmental loss.

Type B a tibial defect that is more than 3 cm long and an intact fibula.

Type C a tibial defect that is more than 3 cm long, involving both the tibia and the fibula.

May classification

May et al. (1989) published a classification system which focused on the status of the tibial bone after soft-tissue and skeletal débridement. It is useful to predict the length of rehabilitation the patient will need. They proposed five categories:

Type I intact tibia and fibula capable of withstanding functional loads (rehabilitation 6–12 weeks).

Type II intact tibia with bone graft needed for structural support (rehabilitation 3–6 months).

Type III tibial defect 6 cm long or less and intact fibula (rehabilitation 6–12 months).

Type IV tibial defect more than 6 cm long and intact fibula (rehabilitation 12–18 months).

Type V tibial defect more than 6 cm long and no functional, intact fibula (rehabilitation 18 months or more).

This classification provides ideal times for rehabilitation and also aids in the management decisions involved in post-traumatic osteomyelitis. However, each patient is different and patient factors influence the management of and the rehabilitation time from osteomyelitis.

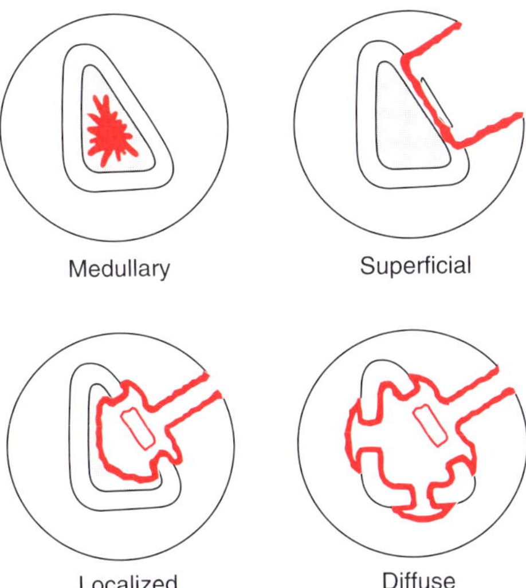

Figure 4.1. The anatomical classification of adult osteomyelitis (reproduced with permission from Evarts, C.M. (1990) *Surgery of the Musculoskeletal System*, Vol. 5).

Cierny and Mader classification

In the classification scheme by Cierny and Mader (1984), host factors are taken into account. Their classification scheme includes four treatment and prognostic factors which must be considered when assessing treatment results and efficacy of treatment alternatives.

1. The degree of necrosis
2. The condition of the host
3. The site and extent of involvement
4. The disabling effects of the disease.

The infection and the host are staged using four anatomical types (Figure 4.1) and three physiological classes (Table 4.10).

1. Medullary osteomyelitis denotes infection confined to the intramedullary surfaces of the bone and includes haematogenous osteomyelitis and infected intramedullary rods.
2. Superficial osteomyelitis which represents a true contiguous focus infection of bone occurs when an exposed infected necrotic area of bone lies at the base of a soft-tissue wound.
3. Localized osteomyelitis is usually characterized by a full thickness cortical sequestration, which can be removed surgically without compromising bone stability.
4. Diffuse osteomyelitis usually requires an intercalary resection of the bone for cure. It includes those infections with a loss of bony stability either before or after débridement surgery.

Table 4.10 Cierny and Mader (1984) classification of osteomyelitis

	Description
Anatomical type	
Stage 1	Medullary osteomyelitis
Stage 2	Superficial osteomyelitis
Stage 3	Localized osteomyelitis
Stage 4	Diffuse osteomyelitis
Physiological class	
A host	Normal host
B host	Systemic compromise (Bs)
	Local compromise (Bl)
C host	Treatment worse than the disease

The host is classified as an A, B or C host. An A host has normal physiological, metabolic and immunological capabilities. The B host is either systemically compromised (s), locally compromised (l) or both (Table 4.11). When the morbidity of treatment is worse than that imposed by the disease itself, the patient is given the C classification.

Thalgott et al. (1990) modified the Cierny and Mader (1984) classification system in order to categorize patients with infections after posterior spinal instrumentation and fusion. Microbiological results were used to indicate the severity of infection rather than a clear anatomical classification. They retained the concept of class A, B and C hosts.

The Severity of Systemic Sepsis

A number of scoring systems for measuring the severity of surgical infections have been published in order

Table 4.11 Cierny and Mader (1984) classification of the host

Systemic (Bs)	*Local (Bl)*
Malnutrition	Chronic lymphoedema
Renal, liver failure	Venous stasis
Diabetes mellitus	Major vessel compromise
Chronic hypoxia	arteritis
Immune deficiency	extensive scarring
Malignancy	radiation fibrosis
Extremes of age	small vessel disease
Immunosuppression	Complete loss of local
Tobacco abuse	sensation
IV drug use	

to facilitate comparison of patient groups. The basis of most scoring systems is the concept of the 'multiple organ failure' syndrome. The published studies have been on a wide variety of surgical and medical patients but the principles apply to multiple trauma patients. These systems are primarily the domain of intensivists but the trauma surgeon may need a working knowledge of scoring systems past and present if planning interventional or descriptive studies of the multiply injured patient.

Fry, Pearlstein and Fulton in 1980 showed that the risk of death after surgery or severe trauma was usually due to infection and increased with the number of failed organs. This was reconfirmed in several later studies including that by Knaus, Draper and Wagner (1985).

A detailed description of the scoring systems that have been proposed since the 1980s is beyond the scope of this book but brief details are given. A review by Dellinger (1988) gives more extensive information and is well worth referring to.

Stevens (1983) defined seven organ systems and assigned a value of 0 to 5 in each system. Scores were calculated by squaring the value assigned to each organ system and adding the highest three scores to get the 'Sepsis Severity Score'. He evaluated this in 35 patients initially and in a further 31 patients later (Stevens, Clemmer and Laub, 1986). The mortality increased significantly in those with a score above 40.

Skau, Nystrom and Carlsson (1985) modified the sepsis severity score of Stevens and compared it with the Acute Physiology Score (APS) (described below). The patient population he used comprised individuals with intra-abdominal sepsis. Elebute and Stoner (1983) divided the clinical features of sepsis into four classes:

Local effects of tissue infection
Scoring of temperature
Scoring of the secondary effects of sepsis
Scoring of laboratory data.

The total score is the sum of individual scores. This system was applied initially to 15 patients by Elebute and Stoner and then in more detail by Dominioni, Diongi and Zanello (1987a,b) on a further 135 patients. Scores ranged from 10 to more than 30. The mortality rate in those with scores of 20 or below was 20% whereas in those with scores greater than 20 the mortality was 89%. A further study by the same authors confirmed this finding in 127 patients.

Dominioni, Diongi and Zanello (1987a,b) also measured acute phase proteins during sepsis and incorporated this into the score. This did not, however, significantly improve the accuracy of the prediction above that described for the severity score alone.

Acute physiology score

Knaus et al. (1981) proposed a two part scoring system called the APACHE score. The first part is an acute physiological assessment (APS-34) based on abnormalities in 34 possible physiological measurements on the first day of admission to intensive care. The second part is a chronic health evaluation (CH) which reviews the patients' medical history. This system was not developed to look specifically at patients with severe surgical sepsis but the initial trial did include such patients. Since APACHE was developed, Knaus et al. (1989) have gone on to modify the scoring system to create the APACHE II and then the APACHE III scores. The purpose of APACHE has been predominantly to predict patient outcomes from key characteristics in populations presenting to intensive care units. In this way variations in the quality of intensive care in different units can be compared as can different management options. These scoring systems are useful therefore for comparing populations of patients but should not be used to aid individual treatment decisions.

Concluding Remarks

Few of the available classification systems described here meet all the required criteria outlined earlier in the book. Nonetheless usable systems exist for the prediction of risk of infection, for the description of postoperative infection and for the classification of chronic osteomyelitis. Large scale validations of these systems in the specific population of trauma patients would be desirable.

References

Ahrenholtz, D.H. (1988) Necrotising soft-tissue infections. *Surg. Clin. North Am.*, **68**(1), 199–214.

American Society of Anesthesiologists (1982) New classification of physical status. *Anesthesiology*, **24**, 111.

Baker, S.P., O'Neill, B., Haddon, W. and Long, W.B. (1974) The Injury Severity Score: A method for describing patients with multiple injuries and evaluating emergency care. *J. Trauma*, **14**, 187–196.

Baxter, C.R. (1972) Surgical management of soft tissue infections. *Surg. Clin. North Am.*, **52**(6), 1483–1499.

Byrne, D.J., Malek, M.M., Davey, P.G. and Cuschieri, A. (1989) Postoperative wound scoring. *Biomed. Pharmacother.*, **43**, 669–673.

Cierny, G. and Mader, J.T. (1984) Adult chronic osteomyelitis. *Orthopaedics*, **7**, 1557.

Cierny, G. (1990) Classification and treatment of adult osteomyelitis. In *Surgery of the Musculoskeletal System* (C.M. Evarts, ed.), Vol. 5. New York: Churchill Livingstone, p. 4337.

Cruse, P.J.E. and Foord, R. (1980) The epidemiology of wound infection. A 10 year prospective study of 62 939 wounds. *Surg. Clin. North Am.*, **60**, 27–40.

Culver, D.H., Horan, T.C., Gaynes, R.P. et al. (1991) Surgical wound rates by wound class, operative procedure and patient risk index. *Am. J. Med.*, **91**, Suppl. 3B, 152s–157s.

Dellinger, P.D. (1981) Severe necrotising soft tissue infections. *JAMA*, **246**, 1717–1721.

Dellinger, P.D. (1988) Use of scoring systems to assess patient with surgical sepsis. *Surg. Clin. North Am.*, **68**, 123–145.

Dominioni, L. and Dionigi, R. (1987) The grading of sepsis and the assessment of its prognosis in the surgical patient: a review. *Surg. Res. Comm.*, **1**, 1–11.

Dominioni, L., Diongi, R. and Zanello, M. (1987) Sepsis score and acute phase protein response as predictors of outcome in septic surgical patients. *Arch. Surg.*, **122**, 141–146.

Elebute, E.A. and Stoner, H.B. (1983) The grading of sepsis. *Br. J. Surg.*, **70**, 29–31.

Freischlag, J.A., Ajalat, G. and Busuttil, R.W. (1985) Treatment of necrotising soft tissue infections. *Am. J. Surg.*, **149**, 751–755.

Fry, D.E., Pearlstein, L. and Fulton, R.L. (1980) Multiple system organ failure. *Arch. Surg.*, **115**, 135–140.

Garner, J.S., Jarvis, W.R., Emori, T.G. et al. (1988) CDC definitions for nosocomial infections. *Am. J. Infect. Control*, **16**, 128–140.

Glenister, H.M., Taylor, L.J., Cooke, E.M. and Bartlett, C.L.R. (1992) A study of surveillance methods for detecting hospital infection. Published by PHLS Central Public Health Laboratory, Division of Hospital Infection.

Gordon, L. and Chiu, E.J. (1988) Treatment of infected nonunions and segmental defects of the tibia with staged microvascular muscle transplantation and bone grafting. *J. Bone Joint Surg*, **70A**, 377.

Haley, R.W., Culver, D.H., Morgan, W.M. et al. (1985) Identifying patients at high risk of surgical wound infection. *Am. J. Epidemiol.*, **121**, 206–215.

Hershman, M.J., Cheadle, W.G., Kuftinec, D. et al. (1988) An outcome predictive score for sepsis and death following injury. *Injury*, **19**, 263–266.

Horan, T.C., Gaynes, R.P., Martone, W.J. et al. (1992) CDC definitions of surgical site infections: A modification of CDC definitions of surgical wound infections. *Am. J. Infect. Control*, **20**, 271–274.

Kelly, P.J. (1984) Infected nonunion of the femur and tibia. *Orthop. Clin. North Am.*, **15**, 481.

Knaus, W.A., Draper, E.A. and Wagner, D.P. (1985) Prognosis in acute organ failure. *Ann. Surg.*, **202**, 685–693.

Knaus, W.A., Draper, E.A., Zimmerman, J.E. and Wagner, D.P. (1989) Development of APACHE. *Crit. Care Med.*, **17**, s181–s158.

Knaus, W.A., Zimmerman, J.E., Wagner, D.P. et al. (1981) APACHE acute physiology and chronic health evaluation: a physiologically based classification system. *Crit. Care Med.*, **9**, 591–597.

Law, D.J.W., Mishriki, S.F. and Jeffery, P. (1990) The importance of surveillance after discharge from hospital in the diagnosis of post operative wound infection. *Ann. R. Coll. Surg. Engl.*, **72**, 207–209.

Lennard, E.S., Hargiss, C.O. and Schwenknecht, F.D. (1985) Post operative wound infection surveillance by use of bacterial contamination categories. *Am. J. Infect. Control*, **13**, 147–153.

Mandell, G., Douglas, L. and Bennett, J. (1995) *Principles and Practice of Infectious Diseases*, 4th edn. Edinburgh: Churchill Livingstone, p. 910.

May, J.W., Jupiter, J.B., Weiland, A.J. and Byrd, H.S. (1989) Current concepts review: Clinical classification of post traumatic tibial osteomyelitis. *J. Bone Joint Surg.*, **71A**, 1422.

National Academy of Sciences, National Research Council, Division of Medical Sciences Ad hoc Committee on Trauma (1964) Postoperative wound infections: The influence of ultraviolet irradiation on the operating room and of various other factors. *Ann. Surg.*, **160** (Suppl. 2), p. 1.

Peel, A.L.G. and Taylor, E.W. (1991) Surgical Infection Study Group. Proposed definitions for the audit of post-operative infection: a discussion paper. *Ann. R. Coll. Surg. Engl.*, **73**, 385–388.

Polk, H.C. Jr (1993) Factors influencing the risk of infection after trauma. *Am. J. Surg.*, **165**, Suppl. 2A, 2S–7S.

Pories, S.E., Gamelli, R.L., Mead, P.B. et al. (1991). The epidemiological features of nosocomial infections in patients with trauma. *Arch. Surg.*, **126**, 97–99.

Skau, T., Nystrom, P.O. and Carlsson, C. (1985) Severity of illness in intra-abdominal infection. *Arch. Surg.*, **120**, 152–158.

Stevens, L.E. (1983) Gauging the severity of surgical sepsis. *Arch. Surg.*, **118**, 1190–1192.

Stevens, L.E., Clemmer, T.P. and Laub, R.M. (1986) Fibronectin in severe sepsis. *Surg. Gynecol. Obstet.*, **162**, 222–228.

Surgical Infection Study Group (1991) see Peel and Taylor (1991).

Thalgott, J.S., Cotler, H.B., Sasso, R.C. et al (1990) Postoperative infections in spinal implants. Classification and analysis — a multicentre study. *Spine*, **16**, 981–984.

Waldvogel, F.A., Medoff, G. and Swartz, M.N. (1970) Osteomyelitis: A review of clinical features, therapeutic considerations and unusual aspects. *N. Engl. J. Med.*, **282**, 198, 260, 316.

Weigelt, J.A. (1985) Risk of wound infections in trauma patients. *Am. J. Surg.*, **150**, 782–784.

Weiland, A.J., Moore, J.R. and Daniel, R.K. (1984). The efficacy of free tissue transfer in the treatment of osteomyelitis. *J. Bone Joint Surg.*, **66A**, 181–193.

Wilson, A.P.R., Sturridge, M.F., Treasure, T. and Gruneberg, R.N. (1986). A scoring method (ASEPSIS) for postoperative wound infections for use in clinical trials of antibiotic prophylaxis. *Lancet*, **i**, 311–313.

5

Union and non-union

S. L. Royston

Introduction

The process of bony regeneration after fracture has rightly been described as one of the most remarkable of all biological repair processes in higher animals. The injured bone is reconstituted, in most cases, by normal bony tissue rather than scar tissue. In those cases in which this regeneration does not proceed normally, factors related to the injured bone, the mechanism and extent of the injury and the methods of treatment may all contribute to this failure. Whilst all orthopaedic clinicians have an intuitive concept of union, a precise workable definition is difficult to find. Definitions based on the absolute restoration of radiographic, mechanical or structural normality are fraught with shortcomings, as many fractured bones will function normally long before regaining pre-injury mechanical strength (Richardson et al., 1994) or all radiographic evidence of the fracture disappearing (Tiedeman et al., 1990). It is the purpose of this chapter to examine classifications which have been used to describe union and non-union. The author will examine union as both a process (of uniting) and as a state (of having united) as well as the state and description of non-union.

Describing the Process of Union

The processes by which bony reconstitution occurs after injury were formalized by McKibbin in 1978. The four types of repair process vary in their speed of onset, ability to bridge gaps, tolerance of movement or rigidity and dependence on intact surrounding soft tissues, as shown in Table 5.1.

The primary callus response occurs almost universally after bony injury but is short-lived. Exuberant callus is formed but the process has little ability to bridge gaps. It is suggested that the primary callus response is initiated by viable cells in the periosteum. Because the primary callus response is so universal, its presence has little relevance to the clinical monitoring of fracture healing.

If bony continuity is not restored by the primary callus response, external bridging callus is formed. The cells responsible for this process are probably derived from the surrounding soft tissues. The callus is formed quickly and is tolerant of movement. It is the predominant form of repair in fracture treated by incomplete immobilization, i.e. cast, elastic external fixation or intramedullary nailing. Total immobilization eradicates the external callus response. Therefore, in fractures treated by rigid internal fixation, the appearance of this form of callus must indicate implant loosening or failure, the so-called 'irritation callus'. However, because of its critical dependence on external soft tissues, this warning sign may be absent if soft-tissue damage has been extensive.

Late medullary callus formation is slow to occur, but has the ability to bridge gaps if reasonable stability is provided. Soft-tissue damage is less important as the callus derives its vascular supply from the intramedullary circulation. In clinical practice, this process is seen when bone gaps remain for a long period after injury, for example, after imperfect reduction and plating or rigid external fixation.

Primary cortical healing is characterized by an absence of callus formation. This is the method by which bone heals after anatomical reduction and rigid plating. In essence, the fracture should progressively disappear on sequential radiographs. No callus formation will be visible radiologically if rigid fixation can be maintained for a sufficient period of time.

Table 5.1 The four types of repair processes

Type of healing	*Speed*	*Bridging ability*	*Motion tolerance*	*Tolerance of rigidity*	*Dependence on soft tissues*
Primary callus response	+ + + +	+	+ + + +	+ + + +	–
External bridging callus	+ + +	+ + +	+ + +	–	+ + + +
Late medullary callus	+ +	+ + + +	+ +	+ + +	–
Primary cortical healing	+	–	–	+ + + +	–

The major significance of McKibbin's system is that, for each method of fracture treatment, clinicians have an indication of the likely sequence of radiographic changes in the fracture. Deviation from this expected sequence implies that union is not progressing satisfactorily and that clinical intervention may be required.

Since the appearance of this seminal article, numerous review papers have been published claiming to examine bone repair in a new light (Simmons, 1985; Hulth, 1989; Cornell and Lane, 1992; Einhorn, 1995). Several have usefully reviewed the microscopic and chemical events underlying this sequence. None, however, has managed to expand or modify McKibbin's original description significantly.

Quantifying the Process of Union

In 1994, Andrew et al. undertook an immunohistochemical study of the role of inflammatory cells in healing human fractures. They describe the histological grading of fracture healing. This is purely descriptive and follows the sequence described by McKibbin.

- 1a Blood clot. Areas of mesenchymal/granulation tissue invasion at periphery of section.
- 1b Most of section is mesenchymal/granulation tissue. May be areas of residual clot. No bone or cartilage formation seen.
- 2a Areas of cartilage formation but no obvious woven bone seen. No remodelling or new lamellar bone.
- 2b Areas of woven bone but no obvious cartilage seen. No remodelling or lamellar new bone.
- 2c Areas of cartilage and woven bone. No remodelling or new lamellar bone.
- 3a Areas of woven and lamellar bone. No cartilage seen. Active bone remodelling.
- 3b Areas of woven and lamellar bone and cartilage. Active bone remodelling.

In clinical practice, union is most often assessed on subjective radiographic and mechanical criteria. Both of these methods have been shown to be fallible (Matthews, Kaufer and Sonstegard 1974). Several studies have examined more objective methods of determining 'mechanical' and 'radiographic' union and, by inference, true bony union.

Hammer (1985 and with others in 1985) examined the process of bony union by analysing sequential radiographs of 127 tibial fractures which were also tested mechanically to determine fracture stiffness. An earlier study had used a testing jig to apply a known force to 157 healing tibial fractures treated conservatively whilst the induced deformation was measured radiographically. The deflection ratio (DR), given by the deformation divided by the applied bending moment, was calculated. It was found that in these 157 fractures, union was sufficiently strong to allow unsupported weight bearing when DR was 0.08 or less and that this policy produced no risk of late deformity or refracture. In the radiological study, this value of DR was used with the measured deflection to divide the 127 fractures into six classes (Table 5.2). The radiographs taken at the time of each measurement, 889 pairs of radiographs in total, were independently assessed by seven experienced radiologists and classified into five groups on the basis of features of bony union (Table 5.3).

When mechanical and radiographic data were compared, poor agreement between mechanical and radiographic data was found, even when the uncertain radiograph group (Group 3) was excluded. 44% of the mechanically stable fractures were classified as radiographic groups 4 or 5, whilst 55% of fractures classed as 'probably unstable' showed radiographic

Table 5.2 Mechanical classification of presumed stability

Class I	Definitely stable
Class II	Probably stable
Class III	Presumably stable
Class IV	Presumably unstable
Class V	Probably unstable
Class VI	Definitely unstable

Table 5.3 Classification of radiographic stage of union

Radiographic assessment	*Callus formation*	*Fracture line*	*Stage of union*
Group 1	Homogenous bone structure	Obliterated	Achieved
Group 2	Massive. Bone trabeculae crossing fracture line	Barely discernible	Achieved
Group 3	Apparent. Bridging of fracture line	Discernible	Uncertain
Group 4	Trace. No bridging of fracture line	Distinct	Not achieved
Group 5	No callus formation	Distinct	Not achieved

evidence of satisfactory union. These reports suggest that even widely accepted radiographic criteria of fracture healing may be unreliable and that mechanical testing may be a more reliable method of classifying union. Interestingly, based on a small subgroup of their 104 patients, the authors felt that their mechanical testing was able to distinguish 14 delayed unions from 13 non-unions. In the delayed unions the mechanical stiffness increased, although more slowly than normal, whereas in the non-unions stiffness measurements failed to improve after reaching a plateau. This supports the comments made below about the importance of identifying 'at risk' slow-to-heal fractures from frank non-union.

Jorgensen (1972) used a Hoffman external fixator fitted with a micrometer to measure the angular deformity produced by a known bending moment applied by a spring balance. He was able to produce a graph of stiffness against time for healing tibial fractures, defining safe healing as occurring if fracture stiffness reached 12–20 Nm/degree. He later retreated from this objective position, in 1979, by suggesting that 'an experienced clinician' could reproduce these findings without a standardized bending moment as long as the induced deflection asymptotically approached 0.25–0.5°.

Richardson et al. (1994) used a force transducer with a flexible electrogoniometer and an instrumented fixator bar to monitor increases in fracture stiffness in tibial fractures treated by monolateral external fixation. In the first cohort of 117 patients, the decision to remove the fixator was made on the basis of clinical and radiographic findings. In the second group of 95 patients, the fixator was only removed when a sagittal plane stiffness of 15 Nm/degree was achieved. In the first group, early refracture occurred in 8 patients, in each of whom fracture stiffness at fixator removal had been less than 15 Nm/degree. There were no refractures in the second group. Although both groups reached a safe stiffness at a similar time after injury (17.6 weeks in group 1; 19.4 weeks in group 2), the fixator was removed at an earlier mean time in group 2 (24 weeks for group 1; 21.7 weeks for group 2). The authors suggest that routine measurement of fracture stiffness will improve the management of externally fixed fractures by defining the optimal time for fixator removal.

Biomechanical measurement of fracture healing has been undertaken using ultrasound transmission across the fracture site and by vibration analysis of the whole tibia (Cunningham, Kenwright and Kershaw, 1990). Both methods are open to the criticism that they assume a consistent relationship exists between the parameter measured and the parameter of interest, i.e. fracture stiffness. As early as 1958, Siegel, Anast and Fields reported a rabbit model in which 'a pattern of sound velocity which seemed to parallel the progress of fracture union was noted.' The authors also noted that in unhealed tibial fractures in humans, ultrasound transmission is slowed. Gerlanc et al. (1975) showed that in 21 tibial fractures treated by casting, the conduction velocity in the fractured tibia approached 80–88% of normal by the time the patients were able to walk comfortably. As part of a larger study, Cunningham showed that, although conduction velocity increased with time, at the time of clinical union, normal velocities were not achieved. They suggested that if ultrasound transmission is used to monitor fracture healing, velocities in the injured limb should exceed 80% of those in the intact bone. Benirschke et al. (1993) and Tower, Beals and Duwelius (1993) described the results of resonant frequency analysis of healing tibial fractures. The resonant frequency (RF) of an intact tibia, a simple beam structure, is proportional to its stiffness. The resonant frequency of a fractured tibia will be proportional to the amount of structural coupling between the fracture fragments. Benirschke et al. noted a consistent rise in RF of 14 healing tibial fractures over time and suggested that, when the RF reached 88% of normal, union could be assumed. Tower's group compared the tibial stiffness index (the ratio of the resonant frequency of the fractured bone to that of the intact tibia) with a 20-point clinical and radiographic score on which a score of 17 or more was taken to indicate union. Despite significant correlation between RF and clinical/radiographic progress, the authors felt they could not recommend the technique for routine monitoring of fracture healing. Whilst mechanical testing has been shown to be useful in fractures treated by casts or external fixation, such techniques have little relevance to fractures treated by other means.

Several attempts have been made to classify the radiological appearances of healing fractures and correlate such appearances to the mineral content of the fracture callus and, by extension, to its mechanical properties. In an effort to formalize interpretation of radiographs, Lane and Sandhu (1987) described a scoring system based on three criteria: bone formation, persistence or disappearance of the fracture line and cortical remodelling (Table 5.4). Whilst this system can be criticized for its assumption that all fractures will eventually fully remodel, which may not be the case after severe fractures, it has been used in parallel with other techniques in a small number of studies with reasonable results.

Table 5.4 The scoring system of Lane and Sandhu (1987)

	Points
Bone formation	
no evidence of bone formation	0
bone formation in 25% of defect	1
bone formation in 50% of defect	2
bone formation in 75% of defect	3
full gap bone formation	4
Fracture line	
full fracture line	0
partial fracture line	2
absent fracture line	4
Remodelling	
no evidence of remodelling	0
remodelling of intramedullary canal	2
full remodelling of cortex	4

Tiedeman et al. (1990) used Lane and Sandhu's classification together with optical densitometry techniques and mechanical testing to examine the healing of both a standardized single cut osteotomy and a 6 mm defect in canine tibiae. Standardized radiographs of the healing osteotomies were analysed using a digital photometer. The densitometry was validated against ash values of eight samples, obtaining a linear relationship between measured density and percentage ash weight ($r = 0.95$ at $p < 0.001$). An exponential relationship between the densitometric evaluation of the healing fracture gap and bending rigidity was found ($r = 0.79$ at $p < 0.001$). The relationship between three-point bending rigidity and scoring according to Lane and Sandhu was much less precise, $r = 0.64$ at $p < 0.001$. When each element of the radiographic scoring was analysed separately against rigidity, it was found that scores for bone formation and union had better correlations than the combined scores and that remodelling had a very low correlation. In fact, it seemed that a combination of bone formation and union gave the closest correlation with rigidity, although this was still lower than for the densitometric data.

The inability to relate radiographic appearances to structural integrity has prompted several authors to investigate the usefulness of less conventional imaging modalities.

In clinical studies of children undergoing limb lengthening by callus distraction (De Bastiani et al., 1987), Eyres and colleagues studied the use of dual-energy X-ray absorptiometry (DEXA) alone (1993a) and in comparison with ultrasound and conventional radiography (1993b) to measure mineralization of the regenerate. Although the healing of fractures is by no means an identical process to limb lengthening, their results suggest that DEXA can accurately estimate the process of mineralization during bone regeneration and may be a tool for fracture management in the future.

Markel et al. (1990) used quantitative computed tomography (QCT), magnetic resonance imaging (MRI), single photon absorptiometry (SPA) and DEXA in a study similar to that of Tiedeman. Invasive techniques were used to measure the torsional and bending stiffness of the healing osteotomies. In general, MRI had poor agreement with local mechanical properties. QCT, SPA and DEXA had broadly similar correlations with torsional stiffness, whilst QCT had the strongest correlation with local material properties such as indentation stiffness and mineral content.

Failure of Union

Fracture healing may be abnormal for a number of reasons, some of which are related to the injured bone itself, some to the nature of the injury and some to the treatment used to treat the fracture (Saleh, 1992). Specific parts of the skeleton are acknowledged to be at increased risk of delayed union or non-union, for example the scaphoid (Herbert and Fisher, 1984) and the femoral neck (Parker and Pryor, 1993). In these cases, the non-union may almost be regarded as inevitable because of the peculiar vascular or mechanical factors existing at these sites. Nonetheless, few studies of non-unions at these sites have attempted a formal classification. For example, in Herbert's description of fixation of scaphoid fractures, he included only established non-unions as opposed to delayed unions, in the 'non-union' subgroup. Fibrous non-unions were classified as Type D1, whilst Type D2 consisted of sclerotic non-unions. No formal details are given of the criteria used to make this distinction. However, in a retrospective study of 90 scaphoid non-unions treated surgically, Cooney, Dobyns and Linscheid (1980) showed that the union rate was higher for undisplaced non-unions (85%) than for those which were displaced (65%), suggesting that even this simple distinction may have prognostic relevance. Marti Schüller and Raaymakers (1989) described a series of 50 patients with non-union of an intracapsular hip fracture treated by Pauwels adduction osteotomy. No description of the non-union itself is included, although avascular changes in the femoral head are listed according to Inoue and Ono's classification (1979).

There are three important determinants of a fracture's ability to heal: the biology of the bone fragments (which is predominantly influenced by vascularity) and the tissues in the fracture gap; the mechanical environment; and the presence or absence of infection. These features are obviously interrelated and each can be influenced by those factors listed in Table 5.5. For example, it was shown by Dickson et al. (1994) that open fractures with a demonstrated vascular lesion had a dramatically increased incidence of non-union and post-traumatic osteomyelitis.

Table 5.5 Features which should be encompassed in an all-inclusive classification system applicable to all non-unions (from Rosen, 1992)

Degree of severity (based in part on time)
- Delayed union (6 months)
- Non-union (6–8 months)
  - mobile
  - taut
  - hypertrophic
  - atrophic
- Synovial pseudarthrosis

Site and displacement
- Diaphyseal
  - non-displaced
  - displaced
- Metaphyseal
  - extra-articular
  - intra-articular

Callus
- Hypertrophic (vascular, reactive)
  - elephant foot
  - oligotrophic
- Atrophic (hypovascular, non-reactive)
  - without gap
  - necrotic butterfly or fragment
  - gap

Infection
- Non-draining (for 3 months)
  - quiescent
  - active
- Draining

Defining Non-union

The important distinction between normal healing, delayed union and non-union has been addressed by several authors, often with conflicting outcomes. In 1954, Urist, Mazet and McLean applied the term delayed union to 'ununited fractures in which (1) the radiographic examination at any time from 4 to 18 months of healing showed inadequate callus and (2) the judgement of the individual surgeon led him to advise a surgical operation to stimulate healing of the fracture.' The authors readily admitted that this distinction was arbitrary but were more specific in their use of the term non-union which was applied to fractures after 18 months of healing if six features were present:

- a bone defect
- false motion
- sclerosis of the bone ends
- rounding, mushrooming or moulding of the fracture surfaces
- sealing of the medullary canal with compact bone
- an apparent arrest of osteogenesis in the fracture gap.

The authors compared 85 ununited fractures, a mixture of delayed unions and non-unions whose exact proportions are not clear, with 100 comparable injuries and suggested that extensive bony and soft-tissue injury and inappropriate open surgery were associated with a high incidence of delayed or non-union.

There have been many attempts to define non-union as existing at a specific time after injury, with different authors suggesting different times, for example 6 months (Crenshaw, 1987), 8 months (Müller et al., 1991) or 12 months (Sharrard, 1984). None of these rigid definitions make allowance for the biological limitations imposed on the fractured bone by the energy and nature of the injury. For example, a high energy, comminuted fracture of the tibial diaphysis with soft-tissue damage may well not be united after 12 months (Blick et al., 1989). Should this failure to unite be regarded as a non-union, a delayed union or the natural course of events for this injury? Non-union is probably more usefully defined on clinical and radiographic grounds. Esterhai et al. (1981) described non-union as 'that condition existing in a fractured bone in which all reparative processes have ceased and yet bony continuity has not been restored.' Nicoll (1964) described non-union as 'a condition in which, in the opinion of the surgeon, the fragments would not have united with further conservative treatment.'

The AO group have adopted a more dogmatic approach to the classification of abnormal union processes. By their definition, any fracture not healed by 6 months is delayed. Non-union exists when there is an arrest of the repair process and the fracture has remained ununited for 6–8 months (Müller et al., 1991).

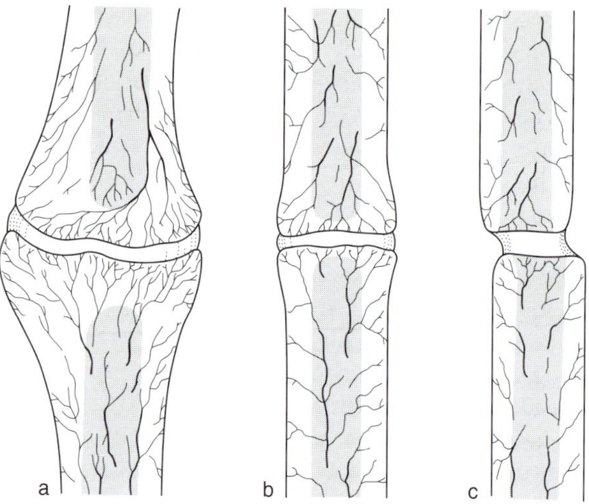

Figure 5.1. Weber and Cech (1976) classification of non-unions: (a) elephant foot; (b) horse's foot; (c) oligotrophic.

Classification of Non-unions

The most widely recognized classification on non-unions is that described by Judet, Judet and Roy-Camille (1958) and by Weber and Cech (1976). Based on radiographic and isotope studies, they classified non-unions into hypertrophic and atrophic. Hypertrophic non-unions are also described as reactive and vascular, whilst atrophic non-unions are said to be non-reactive and frequently avascular, with radiological evidence of osteoporosis. Hypertrophic non-unions have been further divided on the basis of radiographic appearances into three groups (Figure 5.1):

(a) 'Elephant's foot', non-unions in which callus is plentiful.
(b) 'Horse's hoof', non-unions which are less hypertrophic and exhibit less callus formation.
(c) Oligotrophic non-unions which show little, if any, bone formation.

Of non-unions arising from closed fractures treated conservatively, 90% will show hypertrophic appearances. Weber and Cech (1976) examined the so-called avascular non-unions in some detail, identifying the presence of one or more apparently necrotic bone fragments or bone defects as important in their production. Local soft-tissue damage, whether traumatic or iatrogenic, predisposes to atrophic non-union and to infection. There is little doubt that as internal fixation of fractures has been more widely practised, the incidence of atrophic non-unions has risen. Weber and Brunner (1981) report that only 15% of the non-unions treated by them were atrophic. A review of 100 consecutive non-unions from Sheffield (Ribbans, Stubbs and Saleh, 1992) quotes an incidence of at least twice this. Recent histological studies by Andrew et al. (1994) have, however, suggested that this widely-used nomenclature is inaccurate. Biopsies taken from non-unions before or at surgery were examined by an experienced osteo-articular pathologist. Radiologically hypertrophic non-unions often appeared to be histologically inert, whilst samples from some 'atrophic' non-unions showed intense cellular activity. This is perhaps not so surprising. In 1969, Segmüller, Cech and Bekier used isotope scintigraphy to demonstrate that most non-unions, including many with atrophic appearances, are very active in metabolic terms. True synovial pseudarthrosis is a rare form of non-union with specific radiographic, scintigraphic and histological features (Esterhai et al., 1981).

In 1979, the 2nd edition of the AO manual listed non-unions in three groups; non-infected, previously infected, and infected (Müller et al., 1979). Only the non-infected non-unions are then classified as hypertrophic or atrophic, whilst previously infected non-unions are simply grouped into those with bony contact and those with a gap. This would seem to imply that infected non-unions are not capable of division into hypertrophic and atrophic groups, which is unlikely, or that the radiological changes produced by infection somehow mask those occurring at the non-union site *per se*. However, in the 1991 edition, the corresponding section remedies this by mentioning atrophic appearances in the treatment of both infected and previously infected non-unions.

The important implication of the Weber classification is that hypertrophic non-unions are essentially 'trying to heal', bone formation is exaggerated but disorganized and blood supply is excellent. The profuse callus formation represents McKibbin's primary callus response. These non-unions are said, therefore, to require only a change in their mechanical environment to allow the callus to mature and healing to progress. The concept of absolute rigidity and compression as requirements for hypertrophic non-unions to heal has been shown to be invalid. Hypertrophic non-unions are successfully treated by distraction in (elastic) circular frames (Catagni et al., 1994; Saleh and Royston, 1996). Atrophic non-unions, in contrast, are said to require some further stimulus to bone formation as well as stability to allow them to unite. Originally, this stimulus was provided by decortication of the non-union and bone grafting. The introduction of Ilizarov techniques has changed this requirement as well. In bifocal surgery, the production of a lengthening osteotomy dramatically increases blood flow and metabolic activity throughout the operated bone (Sveshnikov et al., 1984; Aronson, 1994). There is still support for Weber's classification in the published results of non-union surgery. The time taken for union to occur after surgical intervention in series with predominantly hypertrophic non-unions (Boyd, Lipinski and Wiley, 1961; Rosen, 1979; Brighton et al., 1981; Gershuni and Pinsker, 1982) tends to be slightly less than for those series with a majority of atrophic cases (Freeland and Mutz, 1976; Green and Dlabal, 1983; Ribbans, Stubbs and Saleh, 1992).

The distinction between sterile and infected non-unions is of major clinical importance. The social and psychological cost of persistent infection at a non-union site is considerable (Toh and Jupiter, 1995) and is often worse than that of a sterile non-union or a below-knee amputation (Lerner et al., 1993).

Those few classifications of non-unions which do exist have tended to concentrate on the tibia and especially on infected tibial non-unions. This is the most common site for non-unions (Ribbans, Stubbs and Saleh, 1992) and sepsis occurs more frequently here than elsewhere in the body (Merle d'Aubigne, 1965; Paley et al., 1989).

Ilizarov's pioneering work from 1950s became known to western surgeons 30 years later. In the late 1970s, Ilizarov, with characteristic simplicity, divided non-unions into two groups on purely clinical grounds. 'Lax' non-unions were those in which any deformity could be easily and immediately corrected. A 'stiff' non-union is one in which deformity was fixed or only partially corrigible by closed means (Ilizarov et al., 1972).

Weiland, Moore and Daniel (1984) described a classification of post-traumatic osteomyelitis based on the extent of the septic lesion: Type 1 in which there is soft-tissue sepsis and exposed bone, but no evidence of bony sepsis; Type 2 in which there is circumferential cortical and endosteal infection; Type 3 in which cortical and endosteal sepsis is associated with segmental bone loss. Chronic osteomyelitis was only present if there had been drainage from a wound with exposed bone for 6 months or more. This classification was applied to a group of 33 patients undergoing 37 free tissue transfers after débridement of septic tissues. Four patients had Type 1 bone lesions, six had Type 2 and 23 had Type 3 lesions. At final review, all four of the patients with Type 1 lesions and all six with Type 2 lesions were able to walk unaided. Only 12 of 20 patients with Type 3 lesions in the lower limb were walking unassisted. Of the remaining eight, five had undergone amputation and three walked with support. Of 27 patients whose free tissue transfer was successful, six had persistent drainage at follow-up. Of these, one had a Type 2 lesion and five a Type 3 lesion. All six had drained for six months or more before treatment.

Gordon and Chiu's classification (1988) of infected tibial non-unions was also based on the use of free tissue transfer and bone grafting. Describing a retrospective series of 14 patients, their diagnostic criteria were less rigid than those of Weiland. Chronic osteomyelitis was diagnosed on the basis of a positive culture from bone and continued drainage for only 2 months or more. Non-union was defined radiographically either if there was no healing at 6 months after injury or if there was a bone defect of more than 3 cm. They describe their classification, which is based on the extent of bony damage after débridement, as having importance in both treatment choice and prognosis. Type A lesions have a tibial defect and non-union but without substantial bone loss (< 3 cm), i.e. that group in whom conventional bone grafting was likely to succeed. Type B lesions have 3 cm or more of bone loss but the ipsilateral fibula is intact. Type C included all lesions in which there is bone loss both in the tibia and the fibula. All six Type A defects healed after free soft-tissue coverage without the need for bone grafting. Of four Type B defects, three underwent vascularized fibula transplantation and one had a proximal and distal tibiofibular synostosis procedure performed. In two of the patients who underwent free fibula transfer, sepsis recurred after the bony procedure, necessitating amputation. The third had a persistent non-union at the proximal end of the transplant. Two patients with Type C defects were treated by tibial shortening, knee arthrodesis and bone grafting after soft-tissue coverage. Both healed well. The remaining two Type C defects, which measured 15 and 17 cm respectively, were treated by free fibula transfer, both of which united successfully. With the exception of the two Type B lesions mentioned above, no patient developed recurrent sepsis after bone and soft-tissue coverage.

May et al. (1989) developed a classification specifically to describe post-traumatic tibial osteomyelitis, based on the extent of tibial loss or débridement and the presence of the ipsilateral fibula suitable as a weight bearing bone after synostosis or as a graft. This classification is accompanied by treatment options available to treat injuries in each group and a likely timescale for these treatments.

Type 1: Intact tibia and fibula capable of withstanding functional loads.
(This represents a soft-tissue injury, such as a burn or degloving injury, with exposed but intact bone. Skeletal involvement is minimal. Limb reconstruction will consist of obtaining soft tissue coverage.)

Type 2: Intact tibia, with bone graft required for structural support.
(Although the tibia is in continuity, it has been so weakened by the débridement that it is likely to fracture under normal loading.)

Type 3: A tibial defect of 6 cm or less, with an intact ipsilateral fibula.

Type 4: A tibial defect of >6 cm with an intact ipsilateral fibula.

Type 5: A tibial defect of >6 cm with no usable ipsilateral fibula.

Based on this classification, the authors make the following treatment recommendations:

Type 1: Soft tissue coverage only. Expected treatment time, 6–12 weeks.

Type 2: Open cancellous bone grafting (Papineau, 1973), posterior bone grafting and tibiofibular synostosis, anterior bone grafting beneath a soft tissue flap. Expected treatment time, 3–6 months.

Type 3: Open cancellous bone grafting, posterior bone grafting and tibiofibular synostosis, anterior bone grafting beneath a soft-tissue flap, distraction osteogenesis (bifocal surgery). Expected treatment time, 6–12 months.

Type 4: Posterolateral grafting, distraction osteogenesis, fibula-pro-tibia transfer, vascularized autogenous bone graft (fibula, iliac crest, rib), non-vascular cortical bone graft, allograft transplant. Expected treatment time, 12–18 months.

Type 5: Contralateral fibula transfer, bi- or tri-focal surgery. Expected treatment time, 18 months or more.

In 1989, Paley et al. described the early Italian experience with the Ilizarov fixator in the treatment of tibial non-unions with bone loss. At the time of publication, the Lecco group had been using these techniques for some seven years and had consolidated on previous publications by Ilizarov to produce a clinical and radiological classification of these non-unions, according to the presence or absence of a bone loss of 1 cm or more, the presence or absence of deformity and the mechanical properties of the non-union site. Type A non-unions, those without appreciable bone loss, were divided into Type A1, in which any deformity is mobile (lax), and Type A2, stiff non-unions. Type A2 is further subdivided into A2-1, stiff without deformity, and A2-2 which has a fixed deformity.

Non-unions with a loss of bone substance of 1 cm or more, whether resulting from previous injury, sepsis or planned débridement of infected bone during treatment, were called Type B. Where bone was lost from the tibia but the overall segment length was normal, usually because of an intact or united fibula, the non-union was of Type B1. In Type B2, the non-union surfaces were in apposition and the limb segment was shortened. Type B3 non-unions had both a bony gap and shortening.

This classification represents the logical summation of the classifications of Ilizarov, Gordon and Chiu and May, as it considers both the mechanical stiffness of the fracture and the presence or absence of an intact fibula. Based, as it was, on the possible Ilizarov strategies available to treat non-unions, it is not surprising that there is an almost direct correlation between this classification system and the treatment used. Although little detail is given, it is stated that 'Type A pseudoarthroses are treated by correction of the deformity, followed by compression. . .' and that 'In Type A1, a deformity should be corrected at the time of treatment because the pseudoarthrosis is mobile.' The paper concentrates on the treatment of Type B non-unions. Type B1 non-unions were treated by bone transport (internal lengthening). Type B2 lesions were treated by monofocal compression-distraction (distraction osteogenesis at the non-union site after initial compression) or bifocal compression-distraction (simultaneous distraction osteogenesis at a distant osteotomy site and compression of the non-union). Type B3 non-unions, with both shortening and a bony defect, were mostly treated by a combination of transport and bifocal compression-distraction. Thus, apart from some minor overlap, there was a clear correlation between their suggested classification and proposed treatment. It is surprising, in the results section of the paper, that there is no mention of the original classifications when considering overall success rates. Indeed, beyond stating that the three hypertrophic non-unions were the only successes in the group of 10 treated by monofocal compression-distraction, and four illustrative figures, no attempt appears to have been made to correlate either the author's non-union types or treatment categories with overall results. The mean treatment time in the fixator for the bifocal treatment group, all of which were atrophic, was 10.6 months. This is comparable to series using other treatment methods. For those patients treated by distraction osteogenesis, the authors quoted a treatment index of 30–40 days per cm of lengthening. They suggested that this index was reduced for the longer lengthenings. These findings, which provide some connection between non-union groups and treatment times, are similar to those used in adult limb lengthening with intact bones.

Two years later, a second paper from the Lecco group (Catteneo, Catagni and Johnson, 1992) described 28 patients with infected non-unions or segmental defects in the tibia, 15 of whom had pre-existing bone loss ranging from 2 to 7 cm. Seven of the remainder had extensive diaphyseal infection and were treated by *en bloc* resection. Despite the apparent similarity between the cases in the 1989 and 1992 articles, and their having two authors in common, one of whom originally described the classification used in 1989 (Catagni, 1986), no mention is made of their grading system in the later article. During the last few years a continuing trickle of papers has described Ilizarov treatments of complex tibial non-unions, all having a number of patients with bone loss and/or sepsis. None of these authors have used the Lecco group's classification in their articles. (Saleh and Royston, 1993; DiPasqualae et al., 1994; Dendrinos, Kontos and Lyritis, 1995; Ebraheim, Skie and Jackson, 1995).

Comment

At present, there exists no unified classification of non-unions which can encompass the heterogeneity seen in clinical practice. Beyond that of Weber, those classifications which have prognostic significance do so only for a limited part of the spectrum of non-unions and are not widely used. Table 5.5 lists some of the clinical, radiological and imaging criteria which can assist in characterizing a non-union. Clearly, the large number of variables suggests that this situation is likely to continue.

References

Andrew, J.G., Andrew, S.M., Freemont, A.J. and Marsh, D.R. (1994) Inflammatory cells in normal fracture healing. *Acta Orthop. Scand.*, **65**, 462–466.

Aronson, J. (1994) Temporal and spatial increases in blood flow during distraction osteogenesis. *Clin. Orthop.*, **301**, 124–132.

Benirschke, S.K., Mirels, H., Jones, D. and Tencer, A.G. (1993) The use of resonant frequency measurements for the non-invasive assessment of mechanical stiffness of the healing tibia. *J. Orthop. Trauma*, **7**, 64–71.

Blick, S.S., Brumback, R.J., Lakatos, R. et al. (1989) Early prophylactic bone grafting of high-energy tibial fractures. *Clin. Orthop.*, **240**, 21–41.

Boyd, H.B., Lipinski, S.W. and Wiley, J.H. (1961) Observations on non-union of the shafts of the long bones with a statistical analysis of 842 patients. *J. Bone Joint Surg.*, **43A**, 159–168.

Brighton, C.T., Black, J., Friedenberg, Z.B. et al. (1981) A multi-center study of the treatment of non-union with constant direct current. *J. Bone Joint Surg.*, **63A**, 2–13.

Catagni, M.A. (1986) Classificazione e trattamento delle pseudartrosi di gamba con perdita di sostanza. In: Proceedings of the 1st National Congress for the Application of the method of Ilizarov to the Tibia. Florence, Italy, October 30–31; pp. 87–90.

Catagni, M.A. Guerreschi, F., Holman, J.A. and Cattaneo, R. (1994) Distraction osteogenesis in the treatment of stiff hypertrophic non-unions using the Ilizarov apparatus. *Clin. Orthop.*, **301**, 159–163.

Catteneo, R., Catagni, M. and Johnson, E.E. (1992) The treatment of infected non-unions and segmental defects of the tibia by the methods of Ilizarov. *Clin. Orthop.*, **280**, 143–152.

Cooney, W.P., Dobyns, J.H. and Linscheid, R.L. (1980) Non-union of the scaphoid: Analysis of the results from bone grafting. *J. Hand Surg.*, **5**, 343–354.

Cornell, C.N. and Lane, J.M. (1992) Newest factors in fracture healing. *Clin. Orthop.*, **277**, 297–311.

Crenshaw, A.H. (1987) Delayed union and non-union of fractures. In *Campbell's Operative Orthopaedics* (A.H. Crenshaw, ed.), Vol. 3, 7th edn. St Louis: Mosby, pp. 2053–2118.

Cunningham, J.L., Kenwright, J. and Kershaw, C.J. (1990) Biomechanical measurement of fracture healing. *J. Med. Eng. Technol.*, **14**, 92–101.

De Bastiani, G., Aldegheri, R., Renzi Brivio, L. and Trivella, G. (1987) Limb lengthening by callus distraction (callotasis). *J. Pediatr. Orthop.*, **7**, 129–134.

Dendrinos, G.K., Kontos, S. and Lyritis, E. (1995) Use of the Ilizarov apparatus for the treatment of non-union of the tibia associated with infection. *J. Bone Joint Surg.*, **77A**, 835–846.

Dickson, K., Katzman, S., Delgado, E. and Contreras, D. (1994) Delayed union and non-unions of open tibial fracture: Correlation with arteriography results. *Clin. Orthop.*, **302**, 189–193.

DiPasquale, D., Ochsner, M.G., Kelly, A.M. and Maloney, D.M. (1994) The Ilizarov method for complex fracture non-unions. *J. Trauma*, **37**, 629–634.

Ebraheim, N.A., Skie, M.C. and Jackson, W.T. (1995) The treatment of tibial non-union with angular deformity using an Ilizarov device. *J. Trauma*, **38**, 111–117.

Einhorn, T.A. (1995) Enhancement of fracture-healing. *J. Bone Joint Surg.*, **77A**, 940–956.

Esterhai, J.L., Brighton, C.T., Heppenstall, R.B. et al. (1981) Detection of synovial pseudarthrosis by ^{99m}Tc scintigraphy: Application to treatment of traumatic non-union with constant direct current. *Clin. Orthop.*, **161**, 15–23.

Eyres, K.S., Bell, M.J. and Kanis, J.A. (1993a) The evaluation of new bone formation during limb lengthening evaluated by dual energy X-ray absorptiometry. *J. Bone Joint Surg.*, **75B**, 96–106.

Eyres, K.S., Bell, M.J. and Kanis, J.A. (1993b) Methods of assessing new bone formation during limb lengthening: Ultrasonography, dual energy X-ray absorptiometry and radiography compared. *J. Bone Joint Surg.*, **75B**, 358–364.

Freeland, A.E. and Mutz, S.B. (1976) Posterior bone grafting for infected un-united fractures of the tibia. *J. Bone Joint Surg.*, **58A**, 653–657.

Gerlanc, M., Haddad, D., Hyatt, G.W. et al. (1975) Ultrasonic study of normal and fractured bone. *Clin. Orthop.*, **111**, 175–180.

Gershuni, D.H. and Pinsker, R. (1982) Bone grafting for non-union of the tibia. A critical review. *J. Trauma*, **22**, 43–49.

Gordon, L. and Chiu, E.J. (1988) Treatment of infected non-unions and segmental defects of the tibia with staged microvascular muscle transplantation and bone grafting. *J. Bone Joint Surg.*, **70A**, 377–386.

Green, S.A. and Dlabal, T.A. (1983) The open bone graft for septic non-union. *Clin. Orthop.*, **180**, 117–124.

Hammer, R.R.R. (1985) Strength of union in human tibial shaft fracture: A prospective study of 104 cases. *Clin. Orthop.*, **199**, 226–232.

Hammer, R.R.R., Hammerby, S. and Lindholm, B. (1985) Accuracy of radiological assessment of tibial shaft fracture union in humans. *Clin. Orthop.*, **199**, 233–238.

Herbert, T.J. and Fisher, W.E. (1984) Management of the fractured scaphoid using a new bone screw. *J. Bone Joint Surg.*, **66B**, 114–123.

Hulth, A. (1989) Current concepts of fracture healing. *Clin. Orthop.*, **249**, 265–284.

Ilizarov, G.A., Kaplunov, A.G., Degtariev, V.E. and Ledvaev, V.I. (1972) Therapy of psuedarthrosis and un-united fractures aggravated by purulent infection by the method of compression-distraction osteosynthesis. *Orthop. Travmatl. Protez.*, **11**, 10–14.

Inoue, A. and Ono, K. (1979) A histological study of idiopathic avascular necrosis of the head of the femur. *J. Bone Joint Surg.*, **61B**, 138–143.

Jorgensen, T.E. (1972) Measurements of stability of crush fractures treated with Hoffman osteotaxis. *Acta Orthop. Scand.*, **34**, 280–291.

Jorgensen, T.E. (1979) A simple mechanical method of assessing fracture healing. In *External Fixation; The Current State of the Art* (A.F. Brooker and C.C. Edwards, eds). Baltimore, London: Williams & Wilkins, pp. 383–390.

Judet, R., Judet, J. and Roy-Camille, R. (1958) La vascularisation des os longs d'après une étude clinique et experimentale. *Rev. Chir. Orthop.*, **44**, 5.

Lane, J.M. and Sandhu, H.S. (1987) Current approaches to experimental bone grafting. *Orthop. Clin. North Am.*, **18**, 213–225.

Lerner, R.K., Esterhai, J.L., Polomano, R.C. et al. (1993) Quality of life assessment of patients with post-traumatic fracture non-union, chronic refractory osteomyelitis and lower-extremity amputation. *Clin. Orthop.*, **295**, 28–36.

McKibbin, B. (1978) The biology of fracture healing in long bones. *J. Bone Joint Surg.*, **60B**, 150–162.

Markel, M.D., Wikenheiser, M.A., Morin, R.L. et al. (1990) Quantification of bone healing: Comparison of QCT, SPA, MRI and DEXA in dog osteotomies. *Acta Orthop. Scand.*, **61**, 487–498.

Marti, R.K., Schüller, H.M. and Raaymakers, E.L.F.B. (1989) Intertrochanteric osteotomy for non-union of the femoral neck. *J. Bone Joint Surg.*, **71B**, 782–787.

Matthews, L.S., Kaufer, H. and Sonstegard, D.A. (1974) Manual sensing of fracture stability: a biomechanical study. *Acta Orthop. Scand.*, **45**, 373–381.

May, J.W., Jupiter, J.B., Weiland, A.J. and Byrd, H.S. (1989) Clinical classification of post-traumatic tibial osteomyelitis. *J. Bone Joint Surg.*, **71A**, 1422–1428.

Merle d'Aubigne, R. (1965) Infection in the treatment of un-united fractures. *Clin. Orthop.*, **43**, 77–82.

Müller, M.E., Allgöwer, M., Schneider, R. and Willenegger, H. (1979) *Manual of Internal Fixation: Techniques recommended by the AO group*, 2nd edn. Berlin: Springer-Verlag.

Müller, M.E., Allgöwer, M., Schneider, R. and Willenegger, H. (1991) *Manual of Internal Fixation:*

Techniques recommended by the AO group, 3rd edn. Berlin: Springer-Verlag, pp. 713–742.

Nicoll, E.A. (1964) Fractures of the tibial shaft. A survey of 705 cases. *J. Bone Joint Surg.*, **46B**, 373–387.

Paley, D., Catagni, M.A., Argnani, F. et al. (1989) Ilizarov treatment of tibial non-unions with bone loss. *Clin. Orthop.*, **241**, 146–165.

Papineau, L.J. (1973) L'excision greffe avec fermeture cutanée dilibérée dans l'osteomyélite chronique. *Nouv. Press. Med.*, **2**, 2753–2755.

Parker, M.J. and Pryor, G.A. (1993) *Hip Fracture Management*. Oxford: Blackwell Science.

Ribbans, W.J., Stubbs, D.A. and Saleh, M. (1992) Non-union surgery: part II — the Sheffield experience. *Int. J. Orthop. Trauma*, **2**, 19–24.

Richardson, J.B., Cunningham, J.L., Goodship, A.E. et al. (1994) Measuring fracture stiffness can define healing of tibial fractures. *J. Bone Joint Surg.*, **76B**, 389–394.

Rosen, H. (1979) Compression treatment of long bone pseudarthroses. *Clin. Orthop.*, **138**, 154–166.

Rosen, H. (1992) Non-union and mal-union. In *Skeletal Trauma* (B.D. Browner, J.B. Jupiter, A.M. Levine and P.G. Trafton, eds). Philadelphia: Saunders.

Saleh, M. (1992) Non-union surgery: part 1 — basic principles of management. *Int. J. Orthop. Trauma*, **2**, 4–18.

Saleh, M. and Royston, S.L. (1993) The circular frame in the management of non-unions: Initial experience with 39 cases. Presented at British Orthopaedic Association, Torquay, England. October.

Saleh, M. and Royston, S.L. (1996) Management of non-union of fractures by distraction with correction of angulation and shortening. *J. Bone Joint Surg.*, **76B**, 105–109.

Segmüller, G., Cech, O. and Bekier, A. (1969) Die osteogene Aktivität im Bereiche der Pseudarthrose langer Röhrenknochen. *Z. Orthop.*, **106**, 599–609.

Sharrard, W.J.W. (1984) Bone and joint. In *Wound Healing for Surgeons* (T.E. Bucknell and H. Ellis, eds). London: Ballière Tindall, pp. 261–279.

Siegel, I.M., Anast, G.T. and Fields, T. (1958) The determination of fracture healing by measurement of sound velocity across the fracture site. *Surg. Gynecol. Obstet.*, **107**, 327–332.

Simmons, D.J. (1985) Fracture healing perspectives. *Clin. Orthop.*, **200**, 100–113.

Sveshnikov, A.A., Barabash, A.P., Chepelenko, T.A. et al. (1984) Radionuclide studies of osteogenesis and circulation in substitution of large defects of the leg bones in experiment. *Ortop. Traumatol. Prutez.*, **11**, 33–37.

Tiedeman, J.J., Lippiello, L., Connolly, J.F. and Strates, B.S. (1990) Quantitative roentgenographic densitometry for assessing fracture healing. *Clin. Orthop.*, **253**, 279–286.

Toh, C-L. and Jupiter, J.B. (1995) The infected non-union of the tibia. *Clin. Orthop.*, **315**, 176–191.

Tower, S.S., Beals, R.K. and Duwelius, P.J. (1993) Resonant frequency analysis of the tibia as a measure of fracture healing. *J. Orthop. Trauma*, **7**, 552–567.

Urist, M., Mazet, R. and McLean, F.C. (1954) The pathogenesis and treatment of delayed union and non-union. *J. Bone Joint Surg.*, **36A**, 931–967.

Weber, B.G. and Brunner, C. (1981) The treatment of non-unions without electrical stimulation. *Clin. Orthop.*, **161**, 24–32.

Weber, B.G. and Cech, O. (1976) *Pseudarthrosis. Pathophysiology, biomechanics, therapy, results.* Bern: Huber.

Weiland, A.J., Moore, J.R. and Daniel, R.K. (1984) The efficacy of free tissue transfer in the treatment of osteomyelitis. *J. Bone Joint Surg.*, **66A**, 181–193.

6

Peripheral nerve injuries

T. E. J. Hems

Introduction

Peripheral nerves can be injured by compression, traction, ischaemia or direct laceration. The management and outcome depend on the extent of damage to the anatomical components of a nerve trunk. During World War II the large number of nerve injuries from battle casualties led to an increased interest in the subject and to the demand for a meaningful classification of nerve injury. Considerable basic scientific and clinical research was carried out over this period. This produced two classifications for nerve injuries which cause loss of function. Both are based on nerve fibre and nerve trunk pathology. These were described by Seddon (1942, 1943) and Sunderland (1951).

Structure of the Peripheral Nerve Trunk

The anatomy of a peripheral nerve trunk is shown in Figure 6.1. Nerve impulses are conducted by axons. The nerve contains many axons which are supported by connective tissue structures. Neurones consist of a cell body, associated dendrites and usually one axon. In order to remain viable an axon must be connected to its cell body. All axons are surrounded by Schwann cells. In myelinated fibres the Schwann cells form an insulating sheath, each Schwann cell being associated with only one axon. Integrity of the myelin sheath is necessary for conduction of nerve impulses in myelinated fibres. Unmyelinated fibres are composed of several axons wrapped by a single Schwann cell. The Schwann cell basement membrane together with endoneurial collagen fibres forms the endoneurial tube. Large numbers of nerve fibres are gathered in fascicles surrounded by a connective tissue sheath called the perineurium. The fascicles are bound together and the whole trunk sheathed by a further connective tissue layer called the epineurium.

Response of a Nerve to Injury

If trauma to a nerve is sufficient to disrupt axons then the distal part of the nerve undergoes Wallerian degeneration. In this process there is lysis of the axoplasm and fragmentation of myelin sheaths leaving an endoneurial tube containing Schwann cells. The axons in the proximal part of the nerve have the potential to regenerate into the endoneurial tubes of the distal segment and subsequently make connections with target organs. The regeneration proceeds slowly with axons growing at only 1–2 mm per day in humans. The quality of recovery depends on the extent of damage to the supporting connective tissue layers. In the event of division of a nerve or severe damage, outcome will be improved by surgical repair.

Seddon's Classification

Seddon (1942, 1943) divided peripheral nerve injuries into three types, neurapraxia, axonotmesis, and neurotmesis based on the study of 650 cases (Table 6.1). This is still the most widely used classification.

Neurapraxia (Nerve not Working)

Neurapraxia is a comparatively mild injury caused by moderate compression such as that caused by a tour-

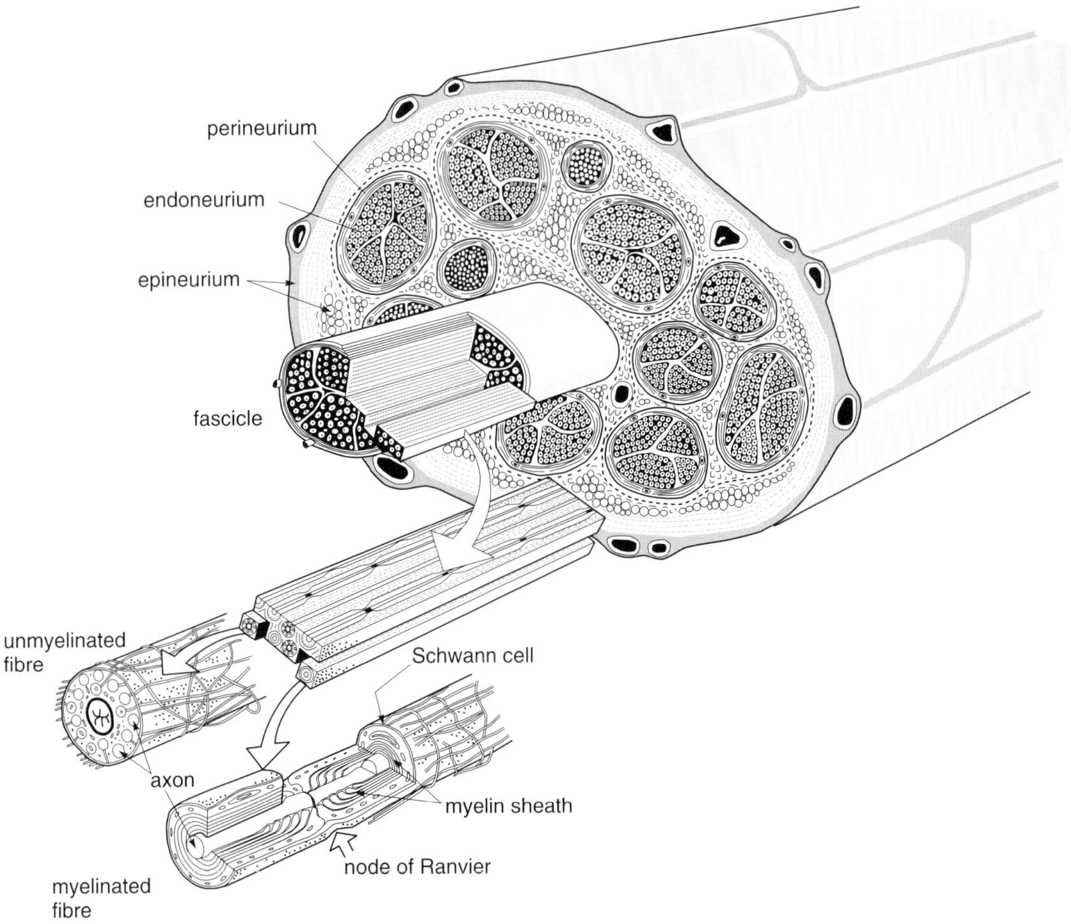

Figure 6.1. Microanatomy of a peripheral nerve trunk (from *Nerve Injury and Repair* (G. Landborg, ed.). Edinburgh: Churchill Livingstone, 1988, p. 33).

niquet, slight stretching, or the passage of a missile near a nerve trunk. Nerve modalities are not equally affected. Motor paralysis is usually complete, while there is only partial sensory loss and little or no disturbance of autonomic function. The cause of neurapraxia is a local block to conduction of nerve impulses at a discrete area along the course of a nerve. The axons are in continuity and there is therefore no Wallerian degeneration; indeed, conduction distal to the site of damage remains normal. Complete recovery occurs providing the cause is removed, but the time for recovery may vary from days to several weeks. Recovery does not follow a proximal-to-distal course as occurs during axon regeneration. Patients commonly complain of spontaneous sensations.

Seddon (1943) did not know the pathological basis of the local conduction block. However, he did note that the order of vulnerability of fibres in cases of neurapraxia is motor, proprioceptive, touch, pain and autonomic, which closely corresponds to fibre size.

The largest diameter myelinated fibres, A alpha and A beta, are more vulnerable than the smaller diameter, A delta and C fibres.

Experimental work (Denny-Brown and Brenner, 1944a,b), in which short segments of peripheral nerve were compressed, revealed that localized demyelination of fibres occurred in the damaged segment of nerve. The lesions created produced complete paralysis without degeneration of axons. Recovery occurred over a period of days. The histological changes noted in the myelin outlasted the return of conductivity. However, they found no difference in the appearance of large and small diameter fibres to account for the greater effect of the lesion on the conduction of motor impulses.

While Seddon's description implies an injury which takes days or weeks to recover, the term 'neurapraxia' has been used to include lesions where recovery is

Table 6.1 Seddon's classification of nerve injury (from Seddon, 1972)

	Neurotmesis	*Axonotmesis*	*Neurapraxia*
Pathological			
Anatomical continuity	May be lost	Preserved	Preserved
Essential damage	Complete disorganization	Nerve fibres interrupted Schwann sheaths preserved	Selective demyelination of larger fibres: no degeneration of axons
Clinical			
Motor paralysis	Complete	Complete	Complete
Muscle atrophy	Progressive	Progressive	Very little
Sensory paralysis	Complete	Complete	Usually much sparing
Autonomic paralysis	Complete	Complete	Usually much sparing
Electrical phenomena			
Reaction of degeneration	Present	Present	Absent
Nerve conduction distal to the lesion	Absent	Absent	Preserved
Motor-unit action-potentials	Absent	Absent	Absent
Fibrillation	Present	Present	Occasionally detectable
Recovery			
Surgical repair	Essential	Not necessary	Not necessary
Rate of recovery	1–2 mm a day after repair	1–2 mm a day	Rapid; days or weeks
March of recovery	According to order of innervation	According to order of innervation	No order
Quality	Always imperfect	Perfect	Perfect

much more rapid. Such lesions are unlikely to have the same pathological basis. Gilliat (1980) reviewed the subject and differentiated between two grades of conduction block:

(1) Rapidly reversible physiological block.
(2) Local demyelinating block.

Rapidly reversible physiological block is the condition when a leg or arm 'goes to sleep' and then recovers quickly when posture is altered and blood supply to the nerve is restored. Investigation of this phenomenon using a pneumatic cuff (Lewis, Pickering and Rothschild, 1931) showed that it results from ischaemia of the compressed nerve. In contrast the acute demyelinating block as described by Denny-Brown and Brenner (1944a,b) results from direct damage to nerve fibres.

Axonotmesis (Axons Divided)

This is the term used for a more severe injury where the axons and their myelin sheaths are broken but the supporting connective tissue structures including the endoneurial tubes, perineurium and epineurium are still intact. Wallerian degeneration occurs distal to the site of injury and hence distal conduction is lost. Clinical examination reveals complete loss of motor, sensory and autonomic function. Recovery occurs by axon regeneration at a rate of 1–2 mm/day. A Tinel sign will be apparent at the level of injury and with time will advance distally. The recovery of neurological function will follow a serial progression from proximal to distal. Recovery is usually good, restoring full and original sensory and motor function.

A common example of axonotmesis is the radial nerve palsy associated with fracture of the humerus which usually recovers to a normal or near normal level. Axonotmesis can be readily produced in experimental animals by crushing a nerve. Not only is regeneration after crush injury more rapid than after nerve division and suture (Gutmann et al., 1942), but functional recovery is also better after crush injury (Gutmann, 1942; De Medinacelli, Freed and Wyatt, 1982).

Neurotmesis (Whole Nerve Divided)

Neurotmesis is the word which describes the state of a nerve that has either been completely severed or is so seriously disorganized by scar tissue that spontaneous regeneration is out of the question. Neurotmesis is the most likely diagnosis in open injuries but high energy traction, injection of noxious drugs and ischaemia can destroy a nerve in this way.

If appropriate repair of the nerve is carried out, recovery may occur by axonal regeneration at a rate of 1–2 mm/day. As with axonotmesis, recovery of function progresses from proximal to distal. However, the quality of recovery is never perfect after neurotmesis. This is thought largely to result from the failure of correct 'rewiring'. The endoneurial tubes and other connective tissue structures have been disrupted. Even with the best surgical repair, regenerated fibres connect with muscles or sensory organs which formerly they did not innervate resulting in loss of function.

Modifications of Seddon's Classification

A subdivision of the neurapraxia group of injuries was described by Birch and St. Clair Strange (1990). This is a type of lesion in which, after a long delay without recovery, surgical exploration and neurolysis are followed by the early onset of recovery, which then progresses in a manner that is identical with that observed after classical neurapraxia. The lesion appears to result from displacement of myelin by continuing local pressure on the nerve. The axons, however, remain intact and are able to start conducting again once the pressure in relieved. Birch and St. Clair Strange coined the term 'Axonamosis' (axons deprived of insulation) for this type of nerve injury.

Limitations of Seddon's Classification

The terms used in the classification have been criticized as they require some knowledge of Greek for their understanding. However, the major limitation of the classification is that it does not distinguish between all grades of intraneural damage. Seddon (1942, 1943) himself noted variation, in the appearance at operation and subsequent recovery, among nerves with injuries classified as axonotmesis. He recognized that in some cases of 'lesion in continuity' there is damage to the endoneurium and perineurium as well as disruption of axons.

Sunderland's Classification

Sunderland (1951) proposed his classification after detailed investigation of the anatomy of peripheral nerves and a long-term study of over 300 cases of peripheral nerve injury. Five degrees of nerve injury were defined on the basis of changes induced in the normal structure of the nerve. The injuries are arranged in ascending order of severity from the first to the fifth degree (Table 6.2).

First degree injury

The basis of a first degree injury is interruption of axonal conduction at the site of injury. It corresponds to neurapraxia in Seddon's classification.

Second degree injury

In a second degree injury the axon is severed or axonal mechanisms are so disorganized that the axon fails to survive below the level of the injury, while

Table 6.2 Summary of the pathological basis for Sunderland's classification of nerve injuries (I–V)

Tissues	*Normal*	*Injury group*				
		I	*II*	*III*	*IV*	*V*
Axon	*	*	#	#	#	#
Myelin sheath	*	#	#	#	#	#
Endoneurial tube	*	*	*	#	#	#
Endoneurium	*	*	*	#	#	#
Perineurium	*	*	*	*	#	#
Epineurium	*	*	*	*	*	#
Blood vessels	*	±	±	±	±	#
Seddon group	Normal	Neurapraxia	Axonotmesis	–	–	Neurotmesis

* = intact, # = severed, ± = intact or severed

the endoneurial tubes remain intact. It therefore corresponds to axonotmesis in Seddon's classification.

Third degree injury

A third degree injury results from more severe trauma which, not only causes axonal disintegration and Wallerian degeneration, but also disorganizes the internal structure of the fasciculi. Continuity of the endoneurial tubes is lost. The perineurium and hence the fasciculi are left in continuity and their general arrangement in the nerve trunk is preserved. Retrograde degeneration of axons is more severe than in second degree injuries particularly when the nerve is injured at a proximal level.

The intrafascicular damage may be further complicated by haemorrhage, oedema, vascular stasis and ischaemia, which favour fibrous tissue formation. Intrafascicular fibrosis then constitutes a serious obstacle to axon regeneration. The quality of regeneration is also adversely affected by loss of continuity of the endoneurial tubes. Regenerating axons, though still confined within the fasciculus, are no longer confined to the tubes which originally contained them. The outcome of the erroneous cross-shunting of axons depends on the fibre composition of the affected fasciculi. The consequences will be minimal when a fascicle contains fibres from one, or two functionally related sources as is usually the case at distal levels. However, at proximal levels fibres from functionally unrelated sources are intermingled, and the consequences of rupture of the endoneurial tubes is greater.

After third degree damage, spontaneous recovery does occur as with second degree injury, but this will not be full and complete. The variation of recovery is broadest in this type of injury and can vary from almost complete recovery to almost no recovery depending on the extent of the factors mentioned above.

Fourth degree injury

In fourth degree injury the perineurium is ruptured and hence the fasciculi are disrupted. The epineurium is intact maintaining continuity of the nerve trunk, though the involved segment is ultimately converted into a tangled mass of ruptured fasciculi, scar tissue and regenerating axons.

Wallerian degeneration occurs as in second and third degree injury but regeneration is further complicated as regenerating axons are no longer confined to their fasciculi and scarring is more severe and extensive. As a result, any spontaneous recovery rarely proceeds to a useful degree. Therefore this type of injury requires excision of the damaged segment and surgical repair of the nerve.

Fifth degree injury

In fifth degree injury there is loss of continuity of the nerve trunk. It is therefore equivalent to neurotmesis in Seddon's classification. Useful recovery is only possible if surgical repair is carried out.

Modifications of Sunderland's Classification

Sunderland (1951, 1991) mentions that some injuries are mixed with different parts of a nerve trunk being affected to a varying degree. This type of injury combines all or many of the five degrees of injury and hence the pattern of recovery will be mixed. Mackinnon and Dellon (1988) classified this situation as a *sixth degree injury*. They also describe it as a neuroma in-continuity, but this term is misleading as it implies some degree of axon regeneration rather than describing the initial state after injury. Mixed injuries are not uncommon and present a particularly difficult clinical problem.

Appraisal of Seddon's and Sunderland's Classifications

Sunderland's classification represents an improvement over that of Seddon in that it more precisely recognizes varying severity of damage to a nerve trunk. However, this is at the expense of greater complexity.

For practical clinical application, the weakness of both systems is their attempt to link clinical features with pathological changes. On initial clinical and neurophysiological assessment of a closed nerve injury, it is not possible to distinguish axonotmesis from neurotmesis or differentiate between Sunderland's second, third, fourth or fifth degree injuries. Therefore, correct classification can only be made retrospectively once recovery has occurred or on surgical exploration. While certainly of pathological and research importance, these classifications may not be useful in determining clinical action.

Birch and Giddins (1995) have suggested a simple distinction between degenerative and non-degenerative lesions which can be made on clinical examination:

Non-degenerative

This is conduction block of short or long duration (neurapraxia or first degree injury). Some modalities of sensation persist. Nerve action potentials are detectable in the distal trunk and there is no spontaneous electrical activity present in muscle.

Degenerative

This encompasses all lesions where there is Wallerian degeneration distal to the site of injury. Prognosis may be favourable or unfavourable. All modalities are lost. There is no distal conduction and fibrillation potentials appear in muscle at about three weeks.

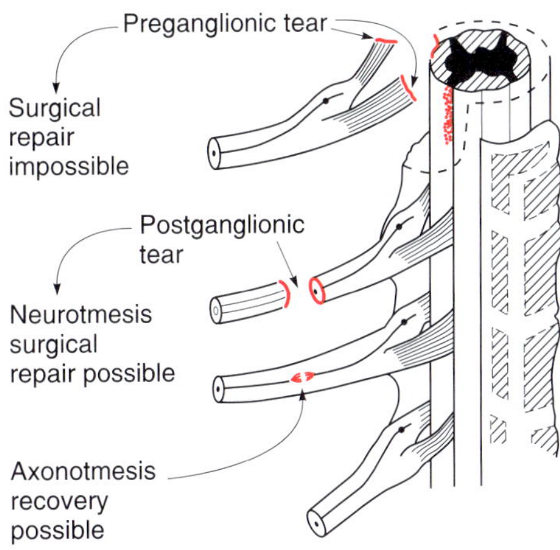

Figure 6.2. Diagram showing the types of injury that can result from traction on the roots of the brachial plexus.

Brachial Plexus Injuries

The general classifications described above can be applied to all peripheral nerves. However, injuries of the brachial plexus represent a particularly complex situation. Many different patterns of injury are possible and no detailed classification has been widely accepted except in the case of obstetric palsy. Although damage may occur at any level in the plexus from the roots to the terminal branches, injuries fall into the two broad categories of supraclavicular and infraclavicular lesions. Both types may coexist in the same patient. In the case of supraclavicular injuries it is important to distinguish between pre- and postganglionic damage to the spinal roots.

Supraclavicular Injuries

Closed traction injuries of the supraclavicular plexus are more common than injuries to the infraclavicular plexus. Birch (1992) has estimated that there are between 300 and 350 such injuries every year in the UK. The consequences of the exact site of damage to the spinal roots are illustrated in Figure 6.2. When damage is distal to the dorsal root ganglion, postganglionic injury, then recovery may occur spontaneously or after surgical repair. However, if roots are ruptured proximal to the dorsal root ganglion, preganglionic injury, then surgical repair is currently thought to be impossible. Experimental work (Carlstedt, 1991, Hems, Clutton and Glasby, 1994) suggests that repair of the ventral roots, but not dorsal roots, may be possible. In view of the implications for management and prognosis it is important to distinguish between pre- and postganglionic injuries. Although, investigations including CT myelography, MRI and electrophysiological examination may be helpful, definite diagnosis is often only made at surgery.

Infraclavicular Injuries

Infraclavicular injury to the brachial plexus corresponds to an injury at the level of the cords or the terminal branches. It is often associated with fracture of the proximal humerus or dislocation of the shoulder. Most authors have reported a more favourable prognosis for infraclavicular compared with supraclavicular injuries and non-operative management was recommended by Leffert and Seddon (1965). However, Burge, Rushworth and Watson (1985) found at least one nerve ruptured in all their cases and on this basis recommended early surgical exploration.

Patterns of Injury to Brachial Plexus

Although many combinations of injury to different parts of the plexus are possible certain patterns are more common. Millesi (1991) classified them into *complete lesions* in which all spinal nerves are involved and *partial lesions* where only some roots are damaged. Five types of partial lesion were distinguished:

1. Upper brachial plexus lesion (C5, C6). The shoulder muscles and the muscles for elbow flexion are paralysed.
2. Extended upper brachial plexus lesion (C5, C6, C7). In addition to shoulder motion and elbow

Table 6.3 Summary of the discussion group's findings

Author	*Good for defining treatment*	*Good for defining outcome*	*Reliability tested?*	*It it sensitive to change?*	*Methods*	*Current use*	*Original purpose?*
Seddon (1942, 1943)	Yes but often only in retrospect	Yes but often only applied in retrospect	No	No	Examination and EMG	Aid prognosis and treatment	War injury
Sunderland (1951)	Yes but often only applied in retrospect	Yes but often only applied in retrospect	No	No	Examination and EMG	Aid prognosis and treatment	To improve Seddon

flexion, the muscles for wrist and finger extension, innervated by the radial nerve, are involved.

3. Peripheral upper brachial plexus lesion. An upper brachial plexus lesion may be imitated by a lesion of the suprascapular, the axillary, and the musculocutaneous nerves. (Many infraclavicular injuries fall into this category.)
4. Intermediate brachial plexus lesion. Mainly C7 is involved. The functional loss corresponds to a radial nerve lesion.
5. Lower brachial plexus lesion (C8, T1). Shoulder muscles, elbow flexion, and wrist and finger extension, as well as flexor carpi radialis and pronator teres muscles are functional, but finger flexion and intrinsic muscles of the hand are completely paralysed.

Although not a widely accepted classification, a similar grouping of brachial plexus injuries has been mentioned by other authors including Seddon (1972). While not including all possibilities it is a useful classification of the pattern of injury in most cases.

However, the prognosis and management in each type will depend on the degree of damage and in the case of the roots whether the lesion is pre- or post-ganglionic.

Discussion (Summarized in Table 6.3)

It was noted that the peripheral nerve injury classifications were pathologically and anatomically based. The original concept of Seddon (1942, 1943) was founded on principles of nerve injury and repair; because of this, it is difficult to define precisely all the classification sub-groups either by examination or by imaging. EMG studies are needed to confirm a diagnosis, although sometimes these tests can not be applied for a number of weeks following an accident. In this respect classification is only of use retrospectively. Common use over a period of years has made the classification of peripheral nerve injuries of some value in guiding surgeons. Classification allows predictions of likely outcome of a particular type of nerve injury at a particular site to be made. Precisely defining a nerve injury according to the classification system of Seddon or Sutherland (1951, 1991) is sometimes difficult at the outset. It was felt that new systems are needed to take into account factors such as age and site of injury. It is also likely that new technologies such as MRI and ultrasound will provide new opportunities for classification.

Acknowledgements

I would like to thank Mr M. Glasby for helpful comments and Mrs J. Hems for careful checking of the manuscript.

References

Birch, R. (1992) Advances in diagnosis and treatment in closed traction lesion of the supraclavicular brachial plexus. In *Recent Advances In Orthopaedics 6* (A. Catterall, ed.). Edinburgh: Churchill Livingstone.

Birch, R. and Giddins, G. (1995) Peripheral nerve injuries. In *Medicolegal Reporting in Orthopaedic Trauma* (M.A. Foy and P.S. Fagg, eds). Edinburgh: Churchill Livingstone.

Burge, P., Rushworth, G. and Watson, N. (1985) Patterns of injury to the terminal branches of the brachial plexus: The place for early exploration. *J. Bone Joint Surg.*, **67B**, 630–634.

Birch, R. and St. Clair Strange, F.G. (1990) A new type of peripheral nerve lesion. *J. Bone Joint Surg.*, **72B**, 312–313.

Carlstedt, T. (1991) Experimental studies on surgical treatment of avulsed nerve roots in brachial plexus injury. *J. Hand Surg.*, **16B**, 477–482.

Denny-Brown, D.E. and Brenner, C. (1944a) Paralysis of

nerve induced by direct pressure and by tourniquet. *Arch. Neurol. Psychiat.*, **51**, 1–26.

Denny-Brown, D.E. and Brenner, C. (1944b) Lesion in peripheral nerve resulting from compression by spring clip. *Arch. Neurol. Psychiat.*, **52**, 1–19.

Gilliat, R.W. (1980) Acute compression block. In *The Physiology of Peripheral Nerve Disease* (A.J. Summer, ed.). Philadelphia: Saunders.

Gutmann, E. (1942) Factors affecting recovery of motor function after nerve lesions. *J. Neurol. Psych.*, **5**, 81–95.

Gutmann, E., Guttmann, L., Medawar, P.B. and Young, J.Z. (1942) The rate of regeneration of nerve. *J. Exp. Biol.*, **19**, 14–44.

Hems, T.E.J., Clutton, R.E. and Glasby, M.A. (1994) Repair of avulsed cervical nerve roots: An experimental study in sheep. *J. Bone Joint Surg.*, **76B**, 818–823.

Leffert, R.D. and Seddon, H. (1965) Infraclavicular brachial plexus injuries. *J. Bone Joint Surg.*, **47B**, 9–22.

Lewis, T., Pickering, G.W. and Rothschild, P. (1931) Centripetal paralysis arising out of arrested blood flow to the limb, including notes on a form of tingling. *Heart*, **16**, 1.

Mackinnon, S.E. and Dellon, A.L. (1988) *Surgery of the Peripheral Nerve*. New York: Thieme.

De Medinacelli, L., Freed, W.J. and Wyatt, R.J. (1982) An index of the functional condition of the rat sciatic nerve based on measurements made from walking tracks. *Exp. Neurol.*, **77**, 634–643.

Millesi, H. (1991) Brachial plexus injury in adults: Operative repair. In *Operative Nerve Repair and Reconstruction* (R.H. Gelberman, ed.). Philadelphia: Lippincott.

Seddon, H.J. (1942) A classification of nerve injuries. *Br. Med. J.*, **ii**, 237–239.

Seddon, H.J. (1943) Three types of nerve injury. *Brain*, **66**, 237–288.

Seddon, H.J. (1972) *Surgical Disorders of the Peripheral Nerves*. Edinburgh and London: Churchill Livingstone.

Sunderland, S. (1951) A classification of peripheral nerve injuries producing loss of function. *Brain*, **74**, 491–516.

Sunderland, S. (1991) *Nerve Injuries and Their Repair, a Critical Appraisal*. Edinburgh: Churchill Livingstone.

7

Physeal injuries

D. M. Eastwood and L. Hamilton

Introduction

Injuries to the immature skeleton can be defined according to the pattern of damage using terms such as torus or greenstick fractures or classified depending on the site of injury such as the physis, metaphysis, diaphysis or epiphysis. Injuries to the physis have been recognized since ancient times and although Hippocrates is often credited with the first written medical account of the condition, it was not until Poland published his book entitled *Traumatic Separation of the Epiphysis* in 1898 that physeal injury was established as a distinct entity. He reviewed all the existing material and was one of the first people to propose a general classification system for such injuries by describing four categories of fracture (Figure 7.1).

Classification systems exist in all branches of medicine and in many cases it would seem that the number of systems described is inversely proportional to their usefulness. A useful system must: (1) be easy and accurate to apply; (2) facilitate exact communication between physicians, and (3) have implications for outcome. Classification systems can be based on various parameters such as anatomical site, mechanism of injury, radiographic features or methods of treatment.

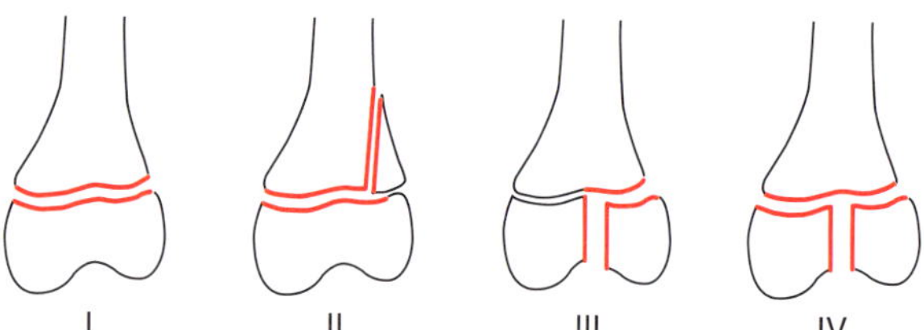

Figure 7.1. Classification system of epiphyseal injuries according to Poland (1898).

In terms of outcome after physeal injury, the two most important issues are joint congruity and growth arrest consequent on damage to the germinal cells of the physis. Classification systems should identify fractures which are associated with a poor prognosis.

Anatomical Classification

Up to the end of the nineteenth century, clinicians such as Poland could only categorize injuries according to site and the obvious clinical deformity or findings on postmortem dissection. Such descriptive attempts at categorization are still valid today and form the basic language of communication between doctors. These descriptions may imply the mechanism of injury but are not usually related to outcome although certain sites of injury are known to be associated with poorer results.

Site

Anatomical site is particularly important when the epiphysis is intra-articular as occurs with the proximal femur and the proximal radius. The blood supply to the intra-articular epiphysis is more tenuous and growth arrest is more common following injury. Physes in the lower limb, particularly around the knee and the distal tibia, are known to be more prone to complications. These physes have irregular shapes and although this increases their strength and hence reduces the risk of injury, once injured the fracture line propagates variably through the different layers of the physis thus increasing the liklihood of damage to the germinal layer of the physis (Lombardo and Harvey, 1977; Riseborough, Barrett and Shapiro, 1983; Petersen, 1984) and hence the risk of growth disturbance.

Open or Closed

Physeal injuries like all other injuries to the musculoskeletal system must also be divided into open or closed categories. Open injuries with tissue loss and/or contamination are associated with a poorer outcome.

Radiographic Classifications

With the development of X-rays the assessment of physeal injuries became more accurate and existing anatomical classification systems were modified and developed according to the radiographic injury patterns seen. It must be remembered that the area of interest (i.e. the physis) is itself radiolucent on such investigations. Interpretation of the injury is indirect and involves assessment of the adjacent radiodense epiphysis and metaphysis. In addition, plain radiographs simply provide a two-dimensional view of the three-dimensional physis and even the use of two radiographs at right angles to each other gives a limited appreciation of the complete physeal injury.

The first significant attempt at developing a classification system after the discovery of the radiograph was by Bergenfeldt in 1933 (Petersen, 1994c). He discussed 310 fractures and divided them into six groups (Figure 7.2). His classification included the first three groups described by Poland although the type 2 fracture (also the Salter–Harris type 2) was subdivided into two sub-groups depending on the size of the metaphyseal fragment (Thurston–Holland fragment). He also described a fracture which traversed the metaphysis, physis and epiphysis (a Salter–Harris type 4). The last fracture type identified an injury through the metaphysis which might cause growth arrest due to periosteal stripping. This work was published originally in German, a fact which may explain why the classification system was never widely adopted in the English speaking world.

Aitken in 1936 wrote on the subject of physeal injuries of the distal tibia and described three patterns of injury. In 1952 he noted similar injuries in the distal femur (Aitken and McGill) and by 1965 it was felt that the three fracture types were applicable to any epiphyseal injury (Figure 7.3).

In 1963, Salter and Harris published their classic descriptive article entitled 'Injuries involving the Epiphyseal Plate' and their classification system is the one which has gained most widespread acceptance. The paper outlined four categories of fracture which had already been described by Poland, Bergenfeldt and Aitken (Rockwood, Wilkins and Beaty, 1996; Petersen 1994c) and added a fifth category in which there was no fracture as such but compression damage to the germinal layer of the physis (Figure 7.4). This system was based on the mechanism of injury, the relationship of the fracture line to the germinal cells of the physis and outcome as regards growth disturbance. It assumed a straight propagation line of the fracture through the physis. In general terms the authors felt that a good prognosis was afforded with fracture types 1, 2 and 3 whereas types 4 and 5 were more likely to be affected by premature growth arrest. Displaced type 3 or 4 fractures would manifest degrees of joint incongruity which could compound the effects of any growth disturbance.

Mercer Rang extended the Salter–Harris classification system in 1969 by adding a type 6 fracture which described a localized injury to the perichondrial ring with an associated high risk of growth disturbance (Figure 7.4). The original injury was described as a direct blow to the perichondrial ring but has been interpreted by many authors as including avulsion injuries of the ring and open injuries such as those associated with lawnmower injuries of the distal tibial physis where often portions of metaphysis, physis and epiphysis are lost or damaged. Such injuries also occur as a result of burns or frostbite.

Although the Salter–Harris system enjoyed popular acclaim for some time, over recent years, several authors have become unhappy with the classification,

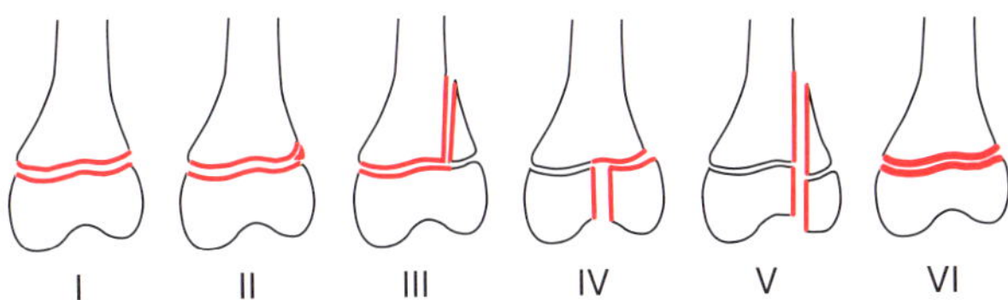

Figure 7.2. Classification system of epiphyseal injuries according to Bergenfeldt (1933).

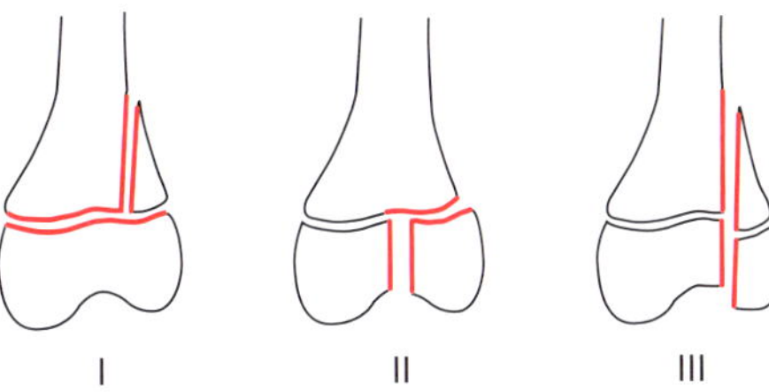

Figure 7.3. Classification system of epiphyseal injuries according to Aitken (1936).

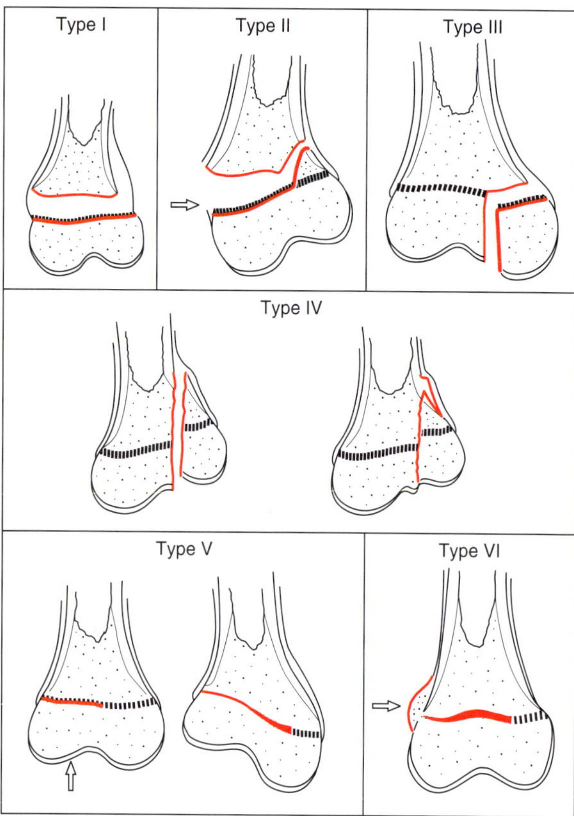

Figure 7.4. Rang's (1969) modification of the Salter-Harris (1963) classification of epiphyseal injuries.

perhaps because its relationship to outcome is less predictable than Salter and Harris had originally implied. It is now well recognized that the determination of outcome for any individual case of physeal injury depends on many factors such as age, site of damage, mechanism of injury, degree of displacement and indeed treatment as well as fracture type (see Eastwood, 1994). There has been considerable controversy over the nature and indeed existence of the type 5 fracture. In this compression injury there is no bony damage and thus the term fracture is a misnomer and the term injury is probably more appropriate. Petersen and Burkhart (1981) doubted the existence of this injury and stated that the prerequisites for diagnosis were: (1) a documented injury with negative radiographs in two planes, (2) no treatment, and (3) growth arrest some time later. Some case reports do exist, one of these documents the development of a valgus deformity of the tibia, secondary to a growth arrest, following an apparently non-physeal fracture of the proximal tibial metaphysis (Canale and Beaty, 1995). Others feel that a crushing element may be asociated with a transphyseal type of injury (e.g. types 2 or 3) and this may account for some of the 'unexpectedly' bad outcomes with these fracture types.

Weber (1980) was unable to identify any Salter–Harris type 5 injuries and chose to develop a classification system, published in 1980, whose aim was to distinguish between fractures with a good prognosis which could be treated by closed means and fractures with a poor prognosis which should be treated more aggressively. His type B fractures were at high risk of developing premature growth arrest. Both types of injury were subdivided giving a total of four fracture types equivalent to the Salter–Harris types 1–4 (Figure 7.5).

Ogden (1981, 1982) proposed an extensive classification system for injuries affecting growth in the limb. He detailed specific subgroups which he felt explained the 'unexpected' cases of growth disturbance with which all clinicians are familiar. He also defined additional fracture patterns (types 7, 8, 9, Figure 7.6) which affected areas of the growth mechanism other than the physeal plate. Whilst this system may be more complete than others it is unwieldy and has not been used regularly in clinical practice.

Shapiro (1982) attempted to correlate the bony injury with the damage to the epiphyseal and metaphyseal blood supply to the osteoprogenitor cells of the physis but he did not describe any new anatomic fracture type.

More recently, Petersen (1994c) described a new classification system following an epidemiological review of 951 physeal fractures over a 10-year period (Figure 7.7). In this system he describes two 'new'

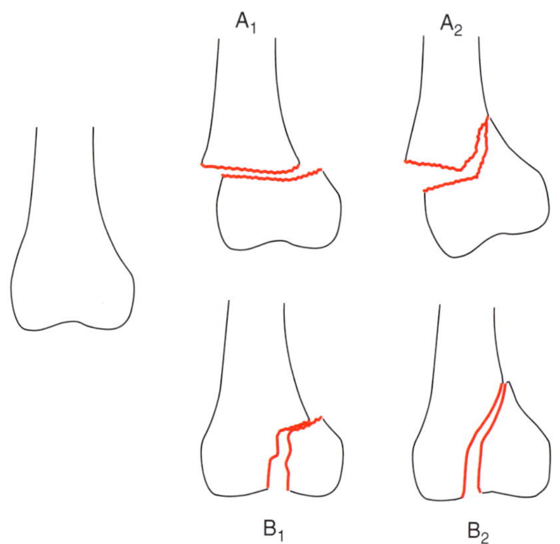

Figure 7.5. Classification system of epiphyseal injuries according to Weber (1980).

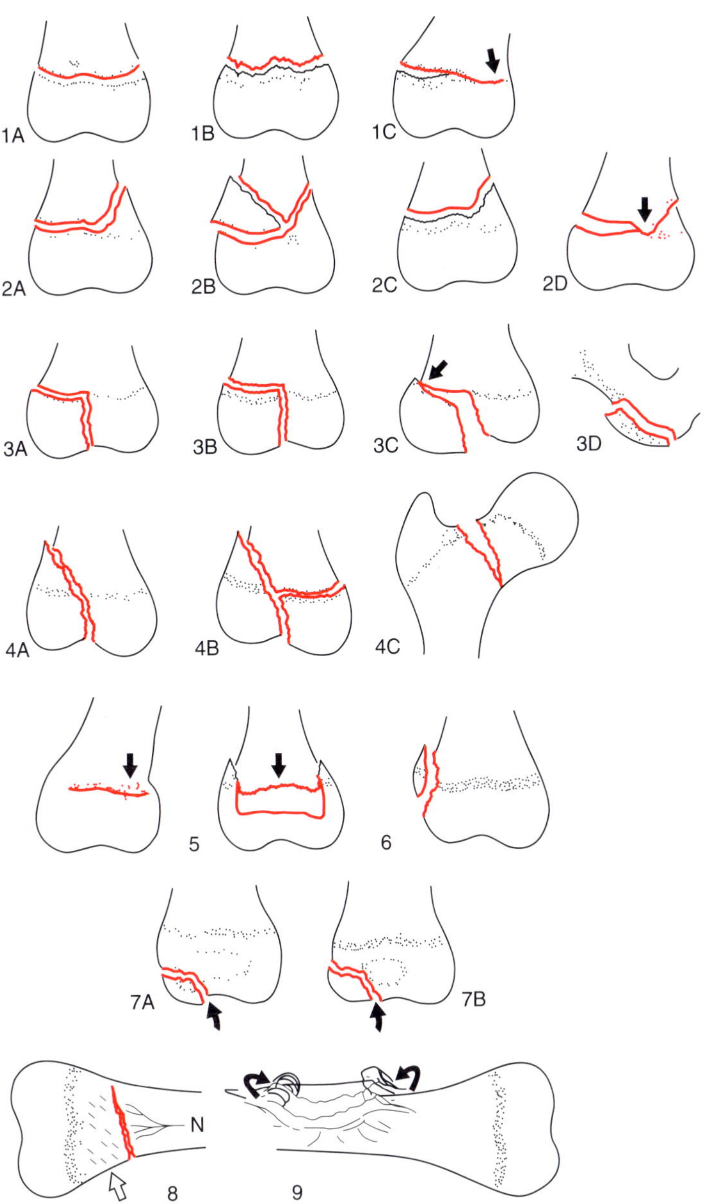

Figure 7.6. Classification system of epiphyseal injuries according to Ogden (1981).

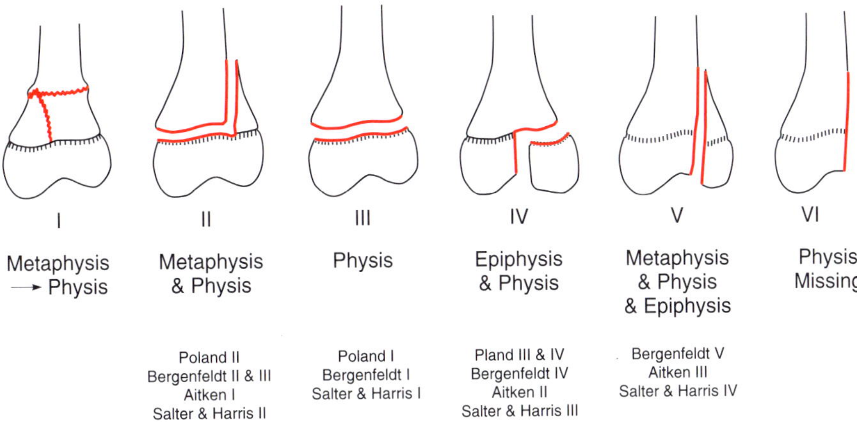

Figure 7.7. Classification system of epiphyseal injuries according to Petersen 1994c (with comparison to other systems).

Table 7.1 Petersen classification system. Number of fractures of each type requiring surgery 1979–1988 (Petersen, 1994c)

	Fracture type						*Total*
	1	*2*	*3*	*4*	*5*	*6*	
Number	147	510	126	104	62	2	951
Immediate surgery	1	23	13	18	12	2	69
Late surgery	0	12	9	15	12	1	49

Table 7.2 Salter–Harris classification system. Number of fractures requiring 'operative' treatment (Mizuta et al., 1987)

Fracture type	*Treatment*		
	No reduction	*Closed reduction*	*Open reduction*
1	13	12	5*
2	187	62	8*
3	19	2	8*
4	26	3*	13
5	1*	0	0

*= growth arrest (total 5 cases)

injury patterns. The first is essentially a transverse fracture of the metaphysis with extension to the physis which many clinicians will have recognized. There may be a small cortical fragment which is detached from both the metaphysis and the epiphysis (and thus distinguishes this injury from a Salter–Harris type 2 fracture). Oblique radiographs are often needed to identify this injury. The other 'new' injury is his type 6 fracture described as a 'part missing' physeal injury in which the part often includes a piece of metaphysis and epiphysis as well as the physis. Many clinicians might see this as an extreme variant of the Salter–Harris type 4 or 6 fractures. Petersen states that his type 6 occurs only in open fractures. This classification system has an anatomical basis and depicts physeal injury as a continuum from relatively minor involvement in a type 1 injury to complete transphyseal involvement (type 3) to longitudinal disruption of the physis with or without loss of some of the physeal cartilage (type 5 or 6). This concept of progressive anatomical damage has led Petersen (1994b) to divide the type 2 injuries into subgroups depending on the size of the metaphyseal fragment. The larger the fragment the less the transphyseal damage to the growth plate.

Epidemiological studies show a good basis for both the Salter–Harris and the Petersen classification systems as with both systems type 2 fractures are most common followed in the Salter–Harris system by types 1, 3, 4, 5 and in the Petersen system by types 1, 3, 4, 5, 6 (although the exact incidences do vary a little from study to study) (Oh, Craig and Banks, 1974; Worlock and Stower, 1986; Mizuta et al., 1987; Petersen, 1994c). As a result of the epidemiological review that he conducted, Petersen also feels that there is a strong prognostic basis for his system. With ascending fracture type the potential for physeal damage increases and so does the need for both early and late surgery (Table 7.1). It was also felt that the complications were less in fracture types 1 and 2 but this would also be true for the first two types in the Salter–Harris classification system. The chance of an injury requiring aggressive treatment increases with increasing fracture type because of the risk of growth arrest and joint incongruity if the fracture is displaced. Results from one such study are shown in Table 7.2.

Some authors have difficulty in applying a general classification system to physeal injuries at different sites of the musculoskeletal system and feel that specific sites deserve specific classifications (e.g. Dias and Tachdjian, 1978; Chadwick and Bentley, 1987). It does become somewhat cumbersome to remember a different system for each anatomical site that may sustain a physeal injury and discussion of such systems is outside the scope of this chapter. Such systems may still have a research use amongst groups of interested workers. Certain injuries such as the triplane ankle fracture were initially considered to be a separate fracture type specific to the distal tibia but most would now agree that this injury can be classified as a Salter–Harris type 4 fracture. A similar fracture pattern has also been described with respect to the distal humeral physis (Petersen, 1992).

The main cause for concern, when dealing with physeal injuries, relates to growth disturbance. It is reasonable to assume that the potential for this occurring must be associated, at least in part, with the damage sustained by the germinal layer of the physis at the time of the injury. Classification systems should, if they are to be useful as prognostic indicators, relate to the degree of physeal damage. Many existing systems have attempted to do this. However, the classification systems outlined above all rely on the accurate interpretation of the radiographic features of the physeal injury. It is well known that not all injuries to the developing skeleton are easy to see on standard plain radiographs. Whenever possible, radiographs should be taken at right angles to the line of the physis (rather than the diaphysis). Oblique views are often required at the ankle and knee to demonstrate particularly the minimally displaced fractures. Careful analysis of the plain films may show small lamellar fragments of bone within the physis. These represent fragments of calcified cartilage, which indicate that the fracture line has crossed the hypertrophic zone and violated at least part of the zone of provisional calcification. Plain radiographs alone will never be able to detail the exact nature of the physeal injury, particularly as the physis itself is a lucent structure. The newer imaging modalities may be able to add more precise information about the injury but, whilst some of these techniques should be in common usage, some remain as research tools awaiting assessment of their value.

Computed Tomography (CT)

Computed tomography is often used to evaluate more fully injuries which have already been identified on plain films and is most commonly used in fractures around the ankle and knee. It is particularly helpful for Salter–Harris type 4 injuries to ascertain the alignment of the articular surface along the entire length of the fracture; a feature which is not easy to determine on the plain film (Rogers and Poznanski, 1994).

Magnetic Resonance Imaging (MRI)

In theory, magnetic resonance imaging has distinct advantages in the evaluation of complex physeal injuries (and any subsequent growth arrest) as it depicts both cartilage and soft tissues as well as bone and displays anatomy in sagittal, coronal and oblique planes. MRI is the method of choice for the assessment of growth arrest, the detection of bony or fibrous bridging of the physis and the identification of vascular damage to the epiphysis and metaphysis following physeal injury. The exact role of MRI in the acute injury has not yet been defined and its implications for prognosis have not been determined. There are, therefore, no specific indications for acute MRI in suspected or actual physeal injury although MRI is particularly useful in the assessment of elbow injuries in young children and in depicting ligamentous injuries which occur commonly in children with epiphyseal separations around the knee. A preliminary report on the value of early MRI in lower limb physeal injuries (Smith et al., 1994) led to a change in Salter–Harris classification in three of four cases when MRI was performed within 10 days of injury. Treatment was not altered as a result of this reclassification but the authors felt that a better understanding of the injury and hence the prognosis was gained. Attempts were made to evaluate the physeal fracture pattern (horizontal or vertical), fracture location (central or peripheral) and level of fracture within the physis. Although it is not possible to delineate the various layers of the physis on MRI, attempts were made to localize the fracture to the epiphyseal or metaphyseal side of the physis in the hope that this might have implications for determining outcome.

In the acute period, MRI defines cartilaginous injury more accurately than plain radiography so that both the physeal and articular extents of the injury can be appreciated. MRI scans also depict the associated soft-tissue insults which may in due course help our understanding of the pathology of growth plate injuries and the potential for adverse effects.

Displacement

Most clinicians feel that fracture outcome is related to the degree of initial displacement and some evidence exists to support this belief (Lombardo and Harvey, 1977; Caterini, Farsetti and Ippolito, 1991). Persistent displacement at the fracture site may lead to immediate joint incongruity or malalignment and later growth arrest. In some studies the phrases 'degree

Table 7.3 Displacement of fractures

NONE	no displacement
MILD	up to one-third diameter displacement
MODERATE	one-third to two-thirds displacement
SEVERE	two-thirds to full diameter displacement
COMPLETE	no contact between fragments

of displacement' and 'accuracy of reduction' are used almost interchangeably which is obviously incorrect and perhaps a universal system for attempting to quantify initial displacement should be encouraged. Displacement is often discussed in terms of the diameter of the bone and many clinicians would agree with the system outlined by Petersen (1994a) (Table 7.3) which could be applicable to certain fracture types such as Salter–Harris types 1 and 2 and possibly 3 and Petersen types 2, 3 and perhaps 4.

Summary

All the radiographic classification systems outlined above describe the same injury patterns. For ease of application and communication the Salter–Harris classification system is understandably popular amongst clinicians although dispute remains about the existence of the type 5 injury. As the type 5 injury can not be diagnosed at the time of presentation but only at the time that the adverse outcome of growth arrest is identified there are reasonable grounds for excluding the category from any clinically useful classification system whilst remembering the pathological basis on which the fracture type was based as this process may occur in other fracture patterns. In terms of outcome, it is now well understood that fracture type is not the only and possibly not the most important factor to be considered. Thus, in order to predict outcome, perhaps a radiographic classification should be combined with an anatomical classification and a measure of the degree of displacement. Other patient variables such as age and mechanism of injury must also be considered.

At the end of the nineteenth century the development of radiographs led to a greater understanding of the nature of physeal injury and perhaps at the end of the twentieth century the advances in musculoskeletal imaging will shed further light on the subject. If more precise identification of the type of injury can be achieved then a better indication of outcome may follow.

Discussion

If we assume that outcome following a physeal injury is related to growth disturbance then a classification system must relate to the degree of physeal damage which has occurred.There is little basic science work which correlates the degree of force applied to the physis with the fracture pattern and which can be extrapolated to apply to physeal injuries at different sites in the skeleton at diferent ages. Whilst Salter and Harris did attempt to relate their classification system to mechanism of injury it would appear that the Petersen system has a more logical progression of physeal damage linked to an increasing need for surgical intervention and, by implication, increasing fracture severity.

Radiographs do not depict physeal damage with any great accuracy and the development of imaging techniques such as MRI may improve our understanding of physeal injuries significantly and with this may come the opportunity to devise the definitive classification system. It remains to be seen whether MRI is sensitive enough to detect damage to the layers of the physis caused by crushing injuries such as suggested in the Salter–Harris type 5 fracture.

At present, we would advocate the use of the Petersen system which identifies the physeal injuries associated with growth disturbance and joint incongruity and describes a progression of injury which helps in defining treatment.

References

Aitken, A.P. (1936) The end result of the fractured distal tibial epiphysis. *J. Bone Joint Surg.*, **18**, 685–691.

Aitken, A.P. (1965) Fractures of the epiphysis. *Clin. Orthop.*, **41**, 19–23.

Aitken, A.P. and McGill, H.K. (1952) Fractures involving the distal femoral epiphyseal cartilage. *J. Bone Joint Surg.*, **34**, 96–108.

Canale, S.T. and Beaty, J.H. (eds) (1995) *Operative Paediatric Orthopaedics*, 2nd edn. St Louis: Mosby, p. 32.

Caterini, R., Farsetti, P. and Ippolito, E. (1991) Unusual physeal lesions of the lower limb. A report of 16 cases with a very long term follow-up observation. *J. Orthop. Trauma*, **5**, 38–46.

Chadwick, C.J. and Bentley, G. (1987) Classification and prognosis of epiphyseal injuries. *Injury*, **18**, 157–168.

Dias, L.S. and Tachdjian, M.O. (1978) Physeal injuries of the ankle in children: classification. *Clin. Orthop.*, **136**, 230–233.

Eastwood, D.M. (1994) Growth plate injuries. In *Outcome measures in Trauma* (P. B. Pynsent, J. C. T. Fairbank and A. J. Carr, eds). Oxford: Butterworth-Heinemann.

Lombardo, S.J. and Harvey, J.P. (1977) Fractures of the distal femoral physis. Factors influencing prognosis: a review of 34 cases. *J. Bone Joint Surg.*, **59A**, 742–751.

Mizuta, T., Benson, W.M., Foster, B.K. and Patterson, D.L. (1987) Statistical analysis of the incidence of physeal injuries. *J. Pediatr. Orthop.*, **7**, 518–523.

Ogden, J.A. (1981) Injury to the growth mechanisms of the immature skeleton. *Skeletal Radiol.*, **6**, 237–253.

Ogden, J.A. (1982) Skeletal growth mechanism injury patterns. *J. Pediatr. Orthop.*, **2**, 371–377.

Oh, W.H., Craig, C. and Banks, H.H. (1974) Epiphyseal injuries. *Pediatr. Clin. North Am.*, **21**, 407–422.

Petersen, H.A. (1984) Partial growth plate arrest and its treatment. *J. Pediatr. Orthop.*, **4**, 246–258.

Petersen, H.A. (1992) Physeal injuries of the distal humerus. *Orthopaedics*, **15**, 799–808.

Petersen, H.A. (1994a) Physeal Fractures: Part 1. Epidemiology in Olmsted County, Minnesota, 1979–1988. *J. Pediatr. Orthop.*, **14**, 423–430.

Petersen, H.A. (1994b) Physeal Fractures: Part 2. Two previously unclassified types. *J. Pediatr. Orthop.*, **14**, 431–438.

Petersen, H.A. (1994c) Physeal Fractures: Part 3. Classification. *J. Pediatr. Orthop.*, **14**, 439–448.

Petersen, H.A. and Burkhart, S.S. (1981) Compression injuries of the epiphyseal growth plate: fact or fiction? *J. Pediatr. Orthop.*, **1**, 377–384.

Rang, M. (1969) *The Growth Plate and its Disorders*. Baltimore: Williams and Wilkins.

Riseborough, E.J., Barrett, I.R. and Shapiro, F. (1983) Growth disturbances following distal femoral physeal fracture-separations. *J. Bone Joint Surg.*, **65A**, 885–893.

Rockwood, C.A., Wilkins, K.E. and Beaty, J.H. (eds) (1996) *Fractures in Children*, 4th edn. Philadelphia: Lippincott Raven.

Rogers, L.F. and Poznanski, A.K. (1994) Imaging of epiphyseal injuries. *Radiology*, **191**, 297–308.

Salter, R.B. and Harris, W.R. (1963) Injuries involving the epiphyseal plate. *J. Bone Joint Surg.*, **45A**, 587–621.

Shapiro, F. (1982) Epiphyseal fracture-separation. A pathophysiological approach. *Orthopedics*, **5**, 720–736.

Smith, B.G., Rand, F., Jaramillo, D. and Shapiro, F. (1994) Early MR imaging of lower extremity physeal fracture-separations: A preliminary report. *J. Pediatr. Orthop.*, **14**, 526–533.

Weber, B.G. (1980) Fracture healing in the growing bone and in the mature skeleton. In *Treatment of Fractures in Children and Adolescent* (B.G. Weber, C. Brunner and F. Freuler, eds). New York: Springer Verlag.

Worlock, P. and Stower, M. (1986) Fracture patterns in Nottingham children. *J. Pediatr. Orthop.*, **6**, 656–660.

8

The spine

M. Scott-Young, H. Hegde and M. F. Brown

Introduction

Classification schemes are devised to assist the surgeon in treatment and decision making. They should:

1. Be simple and easily remembered.
2. Independent of sophisticated diagnostic methods if possible.
3. Provide a guide to the best form of treatment.
4. Give an indication of the prognosis.

Classification of fractures and dislocations of the axial skeleton can be based on any of three variables:

1. Presumptive mechanism of injury.
2. Fracture anatomy.
3. Degree of fracture fragment displacement.

Spinal column fractures may present with two distinct problems:

1. Osteoligamentous disruption.
2. Neurological injury.

A classification system should address both areas. Although no current classification provides a complete description, many advances have been made in classification of spinal fractures over the last 40 years.

Injuries to the Upper Cervical Spine

These injuries have characteristics related to aetiology and treatment that set them apart from those in the rest of the spine. There are nine major groups of injury to the upper cervical spine.

Atlanto-occipital Dislocations (CO/C1):

These injuries are twice as common in children as in adults (Bucholz and Burkhead, 1979). They are often fatal, so that reports on this injury are rare. They are classified into three groups.

A. Anterior
   i. Unilateral
   ii. Bilateral

This is the most common and caused by a hyperextension and distraction force — they are associated with submental lacerations, mandibular fractures and posterior pharyngeal wall lacerations (Eismont and Bohlman, 1978; Collato et al., 1986).

B. Posterior
   i. Unilateral
   ii. Bilateral

These are less common and are associated with cranial nerve injuries (VIn abducens commonly involved, Collato et al., 1986), spinal cord injuries and involvement of upper three cervical nerves (Dublin et al., 1986).

C. Longitudinal (Kaufman, Dunbar and McLaurin, 1982)

This is the least common atlanto-occipital dislocation.

All atlanto-occipital dislocations are extremely rare and most cases are fatal from respiratory arrest with brain stem involvement (Bama et al., 1983). No reliability studies are available.

Fractures of the Lateral Mass and Occipital Condyles (CO)

Anderson and Montisano (1988) described three types of occipital condylar fractures (Table 8.1). Their classification depends on CT scanning. In prac-

Table 8.1 Anderson and Montisano (1988) classification of fractures of the lateral mass and occipital condyles (CO)

Type		*Stability*	*Mechanism of injury (24)*
I	Impaction	Stable	Axial loading
II	Associated with basilar skull fracture	Stable	Direct blow to skull
III	Avulsion fracture of occipital condyle	Unstable	Either rotation or lateral bending or a combination of the two

Table 8.2 The classification of atlas fractures (C1) (Levine and Edwards, 1986; Segal, Grimm and Stauffer, 1987)

Type	*Mechanism of injury*	*Distinctive features*
1. Burst fracture (Jefferson, 1920)	Axial load producing a three or four part fracture	>7 mm atlanto dens interval (19) indicates rupture of the transverse ligament making it unstable
2. Posterior arch fracture, i.e. traumatic spondylolisthesis or dens fracture and nerve deficits (10)	Axial compression and hyperextension, trapping C1 arch between the occiput and C2	50% associated with other cervical fractures of C2
3. Comminuted	Axial	
4. Anterior arch fracture	Dens sheared off fragment in hyperextension	May be associated with fracture of dens
5. Lateral mass fracture	Combined axial loading and lateral surface and sometimes compression through articular mass	Occurs on one side of the neural arch with fracture lines running anterior and posterior to articular surfaces at C1
6. Transverse process fracture		
7. Avulsion of the inferior tubercle	Avulsion force by longus colli muscle due to hyperextension	

tice, all these injuries are treated non-operatively in a halo vest for three months. Type 3 injuries should be carefully evaluated for stability.

Atlas Fractures (C1)

These comprise 25% of all injuries in the C1–C2 complex and 10% of all cervical spine injuries. The original description of these was by Jefferson (1920). They are usually sustained during falls (Han, Witten and Musselman, 1976) or vehicular accidents (Segal, Grimm and Stauffer, 1987). The mechanism of injury is usually axial compression. Neurological deficits are stated to be rare except when the fracture occurs in association with odontoid fractures, rupture of the transverse ligament of the atlas, or intracranial damage (Esses, 1981).

Segal, Grimm and Stauffer (1987) classified these injuries into six groups (expanded from a classification of Gehweiler, Osborne and Becker (1980) and a seventh one was added by Levine and Edwards (1986) (Table 8.2). Landells and van Peteghem (1988) used a three type classification to review 35 cases (Figure 8.1). This is a contraction of the Levine–Edwards

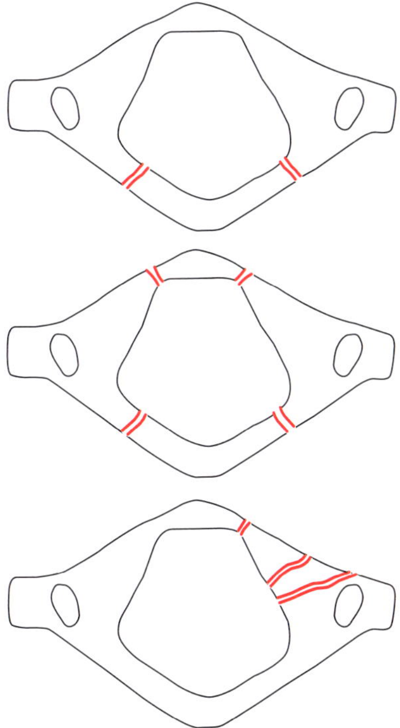

Figure 8.1. Classification of atlas (C1) fractures (from Landells and van Peteghem, 1988).

Table 8.3 The classification of dens fractures (C2) (Anderson and d'Alonzo, 1974)

Type	*Subtype*	*Distinctive features*
II Oblique fracture through the upper part of the odontoid process		Avulsion of alar ligament attachment; stable; rare
II Fracture at the junction of the odontoid and body of axis	i Undisplaced	40% non-union rate
	ii Displaced	30–60% non-union rate
III Fracture extending down into body of C2	i Displaced	Stable with high union rate
	ii Undisplaced	

system. It relates supposed mechanism of injury to prognosis. The series suggests that there is a significant incidence of neurological deficit in these injuries.

Atlanto-axial Subluxation (C1/C2)

These pure ligamentous injuries are rare (De Bur et al., 1988), more often in older age (Hintzer and Schalintzek, 1971), usually fatal (Levine and Edwards, 1989) and are as a result of forced severe flexion (De Bur et al., 1988). There is rupture of the transverse ligament, alar ligaments and disruption of the facet capsules of C1 and C2 resulting in an increase in the atlanto dens interval of more than 3 mm in the flexion extension radiographs done in the awake neurologically intact patient (Fielding et al., 1974).

Atlanto-rotatory Fixation (C1/C2)

This injury is uncommon in adults and is a distinct entity from rotatory subluxation in children. Fielding and Hawkins (1977) have classified these injuries into four types depending on the degree and direction of displacement and the presence or absence of increase in the atlanto dens interval.

Type 1. Simple rotatory displacement without anterior shift. The odontoid acts as a pivot. These are most common and comprise 47% of these injuries and the transverse ligament is intact.

Type 2. Rotatory displacement with anterior displacement of 3.5 mm, the lateral articular process acting as a pivot. These comprise 50% of these injuries and the transverse ligament is deficient.

Type 3. Rotatory displacement with anterior displacement of more than 5 mm. Both the lateral articular processes are subluxed anteriorly one more than the other. The transverse ligament and other supporting ligaments are deficient.

Type 4. Rotatory displacement with posterior shift seen in association with a deficient dens.

Type 5. Frank rotatory dislocation — these are extremely rare (Levine and Edwards, 1989).

Dens Fractures (C2)

Anderson and d'Alonzo (1974) reported on a series of 60 patients with fractures of the odontoid process. Three distinctive patterns of fracture were found (Table 8.3). This classification is widely used and is useful both for prognosis and management. No reliability studies have been published. Clark and White (1985) used the Anderson–d'Alonzo classification in a multicentre study of odontoid fractures. Type I fractures were excluded. They reviewd 96 type II and 48 type III fractures. Their conclusions support the use of this classification but noted that displacement and angulation are important prognostic factors in the type II group.

Burke and Harris (1989) did not believe that type I fractures exist and preferred to divide odontoid fractures into 'high' and 'low'. Hadley et al. (1988) reported 'type II' fractures with comminution at the fracture site which were difficult to reduce. These were designated as IIA. This classification is not widely adopted.

Traumatic Spondylolisthesis of Axis (C2)

This term was introduced by Garber (1964). A classification based on supposed stability was introduced by Francis et al. (1981). It has not been adopted as it reflects neither the mechanism of injury (hyperextension, with or without distraction) nor does it guide prognosis and treatment. Effendi et al. (1981) described a three-part classification which was modified by Levine and Edwards (1985) (Table 8.4).

Table 8.4 The classification of traumatic spondylolisthesis of the axis (C2) (Levine and Edwards, 1985)

Type	*Mechanism of injury*	*Distinctive features*
1. No angulation minimal displacement (3 mm). No shift of C2 on C3	Hyperextension and axial loading	Stable. Neurological deficit absent. Associated spine fractures usually present
2. Displacement of anterior segment with abnormal C2–3 disc	Hyperextension with axial load followed by anterior flexion and compression	Unstable. Associated spine fracture present especially wedge compression fracture of body of C3
2A. Severe angulation of C2 over C3 with minimal displacement more horizontal fracture in C2 arch	Flexion with distraction	Highly unstable
3. Severe angulation and translation with associated unilateral or bilateral facet dislocation of C2 over C3	Flexion and compression forces	Unstable and associated with neurological injury

Lower Cervical Spinal Injuries (C3–C7)

A mechanistic classification of cervical spine injuries has been proposed by Allan et al. (1982) following a review of 165 lower cervical spine injuries in which they identified six common patterns of injury, each of which is subdivided into stages.

Classification System of Allan et al. (1982)

1. Compressive Flexion (CF)

CF stage 1. Blunting of the anterosuperior vertebral margin to a rounded contour with no evidence of failure of the posterior ligamentous complex.

CF stage 2. Obliquity of the anterior vertebral body with loss of some anterior height of the centrum, in addition to the changes seen in stage 1. The antero-inferior body has a 'beak' appearance, concavity of the inferior end plate may be increased, and the vertebral body may have a vertical fracture.

CF stage 3. Fracture line passing obliquely from the anterior surface of the vertebra through the centrum and extending through the inferior subchondral plate and a fracture of the beak, in addition to the characteristics of a stage 2 injury.

CF stage 4. Deformation of the centrum and fracture of the beak with mild (less than 3 mm) displacement of the inferoposterior vertebral margin into the spinal cord.

CF stage 5. Bony injuries as in stage 3 but with more than 3 mm of displacement of the posterior portion of the vertebral body posteriorly into the spinal canal. The vertebral arch remains intact, the articular facets are separated, and the interspinous process space is increased at the level of injury, suggesting a posterior ligamentous disruption in a tension mode. This is grossly unstable.

2. Vertical Compression (VC)

These result from axial injuries to the neutrally aligned spine.

VC stage 1. Fracture of the superior or inferior end plate with a 'cupping' deformity. Failure of the end plate is central rather than anterior, and posterior ligamentous failure is not evident.

VC stage 2. Fracture of both vertebral end plates with cupping deformities. Fracture lines through the centrum may be present, but displacement is minimal.

VC stage 3. Progression of the vertebral body damage described in stage 2. The centrum is fragmented, and the displacement is peripheral in multiple directions. Most commonly the centrum fails with significant impaction and fragmentation. The posterior aspect of the vertebral body is fractured and may be displaced into the spinal canal. The vertebral arch may be intact with no evidence of ligamentous failure, or it may be comminuted with significant failure of the posterior ligamentous complex; the ligamentous disruption is between the fractured vertebra and the one below it. Sometimes known as a 'burst' fracture.

3. Distractive Flexion (DF)
This is the most common pattern and is associated with falls and road traffic accidents. Between 54% and 80% of these injuries are associated with acute disc herniations at the level of the injury.

DF stage 1. Failure of the posterior ligamentous complex, as evidence by facet subluxation in flexion, with abnormal divergence of the spinous process.

DF stage 2. Unilateral facet dislocation (the degree of posterior ligamentous failure ranges from partial failure sufficient only to permit the abnormal displacement to complete failure of both the anterior and posterior ligamentous complexes, which is uncommon). Subluxation of the facet on the side opposite the dislocation suggests severe ligamentous injury. In addition, a small flake of bone may be displaced from the posterior surface of the articular process, which is displaced anteriorly. Widening of the uncovertebral joint on the side of the dislocation and displacement of the tip of the spinous process towards the side of the dislocation may be seen. Beatson (1963) serially divided the posterior interspinous ligaments, facet capsule, posterior longitudinal ligament and found that unilateral facet dislocation can occur with rupture of only the posterior interspinous ligament and the facet capsule.

DF stage 3. Bilateral dislocation. There is usually >50% width displacement of the vertebral bodies.

DF stage 4. Complete displacement of vertebral bodies (>100% width).

4. Compressive Extension (CE)
CE stage 1. Unilateral vertebral arch fracture, with or without antero-rotatory vertebral displacement. Posterior element failure may consist of a linear fracture through the articular process, impaction of the articular process and ipsilateral pedicle and lamina fractures, resulting in the 'transverse facet' appearance on anteroposterior radiographs, or a combination ipsilateral pedicle and articular process fractures.

CE stage 2. Bilaminar fractures without evidence of other tissue failure. Typically the laminar fractures occur at multiple contiguous levels.

CE stage 3. Bilateral vertebral arch fractures with fracture of the articular processes, pedicles, lamina, or some bilateral combination, without vertebral body displacement.

CE stage 4. Bilateral vertebral arch fracture with full vertebral body width displacement anteriorly. The posterior portion of the vertebral arch of the fractured vertebra does not displace, and the anterior portion of the arch remains with the centrum. Ligament failure occurs at two levels: posteriorly between the fractured vertebra and the one above it and anteriorly between the fractured vertebra and the one below it. Characteristically, the anterosuperior portion of the vertebra below is sheared off by the anteriorly displaced centrum. Both stages 3 and 4 are rare injuries.

5. Distractive Extension (DE)
DE stage 1. Either failure of the anterior ligamentous complex or a transverse fracture of the centrum. The injury is usually ligamentous, and there may be a fracture of the adjacent anterior vertebral margin. The roentgenographic clue to this injury is abnormal widening of the disc space.

DE stage 2. Evidence of failure of the posterior ligamentous complex, with displacement of the upper vertebral body posteriorly into the spinal canal, in addition to the changes seen in stage

1 injuries. Because displacement of this type tends to reduce spontaneously when the head is placed in a neutral position, roentgenographic evidence of the displacement may be minimal, rarely greater than 3 mm on initial films with the patient supine.

6. Lateral Flexion (LF)

LF stage 1. Asymmetric compression of the centrum and ipsilateral vertebral arch fracture, without displacement of the arch on the anteroposterior view. Compression of the articular process of comminution of the corner of the vertebral arch may be present.

LF stage 2. Lateral asymmetric compression of the centrum and either ipsilateral displaced vertebral arch fracture or ligamentous failure on the contralateral side with separation of the articular processes. Both ipsilateral and compressive and contralateral disruptive vertebral arch injuries may be present.

No reliability studies of this system has been found. The Allan system is widely acknowledged. It provides a basis for understanding the diversity of these injuries and some guidance to treatment. Considerable controversies remain on the optimum management and prognosis of these injuries outside the scope of this book.

Thoracolumbar Spine Fractures

A useful classification system for these injuries should distinguish between stable injuries and those which require surgical stabilization. It should identify which injuries would benefit from spinal cord decompression. The unstable spine (as defined by White and Panjabi, 1990) has the potential for neurological injury, increasing spinal deformity and long term pain and disability when the spinal column is subjected to physiological loads.

Nicoll (1949) was the first to direct attention to spinal stability. Later Holdsworth (1963) introduced the two-column model of spinal kinematics, and stated that instability was due to disruption of the entire posterior ligament complex. Bedbrook (1971) disagreed with this view, maintaining that the loss of integrity of the disc and anterior longitudinal complex were more important factors in loss of stability. With the advent of CT imaging, Whitesides and Ali Shan (1976) merged the two concepts giving importance to both anterior and posterior restraints, again using Holdsworth's two-column model. This still failed to explain some instances of acute instability.

Denis Classification System

In 1983, Denis introduced the three column model classifying fractures according to the mechanism of injury and the resulting fracture pattern. He described four major and four minor injury patterns (Table 8.5).

1. *Compression fractures*: These occur due to eccentric axial loading resulting in the common anterior wedge fracture or the less common lateral compression fracture. Anterior compression fractures have four subtypes:

A. Both end plates are damaged.
B. The superior end plate alone is fractured.
C. The inferior end plate alone is fractured.
D. The end plates are intact but the anterior vertebral cortex is compressed.

2. *Burst fractures*: These are caused by axial loading leading to damage of the end plate first. There may be:

A. Fractures involving the central portion of the end plate.

Table 8.5 Thoraco-lumbar spine: modes of failure of the three columns for the four major spinal injury types (Denis, 1983)

Type	*Column*		
	Anterior	*Middle*	*Posterior*
1. Compression	Compression	None	None or distraction (severe)
2. Burst	Compression	Compression	None or splaying of pedicles
3. Flexion-distraction	None or distraction	Distraction	Distraction
4. Fracture-dislocation	Compression Rotation Shear	Distraction Rotation Shear	Distraction Rotation Shear

B. Fractures occurring at the periphery of the end plate.
C. Transverse cracks across the end plate.

Denis stated that in these fractures failure occurs through at least the anterior and middle columns. If the posterior elements are also involved the degree of instability is high.

3. *Flexion-distraction injuries*: These were first reported by Chance in 1948. Gertzbein and Court-Brown (1988) subdivided them into five categories related to the posterior element fracture, anterior columns fracture and the state of the vertebral body. Posteriorly the fracture may enter above the spinous process or through it. In a third type it extends symmetrically across, indicative of a rotational element. Anteriorly there are three possibilities:
A. Injury through the disc space.
B. A fracture passing through the anterior vertebral cortex.
C. A fracture exiting through the superior (C1) or inferior end plate (C2)

4. *Fracture-dislocation injuries*: There is failure of all three columns under compression, tension, rotation, shear or a combination of these forces. These are of three types:
A. Flexion-rotation injuries: the posterior and middle columns fail under tension and rotation whereas the anterior column fails under compression and rotation.
B. Shear injuries: they may be both in the antero-posterior or lateral direction.
C. Flexion-distraction burst injuries: a variant of seat belt injury produced by a combination of flexion and distraction.

Magerl et al. (1994) described a classification derived from the mechanism of injury and takes into account the appearance on radiographs and the progressive severity of injury (Table 8.6, Figure 8.2). The classification is based on an analysis of 1445 cases by a group of eminent international spine surgeons who kept to the hierarchy based on '3s' favoured by AO. The main stems are Type A (vertebral body *compression*), Type B (anterior and posterior element injury with *distraction*) and Type C (anterior and posterior injury with *rotation*). Hierarchies of '3s' develop under these headings. Quite how valuable this system is in clinical practice remains to be seen. Studies are required to assess its reliability and relevance to selection of treatment and prognosis.

Other Spinal Injuries

1. Extension fractures. These are rare and are caused by violent extension forces tending to fracture of the posterior elements, retrolisthesis and anterior lip (avulsion) fractures through a vertebral body.
2. Fractures of the articular processes.
3. Fractures of the spinous process.
4. Fractures of the pars interarticularis.

These injuries are stable and are caused by flexion and axial rotation or extension forces or due to blunt trauma. They may be associated with other severe spinal injuries.

Table 8.6 Classification of thoracic and lumbar fractures, the comprehensive classification of Magerl et al. (1994)

Type A injuries: groups, subgroups and specifications

Type A. Vertebral body compression

A1. Impaction fractures
- A1.1. Endplate impaction
- A1.2. Wedge impaction fractures
  - 1 Superior wedge impaction fracture
  - 2 Lateral wedge impaction fracture
  - 3 Inferior wedge impaction fracture
- A1.3. Vertebral body collapse

A2. Split fractures
- A2.1. Sagittal split fracture
- A2.2. Coronal split fracture
- A2.3. Pincer fracture

Table 8.6 (*Continued*)

A3. Burst fractures

A3.1. Incomplete burst fracture
- 1 Superior incomplete burst fracture
- 2 Lateral incomplete burst fracture
- 3 Inferior incomplete burst fracture

A3.2. Burst-split fracture
- 1 Superior burst-split fracture
- 2 Lateral burst-split fracture
- 3 Inferior burst-split fracture

A3.3. Complete burst fracture
- 1 Pincer burst fracture
- 2 Complete flexion burst fracture
- 3 Complete axial burst fracture

Type B injuries: groups, subgroups and specifications

Type B. Anterior and posterior element injury with distraction

B1. Posterior disruption predominantly ligamentous (flexion-distraction injury)

B1.1. With transverse disruption of the disc
- 1 Flexion-subluxation
- 2 Anterior dislocation
- 3 Flexion-subluxation/anterior dislocation with fracture of the articular processes

B1.2. With type A fracture of the vertebral body
- 1 Flexion-subluxation + type A fracture
- 2 Anterior dislocation + type A fracture
- 3 Flexion-subluxation/anterior dislocation with fracture of the articular processes + type A fracture

B2. Posterior Disruption predominantly osseous (flexion-distraction injury)

B2.1. Transverse bicolumn fracture

B2.2. With transverse disruption of the disc
- 1 Disruption through the pedicle and disc
- 2 Disruption through the pars interarticularis and disc (flexion-spondylolysis)

B2.3. With type A fracture of the vertebral body
- 1 Fracture through the pedicle + type A fracture
- 2 Fracture through the pars interarticularis (flexion-spondylolysis) + type A fracture

B3. Anterior disruption through the disc (hyperextension-shear injury)

B3.1. Hyperextension-subluxations
- 1 Without injury of the posterior column
- 2 With injury of the posterior column

B3.2. Hyperextenion-spondylolysis

B3.3. Posterior dislocation

Type C injuries: groups, subgroups and specifications

Type C. Anterior and posterior element injury with rotation

C1. Type A injuries with rotation (compression injuries with rotation)

C1.1. Rotational wedge fracture

C1.2. Rotational split fractures
- 1 Rotational sagittal split fracture
- 2 Rotational coronal split fracture
- 3 Rotational pincer fracture
- 4 Vertebral body separation

C1.3. Rotational burst fractures
- 1 Incomplete rotational burst fracture
- 2 Rotational burst-split fracture
- 3 Complete rotational burst fracture

Table 8.6 ***(Continued)***

C2. Type B injuries with rotation

C2.1 – B1 injuries with rotation (flexion-distraction injuries with rotation)
1 Rotational flexion subluxation
2 Rotational flexion subluxation with unilateral articular process fracture
3 Unilateral dislocation
4 Rotational anterior dislocation without/with fracture of articular processes
5 Rotational flexion subluxation without/with unilateral articular process fracture + type A fracture
6 Unilateral dislocation + type A fracture
7 Rotational anterior dislocation without/with fracture of articular processes + type A fracture

C2.2 – B2 injuries with rotation (flexion distraction injuries with rotation)
1 Rotational transverse bicolumn fracture
2 Unilateral flexion spondylolysis with disruption of the disc
3 Unilateral flexion spondylolysis + type A fracture

C2.3 – B3 injuries with rotation (hyperextension-shear injuries with rotation)
1 Rotational hyperextension-subluxation without/with fracture of posterior vertebral elements
2 Unilateral hyperextension-spondylolysis
3 Posterior dislocation with rotation

C3. Rotational-shear injuries

C3.1. Slice fracture
C3.2. Oblique fracture

Sacral Fracture

These are relatively common injuries (Flurey, 1942; Bonin, 1945); until recently they were overlooked and lumped with fractures of the pelvis. Bonin (1945) was the first to characterize these injuries. They are classified according to the direction of the fracture line which may be vertical, transverse or oblique. Twenty-five per cent of his fractures were associated with neurological deficits.

1. *Vertical fractures*: These are the most common sacral injuries, and have been classified by Schmidek, Smith and Kristiansen (1984) and by Denis, Davis and Comfort (1988).

A. Schmidek classification of vertical fractures of the sacrum (Figure 8.3).

i. Lateral mass fracture
ii. Tuxta articular fracture
iii. Cleaving fracture
iv Avulsion fracture

B. Denis and colleagues classification (Figure 8.4).

i. Zone I Alar region
ii. Zone II Foraminal fractures
iii. Zone III Central sacral canal involvement

2. *Transverse fractures*: They are less common than vertical fractures (Bonin, 1945; Denis, Davis and Comfort, 1988). There are two types:

i. High transverse fractures (Roy-Camille et al., 1985) (Zone III above): These are due to high energy injuries with anterior bending between the superior and inferior fragments, the distal fragment moves vertically and associated with

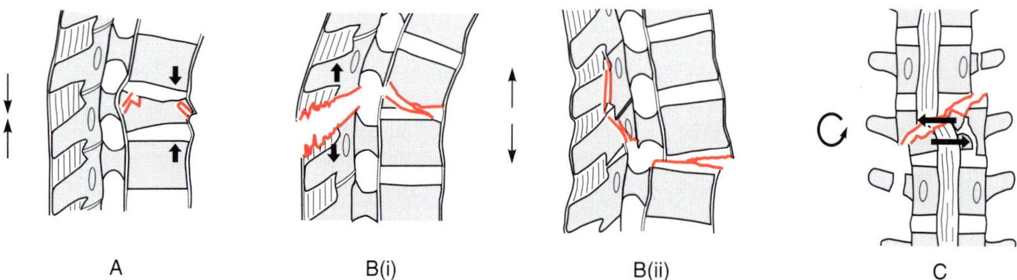

Figure 8.2. Classification of thoracic and lumbar fractures (from Magerl et al., 1994).
a. (A) Compression injuries of the anterior column. (B) Two column injury with either (i) posterior transverse disruption, or (ii) anterior transverse disruption. (C) Two column injury with rotation

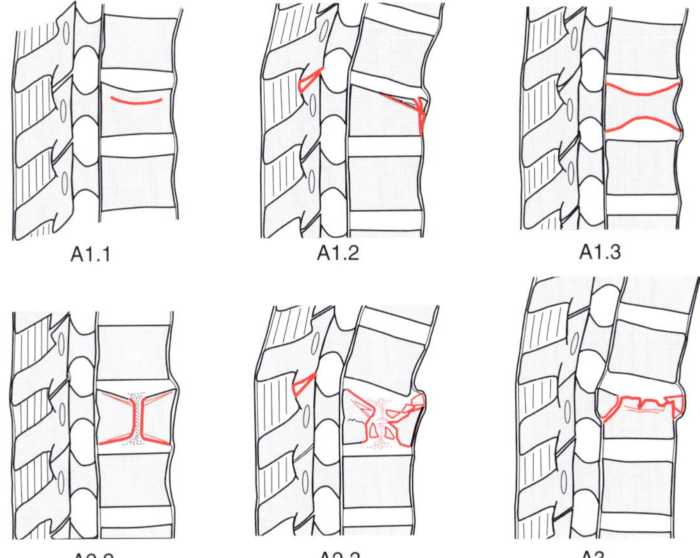

b. 'A' pattern injuries: A1.1 Endplate impaction. A1.2 superior wedge fracture. A1.3 Vertebral body collapse. A2.2 Coronal split fracture. A2.3 Pincer fracture. A3 Superior incomplete burst fracture.

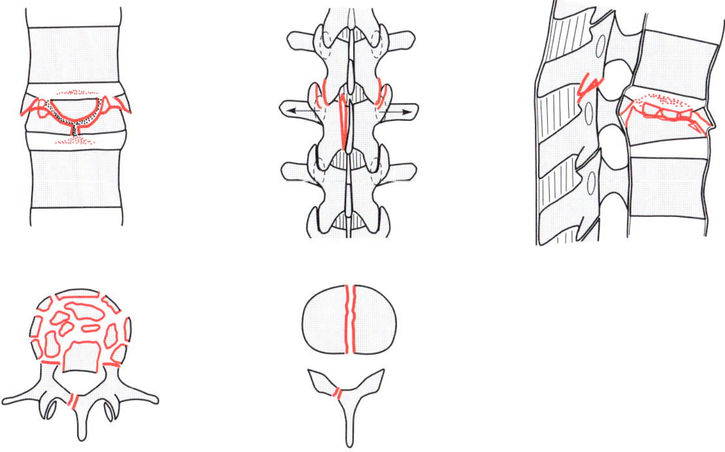

c. Superior burst–fracture (A3.2.1) seen in standard views on plain films and CT.

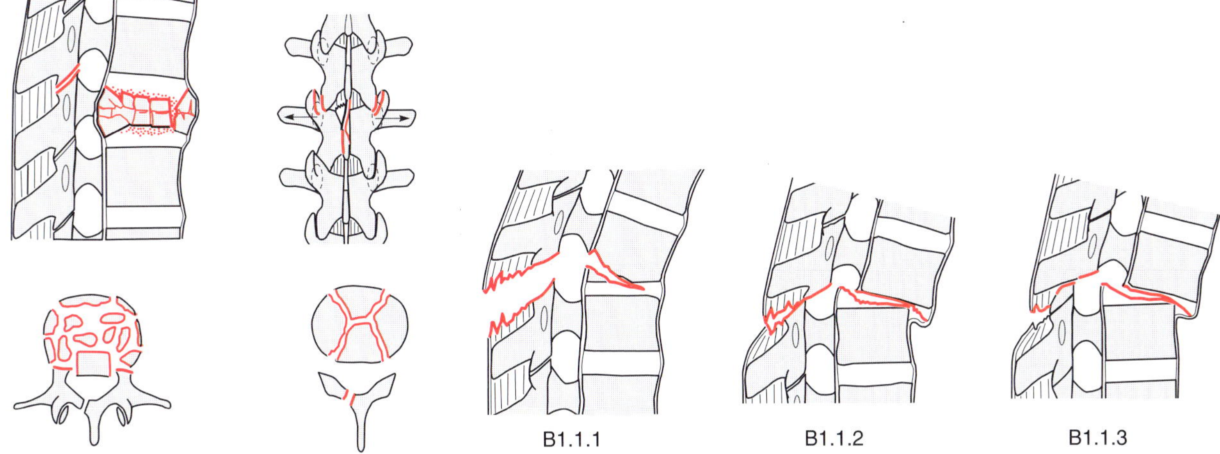

d. Complete axial burst fracture (A3.3.3).

e. 'B' pattern fractures: posterior disruption with anterior lesions.

Figure 8.2. (*Continued*).

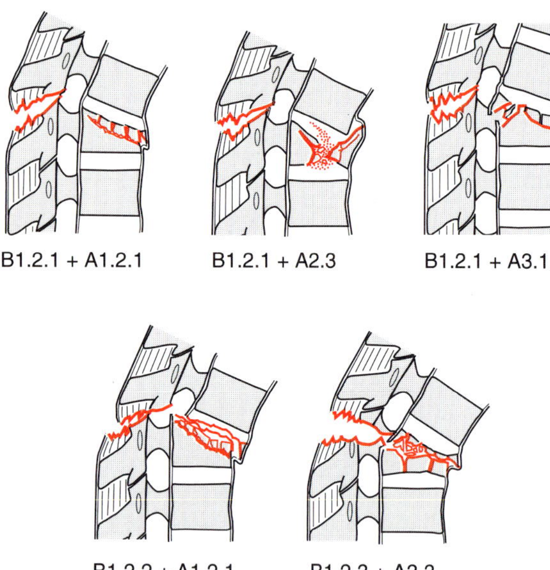

f,g. Posterior disruption (predominantly ligament) with vertebral body fracture.

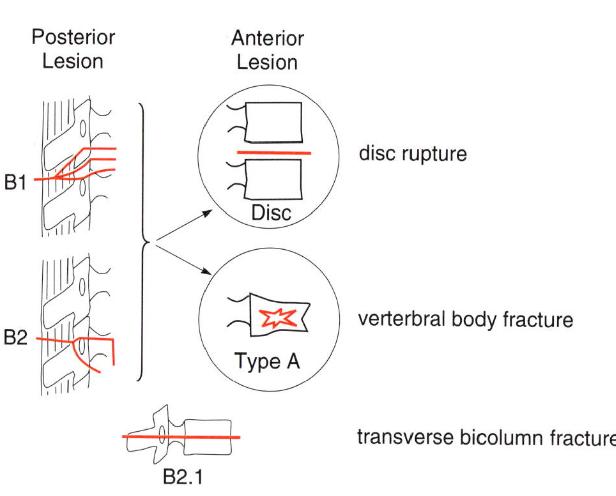

k. Flexion–distraction injuries (B1–B2). Posterior column injuries are associated with disc rupture or vertebral body function.

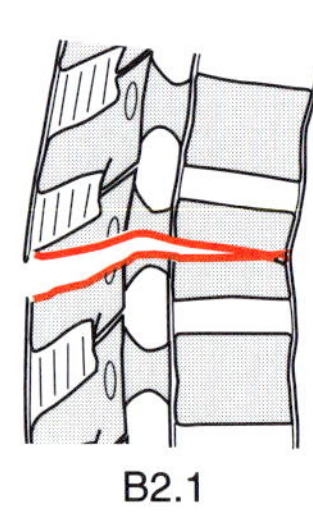

h. Transverse bicolumn fracture.

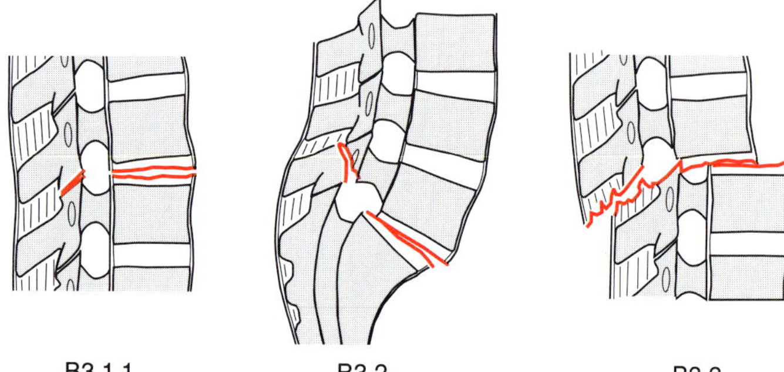

l. Hyper extension–shear injuries with disc disruption.

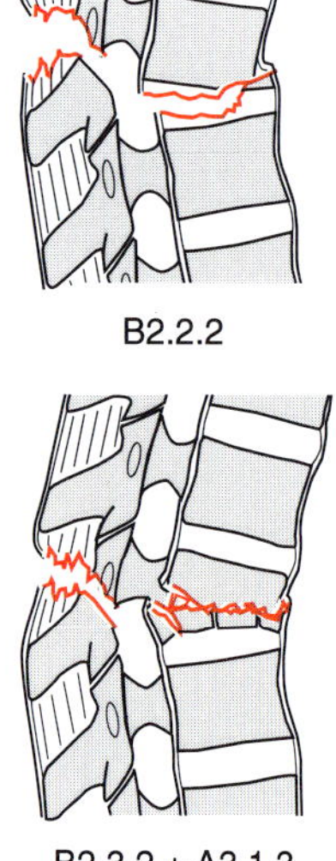

i,j. Flexion spondylolisthesis (B2.2.2, B2.3.2 and A3.1.3).

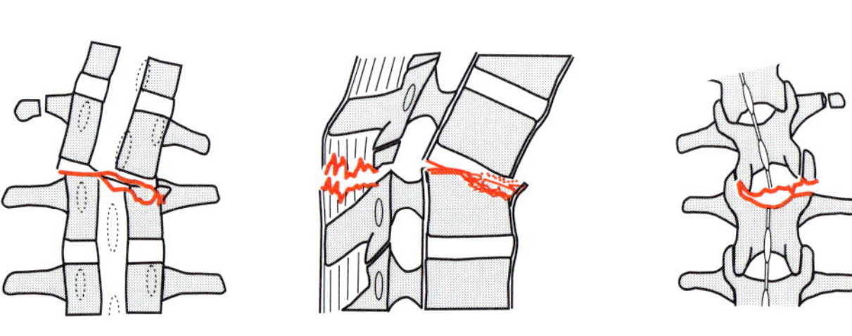

m. Type A fracture with rotation becomes a rotational wedge fracture (C1.1).

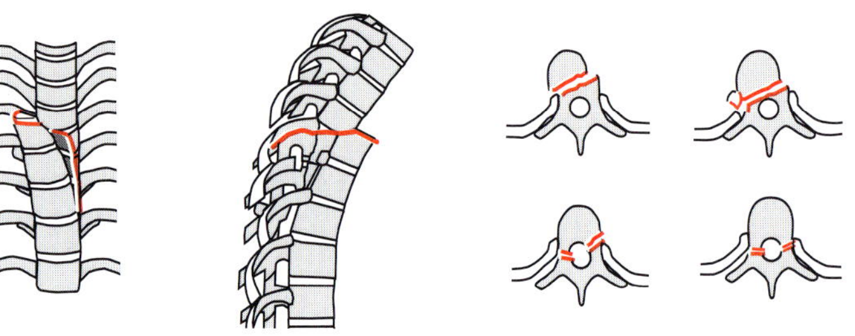

n. Vertebral body separation (C1.2.4).

Figure 8.2. (*Continued*).

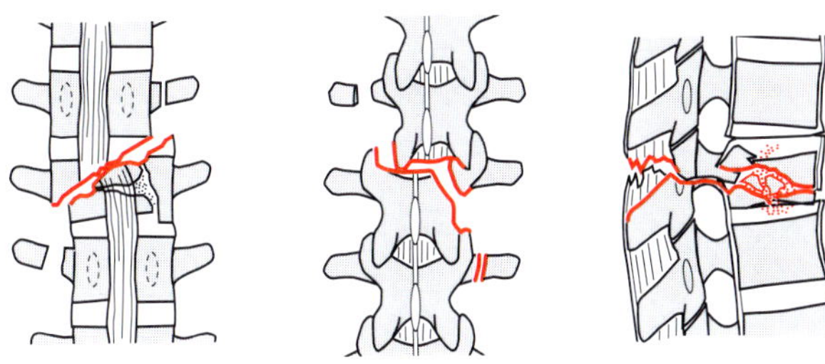

o. Complete burst fracture with rotation (C1.3.3).

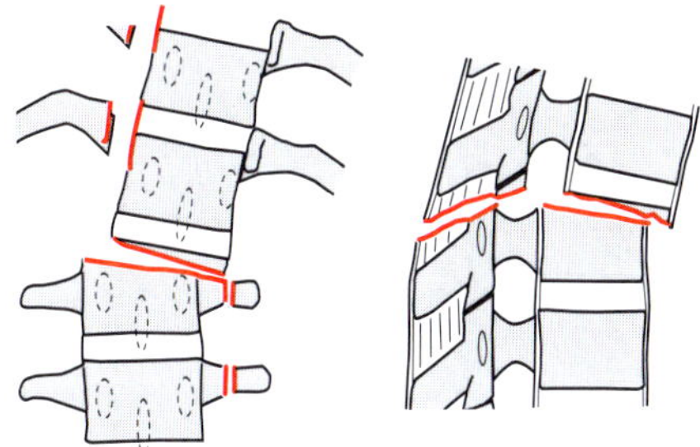

s. Rotational shear injury (C3.1).

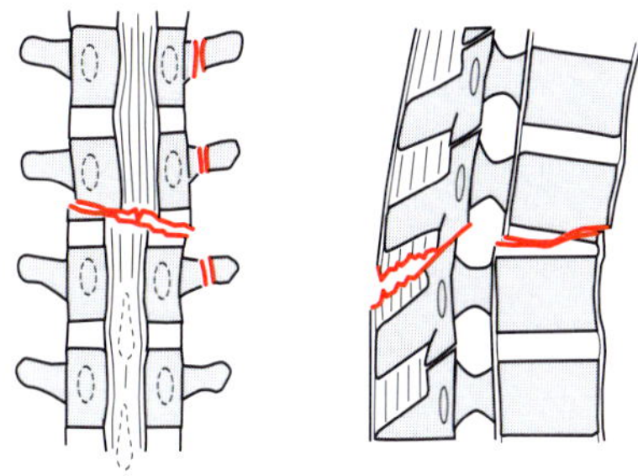

p. 'B' injury with rotation (C2.1.1).

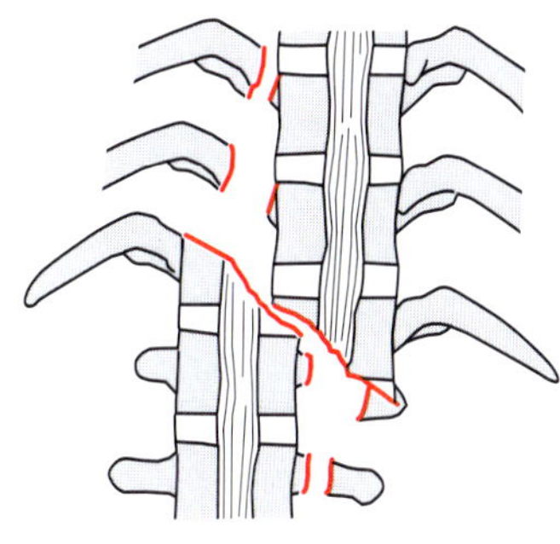

t. Rotation shear injury with oblique vertebral fracture (C3.2).

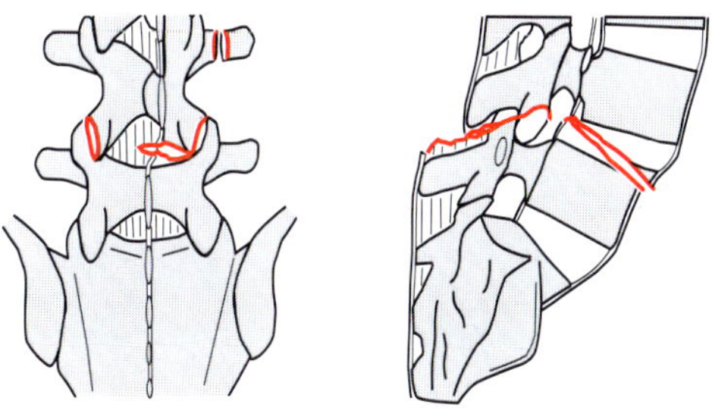

q. 'B' injury with rotation with unilateral dislocation (C2.1.3).

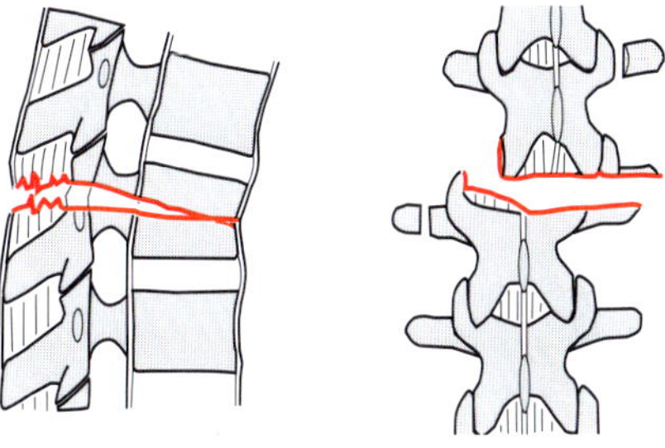

r. Transverse bicolumn injury with rotation (C2.2.1).

Figure 8.2. (*Continued*).

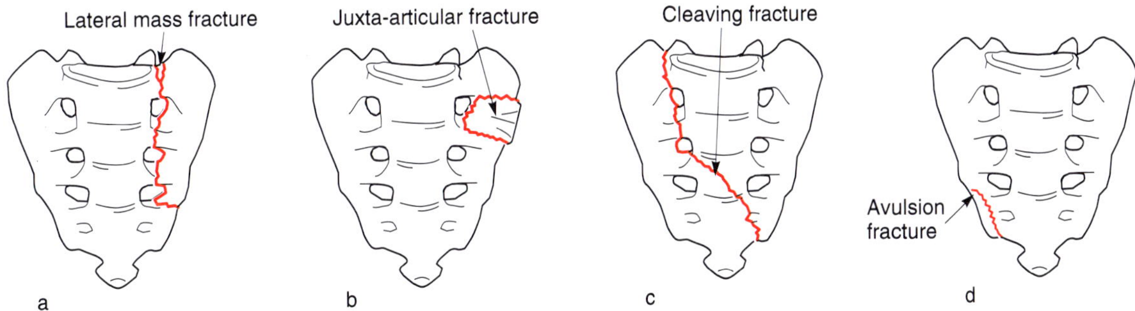

Figure 8.3. Classification of vertical fractures of the sacrum (from Schmidek, Smith and Kristiansen, 1984).

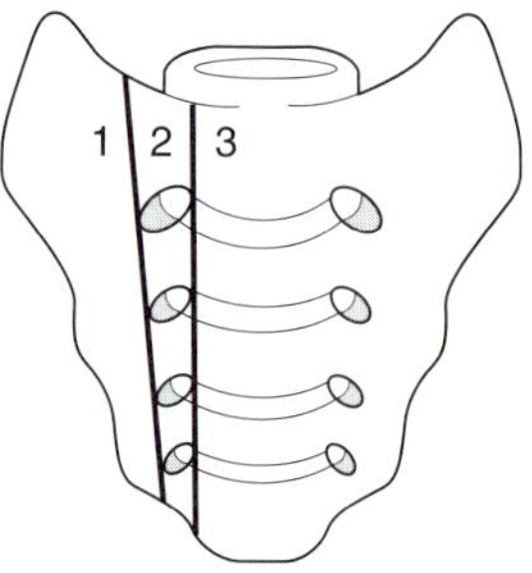

a. Sacral fracture zone.

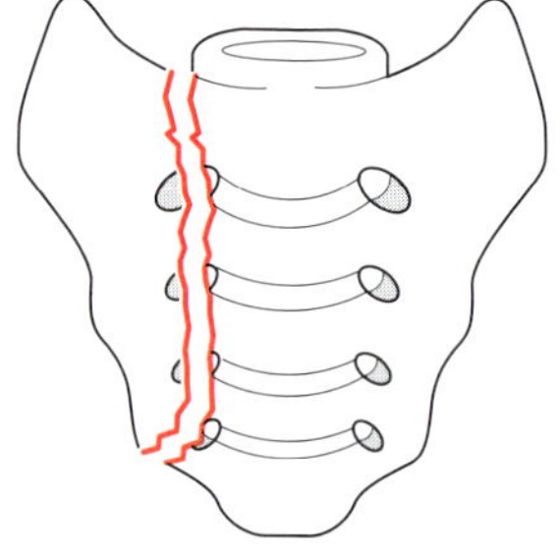

d. Zone 2 fracture.

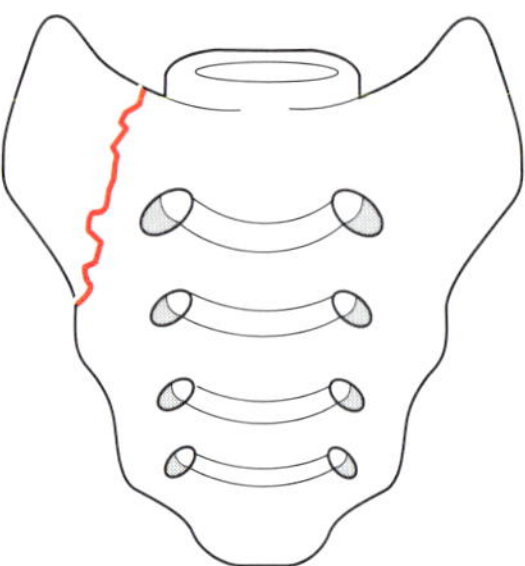

b. Zone 1 fracture.

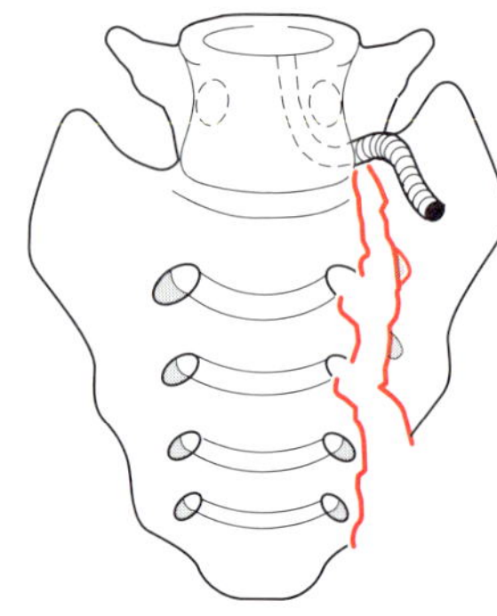

e. Zone 2 fracture with L5 root damage/entrapment.

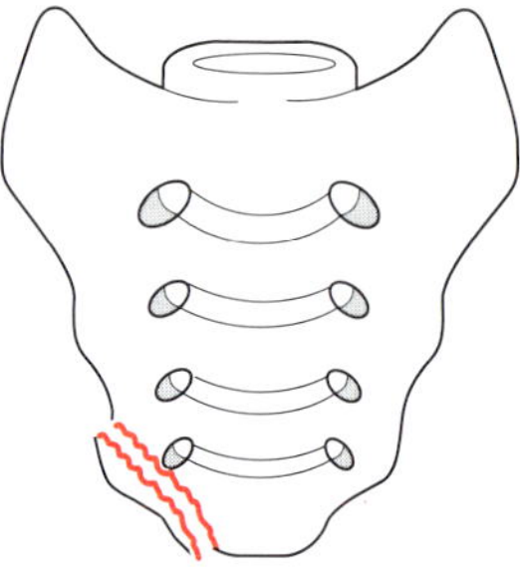

c. Zone 1 fracture with ligament avulsion and severe pelvic instability.

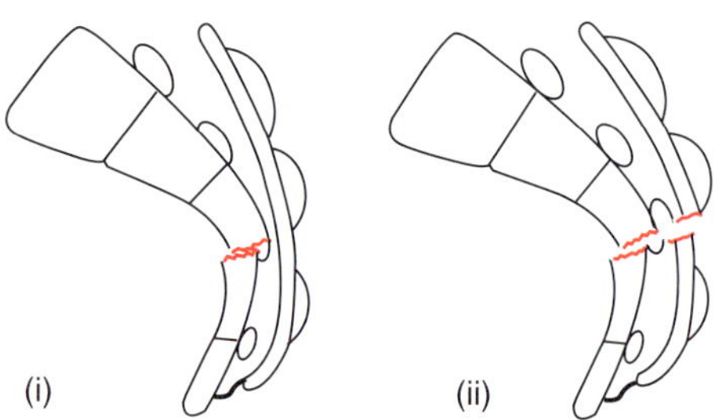

f. Zone 3 fracture: (i) sacral burst; (ii) fracture dislocation.

Figure 8.4. Classification of fractures of the sacrum (from Denis, Davis and Comfort, 1988).

bilateral alar fractures and of the L5 transverse processes. They are frequently associated with neurological deficit including bladder and bowel dysfunction and perineal numbness without motor weakness.

ii. Low transverse fractures (Weaver, England and Richardson, 1981): These are due to a direct blow on the sacrum and they may be associated with rectal preforation, CSF leak or neural deficits.

References

Allan, B.L., Ferguson, R.L., Lehmann, R. and O'Brian, R.P. (1982) A mechanistic classification of closed indirect fractures and dislocation of the lower cervical spine. *Spine*, **7**, 1–27.

Anderson, L.D. and d'Alonzo, R.T. (1974) Fractures of the odontoid process of the axis. *J. Bone Joint Surg.*, **56A**, 1663–1674.

Anderson, P.A. and Montisano, P. (1988) Morphology and treatment of occipital condyle fractures. *Spine*, **13**, 731–736.

Bama, M., Stevenson, G.W., Tumiel, H. and Tumiel, A. (1983) Unilateral atlanto-occipital dislocation complicating an anomaly of the atlas. *J. Bone Joint Surg.*, **65A**, 685–687.

Beatson, T.R. (1963) Fractures and dislocations of the cervical spine. *J. Bone Joint Surg.*, **45B**, 21–35.

Bedbrook, G.M. (1971) Stability of spinal fractures and fracture dislocations. *Paraplegia*, **9**, 23–32.

Bonin, J.G. (1945) Sacral fractures and injuries to the cauda equina. *J. Bone Joint Surg.*, **27**, 113–127.

Bucholz, R.W. and Burkhead, W.Z. (1979) The pathological anatomy of fatal atlanto-occipital dislocations. *J. Bone Joint Surg.*, **61A**, 248.

De Bur, J.D., Thomas, M., Walter, J. and Anderson, P. (1988) Traumatic atlanto axial subluxation *J. Bone Joint Surg.*, **70B**, 652–655.

Burke, J.T. and Harris, J.H. (1989) Acute injuries of the axis vertebrae. *Skelet. Radiol.*, **18**, 335–346.

Chance, C.D. (1948) Note on a type of flexion fracture of the spine. *Br. J. Radiol*, **21**, 452–453.

Clark, C.R. and White, L.A. (1985) Fractures of the dens; A multicentre study. *J. Bone Joint Surg.*, **67A**, 1340–1348.

Collato, P.M., De Muth, W.W., Schwentker, E.P. and Boal, A.K. (1986) Traumatic atlanto-occipital dislocation. *J. Bone Joint Surg.*, **67A**, 1106–1109.

Denis, F. (1983) Three column spine and its significance in the classification of acute thoracolumbar spinal injuries. Spine, **8**, 817–831.

Denis, F., Davis, S. and Comfort, T. (1988) Sacral fractures: an important problem. Retrospective analysis of 236 cases. *Clin. Orthop. Rel. Res.*, **227**, 67–81.

Dublin, A., Marks, W.M., Weinstock, D. and Newton, T.H. (1986) Traumatic dislocation of the atlanto occipital articulation. *J. Neurosurg.*, **52**, 541–546.

Effendi, B., Roy, D., Connigh, B. et al. (1981) Fractures of the ring of the axis. A classification based on the analysis of 131 cases. *J. Bone Joint Surg.*, **63B**, 319–327.

Eismont, F.J. and Bohlman, H.H. (1978) Posterior atlanto occipital dislocation with fracture of the atlas and odontoid process. Report of a case with survival. *J. Bone Joint Surg.*, **60A**, 397.

Esses, S. (1981) Fracture of the atlas associated with fracture of the odontoid process. *Injury*, **112**, 310–312.

Fielding, W.J. and Hawkins, R.J. (1977) Atlanto axial rotatory fixation. *J. Bone Joint Surg.*, **59A**, 37–44.

Fielding, W.J., Hawkins, R.J., Cochran, G.V.B. et al. (1974) Tears of the transverse ligament of the atlas. A clinical and biomechanical study. *J. Bone Joint Surg.*, **56A**, 681–1691.

Flurey, W.W. (1942) Fractures of the pelvis with special reference to associated fractures of the sacrum. *Am. J. Radiol.*, **47**, 89–96.

Francis, W.R., Fielding, J.W., Hawkins, R.J. et al. (1981) Traumatic spondylolisthesis of the axis. *J. Bone Joint Surg.*, **63B**, 313–318.

Garber, J.N. (1964) Abnormalities of the atlas and axis vertebrae: congenital and traumatic. *J. Bone Joint Surg.*, **46A**, 1782–1791.

Gehweiler, J.A., Osborne, R.L. and Becker, R.F. (1980) *The Radiology of Vertebral Trauma.* Philadelphia: Saunders.

Gertzbein, S.D. and Court-Brown, C.M. (1988) Flexion-distraction injuries of the lumbar spine. Mechanisms of injury and classification. *Clin. Orthop.*, **227**, 52–60.

Hadley, M.N., Browmer, C.M., Liu, S.S. and Sonntag, V.K. (1988) New subtype of acute odontoid fractures (type IIA). *Neurosurg.*, **22**, 67–71.

Han, S.Y., Witten, D.M. and Musselman, J.P. (1976) Jefferson fracture of the atlas. Report of six cases. *J. Neurosurg.*, **44**, 368–371.

Hintzer, L. and Schalintzek, M. (1971) Fractures and subluxations of the atlas and axis. *Acta Orthop. Scand.*, **42**, 251–258.

Holdsworth, F.W. (1963) Fractures, dislocations and fracture-dislocations of the spine. *J. Bone Joint Surg.*, **45B**, 6–20.

Jefferson, G. (1920) Fracture of the atlas vertebra. Report of four cases and a review of those previously recorded. *Br. J. Surg.*, **7**, 407–422.

Kaufman, R.A., Dunbar, J.A. and McLaurin, R.L. (1982) Traumatic longitudinal atlanto-occipital distraction injuries in children. *Am. J. Neuroradiol.*, **3**, 415–419.

Landells, C.D. and van Peteghem, P.K. (1988) Fracture of the atlas: classification, treatment and morbidity. *Spine*, **13**, 450–452.

Levine, A.M. and Edwards, C.C. (1985) The management of traumatic spondylolisthesis of the axis. *J. Bone Joint Surg.*, **67A**, 217–226.

Levine, A.M. and Edwards, C.C. (1986) Treatment of injuries in the C1-C2 complex. *Orth. Clin. North Am.*, 31–44.

Levine, A.M. and Edwards, C.C. (1989) Traumatic lesions of the occipital atlanto axial complex. *Clin. Orthop.*, **239**, 53–68.

Margerl, F., Aebi, M., Gentzbein, S.D. et al. (1994) A comprehensive classification of thoracic and lumbar injuries. *Eur. Spine J.*, **3**, 184–201.

Nicoll, E.A. (1949) Fractures of the dorso-lumbar spine. *J. Bone Joint Surg.*, **31B**, 276–294.

Roy-Camille, R., Saillant, G., Gagna, G. and Hazel, C. (1985) Transverse fracture of the upper sacrum, suicidal jumpers fracture. *Spine*, **10**, 838–845.

Schmidek, H.H., Smith, D. A. and Kristiansen, T.K. (1984) Sacral fractures. *Neurosurg.*, **15**, 735–746.

Segal, L.S., Grimm, J.O. and Stauffer, E.S. (1987) Non-union of fractures of the atlas. *J. Bone Joint Surg.*, **69A**, 1423–1434.

Weaver, E.N., England, G.D. and Richardson, D.E. (1981) Sacral fracture: Case presentation and review. *Neurosurg.*, **9**, 725–728.

White, A.A. and Panjabi, M.M. (1990) *Clinical Biomechanics of the Spine*, 2nd edn. Philadelphia: Lippincott.

Whitesides, T.E. and Ali Shan, S.G. (1976) On the management of unstable fractures of the thoracolumbar spine. Rationale for the use of anterior decompression and fusion and posterior stabilisation. *Spine*, **1**, 99.

9

Spinal cord injuries

H. Hegde, M. F. Brown, M. Scott-Young and P. Sett

Introduction

Injuries to the spinal cord are often associated with injuries of the spinal column, but may also occur without obvious bony injury. They result in varying degrees of neurological deficit, covering a spectrum from 'mild and temporary' to 'severe and permanent'. Early examination is of prime importance to assess injury to the spinal cord and nerve roots — in particular whether the cord lesion is complete or incomplete. A classification of the extent and type of neurological deficit should assist with giving an accurate prognosis and may guide treatment. It is essential for research and audit.

Classification

Complete Spinal Cord Injury

Traditionally complete anaesthesia and complete motor paralysis distal to the level of an injury is known as a complete spinal cord injury. In these cases there is, by definition, no evidence of sacral sparing, this is an important sign of potential recovery. Unfortunately, the majority with sacral sparing do not show recovery. During the period of 'spinal shock', a definite diagnosis of a complete, or incomplete, lesion can usually be made (Holdsworth, 1970).

Spinal shock may be considered to be the spinal equivalent of cerebral concussion. It is a reversible condition defined as spinal cord nervous dysfunction based on a physiological, rather than a structural, disruption (Waters, Adkins and Yakura, 1991). It can be shown to have resolved when reflexes below the level of injury can be elicited. Because of the difficulty in defining the level of injury in a patient with possibly resolving spinal shock, the lowest sacral reflexes (bulbocavernosus and anal wink) are conventionally used. The presence of the bulbocavernosus reflex indicates that the sacral roots and sacral spinal cord segments are functioning. It is, of course, normal to have preserved reflexes with a rostral cord injury, but in some cases there may be a mixture of cord and root (cauda equina) injury, and in such cases these reflexes may remain absent. It has been found that 99% of patients with complete spinal cord lesions have return of the bulbocavernosus reflex within 24 hours (Morgan, Wharton and Austin, 1971; Bohlman, 1985).

Complete spinal cord injury is then classified according to the level of the lesion. The problem with this approach has been lack of agreement as to what constitutes a level and at least three classifications have been used (Michaelis, 1968; 1969):

1. The neurological level of the injury is the lowest level where motor function is present (motor at least 3 out of 5 on the MRC scale) (Stoke Mandeville and Long Beach systems).
2. The first 'affected' level (French system).
3. The level of the damaged vertebra (bony level — limited usage except for planning surgery).

Incomplete Spinal Cord Injury

By contrast, the presence of some neurological function below the level of injury is known as an incomplete lesion. The pattern of preserved sensory and motor functions differentiates the various incomplete cord lesion syndromes.

Preferred Nomenclature

A recent joint working party of the International Medical Society of Paraplegia (IMSOP) and the American Spinal Injuries Association (ASIA) have clarified the nomenclature (Maynard et al., 1997). It must also be recognized that whilst all bony injury below the conus medullaris (effectively all lumbar and sacral injury) tends only to cause root damage, paraplegic patients with such injury are commonly said to suffer from spinal cord injury. This is, of course, not strictly accurate.

The important points about the ASIA/IMSOP classification are as follows:

- Motor and sensory modalities are separately described at each level and for each side.
- The definition of 'complete' is clarified as absence of function in the lowest segments testable, i.e. no sacral sparing (Waters, Adkins and Yakura, 1991).
- The level of the injury is the lowest *normal* level of motor and sensory function. These are given separately and given on left and right separately.
- It acknowledges a 'zone of partial preservation' in which there is incomplete motor and sensory function. Again, this is described separately for each side.

This classification has been investigated and seems to have face validity (Ditunno et al., 1994). Two cases were analysed in detail by five international experts (Donovan et al., 1997). There were difficulties in agreeing motor scores and levels, because of pain inhibition, and the zone of partial preservation.

There are five basic types of lesion, excluding cauda equina and root lesions which are dealt with separately below, which apply to incomplete lesions. These are:

1. Central cord syndrome (Schneider, Cherry and Pantek, 1954; Schneider, Thompson and Bebin, 1958)

Central cord syndrome is the most common incomplete cord syndrome. It represents severe damage to the central grey matter with preservation of the peripheral spinal cord structure. The patient presents with an incomplete tetraplegia involving the upper extremities (mainly the hands) to a greater extent than the lower. Sensory sparing is variable but usually sacral pinprick is preserved. There is an early return of bladder and bowel function, although these usually remain abnormal. Motor recovery, if any, starts in the most distal sacral elements (toe flexors and then extensors), followed by the lumbar elements of the ankle, knee and hip. Upper extremity functional recovery is usually minimal depending on the extent of central grey matter damage. Recovery from central cord syndrome is variable, a 75% chance of functional motor recovery was observed by Schneider et al. (1954). More recent figures are not available.

2. Anterior cord syndrome (Schneider, 1955)

This is characterized by complete or 'incomplete' motor loss and loss or 'diminution' of pain and temperature sensation below the level of the lesion with preservation of posterior column sensation, i.e. proprioception and vibration senses. Prognosis of significant recovery is variable. In those who have complete motor loss prognosis is poor, with a 10% functional motor return (Bosch, Stauffer and Nickel, 1971).

3. Posterior cord syndrome (Bosch, Stauffer and Nickel, 1971)

A rare syndrome, consisting of loss of deep pressure, deep pain and proprioceptive sensation. The patient walks with a foot-slapping gait similar to that of someone afflicted with tabes dorsalis (the loss of neurological function is very similar).

4. The Brown–Séquard syndrome

The Brown–Séquard syndrome is an injury to one-half of the spinal cord producing ipsilateral motor loss, loss of vibration and position sense. There is contralateral loss of pain and temperature sensibility. There is usually acute loss of bladder and bowel function and the ability to walk. The prognosis for the recovery of these latter functions is good (Bosch, Stauffer and Nickel, 1971).

5. Conus medullaris syndrome

Injuries at the level of the thoracolumbar junction are very common and frequently affect the conus. In such traumatic cases there is usually an associated nerve root injury leading to lower limb dysfunction and flaccid paralysis. There is loss of all bladder and peri-anal muscle control. The bulbocavernosus reflex and anal wink are usually absent. Motor loss is variable, depending on the number of nerve roots spared.

Table 9.1 The Frankel grading system (Frankel, Hancock and Hyslop, 1969)

Grade A	Complete injury, no motor or sensory function below the level of injury
Grade B	Sensation only, some preserved function below the level of injury; this does not apply to a slight discrepancy between the motor and sensory level, but does apply to sacral sparing
Grade C	Motor function useless, preserved motor function below the level of injury, but it is of no practical use to the patient
Grade D	Motor function useful, preserved useful motor function below the level of injury, patients in this group can walk with or without aids
Grade E	Recovery, normal motor and sensory function, abnormal reflexes may be present

Table 9.2 Sunnybrook Cord Injury Scale (Tator, 1982)

Grade	*Description*	*Corresponding Frankel grade*
1	Complete motor and sensory loss	A
2	Complete motor loss; incomplete sensory loss	B
3	Incomplete motor but unclear	C
4	Incomplete motor loss; incomplete sensory loss	
5	Incomplete motor but useless; normal sensory	C
6	Incomplete motor, complete sensory	D
7	Incomplete motor, incomplete sensory	D
8	Incomplete motor, normal sensory	D
9	Normal motor, incomplete sensory	D
10	Normal motor, normal sensory	E

Root and Cauda Equina Injuries

Whilst these are not strictly speaking spinal cord injuries, they are often associated with spinal fractures.

Cauda equina syndrome

This is an injury below the conus involving the lumbosacral nerve roots (cauda equina) within the vertebral canal. This results in an areflexic bladder and bowel. There is also lower limb weakness with lower motor neurone signs.

Nerve root injury

Injury to nerve roots may occur either in isolation or in association with spinal cord injuries. The prognosis of motor recovery in nerve root injury is favourable.

Neurological Grading

Frankel, Hancock and Hyslop (1969) were the first to attempt a grading system to define the severity of neurological injury. This is outlined in Table 9.1. The Frankel grading system has been widely used to measure neurological deficit. It can be applied in a variety of conditions to measure neurological change. The advantage of clear cut grades make the instrument robust as an outcome measure. The disadvantage of the system lies in its inability to detect and define those who show changes in their neurological function within a single group.

Tator (1982) expanded the Frankel system and introduced the Sunnybrook Cord Injury Scales to assess the degree of neurological injury and recovery. The system used 10 grades to define severity of injury. Grade 1 was defined as complete motor and sensory loss and Grade 10 represented normal (Table 9.2). It allows for complete and incomplete sensory loss and has a second code for changes in neurological status. This system has not proved popular in Europe. A study from Canada of incomplete thoracolumbar and lumbar injuries favoured the reliability of the Frankel scale over the Sunnybrook Scale (Davis et al., 1993).

ASIA has recently modified the Frankel grading to the one shown in Table 9.3.

Motor and Sensory Scoring Systems

The first attempt at motor scoring was made by Lucas and Ducker (1979). They proposed a Neuro Trauma Motor Index score (Table 9.4). In this

Table 9.3 The ASIA motor impairment scale

A	Complete	No sensory or motor function is preserved in the sacral segments S4 and S5
B	Incomplete	Sensory function is preserved below the neurological level and extends through the sacral segments S4 and S5
C	Incomplete	Motor function is preserved below the neurological level, and the majority of key muscles below the neurological level have muscle grade less than 3
D	Incomplete	Motor function is preserved below the neurological level, and the majority of key muscles below the neurological level have a muscle grade greater than or equal to 3
E	Normal	Sensory and motor function normal

Table 9.4 The Neuro Trauma Motor Index: Each muscle is graded according to a functional grade (0–5), and these grades summated. The grades are as follows: 0 = absent, 1 = trace, 2 = poor, 3 = fair, 4 = functional, and 5 = normal

Muscle	*Right*	*Midline*	*Left*
Diaphragm		2	
Deltoid	5		5
Biceps	5		5
Triceps	5		5
Flexor digitorum	5		5
Abductor			
Digiti minimi	5		5
Intercostals		2	
Upper abdominals		2	
Lower abdominals		2	
Iliopsoas	5		5
Quadriceps	5		5
Extensor digiti	5		5
Gastrocnemius	5		5
Anal		2	

Normal = 100

index the maximum score was 100. The system graded muscles on a scale 0–5 according to function (0 = absent, 1 = trace, 2 = poor, 3 = fair, 4 = functional and 5 = normal) (Table 9.5). This grading was then applied to 14 muscles to give the motor index as shown in Table 9.6.

This system is subjective and attempts to generate an ordinal scale which contravenes basic statistical techniques. The scale has been used by Bondurant et al. (1990) and Cotler et al. (1990).

Table 9.5 The MRC Grading of motor weakness

0	Total paralysis
1	Palpable or visible contraction
2	Active movement, full range of motion (ROM) with gravity eliminated
3	Active movement, full ROM against gravity
4	Active movement, full ROM against active resistance
5	Normal

Table 9.6 Motor Index Score adapted by ASIA

Grade on right	*Muscle*	*Grade on left*
0–5	C5 (elbow flexors)	0–5
0–5	C6 (wrist extensors)	0–5
0–5	C7 (elbow extensors)	0–5
0–5	C8 (finger flexors)	0–5
0–5	T1 (finger abductors)	0–5
0–5	L2 (hip flexors)	0–5
0–5	L3 (knee extensors)	0–5
0–5	L4 (ankle dorsiflexors)	0–5
0–5	L5 (great toe extensors)	0–5
0–5	S1 (ankle plantar flexors)	0–5
0–50		0–50

The motor index of Lucas and Ducker (1979) has been modified by the American Spinal Injury Association into the Motor Index Score, where 100 represents the perfect scale. This scale is more precise and the score is calculated by summing separate measurements of the individual muscles with each individual measure, easily categorized by the usual MRC 0–5 rating. Again, its deficiency is that bowel and bladder function are not evaluated. A muscle with grade 3 power is deemed to have its rostral nerve supply intact. For an example of applying this convention, if the finger flexors (C8) has power 3 and the elbow extensor (C7) is power 4 or 5, then the motor level is graded as C8

Manabe scale

In order to overcome the deficiencies of the Frankel grading system, which was originally devised using metastatic disease rather than trauma, and that of the Motor Index Score, Manabe et al. (1989) combined the two to produce their own grading. This is shown in Table 9.7. The Manabe scale has not been used by spinal injury units but, in a modified form (Table 9.8), has been employed in a multicentre spine fracture study carried out by the Scoliosis Research Society (Gertzbein, 1992).

Table 9.7 The Manabe scale of change (Manabe et al., 1989)

Satisfactory result		
Grade I	Excellent	Complete or marked recovery from pain and neurological deficits
Grade II	Good	Neurological deficits improved by one Frankel grade. No pain
Unsatisfactory result		
Grade III	Fair	Slight recovery of neurological function within the same Frankel grade but relief from pain
Grade IV	Poor	No improvement or worsening of neurological deficits

Table 9.8 The Manabe scale of change as used by the Scoliosis Research Society (Gertzbein, 1992)

Grade I	Excellent	Improvement by more than one Frankel grade
Grade II	Good	Improvement by only one Frankel grade
Grade III	Fair	Improvement of function within the same Frankel grade
Grade IV	No change	No change in Frankel grade or motor score
Grade V	Poor	Deterioration of Frankel grade or motor score

In the second National Acute Spinal Cord Injury Study (NASCIS II), Bracken et al. (1990) used its own scoring systems as an outcome measure. For sensory scoring, 29 segments from C2 to S5 were tested bilaterally for pin-prick and light touch. Sensation was scored as: absent = 1, decreased = 2 and normal = 3. Hence the maximum sensory score is 174, i.e. 87 per side. The motor score utilized the MRC grading system (vide infra) and applied this to test 14 muscles on each side to give a total motor score = 140. The NASCIS II trial showed that there was improvement in the motor score by 7 points at the end of one year in patients who were given high dose methylprednisolone (this was a statistically significant improvement).

ASIA scoring system

ASIA have now combined and modified these two systems and issued a revised motor and sensory scoring system. This is clarified in the ASIA classification (Figure 9.1). For sensory scoring ASIA have reduced the number of sensory dermatomes to 28 by combining S4 and S5 into one segment. The scoring is altered to: 0 = absent, 1 = impaired, and 2 = normal. Pin-prick and light touch are separately tested and scored. The maximum score for both pin-prick (both sides) and for light touch (both sides) is 112.

ASIA recommends that sensation in each dermatome be tested at the key sensory points for that dermatome described by Austin (1972). These points are shown as black dots over each dermatome in Figure 9.1. The ASIA and NASCIS motor scores have been validated by El Masry et al. (1996).

Neurophysiological Studies

Neurophysiological studies are based on establishing intact fibre tracts within the damaged cord. Motor and sensory evoked studies have shown conclusively that a patient with a clinically complete lesion could still have intact motor and sensory conducting pathways. Such investigations can help in assessment and in prognosis of spinal injury. Much of this work has been done by Dimitrijevic (1988).

Tsubokawa (1987) has evaluated the role of evoked potential in spinal cord injuries and has found it useful in localizing and determining the extent of cord injury and in defining prognosis. It is now possible to elicit evoked responses in the peripheral muscles by transcranoid and plexus magnetic stimulation. It is likely that such techniques will have a substantial role in neurophysiological studies in the future.

Magnetic Resonance Imaging (MRI)

MRI has made it possible for the first time to visualize the spinal cord directly. It is certainly able to show the structural damage sustained and can help to establish whether the cord has been transected or not. It can play a role in defining prognosis. The role of MRI in spinal injuries has been discussed by Sett and Crockard (1991). They have, for example, demonstrated an incidence of post-traumatic syrinx of 20%. MRI can also be used to demonstrate disc degeneration adjacent to fractured vertebrae as well as compression of the spinal cord, conus or cauda equina. Demonstration of disc damage is likely to

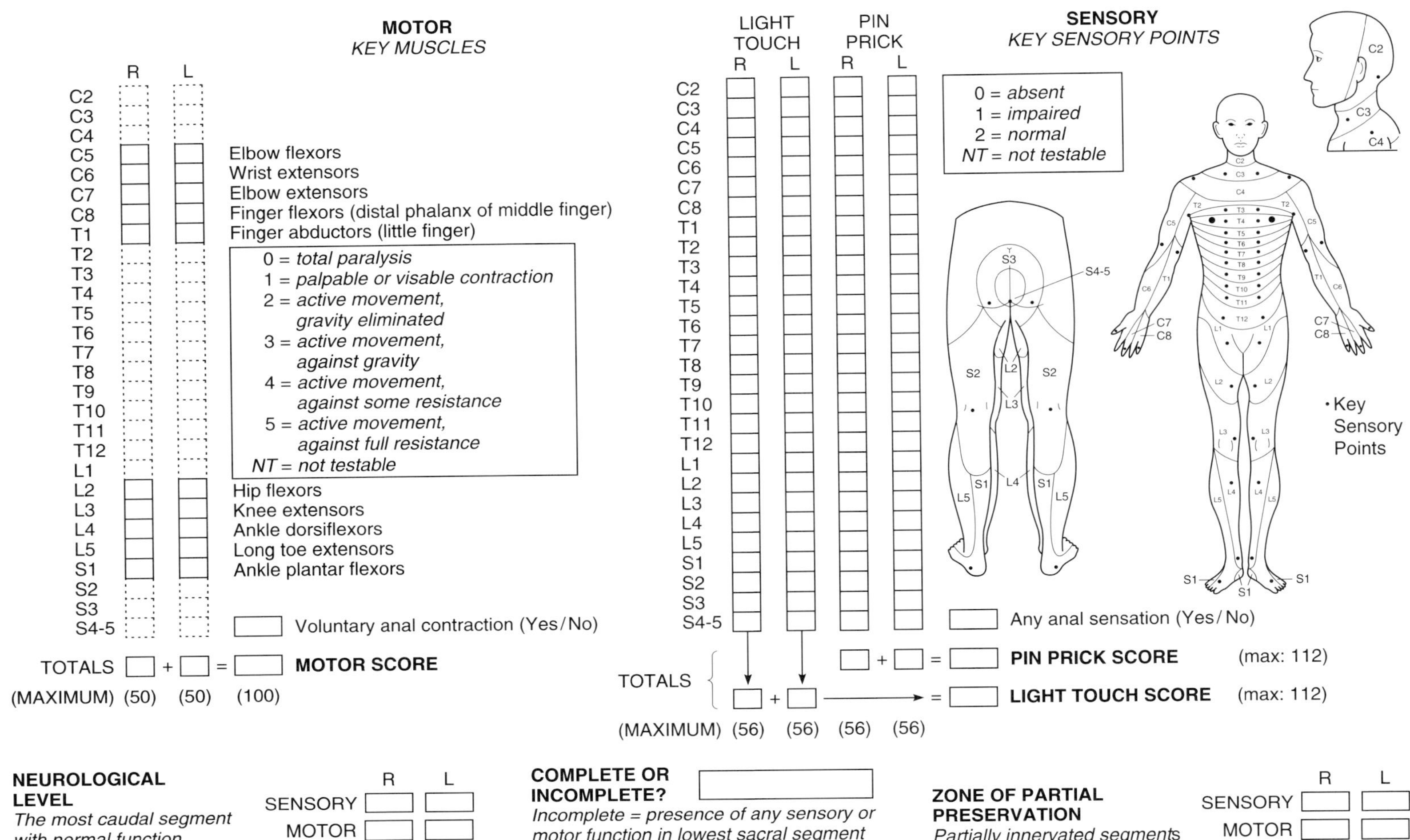

Figure 9.1. Revised motor and sensory score (ASIA classification). ASIA Motor score: ASIA motor score uses the MRC grades (0–5) and extends this to the 10 key muscles already mentioned (Table 9.2) to give a maximum score of 50 on one side and a total of 100. Sensory score: For sensory scoring ASIA have reduced the number of sensory dermatomes to 28 by combining S4 and S5 into one segment. The scoring is altered to: 0 = absent, 1 = impaired and 2 = normal. Pin prick and light touch are separately tested and scored. The maximum score for both pin prick (both sides) and for light touch (both sides) is 112.

be of great value in planning surgical treatment, especially in the choice of anterior or posterior fixation (Bondurant et al., 1990).

MRI can be used to assess the spinal cord following injury. To our knowledge there is no validated classification system although there are a number of publications describing different patterns. It is said that MRI can distinguish compression from contusion and can discriminate oedema from haemorrhage. Kulkarni et al. (1987) report three patterns of intramedullary signal changes in patients with acute trauma:

1. Acute foci of hyperintensity on T1- and T2-weighted images. These are said to be consistent with cord contusion with intramedullary haemorrhage and suggest poor prognosis. These may change at a later stage producing high intensity on T1 images and low signal on T2-weighted images. Subsequently this evolves to a high signal on both T1- and T2-weighted images.
2. A low or normal signal on T1-weighted images and a spindle shape region of high signal on T2-weighted images. These are said to be consistent with spinal cord oedema and carry a better prognosis.
3. A mixed pattern with low signal centre with high signal peripheral lesion on T2-weighted images. These are said to have improvement with appropriate treatment.

Discussion

There has been the need to standardize and simplify neurological examination and scoring in spinal cord injury and the recent revised edition of the ASIA classification is simple and practical. It is hoped that spinal and trauma units will use it to grade their patients. Motor and sensory scoring on the ASIA system is now widely practised in the USA and has been helpful in standardizing neurological examination.

The measurement of neurological outcome is still mainly clinical. With the simplified scoring system of the ASIA classification it is expected that most clinicians will now grade and score spinal injury patients. It will then be possible to measure outcome and prognosis. The neurophysiological studies are improving continuously and will probably have a greater role to play in future. At the present time, they are used mainly in cases where controversy exists. Examples are: (i) the differentiation between organic spinal cord injury and functional paraplegia; (ii) definition of the degree of cord transection; (iii) those patients with high cord lesions requiring respiratory support (Jellinek et al., 1992).

The value of acute MRI lies in its ability to depict the damage to bone, soft tissue and neurological structures. This can help in defining management and predicting outcome. Another important role of MRI is in detecting and measuring the progress of post-traumatic syringomyelia. It is now well established that the incidence of asymptomatic syrinxes is far more common than hitherto believed. Symptoms of a developing syrinx can be protean. They may precede neurological change by months or years. In order to plan treatment and measure outcome, regular MRI is necessary. Shunting procedures to relieve syrinx must be followed up regularly with MRI scans (Sett and Crockard, 1991).

Recommendations

The new ASIA/IMSOP classification offers the most valuable method of classifying spinal cord injuries and has served to clear up much of the confusion that existed before. There seems no virtue in continuing to use the previous systems. This system does require further efforts to validate it and to look at its repeatability. An MRI classification needs to be established and validated.

References

Austin, G.M. (1972) *The Spinal Cord: Basic Aspects and Surgical Considerations*. Springfield, IL: Thomas, p. 762.

Bohlman, H. (1985) Surgical management of cervical spinal fractures and dislocations. *Course Lectures*, **34**, 163–187.

Bondurant, F.J., Cotler, H.S., Kulkarni, M.V. et al. (1990) Acute spinal cord injury — a study using physical examination and magnetic resonance imaging. *Spine*, **15**, 161–168.

Bosch, A., Stauffer, E.S. and Nickel, V.L. (1971) Incomplete traumatic quadriplegia: A ten year review. *JAMA*, **216**, **3**, 473–478.

Bracken, M.B., Shepard, M.J., Collins, W.F. et al. (1990) A randomised, controlled trial of methylprednisolone or naloxone in the treatment of acute spinal-cord injury. Results of the Second National Acute Spinal-Cord Injury Study. *N. Engl. J. Med*, **322**(20), 1405–1411.

Cotler, H.B., Cotler, J.M., Alder, M.E. et al. (1990) The medical and economic impact of closed cervical spine dislocation. *Spine*, **15**, 448–452.

Davis, L.A., Warren, S.A., Reid, D.C. et al. (1993) Incomplete neural deficits in thoracolumbar and lumbar spine fractures. Reliability of Frankel and Sunnybrook scales. *Spine*, **18**(2), 257–263.

Dimitrijevic, M.R. (1988) Clinical neurophysiological evaluation of motor and sensory functions in chronic spinal cord injury patient. In *Spinal Cord Dysfunction* (L.S. Illis, ed.) Oxford: Oxford University Press, pp. 274–286.

Ditunno, J.F., Young, W., Donovan, W.H. and Creasy, E. (1994) The International Standards booklet for neurological and fucntional classification of spinal cord injury. *Paraplegia*, **32**, 70–80.

Donovan, W.H., Brown, D.J., Ditunno, J.F. et al. (1997) Neurological issues. *Spinal Cord*, **35**, 275–281.

El Masry, W.S., Tsubo, M., Katoh, S. et al. (1996) Validation of American Spinal Injury Association (ASIA) Motor Score and the National Acute Spinal Cord Injury Study (NASCIS) Motor Score. *Spine*, **21**, 614–619.

Frankel, H.L., Hancock, D.O. and Hyslop, G. (1969) The value of postural reduction in the initial management of closed injuries of the spine with paraplegia and tetraplegia. *Paraplegia*, **7**, 179–192.

Gertzbein, S.D. (1992) Scoliosis Research Society: Multicentre Spine Fracture Study. *Spine*, **17**, 528–540.

Holdsworth, F. (1970) Fractures, dislocations and fracture dislocations of the spine. *J. Bone Joint Surg.*, **52A**, 1534–1551.

Jellinek, D.A., Bradford, R., Bailey, I. and Symon, L. (1992) The role of motor evoked potentials in the management of hysterical paraplegia: case report. *Paraplegia*, **30**, 300–302.

Kulkarni, M.V., McArdle, C.B., Kopanicky, D. et al. (1987) Acute spinal cord injury: MR imaging at 1.5T. *Radiology*, **164**, 837–843.

Lucas, J.T. and Ducker, T.B. (1979) Motor classification of spinal cord injuries with mobility, morbidity and recovery indices. *Am. J. Surg.*, **45**, 13–20.

Manabe, S., Tateishi, A., Abe, M. and Ohno, T. (1989) Surgical treatment of metastatic tumours of the spine. *Spine*, **14**, 41–47.

Maynard, F.M., Bracken, M.B., Creasey, G. et al. (1997) International standards for neurological and functional classification of spinal cord injury. *Spinal Cord*, **35**, 266–274.

Michaelis, L.S. (1968) Discussion on classification of neurological lesions. *Paraplegia*, **6**, 46–53.

Michaelis, L.S. (1969) International inquiry on neurological terminology and prognosis in paraplegia and tetraplegia. *Paraplegia*, **7**, 179–192.

Morgan, T.H., Wharton, G.W. and Austin, G.W. (1971) The results of laminectomy in patients with incomplete spinal cord injuries. *Paraplegia*, **9**, 14–23.

Schneider, R.C. (1955) The syndrome of acute anterior cervical spinal cord injury. *J. Neurosurg.*, **12**, 95–122.

Schneider, R.C., Cherry, G. and Pantek, H. (1954) The syndrome of acute central spinal cord injury with special reference to the mechanisms involved in hyperextension injuries of the cervical spine. *J. Neurosurg.*, **11**, 546–577.

Schneider, R.C., Thompson, J.M. and Bebin, J. (1958) The syndrome of acute central cervical spinal cord injury. *J. Neurol. Neurosurg. Psychiatry*, **21**, 216–227.

Sett, P. and Crockard, H.A. (1991) The value of magnetic resonance imaging in the follow-up management of spinal injury. *Paraplegia*, **29**, 396–410.

Tator, C.H. (1982) *Early Management of Acute Spinal Cord Injury*. New York: Raven Press.

Tsubokawa, T. (1987) Clinical value of multimodality spinal cord evoked potential in the prognosis of spinal cord injuries. In *Thoracic and Lumbar Spine and Spinal Cord Injuries* (P. Harris, ed.). New York: Springer Verlag.

Waters, R.L., Adkins, R.H., and Yakura, J.S. (1991) Definition of complete spinal cord injury. *Paraplegia*, **29**, 573–581.

10

Pelvis and acetabulum

S. Chugh

Pelvis

Introduction

Fractures of the pelvis account for between 2 and 5% of all fractures treated in a hospital. In younger patients these are associated with severe injuries. These carry a significant mortality rate which increases with associated soft-tissue injuries and/or open wounds. With the advancement of resuscitation techniques, mortality rates are improving. The focus of assessment is shifting from death to morbidity in the long term. Treatment options have increased from single pelvic slings and postural reduction and skeletal traction to the more complex internal or external fixation.

Pelvic haemorrhage is a major cause of death in these patients. There is increasing realization that early effective management of severe pelvic trauma contributes significantly to the survival of the individual. Modern techniques allow early skeletal stabilization to proceed side-by-side with general resuscitative measures, allowing haemostasis and thereby early reduction of the total blood loss.

An effective classification system should be based on the understanding of the fracture with respect to the mechanism of injury, the direction and magnitude of the forces involved and predict the potential problems. To understand the classification of fractures of the pelvis, it is essential to understand the anatomy of the pelvis and the concept of pelvic stability.

Anatomy

The pelvis is a ring formed by the sacrum and the two lateral components, each composed of ilium, ischium and pubis. These components have no inherent stability and rely on ligamentous support for their integrity (Figure 10.1).

The pelvis can be divided into two halves, anterior and posterior, relative to the acetabulum. The anterior half comprises the pubic symphysis, pubic bone and the rami and acts as a supporting strut to prevent anterior collapse on normal weightbearing. The posterior arch is comprised of the sacrum, the sacroiliac joints and the ilium. It is the weight-bearing component of the pelvic ring.

The stability of the sacroiliac joint is not through the bony articulation between the sacrum and the ilium but through the ligamentous attachments between them. The anterior sacroiliac ligaments are flat and strong in order to resist the external rotation and shearing forces. However, the posterior ligamentous structures form the major support. These comprise the iliolumbar ligament, the transverse fibres of

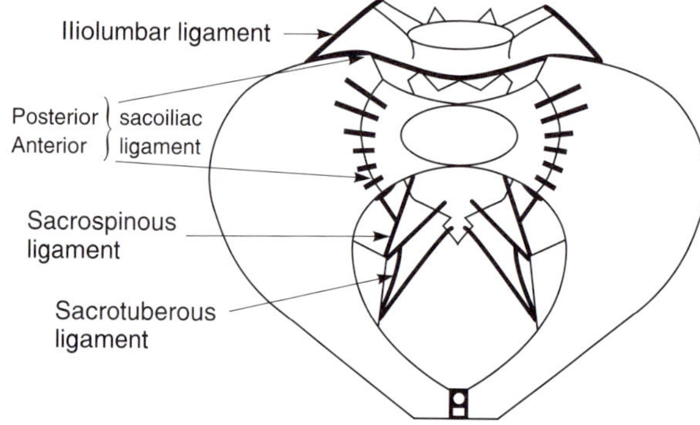

Figure 10.1. Diagrammatic representation of the major ligaments of the pelvis: the strong pubic symphysis anteriorly and the posterior tension bond of the pelvis, including the iliolumbar, posterior sacroiliac, sacrospinous and sacrotuberous ligaments (Tile, 1984).

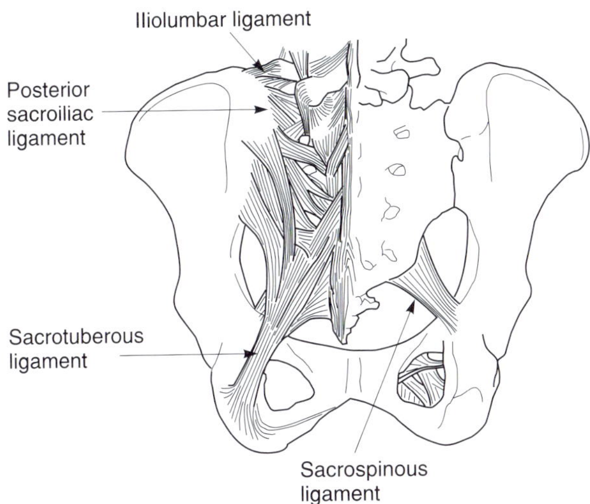

Figure 10.2. The major posterior stabilizing structures of the pelvic ring (from Tile, 1984).

the interosseous sacroiliac ligaments and the posterior sacroiliac interosseous ligament which is the strongest ligament in the body (Figure 10.2).

The posterior ligament complex prevents anterior displacement of the sacrum or posterior displacement of the pelvic ring on the sacrum but allows some rotation between the sacrum and the ilium on weight-bearing.

The superior aspect of the posterior pelvic ring is stabilised by the posterior ligament complex, while the inferior portion is dependent on the sacrospinous and the sacrotuberous ligaments. These are placed perpendicular to each other. The sacrospinous ligament runs from the lateral edge of the sacrum to the ischial spine and resists external rotatory forces. The sacrotuberous ligaments resist vertical shear forces and run from the sacroiliac complex to the ischial tuberosity (Figure 10.3).

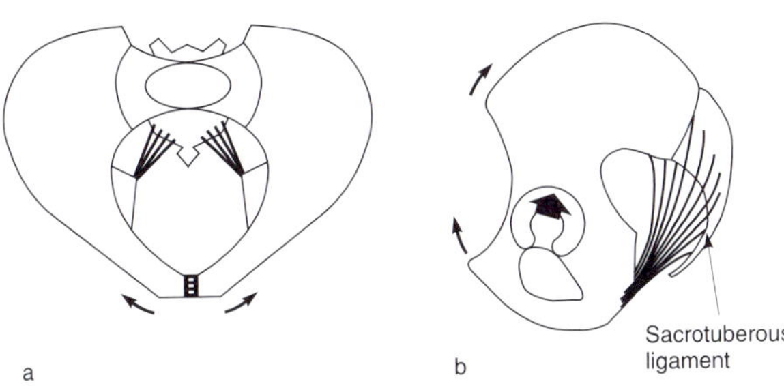

Figure 10.3. (a) The sacrospinous ligaments, resist external rotatory forces. (b) The sacrotuberous ligaments, resist a shearing rotatory force. (Tile, 1984).

Stability of the Pelvis

Stability is defined as the ability to withstand physiological forces without abnormal deformation. The stability of the pelvis depends on ligamentous support. Tile demonstrated that the division of the symphysis ligaments while the posterior ligaments were intact allowed the pubic symphysis to open only up to 2.5 cm. Additional division of the sacrospinous and sacrotuberous ligaments leads to greatly increased pelvic instability.

Radiological Assessment

Prior to discussing the classification in more detail, it is important to visualize the fractures present in order to understand the type and pattern of injury. It is equally important to understand that the pelvic ring has to fracture in at least two sites, as described by Gertzbein and Chenoweth (1977). At least three radiological views of the pelvis are necessary:

1. Anteroposterior.
2. Inlet view.
3. Outlet view.

Fractures of the posterior part of the pelvis may require CT scanning.

Classification

Peltier (1965) classified pelvic fractures into injuries involving the weight-bearing part of the pelvis and those involving the non-weight-bearing segment. He demonstrated that the highest concentration of serious complications occurred in a group of patients with fractures occurring in the weight-bearing portion of the pelvis.

Trunkey et al. (1974) attempted to classify pelvic fractures in relation to the degree of injury.

Type I: Comminuted or crush injury with three or more major components involved.
Type II: Unstable pelvic fractures involving the weight-bearing portion of the pelvis.
Type III: Stable group of fractures.

The mortality and morbidity rates were highest in type I and lowest in type III.

Pennal et al. (1980) drew attention to the importance of classifying fractures on the basis of the forces involved. The main forces as recognized by Pennal were external rotation, lateral compression (internal rotation) and vertical shear.

Tile (1984) recognized that in major trauma, the forces may be too complex to be easily classified. They added a fourth category, 'The Complex Group'. Today, this is the most widely used classification and is described below.

Tile's Classification (Tile, 1984)

Pelvic ring injuries have classically been divided into two groups, stable and unstable. Tile studied the ligamentous contributions to stability by sectioning ligaments of anatomical specimen pelves. He found that when pubic symphysis was sectioned without injury to other ligamentous structures diastasis of no more than 2.5 cm was seen on pelvic loading. If the sacrospinous ligaments were also sectioned, then more than 2.5 cm diastasis was observed. Impingement of the posterior iliac spines against the sacrum eventually limited the total external rotation. This indicates that the pelvis is rotationally unstable in the horizontal plane and the restoration of the anatomical integrity is possible by using the intact posterior ligament complex as a hinge or a tension band. Sectioning of the pubic symphysis, sacrospinous ligament and the sacrotuberous ligament created an extremely unstable situation allowing posterior and superior translation of the ilium in relation to the sacrum indicating instability in the vertical plane in addition to rotational instability.

The fractures are classified into three groups as shown in Table 10.1 and Figure 10.4.

Group 1. Stable pelvic fractures

These have an intact posterior ligamentous complex or impaction at the base of the fracture site and are subdivided according to the direction of impact.

Anteroposterior compression injury

Type I Open book type (Figure 10.4a).
Force applied to the anterior superior iliac spines of a fixed pelvis or via the externally rotated femora can produce an open book type of injury. Similarly, a blow to the posterior superior iliac spine may also produce this injury. This leads to separation of the pubic symphysis with disruption of the anterior sacroiliac ligament. The posterior ligament complex is left intact. This injury pattern can be staged according to severity:

Stage I. There is pubic symphysis interruption but the sacrospinous ligaments are intact. The disruption is less than 2.5 cm.

Stage II As the forces continue, there is disruption of sacrospinous ligaments and anterior sacroiliac ligaments with a gap of more than 2.5 cm at the pubic symphysis.

Stage III There is complete disruption of the anterior sacroiliac liga-

Table 10.1 Classification of Tile (1984)

	Direction of injurious force	*State of posterior sacroiliac complex*
Stable	Anteroposterior compression *Type I*: Open book (Stages I to III) *Type II*: Four rami fracture	Intact
Stable	Lateral compression *Type I*: Ipsilateral anterior and posterior injury *Type II*: Bucket handle *Type III*: Four rami type	Impacted
Unstable	Vertical shear	Disrupted • unilateral • bilateral
Miscellaneous	Complex Bilateral sacroiliac joint disruption with an intact anterior arch Acetabular fracture associated with pelvic ring disruption	

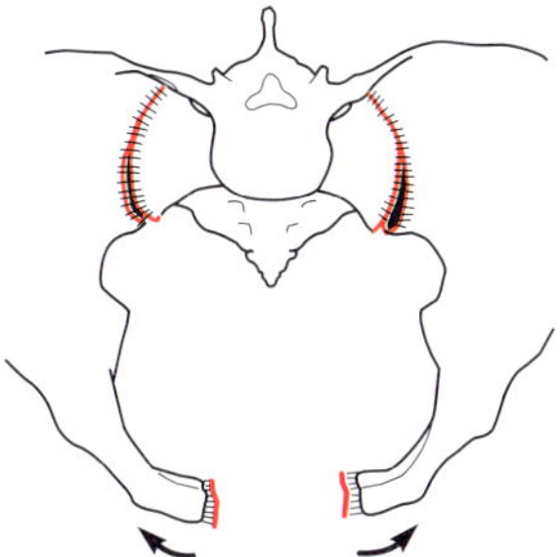

a. Anteroposterior compression (Type I): open book pelvic fracture.

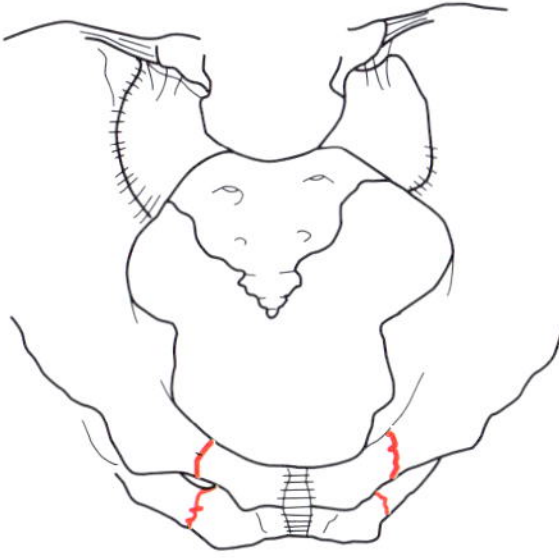

b. Anteroposterior compression (Type II): four rami fracture.

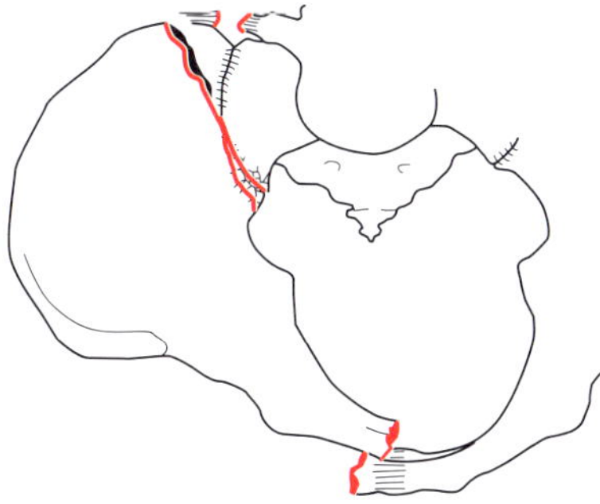

c. Lateral compression (Type I): ipsilateral anterior and posterior injury.

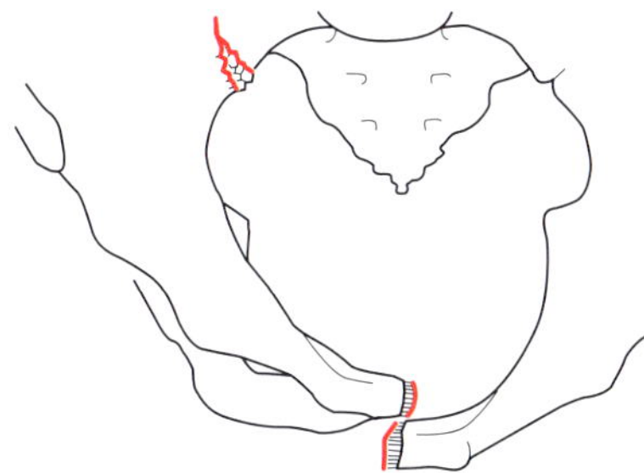

d. Lateral compression (Type I): locked symphysis (variation in anterior injury pattern).

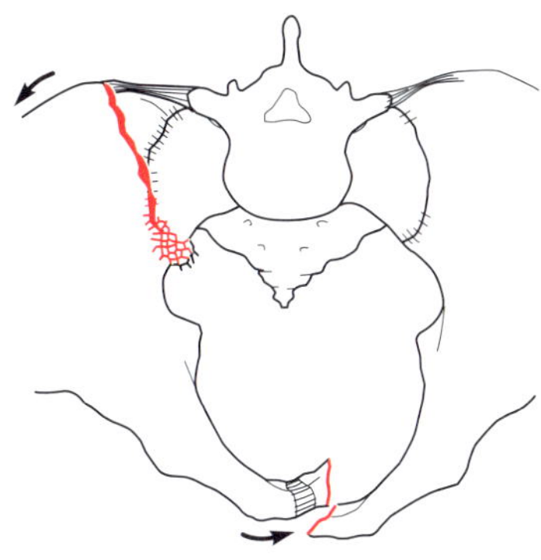

e. Lateral compression (Type II): contralateral anterior and posterior injury (bucket handle).

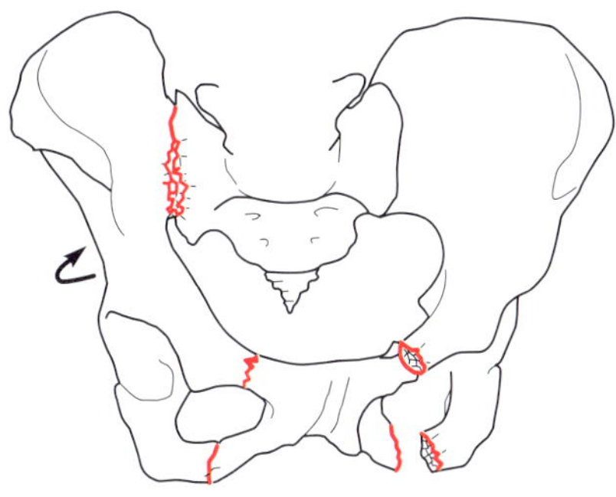

f. Lateral compression (Type III): four rami type (Straddle or butterfly fracture).

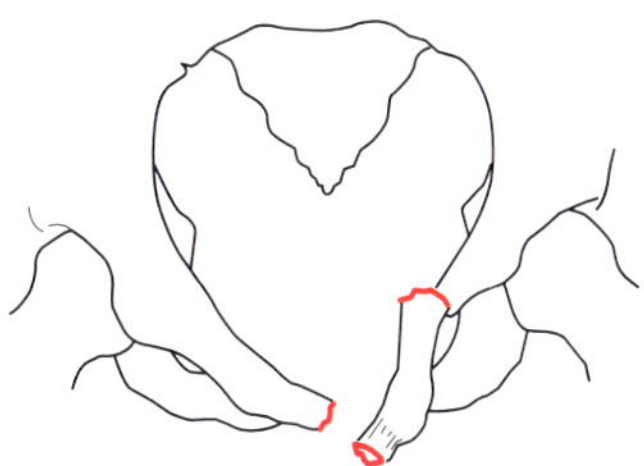

g. Lateral compression (Type III variation): superior pubic rami fracture involving the anterior column of the acetabulum.

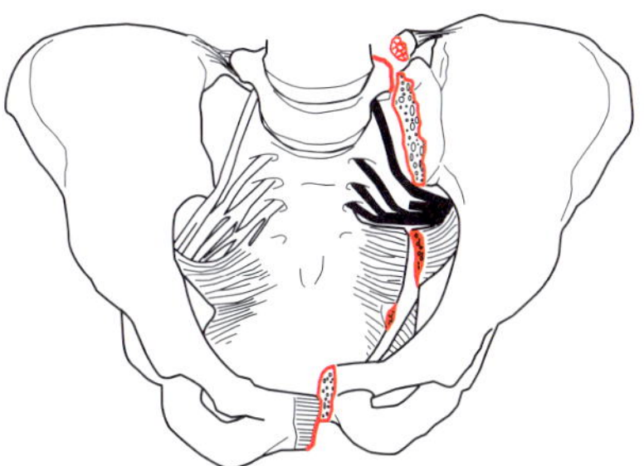

h. Unstable pelvic fracture: vertical shear (Malgaigne fracture).

Figure 10.4. Classification of Tile (1984).

ments. This can be associated with injuries to the urethra, vagina, bladder, rectum and the neurovascular structures.

This injury is stable in the vertical plane as the posterior ligament complex is maintained and can be managed by closing the pelvic ring as the pelvic volume is increased.

Variations are common. The posterior lesion can be a fracture of the ilium rather than disruption of the sacroiliac joint which is either unilateral or bilateral. In some cases, there may be vertical instability. This emphasizes the importance of assessing the posterior complex. The management of these injuries is different.

Type II Four rami fractures (Figure 10.4b). These are due to direct injury.

Usually these are due to lateral compression or vertical shear forces, and indicate that significant forces have been involved and stability needs to be assessed. Isolated fractures of the anterior pubic rami are not generally associated with direct injury.

Lateral compression injury

These are the most common injuries and may be of many types. They tend to be stable as the posterior complex impacts. If the application of force continues, it may disrupt the posterior ligamentous complex and then the pelvis becomes unstable. In this situation the pelvic volume is reduced.

Type I Ipsilateral anterior and posterior injury. As lateral (internal rotation) force is applied, the pubic ramus fractures on the side of the applied abnormal force and, as the force continues, the anterior sacrum is impacted but posterior ligaments remain intact (Figure 10.4c). When the force ceases, there is recoil of the fracture fragments and the fracture may reduce spontaneously. The result may be that the posterior lesion is no longer obvious. However, if a pelvic sling or an external fixator is applied with a compressive force, the fracture may re-displace. The original displacement may be associated with injury to the pelvic viscera. Rarely, there may not be a fracture of the pubic rami but the pubic symphysis can overlap and impact (locked symphysis) (Figure 10.4d).

Type II Contralateral (bucket handle). When the lateral compressive force is coupled with a rotatory force, there is either disruption of the pubic symphysis or fracture of both pubic rami with injury to the posterior complex on the contralateral side (Figure 10.4e). The posterior complex is impacted and the ligaments can be disrupted. This leads to a shortened and internally rotated lower extremity as the side of the posterior complex injury is shifted superiorly.

Type III Four rami type. The lateral compression force disrupts all four pubic rami, leading to the classical 'straddle' or 'butterfly' fracture (Figure 10.4f) or can lead to two rami fractures and pubic symphysis disruption. These are frequently associated with an underlying pelvic visceral injury. The posterior injury is impaction with or without involvement of the ligamentous complex. A variation of this injury is that the superior pubic rami fracture also involves the anterior column of the acetabulum (Figure 10.4g).

Group 2. Unstable pelvic fractures

Vertical shear: Malgaigne fracture (Figure 10.4h). These are caused by severe trauma and can be unilateral or bilateral. Anteriorly there can be disruption of the pubic symphysis or two or four rami fractures. Posteriorly the ligament complex is disrupted leading to a hemipelvic shift and gross displacement. These are very unstable injuries with disruption of the posterior ligamentous complex, sacrotuberous and sacrospinous ligaments. They have the highest incidence of genito-urinary, gastro-intestinal, neurological and vascular injuries of all pelvic fractures.

The posterior lesion can be unilateral or bilateral. Bilateral lesions are more unstable.

Group 3. Miscellaneous group of pelvic fractures

Complex

These usually follow major trauma and are unstable. They lie outside the groups described above. Examples are:

1. Bilateral sacroiliac dislocation with an intact anterior arch. This is rare and seen in young women when it occurs.

2. Pelvic disruption associated with acetabular fracture. The prognosis in these depends more on the acetabular component of the injury than the pelvic ring fracture.

Young's Classification (Young et al. 1986)

Young et al. (1986) refined the Pennel and Tile classifications. They developed a classification by injury and included a fourth category, the combined mechanical (CM) injury (Table 10.2 and Figure 10.5). They ascertained that the accurate identification of force with a radiographic quantification, coupled with physical assessment of the patient, gives a valuable insight into treating both the pelvic injury and the whole patient.

This classification enables clinicians to develop a system of pattern recognition based on a radiographic protocol. A standard series of three films (lateral cervical spine, chest and pelvis) supplemented by inlet and outlet views are used. This is recommended because:

1. It increases the clinician's ability to detect frequently missed injuries particularly of the posterior ring.
2. It provides a predictive index for both local and distal associated injuries, and hence the necessary resuscitative measures (validated by Dalal et al., 1989).
3. It enables the clinician to reduce the morbidity and mortality.

AO Classification (Müller, 1991)

The AO group has formulated its own classification, dividing pelvic fractures into A, B and C groups (Figures 10.6 and 10.7).

Type A Stable, minimally displaced.

A1 Pelvic ring is not involved. This includes avulsion of anterior superior iliac spines, anterior inferior iliac spines or ischial tuberosity.

A2 Pelvic wing fracture with or without involvement of the pelvic ring, which, if fractured, is not displaced.

Table 10.2 Young's classification (Young et al. 1986)

Injury classification keys *Category*	*Characteristics*	*Common differentiating features*
LC-I (Figure 10.12)	Anterior transverse fracture (pubic rami)	Sacral compression on the side of impact
LC-II (Figure 10.13)	Anterior transverse fracture (pubic rami)	Crescent (iliac wing) fracture on the side of the impact
LC-III (Figure 10.14)	Anterior transverse fracture (pubic rami)	Contralateral open book (APC) injury
APC-I (Figure 10.15)	Symphyseal diastasis	Slight widening of pubic symphysis and/or sacroiliac joint; stretched anterior and posterior ligaments
APC-II (Figure 10.16)	Symphyseal diastasis or anterior vertical fracture	Widened sacroiliac joint; disrupted anterior ligament; posterior ligament intact
APC-III (Figure 10.17)	Symphyseal diastasis or anterior vertical fracture	Complete hemipelvis separation, but no vertical displacement; complete sacroiliac joint disruption; complete anterior and posterior disruption
VS (Figure 10.18)	Symphyseal diastasis or anterior vertical fracture	Vertical displacement anteriorly and posteriorly, usually through sacroiliac joint, occasionally through iliac wing and/or sacrum
CM (Figure 10.19)	Anterior and/or posterior, vertical and/or transverse components	Combination of other injury patterns: LC/VS or LC/APC

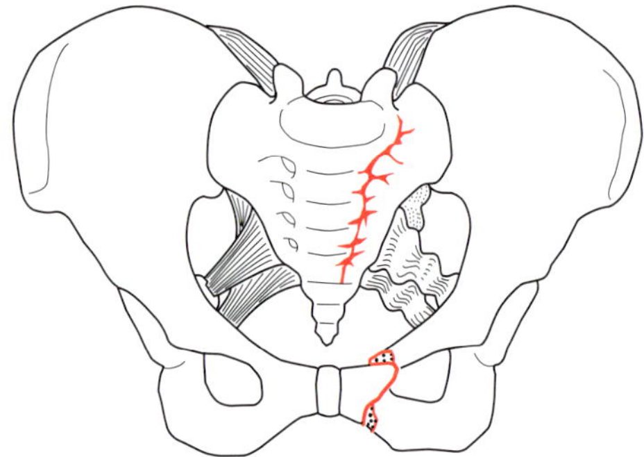

a. LC-I: Inlet view of impaction injury of the sacrum and transverse anterior injury.

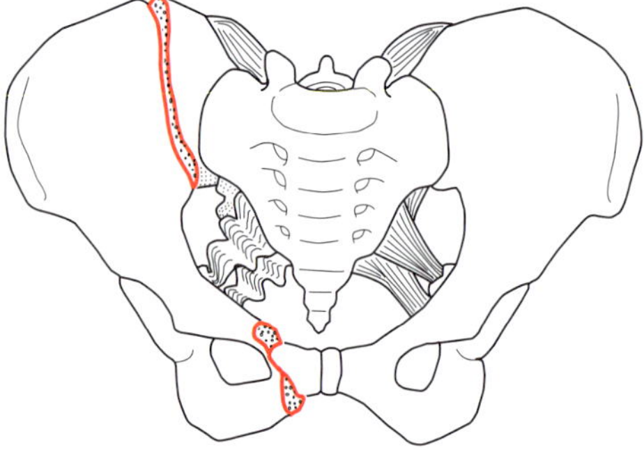

b. LC-II: ipsilateral crescent fracture of the ilium and anterior transverse injury.

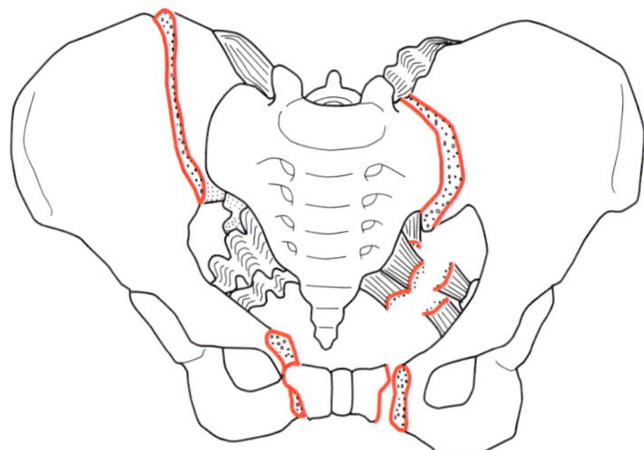

c. LC-III: anterior transverse fracture and contralateral open book injury.

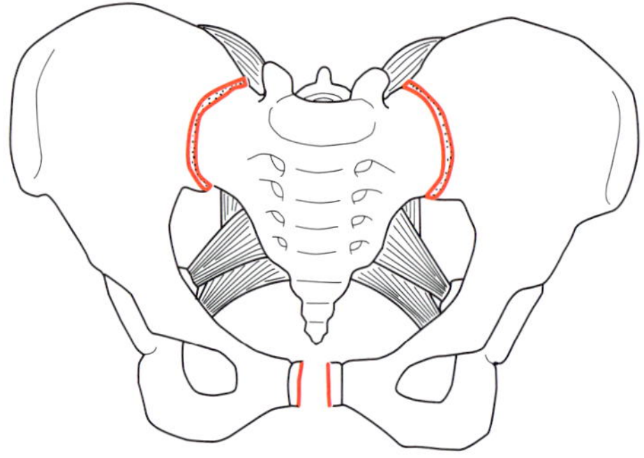

d. APC-I: slight widening of the pubic symphysis and stretched anterior ligaments of the SI joints.

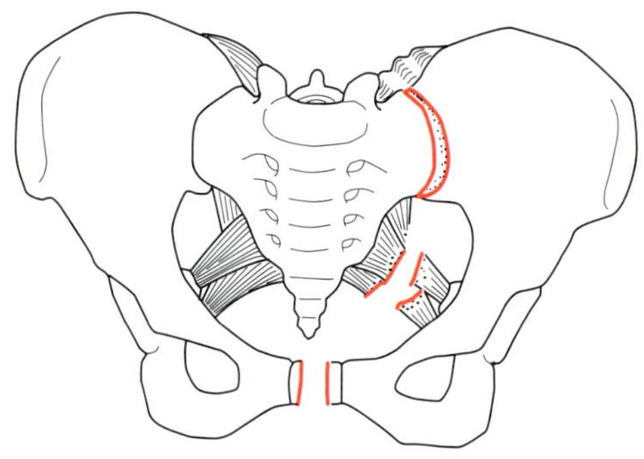

e. APC-II: injury to the sacrotuberous and sacrospinous ligaments.

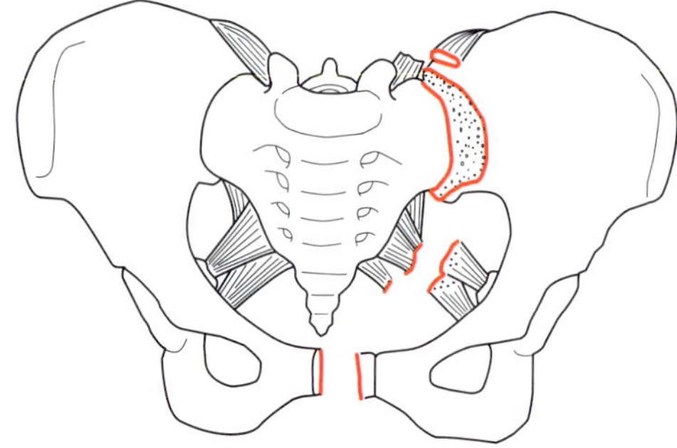

f. APC-III: disruption of all ligamentous structures on one side.

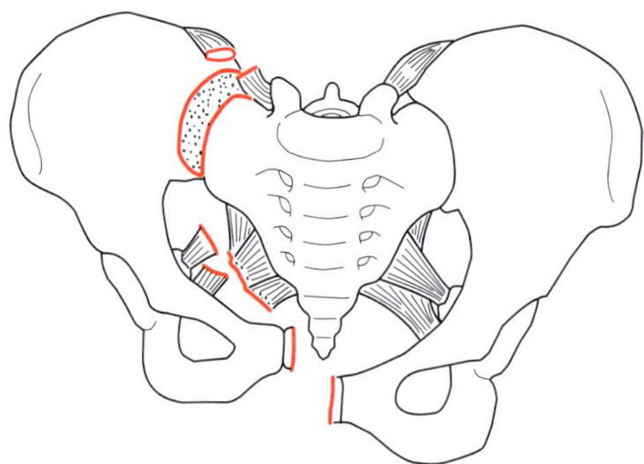

g. VS: vertical displacement of the hemipelvis.

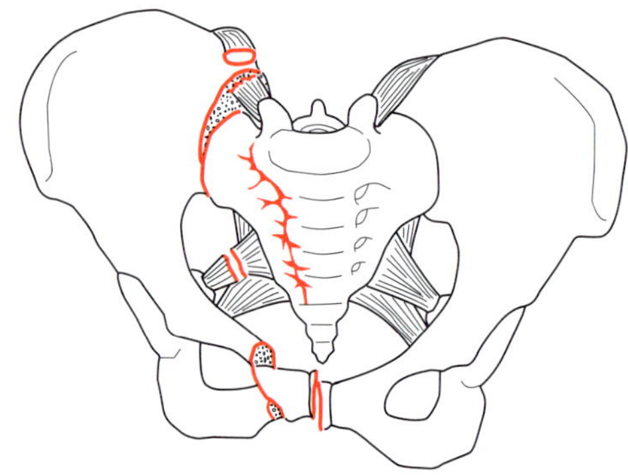

h. Combined lateral compression and vertical shear fracture.

Figure 10.5. Young's classification. From Burgess and Jones, 1996.

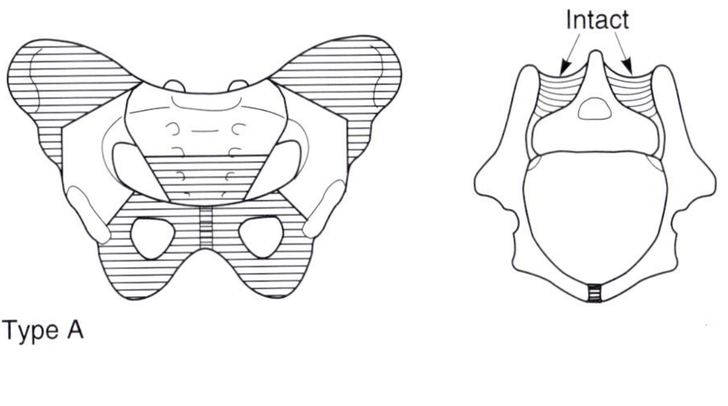

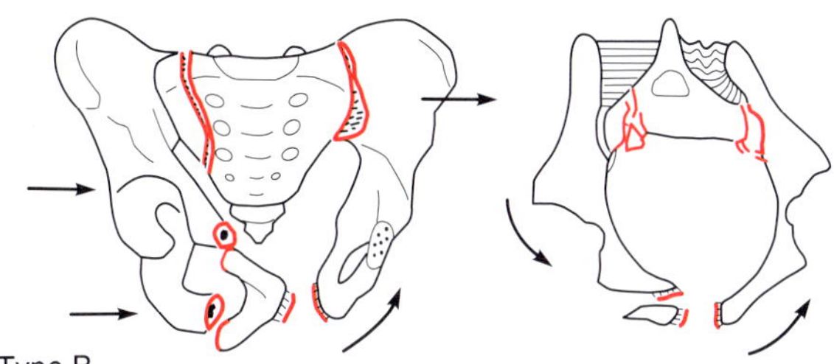

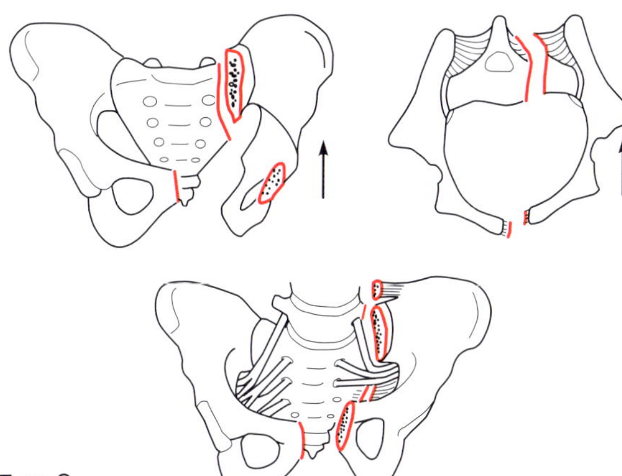

Figure 10.6. AO/ASIF classification of pelvic fractures (AO Manual, Müller et al., 1991).

A3 Transverse fractures of the sacrum and coccyx, not involving the pelvic ring.

Type B: Rotationally unstable, vertically stable. This corresponds to the stable group in Tile's classification.

B1 Open book injury, external rotation.

B2 Lateral compression injury.

B2.1 Ipsilateral type.

B2.2 Contralateral type (bucket handle).

B3 Bilateral type B injury.

Type C: Rotationally and vertically unstable. Vertical shear. This is Tile's 'unstable' group.

C1 Unilateral: posterior lesion can be ligamentous disruption, fracture of the ilium or sacrum.

C2 Bilateral.

C3 Pelvic ring fractures associated with fractures of the acetabulum.

The above scheme has been further developed into a comprehensive classification (Isler and Ganz, 1996) in an attempt to identify all the possible variations.

Discussion

Any of these classifications may serve only as a guide to selecting an appropriate method of treating the injury. The magnitude and the complex nature of the forces involved in producing the injuries are such that there are always fractures which are unclassifiable. There are no published figures on the reliability of these systems.

Acetabulum

Introduction

Acetabular fractures are associated with high energy injuries. With the increase in vehicular traffic there are more acetabular fractures. Even so, the frequency of acetabular fractures remains low compared to other injuries. The average orthopaedic surgeon in a district general hospital is unlikely to have a wide experience of them.

The acetabulum is a major weight-bearing joint of the body and the principle of management should be as for any other displaced intra-articular fracture of the lower extremity. Displaced fractures of the acetabulum usually lead to a rapidly disabling post-traumatic arthritis of the hip joint if they are left unreduced.

Ideal treatment consists of anatomical reduction and early motion to preserve the function of the joint. Provided that joint congruity can be achieved and maintained the result should be satisfactory. There are many reports comparing operative and non-operative methods of treatment. Critical appraisal of the literature suggests that like has not been compared with like and hence the conclusions must be drawn with caution.

The complexity of acetabular fractures makes their classification difficult. The variations are so numerous that if they were all included the classification would be cumbersome and confusing. But it is obvious that a meticulous assessment is necessary to understand the fracture geometry. A detailed history, when obtainable, of how the injury occurred will give an idea of the magnitude and direction of the forces

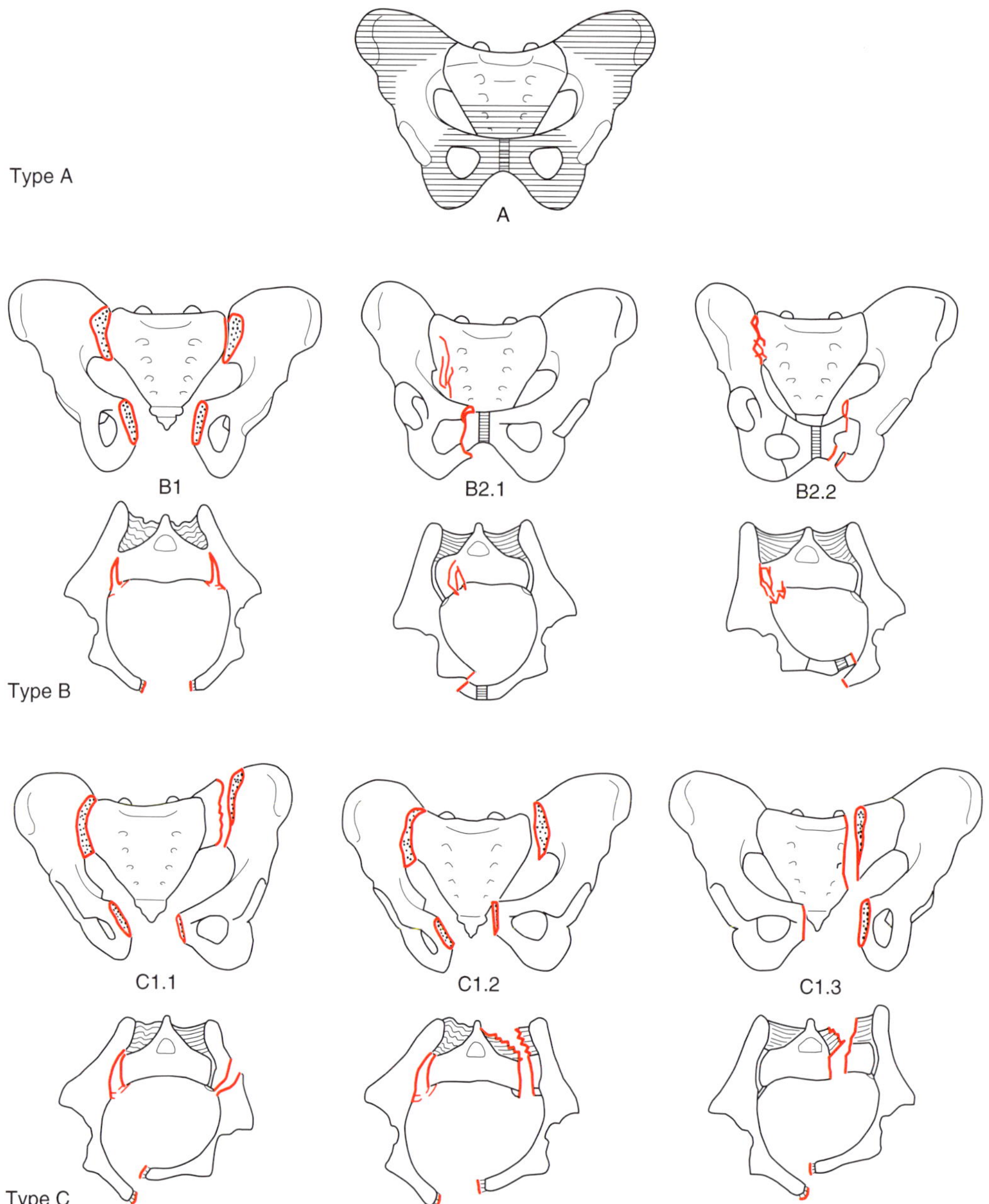

10.7. Classification of pelvic fractures: subgroups (AO Manual, Müller et al., 1991).

involved. Radiological assessment must be thorough and should include:

1. Pelvic views
   Anteroposterior view
   Inlet view } Only if there is an associated pelvic
   Outlet view } ring fracture
2. Acetabular views
   Anteroposterior view
   Obturator oblique view } 'Judet views'
   Iliac oblique view } (Figure 10.8)
3. Computerized tomography:
   Conventional and 3-D reconstructions

Classification

Early literature grouped fractures of the acetabulum into two broad categories:

- Central dislocation
- Posterior dislocation of the hip with a fracture of the acetabulum.

Cauchoix and Truchet (1951) described intermediary forms.

Creyssel and Schnepp (1960) tried to define fractures using these divisions distinguishing principal

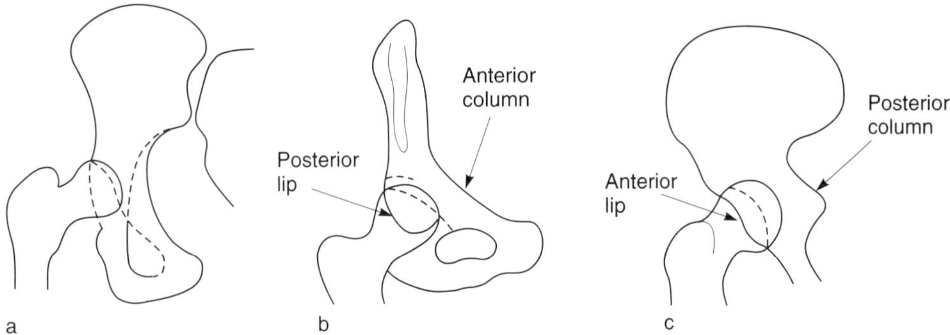

10.8. Judet views (McLaren, 1990): (a) anteroposterior; (b) obturator oblique; (c) iliac oblique.

and accessory fractures. However, Judet et al., 1964 maintained that all the fracture lines traversing the acetabulum were of similar significance; while recognizing that some patterns are more common, there was no proof that a zone of acetabulum was less useful than another as far as the long term prognosis was concerned.

Rowe and Lowell (1961) used the radiological dome (Figure 10.9) as their basis of classifying the fracture. This is unhelpful in assessing the extent of post-traumatic congruity between the femoral head and the acetabular articular surface.

Judet and Letournel (1964) proposed a classification system which is the most widely accepted. They suggested two major types, elementary fractures and associated fractures with five subgroups in each (Table 10.3).

Emile Letournel first published this in his thesis in 1961 and the scheme underwent further modifications until 1965. Subsequently it has remained largely unchanged. The basis of this classification is the understanding of the acetabulum as being supported by two columns, the anterior iliopubic column and the posterior ilioischial column (Figure 10.10). It does not take into account the direction of displacement of the femoral head, although this was recognized as the second anatomical lesion and of great importance in the prognosis as it is primarily responsible for osteonecrosis of the femoral head, one of the most significant complications of this injury.

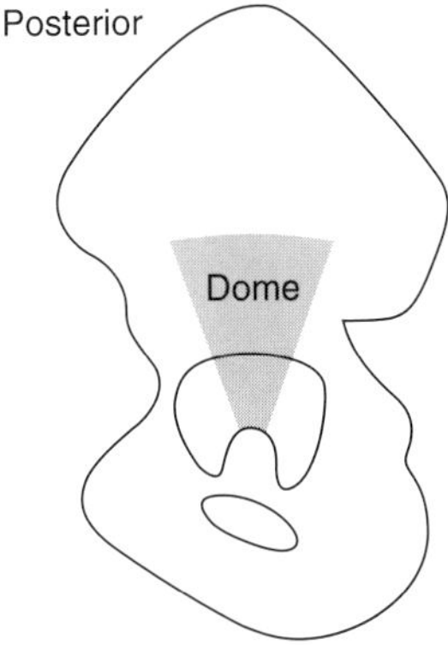

Figure 10.9. The radiological dome (McLaren, 1990).

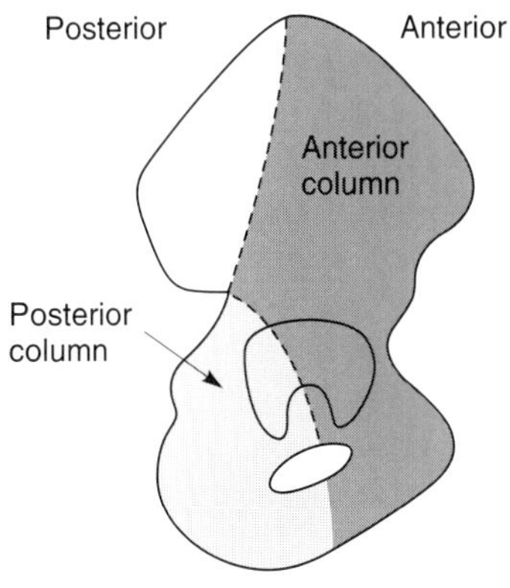

Figure 10.10. The two columns of the pelvis: anterior and posterior (McLaren, 1990).

In fact the series constituted a continuous spectrum of fractures, although each of them could be described individually. The fractures were divided into two broad groups.

Letournel and Judet's (1994) Classification (Figure 10.11)

I. Elementary fractures (five subgroups)

In these all or a part of one of the columns supporting the acetabulum has been detached. The transverse fracture is included in this group.

- i. Posterior wall fractures (Figure 10.11a)
- ii. Posterior column fractures (Figure 10.11b)
- iii. Anterior wall fractures (Figure 10.11c)
- iv. Anterior column fractures (Figure 10.11d)
- v. Transverse fractures (Figure 10.11e)

Table 10.3 Letournel and Judet's classification of acetabular fractures (1994)

Elementary fractures	*Associated fractures*
Posterior wall	T-fracture
Posterior column	Posterior column + posterior wall
Anterior wall	Transverse + posterior wall
Anterior column	Anterior column (or wall) with posterior hemitransverse
Transverse	Double column

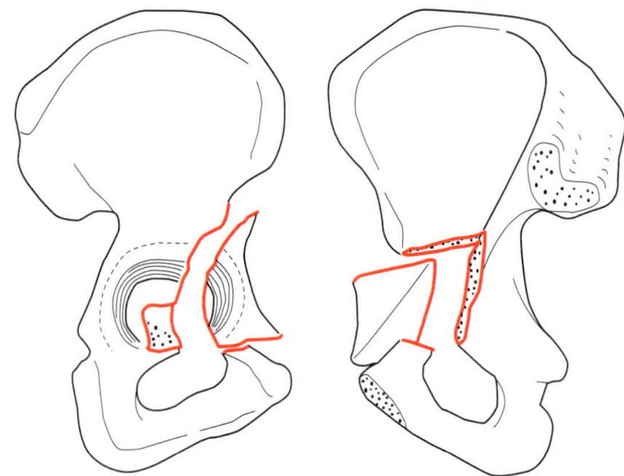

c. Anterior wall fracture.

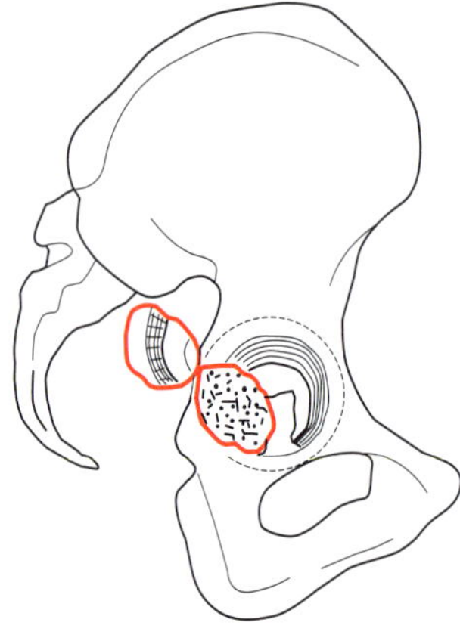

a. Posterior wall fracture.

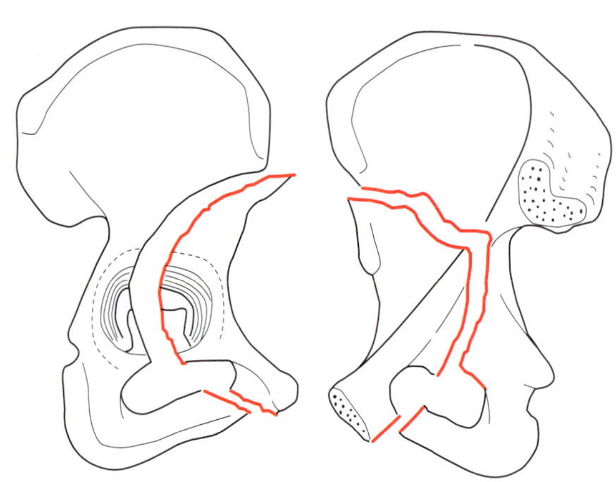

d. Anterior column fracture.

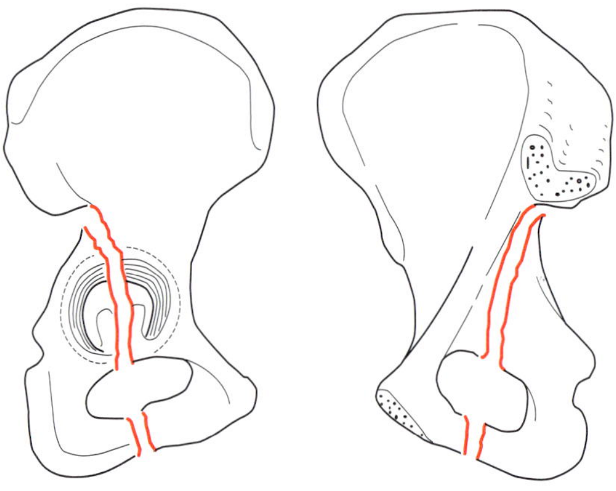

b. Posterior column fracture.

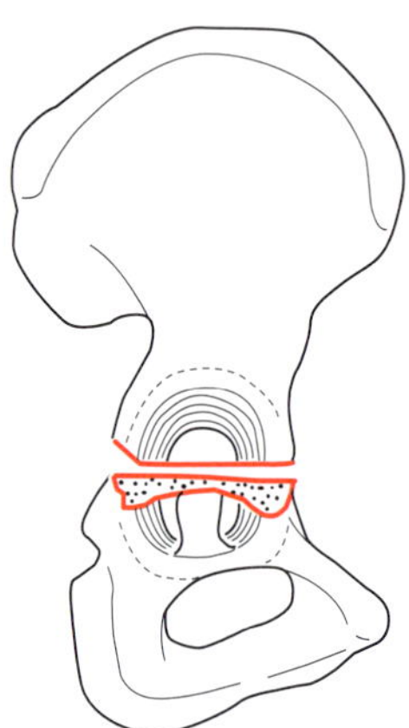

e. Transverse fracture.

Figure 10.11. Classification of elementary fractures (Letournel, 1965).

II. Associated fractures (five subgroups)
These include at least two of the elementary forms and can have five principal associations.
  i. T-shaped fractures (Figure 10.12a)
  ii. Posterior wall and column fractures (Figure 10.12b)
  iii. Transverse and posterior fracture (Figure 10.12c) with dislocation either posteriorly or anteriorly
  iv. Anterior wall or column fracture with hemi-transverse fracture posteriorly (Figure 10.12d)
  v. Both columns fracture (Figure 10.12e). (In this, the two columns become mutually isolated from each other. The only part that remains attached to the axial skeleton is a piece of the iliac wing.)

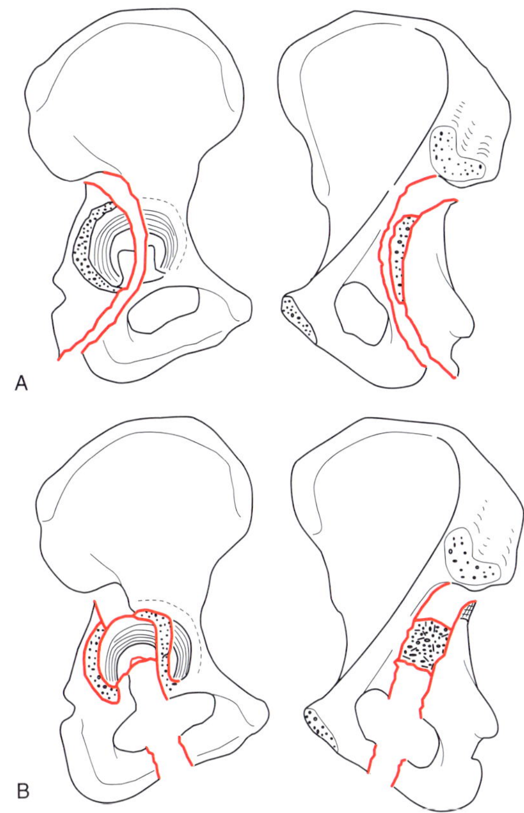

b. Posterior wall (A) and posterior column (B) fracture.

Pennal's Classification

Pennal et al. (1980) added two further features to the classification of these fractures:

1. The amount of residual acetabular displacement at the conclusion of treatment.
2. The amount of comminution affecting the weight-bearing surface of the superior acetabulum.

They proposed:

I. Single-column fractures
  i. Anterior
  ii. Posterior
II. Two-column fractures
  i. Transverse
  ii. Oblique
  iii. T-shaped

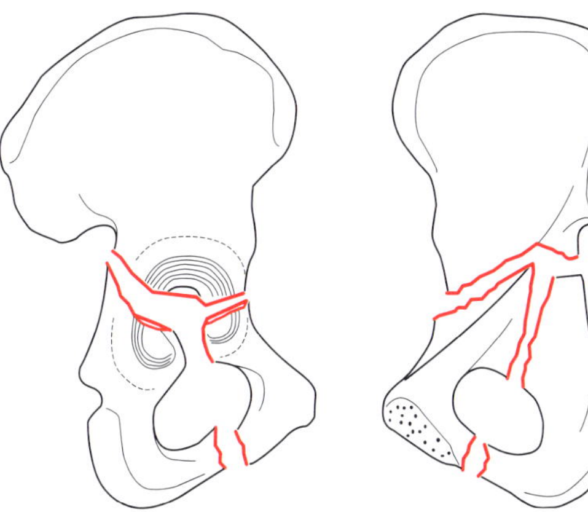

a. T-shaped fracture.

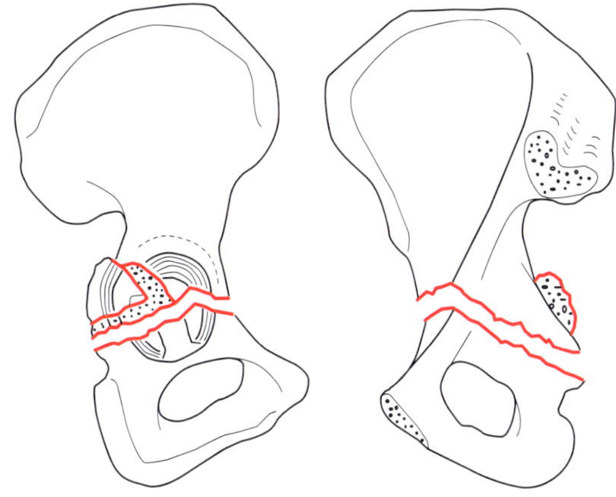

c. Transverse and posterior wall fracture.

Figure 10.12. Classification of associated fractures (Letournel, 1994).

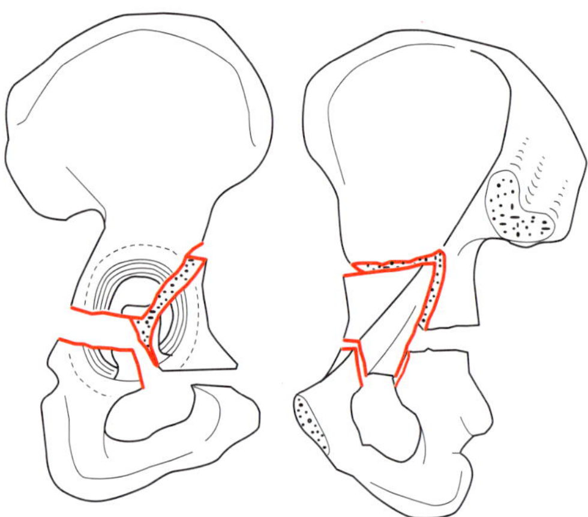

d. Anterior wall and column fracture with posterior hemi-transverse fracture.

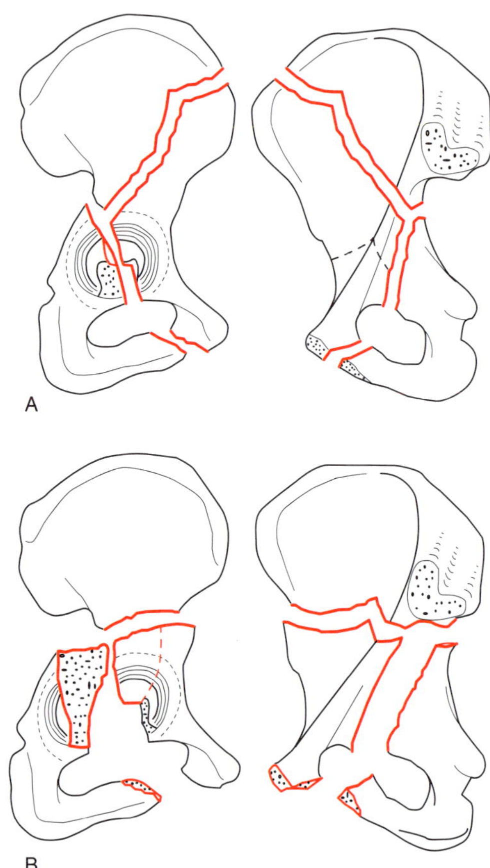

e. Both-column fracture: (A) extending to the iliac crest; (B) extending to the anterior border of the ilium.

Figure 10.12. (*Continued*).

These, in addition, were graded by displacement and by comminution:

I. Undisplaced
II. Moderately displaced
III. Severely displaced, defined as separation of the fracture by more than half the diameter of the femoral head
IV. Fractures of the femoral neck

And further graded by comminution of the superior articular surface:

I. No comminution
II. Moderate comminution
III. Severe comminution

However, these subdivisions were arbitrary and subjective.

Tile's Classification (Table 10.4)

Tile (1984) felt the above classification did not address many of the factors such as the degree of displacement, comminution and the integrity of the superior weight-bearing surface. His scheme incorporates all of the Judet and Letournel types in a different format.

The fracture was divided into two groups, undisplaced and displaced. The displaced group was subdivided into three types:

Table 10.4 Tile's classification of acetabular fractures

- Undisplaced fractures
- Displaced fractures

  Type I
  Posterior types + posterior dislocation
    A. Posterior column
    B. Posterior wall
      – associated with posterior column
      – associated with transverse fracture

  Type II
  Anterior types + anterior dislocation
    A. Anterior column
    B. Anterior wall
    C. Associated anterior and transverse fracture

  Type III
  Transverse types + central dislocation
    A. Pure transverse
    B. T-fractures
    C. Associated transverse and acetabular wall fractures
    D. Double column fractures

Type I Posterior types and posterior dislocation

A. Posterior column (Figure 10.13a)

B. Posterior wall (Figure 10.13b)

i. Associated with posterior column (Figure 10.13c)

ii. Associated with transverse fracture (Figure 10.13d)

Type II Anterior types and anterior dislocation

A. Anterior column (Figure 10.13e)

B. Anterior wall (Figure 10.13f)

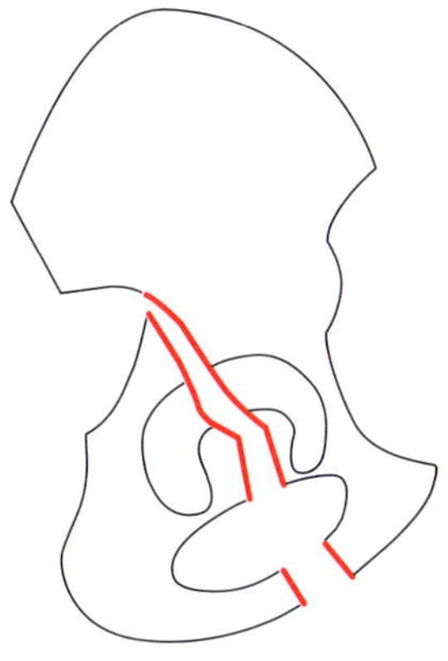

a. Posterior column fracture.

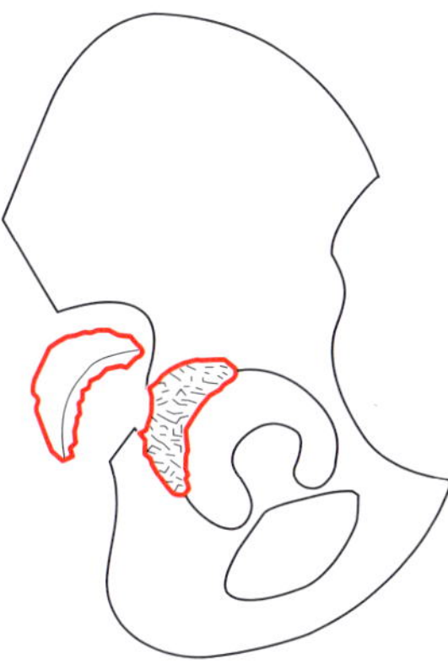

b. Posterior wall fracture.

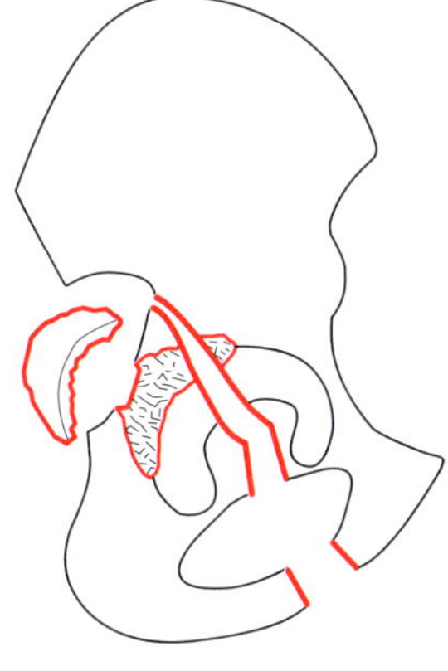

c. Posterior wall associated with posterior column fracture.

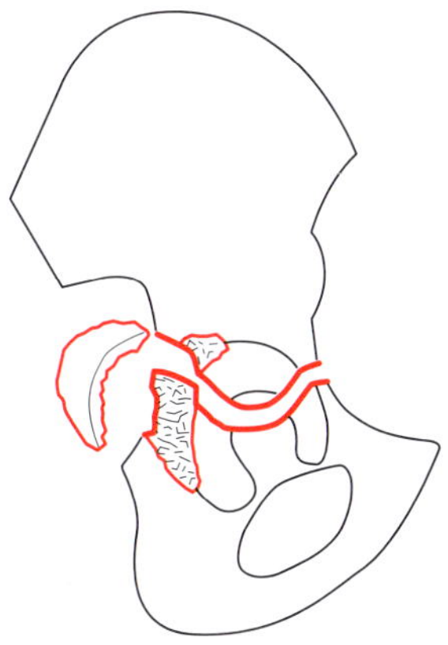

d. Posterior wall associated with transverse fracture.

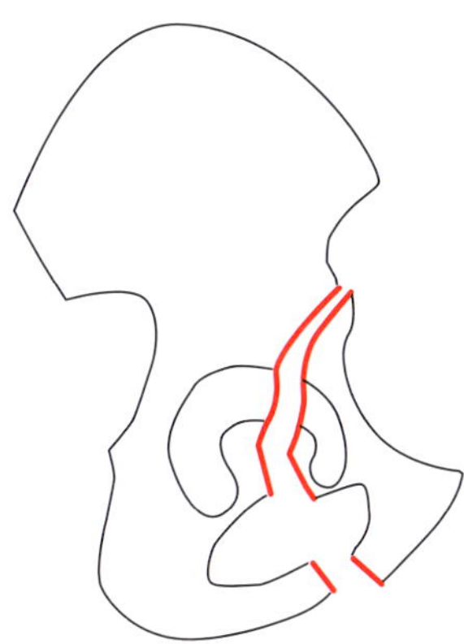

e. Anterior column fracture.

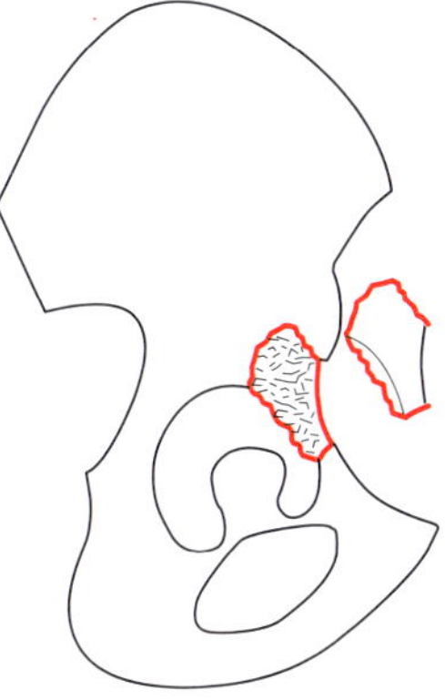

f. Anterior wall fracture.

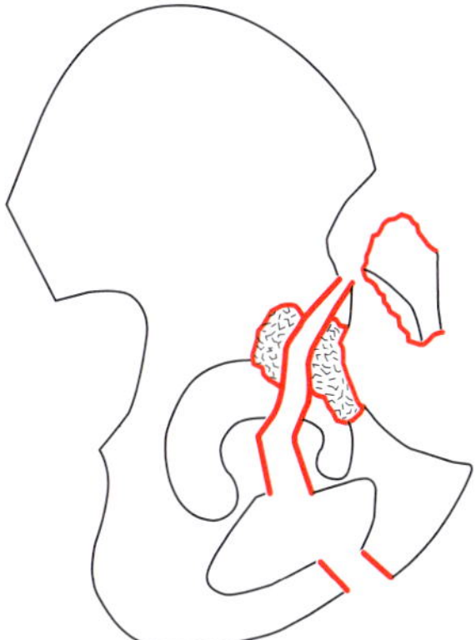

g. Anterior wall associated with anterior column fracture.

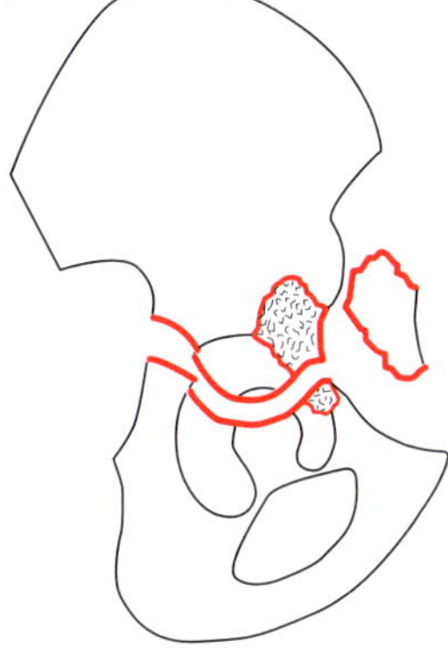

h. Anterior wall associated with transverse fracture.

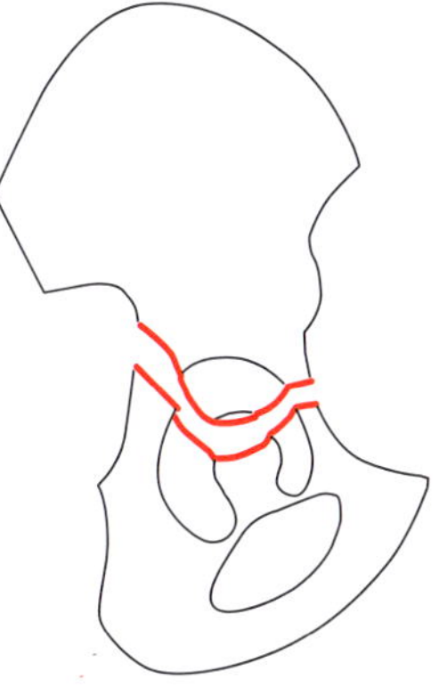

i. Transverse fracture.

Figure 10.13. Classification of acetabular fractures (Tile, 1984).

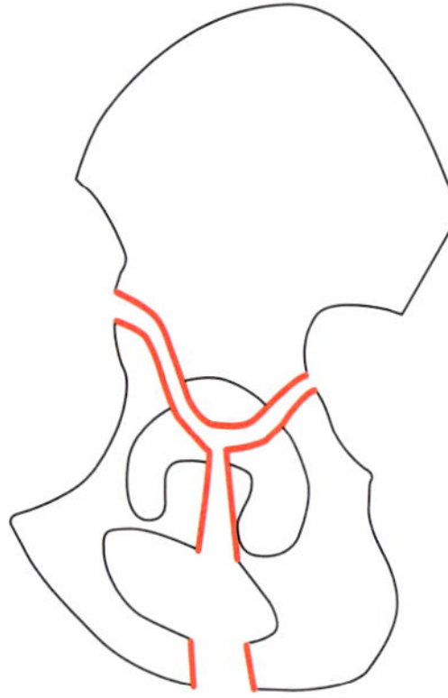

j. T-fracture.

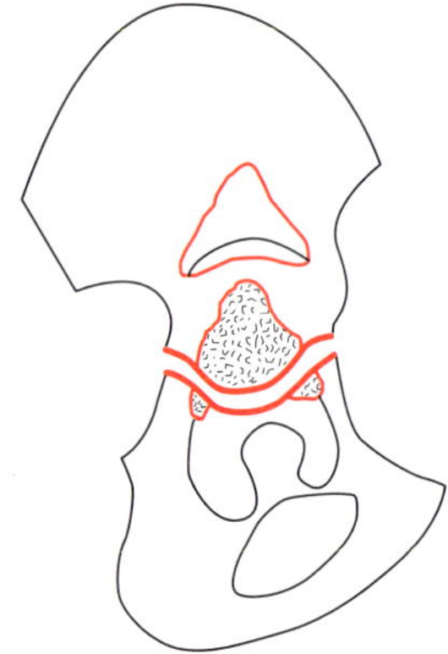

k. Transverse and acetabular wall fracture.

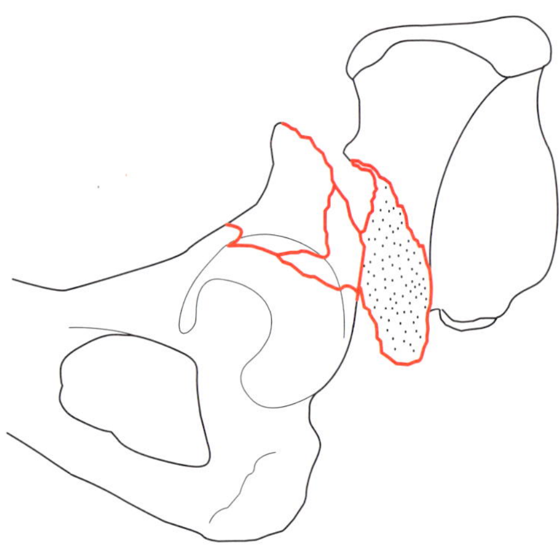

l. Double column fracture.

Figure 10.13. (*Continued*).

i. Associated with anterior column (Figure 10.13g)
ii. Associated with anterior and transverse fracture (Figure 10.13h)

Type III Transverse types and central dislocation

A. Pure transverse (Figure 10.13i)
B. T-fractures (Figure 10.13j)
C. Associated transverse and acetabular wall fractures (Figure 10.13k)
D. Double column fractures (Figure 10.13l)

AO Classification (Müller et al., 1991)

The AO documentation centre modified Letournel's classification to fit into the three groups of the AO classification (Table 10.5 and Figure 10.14).

Type A: Only one column of the acetabulum involved, the other column remaining intact.

A1. Posterior wall fracture and variations.
A2. Posterior column fracture and variations.
A3. Anterior wall and anterior column fractures.

Type B: Characterized by transverse fracture component, where a portion of the roof remains attached to an intact ilium.

B1. Transverse fracture and transverse plus posterior wall fracture.
B2. T-shaped fracture with variations.
B3. Anterior wall or column plus posterior hemitransverse fracture.

Type C: Fractures involving both columns: characterized by fracture lines involving anterior as well as posterior column, but differs from type B in that all articular segments including the roof are detached from the remaining segment of the intact ilium.

C1. Anterior column fracture line extending to the iliac crest.
C2. Anterior column fracture line extending to the anterior border of the ilium.
C3. Fracture line which enters the sacro-iliac joint.

Table 10.5 AO classification of acetabular fractures (Müller et al. 1991)

A	Fractures involving one column or wall
B	Fractures involving both columns with a portion of the dome attached to the axial skeleton
C	Fractures involving both columns with no portion of the dome attached to the axial skeleton

They add a list of qualifiers relating to the quality of the articular surface and the direction of displacement of the femoral head.

The concepts of all the aforementioned schemes have now been incorporated by the AO group in a comprehensive classification of fractures of the acetabulum (Tile et al., 1995).

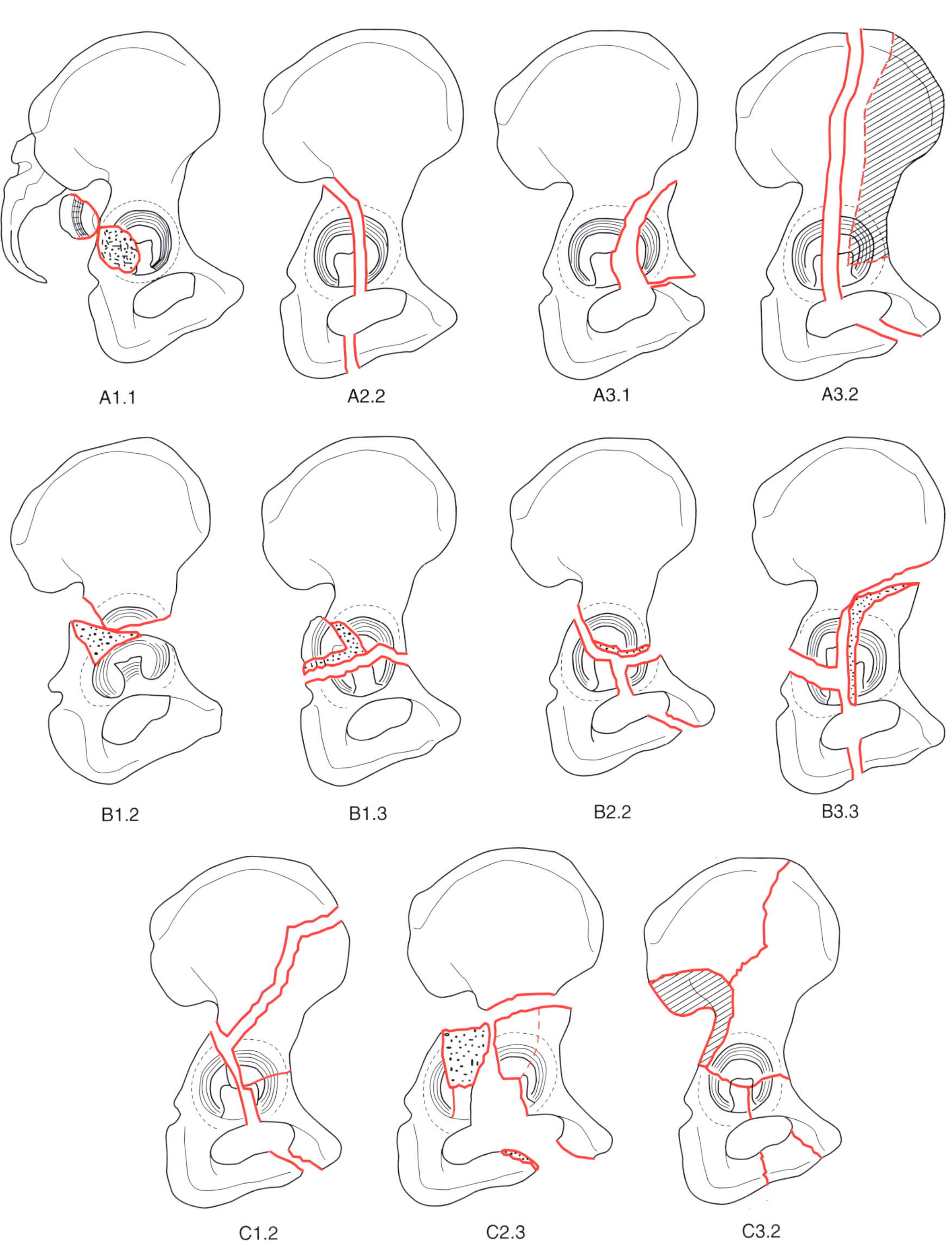

10.14. AO classification of acetabular fractures (Müller et al., 1991).

Discussion

It is important to realize that all these classifications are only a guide to treating a particular patient. The 'personality' of the fracture must be recognized, along with the factors that will affect prognosis, including the presence or absence of dislocation or fracture of the femoral neck. However, the surgeon must not lose sight of the individual patient nor of the limitations of his own experience and expertise in coming to a logical decision regarding the management of these fractures.

Matta, Mehne and Roffi (1986) undertook a retrospective analysis followed by a prospective study of the fractures of the acetabulum and quantified the weight-bearing dome with three measurements. The medial, anterior, and posterior roof arcs were obtained from standard radiographs and emphasized that most displaced fractures involved the weight-bearing zone and required surgery. With an intact weight-bearing dome, non-operative treatment may be considered.

There is general recognition that the management of acetabular fractures is an extremely specialized field and should probably be undertaken by a limited number of surgeons who do this work on a regular basis.

References

Bucholz, R.W. (1981) The pathological anatomy of Malgaigne fracture dislocation of the pelvis. *J. Bone Joint Surg.*, **63A**, 400–404.

Burgess, A.R. and Jones, A.L., (1996) Fractures of the pelvic ring. In *Rockwood and Green's Fractures in Adults* (C.A. Rockwood, Jr, D.P. Green, R.W. Bucholz and J.D. Heckman, ed). Philadelphia: J.B. Lippincott, vol. 2, 1575–1615.

Burgess, A.R., Eastridge, B.J., Young, J.W.R. et al. (1990) Pelvic ring disruptions: Effective classification system and treatment protocols. *J. Trauma*, **30**, 848–856.

Cauchoix and Truchet (1951) Articular fractures of the hip (excluding neck of femur). *Rev. Chir. Orthop.*, **37**, 266–332, (quoted by Judet, R. and Letournel, E. (1993)).

Creyssel, J. and Schnepp, J. (1960) On the utilisation of external fixators in certain fractures of the pelvis. *Lyon Chir.*, **56**, 584–586. (Quoted by Judet, R. and Letournel, E. (1993)).

Dalal, S.A., Burgess, A.R., Siegel, J.H. et al. (1989) Pelvic fractures in multiple trauma: Classification by mechanism is key to pattern of organ injury, resuscitative requirements and outcome. *J. Trauma*, **29**(7), 981–1002.

Gertzbein, S.D. and Chenoweth, D.R. (1977) Difficult injuries of the pelvic ring. *Clin. Orthop.*, **128**, 201–207.

Isler, B. and Ganz, R. (1996) Classification of pelvic ring injuries. *Injury*, **27**, (Suppl 1), S-A3-12.

Judet, R., Judet, J. and Letournel, E. (1964) Fractures of the acetabulum: classification and surgical approaches for open reduction. *J. Bone Joint Surg.*, **46A**, 1615.

Letournel, E. and Judet, R. (1994) *Fractures of the Acetabulum*, 2nd edn. Berlin: Springer-Verlag.

Knight, R.A. and Smith, H. (1958) Central fractures of the acetabulum. *J. Bone Joint Surg.*, **40A**, 1–16.

Letournel, E. (1980) Acetabulum fractures: classification and management. *Clin. Orthop.*, **151**, 81–106.

Matta, J.M., Anderson, L.M., Epstein, H.C. and Hendricks, P. (1986) Fractures of the acetabulum: early results of a prospective study. *Clin. Orthop.*, **205**, 241–250.

Matta, J.M., Mehne, D.K. and Roffi, R. (1986) Fractures of the acetabulum: a retrospective analysis. *Clin. Orthop.*, **205**, 230–240.

McLaren, M.I. (1990) Fractures of the acetabulum. *Surgery*, **82**, 1948–1953.

Müller, M.E., Allgöwer, M., Schneider, R. and Willenegger, H. (1991) *Manual of Internal Fixation Techniques Recommended by the AO-ASif Group*, 3rd edn. Berlin: Springer-Verlag.

Peltier, L.F. (1965) Complications associated with fractures of the pelvis. *J. Bone Joint Surg.*, **47A**, 1060–1069.

Pennal, G.F., Tile, M., Waddell, J.P. and Garside, H. (1979) Pelvic disruption: assessment and classification. *Clin. Orthop.*, **151**, 12–21.

Pennal, G.F., Davidson, J., Garside, H. and Plewes, J. (1980) Results of treatment of acetabular fractures: *Clin. Orthop.*, **151**, 115–123.

Rowe, C.R. and Lowell, J.D. (1961) Prognosis of fractures of the acetabulum. *J. Bone Joint Surg.*, **43A**, 30.

Rockwood, C.A. Jr, Green, D.P. and Bucholz, R.W. (1991) *Fractures in Adults*. Philadelphia: Lippincott, pp. 1414–1421.

Schatzker, J., Tile, M., Axelrod, T.S. et al. (1996) *The Rationale of Operative Fracture Care*. Berlin: Springer-Verlag.

Tile, M. (1984) *Fractures of the Pelvis and Acetabulum*. Baltimore: Williams and Wilkins.

Tornetta, P. (1996) Techniques and outcomes in pelvic fractures. Symposium. *Clin. Orthop.*, **329**, Section I, 2–206.

Trunkey, D.D., Chapman, M.W., Lim, R.C. Jr, and Dunphy, J.E. (1974) Management of pelvic fractures in blunt trauma injury. *J. Trauma*, **14**(11), 912–932.

Young, J.W.R., Burgess, A.R., Brumback, R.J. and Poka, A. (1986) Pelvic fractures: value of plain radiography in early assessment and management. *Radiology*, **160**, 445–451.

11

The shoulder, scapula and proximal humerus

R. Y. L. Liow

Introduction

The classification of injuries occupies an important role in the practice of orthopaedic surgery. Such schemes provide a means of describing the injuries, enabling accurate communication between practitioners. Their natural histories can be predicted and rational guidelines for management can thus be formulated. Comparison between various treatments can also be made on the assumption that injuries of comparable severity are being treated.

The shoulder is a complex joint that demonstrates remarkable mobility, much of which is attributable to the musculotendinous and ligamentous structures that surround the shoulder girdle. Injuries to the skeleton of the pectoral girdle and shoulder joint tend to coexist with a variable degree of soft-tissue disruption. A comprehensive classification should therefore include the assessment of soft-tissue disruption as well as osseous fractures.

Advancement in the imaging of the shoulder joint has also increased the understanding of injuries in this region. Radiographic assessment of osseous abnormalities can be further assisted by CT scans, with or without three-dimensional reconstruction. Rotator cuff pathology has been most frequently evaluated with arthrograms but this is now superseded by ultrasound and magnetic resonance imaging. An ideal classification system should be applicable to all methods of imaging.

It is surprising that the validity and reliability of the existing classification schemes are infrequently evaluated. Fractures of the proximal humerus is the only classification that has been assessed for interobserver reliability and intra-observer reproducibility (Siebenrock and Gerber, 1992; Sidor et al., 1993). The disappointing reliability found in these studies has important implications in the evaluation of published reports of such injuries and their treatments. The need for classification systems that are easy and accurate to apply cannot be over-emphasized.

This chapter will consider classification systems for the scapula including the coracoid process and glenoid fossa, the clavicle, the sternoclavicular joint, the acromioclavicular joint, the rotator cuff and finally, the proximal humerus. Specific classifications for the immature skeleton are considered for the clavicle and proximal humerus as they differ considerably from their adult counterparts.

Classification of Fractures of the Scapula

Complete appraisal of fractures of the scapula requires several radiographic views. These include a 'true' anteroposterior view orientated at a plane perpendicular to the scapula, a lateral view that is parallel to the scapular plane, and an axillary view. The axillary view is particularly useful in identifying acromial fractures and subluxation associated with intra-articular fractures of the glenoid. Other supplementary views with the arm in varying degrees of rotation or abduction can provide more useful information and are recommended by some authors (Imatani, 1975). A CT scan is useful in evaluating intra-articular glenoid fractures, particularly in providing an accurate assessment of the amount of articular surface affected as well as the degree of displacement of the humeral head. Three-dimensional CT reconstruction provides excellent reproduction of the fracture patterns but has yet to be proven beneficial in determining a treatment plan.

Classifications of fractures of the scapula are based on the anatomical site of the fractures. Desault (1805) identified two distinct types of fractures to the scapula: fractures of the acromion and fractures of the inferior angle of the scapula. Modern authors tend to group these fractures into three anatomical categories: (1) fractures of the glenoid and neck, (2) fractures of the apophyses — the acromion and coracoid processes, and (3) fractures of the muscle-covered body of the scapula (Zdravkovic and Damholt, 1974; Hardegger, Simpson and Weber, 1984; Thompson et al., 1985; Ada and Miller, 1991; Euler et al., 1992). Fractures of the scapular body are the most common followed by fractures of the neck, glenoid, acromion, spine and coracoid process. Combinations of these fractures can occur, the commonest involving the body with an extra-articular extension into the neck, or intra-articular glenoid fractures that extend into the neck and body. Examples of these classifications are given in Table 11.1 and Figure 11.1a and b.

In the AO classification of scapular fractures, the stability of the fractures is taken into account (Ruedi and Schweiberer, 1991) (Figure 11.2). Extra-articular fractures of the body and processes of the scapula, single or combined are considered to be stable. Most fractures of the neck, despite some displacement, are usually stable and included in this category. Unstable extra-articular fractures of the neck tend to occur with fractures of the clavicle, coracoid process or acromion. The combination of a neck of scapula fracture with a clavicular fracture renders the whole shoulder mobile with a floating glenoid fragment. Inferior displacement of the glenoid due to the weight of the arm frequently occurs. Intra-articular fractures are rarer and usually include a transverse fracture through the glenoid fossa. Large lip fractures of the glenoid rim are usually associated with a subluxation or partial dislocation of the head of humerus.

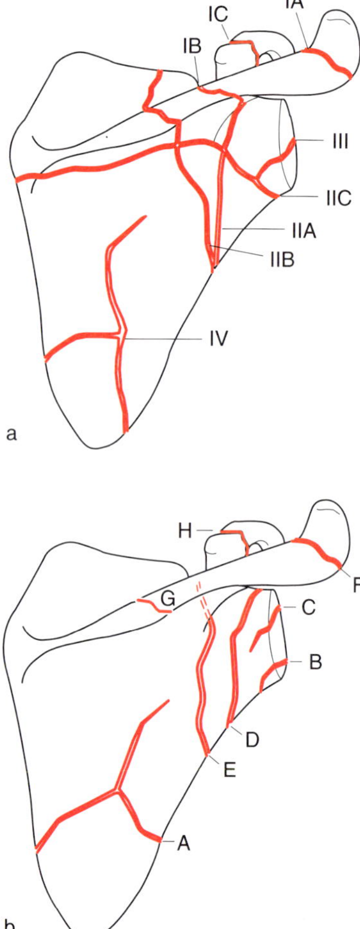

Figure 11.1. Classification of scapular fractures. These are based on anatomical sites (a) (Ada and Miller, 1991). IA, acromion; IB, spine; IC, coracoid; IIA, IIB and IIC, neck; III, glenoid and IV body. (b) (Hardegger, Simpson and Weber, 1984). A, body; B, glenoid rim; C, glenoid fossa; D, anatomical neck; E, surgical neck; F, acromion; G, spine; H, coracoid process.

Isolated fractures of the coracoid process are rare. They are caused by direct trauma or avulsion by either the coracoclavicular ligaments in acromioclavicular dislocations or by violent contractions of the conjoint tendon. Such fractures may also occur, albeit rarely, in association with dislocations of the shoulder or fractures of the clavicle. A classification based on the sites of the fractures and the patho-

Table 11.1 Classifications of scapular features

Zdravkovic and Damholt (1974)		
	Type I	Fracture of the body
	Type II	Fracture of apophysis
	Type III	Fracture of the superolateral angle including the neck and glenoid
Thompson et al. (1985)		
	Class I	Coracoid and acromion and small fractures of the body of scapula
	Class II	Fractures of the glenoid and neck of scapula
	Class III	Major scapular body fractures

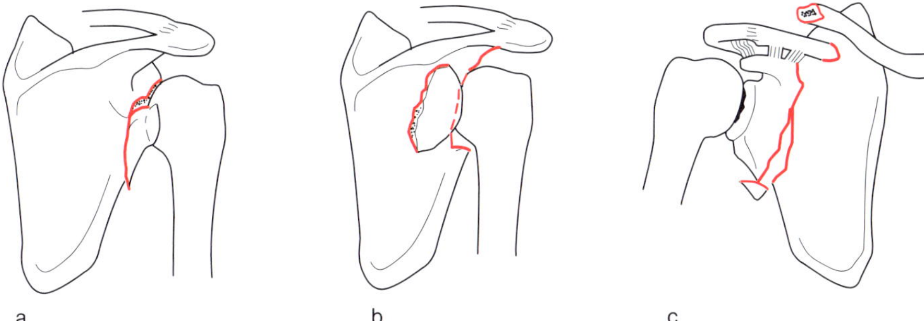

Figure 11.2. AO/ASIF classification of scapular fractures. (a) Intra-articular fracture of the glenoid; (b) Extra-articular fracture of the glenoid; (c) Combined fractures of the glenoid, spine and clavicle. This results in an unstable joint with a tendency of secondary caudal displacement of the whole shoulder. Associated injuries to the thorax and brachial plexus are quite common.

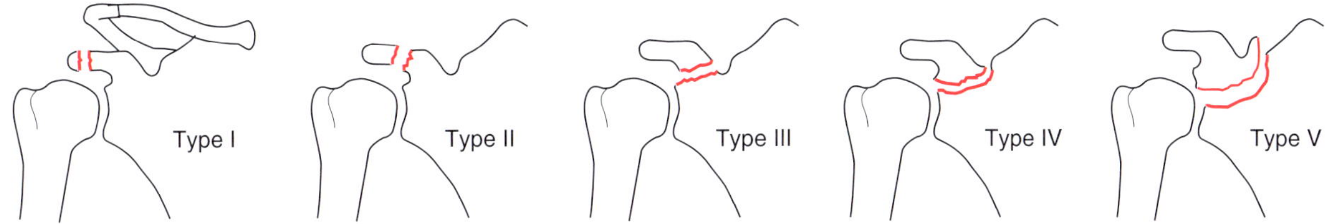

Figure 11.3. Classification of coracoid fractures (Eyres, Brooks and Stanley, 1995). Type I, coracoid tip or epiphyseal fracture; Type II, mid-process; Type III, basal fracture; Type IV, involvement of the superior body of scapula; Type V, extension into the glenoid fossa. The suffix A or B can be used to record the presence or absence of damage to the clavicle or its ligamentous connections to the scapula.

mechanism has been proposed by Ogawa et al. (1990). The type I fractures occur at the apex, and tye II at the mid-portion of the coracoid process. Both are due to traction force from the attached tendons. The type III fractures are at the base of the coracoid process whilst the type IV involve the upper quarter to a third of the glenoid cavity. Type III and IV fractures occur as a result of a medially-directed force, producing a shearing injury as the scapula shifts medially in relation to the clavicle. The connection between the scapula and the clavicle is preserved in types I and II coracoid fracture, but is lost in types III and IV fracture. Such injuries constitute a functional acromioclavicular joint separation. If the coracoclavicular ligaments are attached to the fracture fragment and the acromioclavicular ligaments are torn, a complete acromioclavicular dislocation will result.

An almost identical classification scheme was proposed by Eyres, Brooks and Stanley (1995), based on the anatomical site and the pathomechanism (Figure 11.3). Types I, II and III are identical to the description of Ogawa et al., while type V is equivalent to Ogawa's type IV. Operative stabilization is necessary to maintain scapuloclavicular integrity and to restore the congruity of the glenoid fossa in Ogawa types III and IV and Eyres' types IV and V.

Intra-articular Fractures Involving the Glenoid Fossa

The most comprehensive and widely accepted classification scheme for these fractures is proposed by Ideberg (1984) (Figure 11.4). This classification is based on a review of 200 fractures of the glenoid cavity and the rim of the glenoid fossa. Five patterns of fractures are described. The type I fracture is the most common variety and involves avulsion of the anterior margin of the glenoid fossa. This is associated with a dislocation of the glenohumeral joint in more than half of the type 1 fractures in the Ideberg series. Not surprisingly, persistent subluxation is a serious complication and can lead to the development of secondary osteoarthritis. This is the only type of glenoid fracture caused by indirect trauma to the shoulder.

The type II fracture results from trauma to the lateral aspect of the shoulder with significant violence. The fracture line is oriented transversely in the glenoid fossa, separating the inferior part of the glenoid and neck as a triangular fragment which is displaced with the humeral head. The humeral head slips inferiorly as it lacks support of the glenoid. The type III fracture is often combined with either a fracture of the acromion, a fracture of the clavicle or with

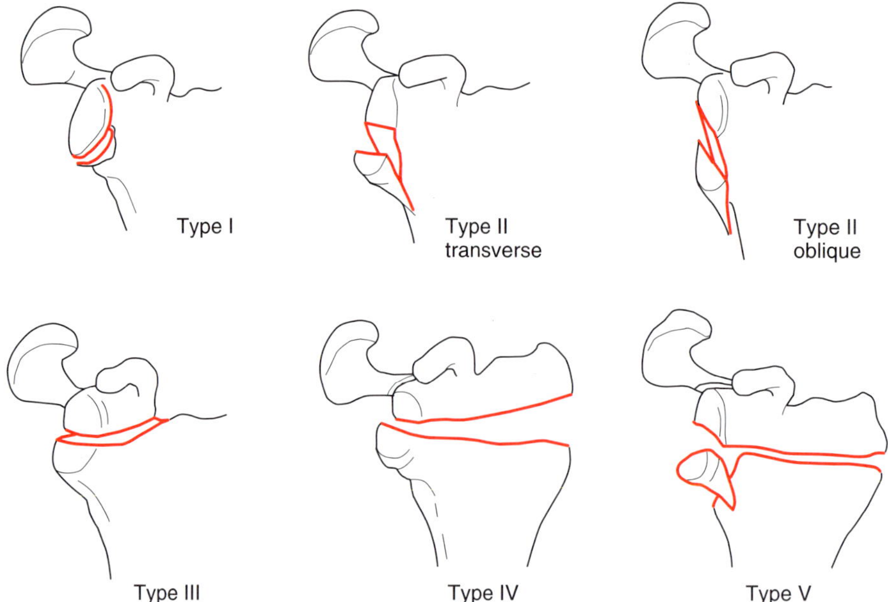

Figure 11.4. The Ideberg (1984) classification for fractures of the glenoid cavity. Type I, avulsion fracture of the anterior rim of the glenoid, and can be associated with humeral dislocation. Types II–V, glenoid fossa fractures (see text).

dislocation of the acromioclavicular joint. The fracture runs obliquely from the glenoid cavity, through the neck and to about the middle of the superior margin of the scapula.

The type IV fracture is transversely orientated in the glenoid fossa, extending horizontally through the neck and blade. Severe displacement of the fragments can be encountered, resulting in incongruity of the articular surface of the glenohumeral joint. The type V fracture is a combination of type IV with a transverse component at the scapular neck which separates the inferior half of the glenoid articular surface.

Several other patterns of glenoid fractures have been reported. Goss modified the Ideberg classification to include these (Goss, 1992) (Figure 11.5).

Nearly 90% of fractures of the glenoid cavity can be treated conservatively with good or excellent results. Indications for open reduction and internal fixation of these fractures are based on the amount of articular step-off (more than 5 mm), the size of the fragment (more than 25% of the articular surface), the amount of fragment displacement and the presence of humeral head subluxation. Any disruption of the superior suspensory complex of the shoulder (i.e. osseoligamentous linkage between the glenoid cavity–coracoid process–coracoclavicular ligaments–clavicle–acromioclavicular joint) must be taken into account in any decision to operate (Goss, 1992). Although Ideberg discussed some factors which influence the indication for operative treatment, the classification is a description of the pattern of fracture and does not provide a guide for operative treatment of these injuries. Nor does it clearly define prognosis. There is no study in the literature that documents the results of treatment in the markedly displaced fractures. This will require a multicentre study because of the rarity of these injuries.

Scapular Fracture in Children

These fractures are rare and no specific classification scheme has been described. Anatomical description of fracture sites, similar to the adult classification, can be employed.

Classification of Fractures of the Clavicle

The clavicle is the commonest bone to be fractured in the body. Allman (1967) categorized three groups of fractures, based on the anatomical site. Group I are fractures of the middle third, where the clavicle is unsupported by ligaments. Group II fractures are lateral to the coracoclavicular ligament and non-union is a frequent complication. The Group III fractures involve the proximal end of the clavicle where displacement and non-union are rare. Within Groups

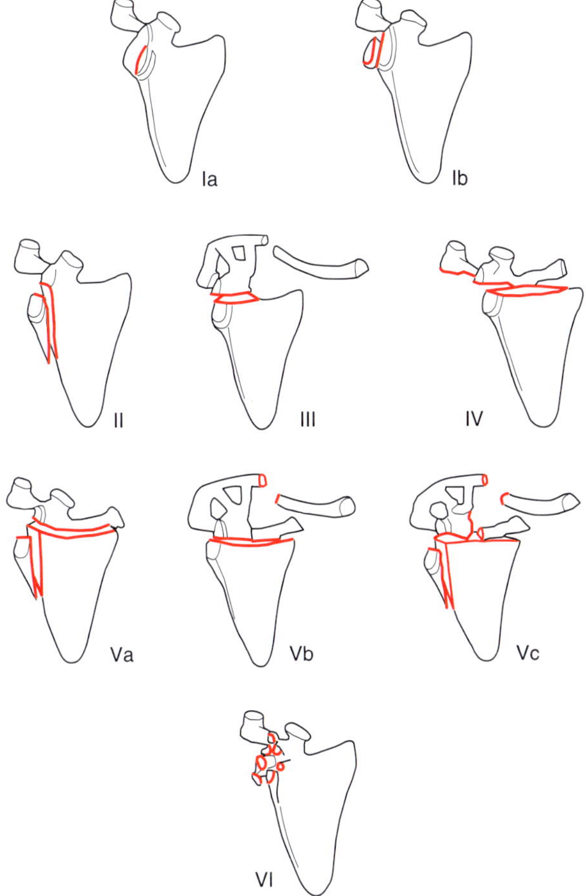

Figure 11.5. The Goss (1992) modification to the Ideberg classification. Type 1, glenoid rim fractures, subgrouped a or b depending on whether anterior or posterior glenoid rim is involved; Type II, the fracture line runs from the glenoid fossa to the lateral border of the scapula; Type III, transverse fracture of the glenoid fossa extending to the superior border of the scapula, and may involve the coracoid process and the suspensory complex of the scapula; Type IV, the fracture extends from the glenoid fossa and exits the medial border; Type V, combines the features of Types II, III and IV fractures; Type VI, severely comminuted glenoid fossa fractures.

II and III further subdivision of fracture types has been described by other authors.

Group 1, Middle Third Fractures of the Clavicle

Group I fractures (middle third) are the commonest in adults and children, accounting for 80% of all clavicular fractures. Several inherent features of the clavicle have been identified as the cause for this. When falling, the force of impact is transmitted from the shoulder along the curve of the clavicle and dissipated upon reaching the lateral curve. The change of the cross-sectional appearance from a prismatic to a more flattened shape has also been implicated as a cause of frequent fractures at this point. In addition, the medial and lateral parts of the clavicle are relatively secured by ligamentous attachments, leaving the central segment relatively unprotected and free. The area just proximal to the attachment of the coracoclavicular ligaments is particularly vulnerable. There is no subclassification for this group of fractures.

Group II, Distal Third Fractures of the Clavicle

Group II or distal third fractures are subclassified according to the location of the coracoclavicular ligaments relative to the fracture fragments. Neer (1968) initially described two types of distal clavicular fractures (Figure 11.6). In the type I fracture, the ligaments remain intact and attached to the medial bone segment. This limits the tendency towards displacement. The type I fracture is a interligamentous fracture at the site between the conoid and trapezoid or the coracoclavicular and the acromioclavicular ligaments. This is the most frequent type of distal clavicle fracture and carries a good prognosis. In Type II distal clavicle fractures, the coracoclavicular ligaments are ruptured and detached from the medial fragment which, in turn, is often displaced upwards. These fractures are associated with a high incidence of non-union due to the forces acting on the fracture, resulting in displacement and impairment of healing. in the type III distal clavicle fracture was subsequently added by Neer (Neer and Rockwood, 1984) and the fracture line involves the articular surface of the acromioclavicular joint without ligamentous damage. These fractures can be subtle and may be confused with first degree acromioclavicular injury. Post-traumatic osteoarthritis may develop as a complication later.

Buschle, von Breitner and Jager (1983) proposed a classification system for distal clavicular fracture which, like the Neer classification, is based on the fracture location and the degree of injury to the ligaments (Figure 11.7). Type Ia are avulsion fractures distal to the coracoclavicular attachments. The distal fragment remains attached to the acromioclavicular ligament and tend to be displaced laterally and superiorly. The fracture line in the type Ib fracture is oriented obliquely. The acromioclavicular and coracoclavicular ligaments are both intact. The medial fragment is often pulled superiorly by the action of the trapezius muscle. Type Ic occurs in children and teenagers where the periosteum is stripped of the clavicle. This is sometimes termed a 'pseudodislocation' and is equivalent to the Group II type IV as described by Craig (see below). Type II fractures occur at the

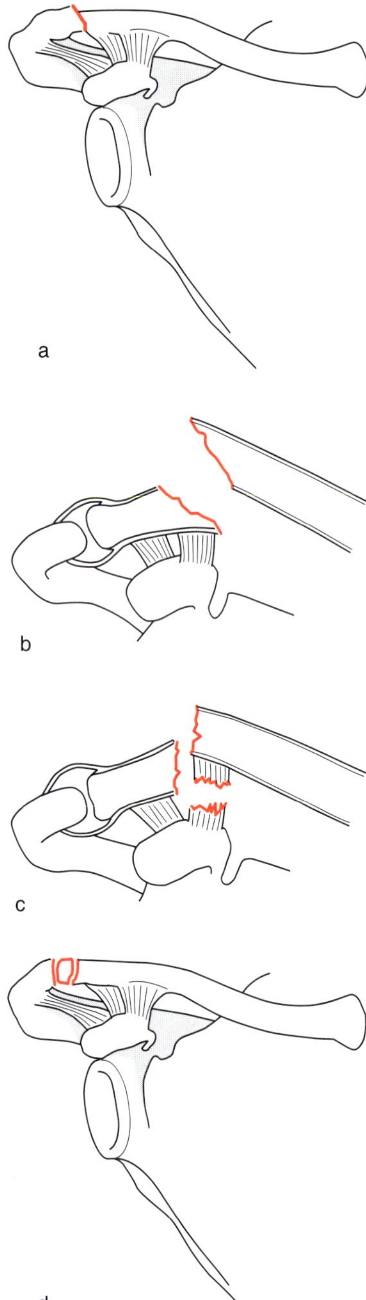

Figure 11.6. Fractures of the distal third of the clavicle. Neer described three types of fractures (Neer, 1968; Neer and Rockwood, 1984). Craig (1991) modified this classification by introducing the subtype A or B. Two other fracture patterns were also added (not illustrated, see Table 11.2). (a) Type I, fracture between the attachments of the conoid and trapezoid ligament, held in place by the intact ligaments; (b) Type IIA, the distal segment is attached to the coracoid process but the intact conoid and trapezoid ligament, while the proximal segment is displaced; (c) Type IIB, the conoid ligament is ruptured while the distal segment remains attached to an intact trapezoid ligament. The proximal segment is displaced; (d) Type III, there is no ligamentous disruption or displacement. The fracture involves the articular surface of the acromioclavicular joint.

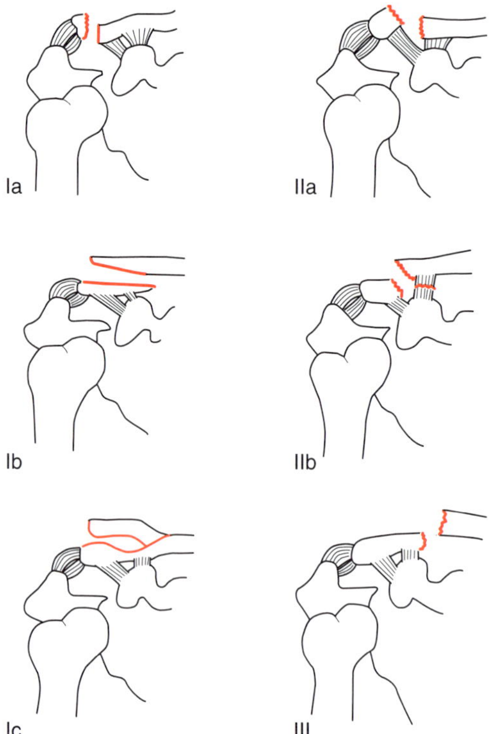

Figure 11.7. Classification of the fractures of the distal clavicle (Buschle, von Breitner and Jager, 1983). Type I, the coracoclavicular ligaments are intact and fracture pattern dictates the subclassification (N.B. the Type Ic is also known as a 'pseudodislocation'); Type II, disruption of either the conoid or trapezoid ligament; Type III, the fracture is proximal to the coracoclavicular ligament.

site between the conoid and trapezoid ligaments with ruptures of either ligaments. In type IIa fractures, the trapezoid ligament is ruptured, resulting in superior displacement of the distal fragment, while type IIb involves a ruptured conoid ligament with superior migration of the proximal clavicular fragment. The fracture line in type III injury is medial to the intact coracoclavicular ligament, and results in a superior migration of the medial fragment.

Group III, Fractures of the Medial Third of the Clavicle

Fractures of the medial end of the clavicle are the least common. If the supporting costoclavicular ligament remains intact and attached to the outer fragment, little or no displacement is seen.

Craig (1991) based his classification system on the three Allman groups. He modified the Neer classification of distal clavicular fractures (Table 11.2). If the conoid and the trapezoid are attached to the distal fragment, the fracture is subtyped IIa. If the conoid

Table 11.2 Classification of fractures of the clavicle (Craig, 1991)

Group I	Middle third fractures	
Group II	Distal third fractures	
	Type I	Minimal displacement (interligamentous)
	Type II	Displacement secondary to fracture medial to coracoclavicular ligaments A. Conoid and trapezoid attached B. Conoid torn, trapezoid attached
	Type III	Articular surface fractures
	Type IV	Ligaments intact to periosteum (children), with displacement of the proximal fragment
	Type V	Comminuted, with no ligamentous attachment proximally nor distally, but to an inferior, comminuted fragment
Group III	Proximal third fractures	
	Type I	Minimal displacement
	Type II	Significant displacement (ligaments ruptured)
	Type III	Intra-articular
	Type IV	Epiphyseal separation (children and young adults)
	Type V	Comminuted

ligament is ruptured whilst the trapezoid ligament remains intact and attached to the distal segment, the fracture is subtyped type IIb. Two other fracture patterns are described.

Type IV fractures occur in children under the age of 16 years and can be confused with complete acromioclavicular separation. The distal clavicle fractures but the acromioclavicular joint remains intact. The proximal fragment ruptures through the thin periosteum, and may be displaced upward by the muscular forces. The coracoclavicular ligaments remains attached to the periosteum or are avulsed with a small fragment of bone. These injuries are also termed the 'pseudo dislocation' of the acromioclavicular joint. Indeed, clinical and radiological differentiation of type IV from type II fractures of the distal clavicle and grade III acromioclavicular dislocation can be very difficult.

Type V fractures occur in adults where neither of the main fragments has any functional coracoclavicular ligament attachments. These fractures are displaced by trapezius and deltoid muscles. The coracoclavicular ligaments remain intact and are attached to a small, third intermediary segment of this comminuted fracture. This is a more unstable fracture when compared to type II distal clavicle fracture.

Craig also subclassified the Group III (proximal third) clavicle fractures as shown in Table 11.2.

Thompson (1989) developed a classification scheme for adult fractures, which was designed to identify fractures at risk of non-union. Five anatomical locations are identified, and within each segment, three categories of subclassification are applied. (Figure 11.8).

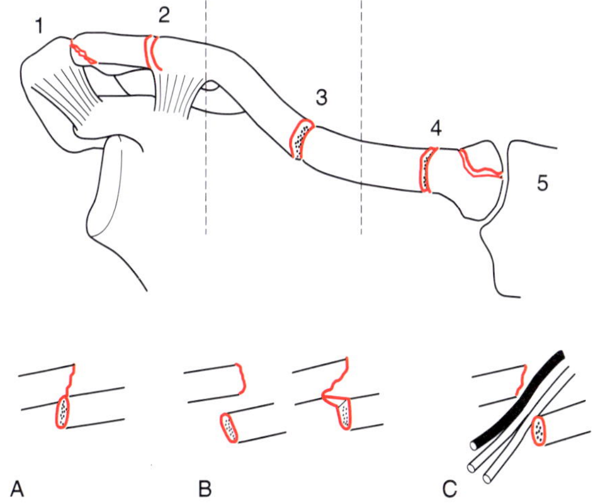

Figure 11.8. The Thompson classification of clavicular fractures. Five locations are identified: Location 1, the acromioclavicular joint; location 2, the lateral one-third of the clavicle; location 3, the middle one-third; location 4, the medial one-third; location 5, the sternoclavicular joint. Displacements are graded as: A, indicating < 100% displacement; B indicating > 100% displacement; C indicating neurovascular compression.

Non-union is an unusual complication when fractures of the clavicle are treated non-operatively. The acceptance of non-union in the elderly or patients who are of poor surgical risk is compatible with good function and comfort. The Thompson classification, the Neer classification and the Buschle, von Breitner and Jager scheme all identify problem fractures. Thompson found that type 3B mid-third fractures of the clavicle, although accounting for only 3% of all clavicular fractures, account for 90% of non-unions. This led him to recommend primary open reduction and internal fixation for this fracture if it could not be converted to a type 3A (Thompson, 1989). The Neer type II fracture of the distal clavicle (Group II) and its equivalent, the Bruschle, von Breitner and Jager type IIa, cannot be satisfactorily controlled by external immobilization. Therefore, open reduction and internal fixation is recommended to pre-empt non-union or delayed-union (Kavanagh, Sarkar and Phillips, 1985; Post, 1989; Edwards, Kavanagh and Flannery, 1992; Brunner, Habermeyer and Schweiberer, 1992).

Sternoclavicular Injuries

Injury to the sternoclavicular joints are rare. The affected anatomical structures are the capsule of the joint, the sternoclavicular or costoclavicular ligaments in isolation or in combination. Allman classified these injuries into three categories (Allman, 1967). Grade I injuries are sprains of the sternoclavicular joint from a medially-directed force applied to the shoulder or from a sudden, anteriorly-directed force on the shoulder. There is no joint laxity and minimal pain. Grade II injuries involve ruptures of the sternoclavicular ligament but with preservation of the costoclavicular ligament. A mild deformity may be present. Grade III injuries are complete ruptures of the sternoclavicular and costoclavicular ligaments with either anterior or retrosternal displacement.

Acromioclavicular Injuries

Patients with severe pain, with or without obvious dislocation, should have radiographic assessment of their injuries. These should be specific for the acromioclavicular joint to ensure appropriate orientation and penetration of the beam. Comparison with the normal uninjured side may be necessary (Zanca, 1971). Stress radiographs can help distinguish between different grades of injury.

Urist (1946) reported on 41 cases of complete acromioclavicular dislocations. He noted that the successes reported by several papers in the literature may be due to the treatment of minor acromioclavicular joints. The importance of a classification scheme to enable comparison of different types of treatment was stressed. Subsequent authors, however, merely grouped these injuries as either complete dislocations or subluxations.

Tossy, Mead and Sigmond (1963) graded these injuries into three stages based upon extent of disruption to the ligamentous structure that surrounds the acromioclavicular joint, namely the capsular (acromioclavicular ligaments) and the extra capsular ligaments (coracoclavicular ligament). This is ascertained by the amount of displacement of the bony structures. The Allman classification (1967) is very similar to that of Tossy's (Tables 11.3 and 11.4) (Figure 11.9). Several authors have described other patterns of injuries and extended the existing grades as

Table 11.3 The classification of acromioclavicular injuries (Tossy, Mead and Sigmond, 1963)

Stage I	Stretching or partial rupture of the acromioclavicular ligament. No gross abnormality on plain radiograph This represents a sprain
Stage II	Rupture of the acromioclavicular ligament and partial tear of the conoid and trapezoid ligament. Cephalad displacement of distal clavicle about half the acromioclavicular joint depth is seen. In addition the distance between the inferior cortex of the clavicle and the superior tip of the coracoid process is increased This represents a subluxation
Stage III	Rupture of the acromioclavicular and coracoclavicular (conoid and trapezoid) ligaments. Displacement of greater than half acromioclavicular joint depth This represents a dislocation

Table 11.4 The Allman classification (1967) of acromioclavicular separation

Grade I	Sprains of the acromioclavicular joint with minor ligament and capsule damage. No laxity of the acromioclavicular joint is observed
Grade II	Moderate force resulting in rupture of the capsule and acromioclavicular ligament. There is no rupture of the coracoclavicular ligament. The injury is often referred to as a subluxation
Grade III	Rupture of both acromioclavicular and coracoclavicular ligaments. Referred to frequently as a dislocation

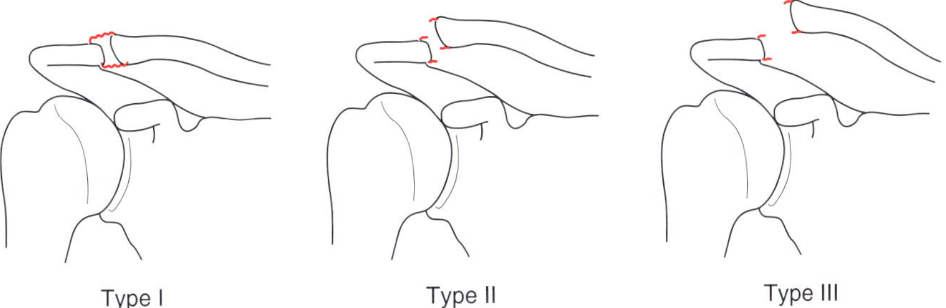

Figure 11.9. The classification of acromioclavicular joint injuries: Type I, a sprain; Type II, a subluxation; and Type III, a dislocation (Tossy, Mead and Sigmond, 1963; Allman,1967).

described by Tossy and Allman. Post (1985) grouped avulsion fractures of the inferior lateral clavicle and severe tears of the soft tissue envelope around the lateral end of the clavicle including button-hole injury of the clavicle into the trapezius, as grade IV. Posterior displacement of the lateral clavicle and the rarer posterior acromioclavicular dislocation are assigned grade V. An inferior dislocation of the distal clavicle trapped beneath the coracoid process of the scapula behind the conjoined tendon is described as a grade VI injury.

Rockwood (1984) identified three additional types of injuries which have been added to the classification as described previously by Tossy and Allman (Figure 11.10). Type IV acromioclavicular injury occurs when the distal clavicle is driven posteriorly in relation to the acromion. There is usually significant damage to the deltoid and trapezius musculature and fascia. The anteroposterior radiograph may appear normal, the displacement is only apparent in the trans-axillary view. The type V injury is similar to the type III but the acromion is driven distally, giving the appearance of a superiorly driven distal clavicle. The coracoclavicular distance is increased by 100 to 300%. The type VI is the least common, and, like the grade VI described by Post (1985), is an inferiorly dislocated distal clavicle trapped beneath the coracoid or the acromion.

Injuries to the acromioclavicular joint also disrupt the deltoid and trapezius. The degree of muscular and ligamentous disruption in the Tossy type III acromioclavicular dislocation can be further assessed with stress radiographs. Bannister et al. (1983,1992) observed three patterns of radiographic features when pre-operative weight-bearing and weight-lifting views are compared (Figure 11.11). These stress radiographs are taken after infiltration of tender areas with local anaesthetic. The patient is then asked to bear 4 kg weights to demonstrate the amount of laxity, and to lift the same weight to contract the anterior deltoid and to assess residual muscle power. The type A dislocation partly reduced, indicating that the deltoid attachment to the clavicle is essentially intact. In the type B injury the dislocation remained unchanged on weight bearing. The dislocation in type C was accentuated in the weight-bearing view, suggesting that the clavicular attachment of the deltoid is stripped off. This type of injury could also be identified by acromioclavicular displacement of >2 cm on plain anteroposterior radiographs.

The authors found all type C injuries to have avulsion of the anterior origin of the deltoid at exploration. Non-operative management resulted in painful subluxation in one-fifth of type A injuries while two-thirds of type B dislocations remain persistently dislocated but with sufficient muscle attachment to avoid fatigue on activity. The type C dislocations leave a weak and cosmetically unacceptable shoulder when treated non-operatively. The authors recommend that types A and C dislocations should be offered early operations as they are at risk of late complications.

The six-part classification system described by Rockwood facilitates the understanding of the pathological anatomy. Furthermore, the ascending grades of acromioclavicular separation are related to an increasing severity of injury, thus affecting the treatment recommendations and prognosis. Types I and II acromioclavicular separation need not be treated surgically and carry good to excellent long-term results while type IV, V and VI injuries require surgical treatment (Richards, 1993). Controversy surrounds type III acromioclavicular separation and there is no current consensus on management of such injuries. Non-surgical treatment can produce good results (Bjerneld, Hovelius and Thorling, 1983; Dias et al., 1987). In a prospective, randomized controlled trial, conservatively managed patients were found to regain movement more rapidly and fully, returned to work and sporting activities earlier and had fewer unsatisfactory results (Bannister, 1983). Although significant failure rates and a high incidence of complications

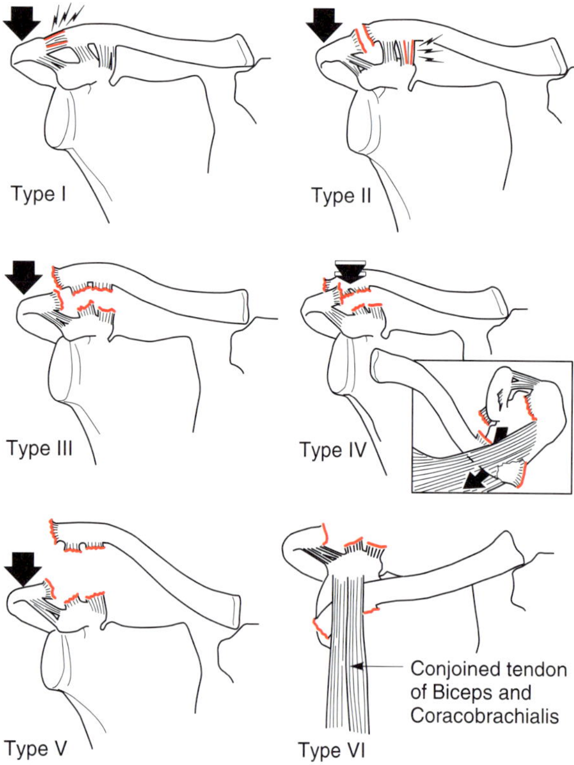

Figure 11.10. The Rockwood (1984) classification of acromioclavicular injuries. This is a modification of the Tossy classification with the addition of three other injury patterns. Type I, sprain of the acromioclavicular ligament; Type II, complete disruption of the acromioclavicular ligament and partial tears of the conoid and trapezoid ligaments. This may increase the coracoclavicular interspace; Type III, complete rupture of the conoid and trapezoid ligaments with disruption of the acromioclavicular ligaments. The coracoclavicular interspace increases by 25–100%. The deltoid and trapezius muscles are usually detached from the distal clavicle. Type III variants include the 'pseudodislocation' through the periosteal sleeve, physeal injury and coracoid process fracture; Type IV, the distal clavicle is driven posteriorly in relation to the acromion (or vice versa).The coracoclavicular ligament and the acromioclavicular ligament are disrupted. The acromioclavicular joint is dislocated. There is often significant detachment of the deltoid and trapezius musculature from the distal clavicle. The coracoclavicular interspace may appear normal; Type V, the pattern of injury is similar to Type III. The acromion is driven distally and the distal clavicle has the appearance of being driven superiorly. Acromioclavicular and coracoclavicular ligaments are ruptured. The coracoclavicular distance is increased by 100–300%. The deltoid and trapezius are detached from the clavicle; Type VI, the distal clavicle is driven beneath either the coracoid or the acromion. Acromioclavicular and coracoclavicular ligaments are ruptured and the deltoid and trapezius detached from the clavicle.

can be expected following surgery, operative treatment is thought to be indicated in severe type III injuries, e.g. in the Bannister type IIIc group, thus emphasizing the value of this classification system in determining the management.

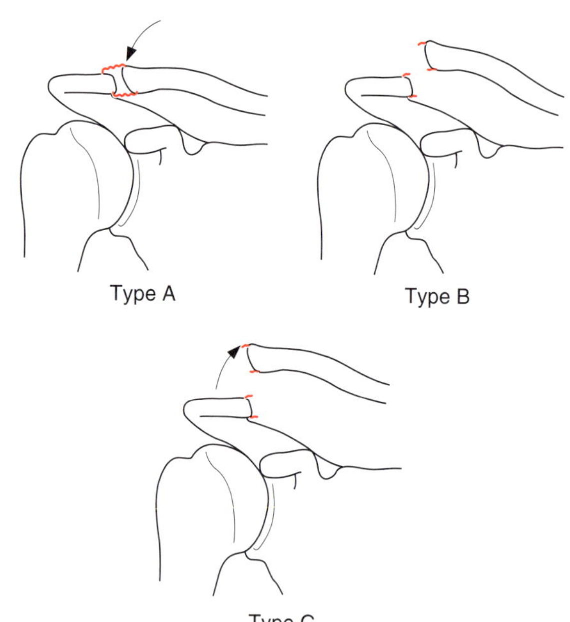

Figure 11.11. Classification of the acute acromioclavicular dislocation (Tossy Type III) by Bannister et al. (1992). A comparison made between the weight-bearing and weight-lifting views is thought to determine the degree of deltoid detachment from the distal clavicle. Type A, the dislocation partly reduced on weight-bearing; Type B, the dislocation remains unchanged on weight-bearing; Type C, the dislocation increased on weight-bearing.

Classification of Fractures of the Clavicle in Children

The clavicle is perhaps the most commonly fractured bone in children with more than half occurring before the age of 10. Clavicular fractures are the most common musculoskeletal birth injury. Like their adult counterparts, these injuries are first divided according to the sites of the fracture, i.e. fractures of the medial end of the clavicle including sternoclavicular joint injuries, fractures of the shaft and fractures involving the distal clavicle including acromioclavicular injuries.

Fractures of the medial clavicle are rare injuries, accounting for only 6% of clavicular fractures in all age groups. Most fractures involve the medial growth plate which does not fuse until early adulthood. Medial shaft fractures are rare, as are true sternoclavicular dislocations. The most common fractures occur through the medial epiphyseal growth plate and are thus categorized according to the Salter–Harris classification (Salter and Harris, 1963) (see Chapter 7). Most injuries are type I or type II, with the epiphyseal fragment attached to the sternum with or without a fragment of the metaphysis. The shaft fragments are displaced either anteriorly or posteriorly. Anterior displacement is more common but posterior displacement is the more important owing to the potential damage to the neurovasular structures emerging from the thoracic outlet and mediastinum. Medial shaft fractures and sternoclavicular dislocations are classified by the direction of fragment displacement as either anterior or posterior.

The thick periosteal coat surrounding the clavicle in children protects the shaft of the clavicle from displacement in the event of a fracture. Fractures of the shaft of the clavicle are thus minimally angulated greenstick fractures or undisplaced. With uncomplicated and rapid union being the most usual outcome, it is thus not surprising that these fractures are not formally classified.

Injuries of the distal clavicle and acromioclavicular joint in children over the age of 15 years can result in acromioclavicular dislocation similar to adults. An adult classification is therefore appropriate. In younger children, however, the same injury usually results in a fracture of the distal clavicle. Disruption of the thick periosteal tube which envelopes the clavicle allows the clavicle to be displaced while the coracoclavicular ligaments remain intact. The Dameron and Rockwood (1984) classification is based upon the position of the distal clavicle and accompanying injury to the periosteal covering (Figure 11.12).

Type I injuries are mild sprains of the acromioclavicular ligaments without disruption of the periosteal tube. The distal clavicle is stable on examination, and radiographs are normal compared with the uninjured side. Type II injuries include partial disruption of the dorsal periosteal envelope with some instability at the distal clavicle on examination. Radiographs show slight widening of the acromioclavicular joint but no change in the coracoclavicular interval. Type III injuries have a large dorsal longitudinal split in the periosteal tube with gross instability of the distal clavicle. Radiographs reveal superior displacement of the clavicle in relation to the coracoid and acromion. The coracoclavicular interval is between 25% to 100% greater than that of the normal shoulder.

Type IV are similar to type III injury but with distal clavicle displaced posteriorly and buttonholed through the trapezius muscle, with the distal end completely buried in the muscle. Widening of the acromioclavicular joint is seen on radiography but there is little in the way of superior migration, similar to type II injury. The axillary film shows posterior displacement of the distal clavicle in relation to the acromion. In type V injury, the periosteum is completely split on its dorsal surface with superior subcutaneous displacement of the clavicle. The deltoid origin and the trapezius insertion are often split. The coracoclavicular distance is greater than 100% when compared with the uninjured side on radiography. Type VI injury involves an inferiorly dislocated clavicle which is lodged beneath the coracoid process.

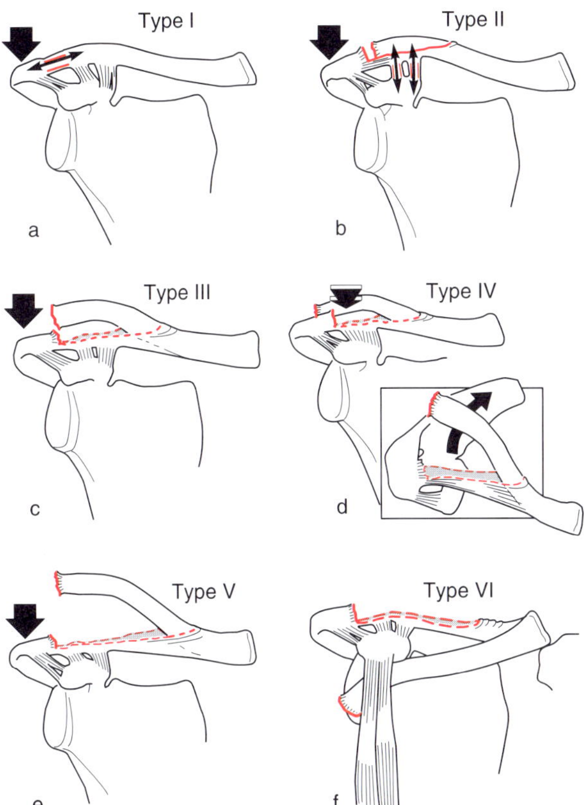

Figure 11.12. Rockwood classification of clavicular and acromioclavicular joint injuries in children (see text for description).

Classification of Rotator Cuff Injuries

Injuries may affect any part of the rotator cuff but the supraspinatus tendon appears to be the most frequently affected. These lesions are also the most persistent and the most disabling (Kessel and Watson, 1977). Degeneration and thinning of the rotator cuff becomes more common with advancing age and injuries in such individuals can be quite extensive. On the other hand, young individuals with little or no previous shoulder symptoms usually suffer a strain lesion of the healthy rotator cuff tissue and it is rare to encounter severe tears with inability to raise the arm in such fit, young adults. The severe acute tear is different from a tear due to chronic attrition as it carries a poorer prognosis. The fraying of the torn edge of tendon and muscle retraction enlarge the defect over time, and the results of early surgery in acute tears are better than those of repair of a chronic lesion (Wolfgang, 1974; Cofield, 1985).

Ultrasonography and arthrography are of limited value in the assessment of rotator cuff injuries. A number of studies have documented the accuracy of MRI in the diagnosis of rotator cuff lesions and have also demonstrated the excellent correlation for the size of tears and the quality of the torn edges with the findings at the time of operation (Recht and Resnick, 1993). The use of MRI to assess the shoulder postoperatively has not been fully assessed. Arthroscopy of the shoulder allows visualization of the rotator cuff and often provides the definitive diagnosis. This is supplemented by further assessment by using a suture marker, and by assessing the same lesion from the bursal side of the shoulder joint. The thickness and quality of the remaining rotator cuff is 'palpated' with the use of a probe (Snyder, 1993).

Classification of Rotator Cuff Lesions

Characterization of rotator cuff lesions according to the macroscopic pathological changes was first described by Bosworth in 1940. The Neer classification for rotator cuff changes in impingement is not dissimilar (1983) (Table 11.5). Pathological classifications are somewhat limited in their use as they merely describe the changes to the rotator cuff tendons and are not applicable to acute injuries.

The dimensions and extent of the rotator cuff tears appear to be important in determining the outcome. It has been reported that repairs of small tears yield predictably less pain and better function than repairs of larger tears (Godsil and Linscheid, 1970; Post, Silver and Singh, 1983) and the strength of post-operative abduction and external rotation is also thought to be related to the size of the lesions found intra-operatively (Hawkins, Misamore and Hobeika, 1985; Gore et al., 1986; Ellman, Hanker and Bayer, 1986).

Post, Silver and Singh, in 1983, classified these lesions on the basis of the thickness of the tears in the rotator cuff and the dimensions of the lesions. Incomplete tears do not involve the entire thickness of the rotator cuff tissue and may be present within the substance of the cuff or the inner surface. Surface tears may cause irregularities and irritation of the bursal floor. Direct communication between the subdeltoid or subacromial bursae and the joint cavity constitutes a complete tear. The tears may be further divided into different sizes. Measured in its longest diameter, a small tear is under 1 cm, a medium-sized tear is less than 3 cm, a large tear under 5 cm and a massive tear measuring more than 5 cm. Several other examples are given in Table 11.6.

An accurate appraisal of the nature of rotator cuff lesions is essential and an ideal classification system should also be applicable to all the imaging modalities as well as arthroscopic and intra-operative findings. The existing classification systems fail to differentiate between a tear in the previously healthy rotator cuff from a degenerate one. The management of tears of the rotator cuff remains a major problem and evaluation of various reported treatments has been hampered by the absence of a unified classification system.

Classification of Glenohumeral Instability

Glenohumeral instability can be categorized in a number of ways. The degree of instability is usually classified into either a dislocation or subluxation depending upon the amount of displacement of the humeral head from the glenoid fossa. Further classification of glenohumeral instability is generally determined by the direction of the displacement (i.e. anterior, posterior or inferior) or by the chronology of the injury (i.e. acute, recurrent or chronic instability). The amount of force initiating the process and the intention of the patient to contribute to the instability are other factors that can also be taken into account.

Table 11.5 Pathological stages of rotator cuff damage in impingement (Neer, 1983)

Stage I	Oedema and haemorrhage from excessive overhead use in sports and occupation <25 years, conservatively managed
Stage II	Fibrosis and thickening of supraspinatus tendon, biceps tendon and subacromial bursa due to chronic inflammation or repeated impingement 25–40 years, operation required if conservative treatment fails
Stage III	Partial or complete rotator cuff tear with degeneration or tear (or both) of biceps tendon secondary osseous change to under surface of anterior acromion and greater tuberosity

Table 11.6 Classifications of rotator cuff tears

Post, Silver and Singh (1983)			
	Incomplete tears		
	Complete tears: direct communication between subdeltoid and subacromial bursa		
		Small tear	< 1 cm
		Medium-sized	1 to 3 cm
		Large tear	3 to 5 cm
		Massive tear	> 5 cm
Harryman et al. (1991)			
	Type O	Intact cuff	
	Type 1A	Thinning or partial thickness defect of supraspinatus tendon	
	Type 1B	Full thickness defect of supraspinatus tendon	
	Type 2	Full thickness defect in supraspinatus and infraspinatus tendons	
	Type 3	Full thickness defect involving supraspinatus, infraspinatus and subscapularis	
Watson described size of full thickness defect in terms of 'tenths' (1985)			
	Subscapularis, supraspinatus and infraspinatus tendons each constitute 3/10 of whole rotator cuff width		
	Biceps long head tendon forming the other 1/10		
The Snyder classification (1993) is based on the dimensions as well as the location of the tears found during arthroscopy of the shoulder.			
	Location of tear		
	A	Articular surface	
	B	Bursal surface	
	C	Complete tear, connecting A and B sides	
	Severity of tear		
	0	Normal cuff, with smooth coverings of synovium and bursa	
	I	Minimal superficial bursal or synovial irritation or slight capsular fraying in a small localized area, usually < 1 cm	
	II	Actual fraying or failure of some rotator cuff fibres in addition to synovial, bursal, or capsular injury, usually < 2 cm in size	
	III	More severe rotator cuff injury, including fraying and fragmentation of the tendon fibres, often involving the whole surface of a cuff tendon (most often the supraspinatus), usually < 3 cm in size	
	IV	Very severe partial rotator cuff tear that usually contains, in addition to fraying and fragmentation of tendon tissue, a sizable flap tear, usually larger in size than grades I–III and often encompass more than a single tendon	

While classification systems exist for chronic instability of the shoulder, there is no specific classification for acute glenohumeral instability. Glenohumeral separation in association with either glenoid or proximal humeral fractures are encompassed within the Ideberg and Neer classifications respectively.

Classification of Fractures of the Proximal Humerus

Generally, fractures of the proximal end of the humerus are not displaced badly and non-operative treatment yields good functional results. There is, however, significant fragment displacement in about 15–20 % of these fractures. These pose a therapeutic challenge. The management of the displaced fractures of the proximal humerus has been a matter of considerable dispute. A variety of classification systems have been proposed.

Early Classifications for Proximal Humeral Fractures

In 1855, Malgaigne identified proximal humeral fractures as either intra-articular or extra-articular. Kocher (1896) and Böhler (1929), classified these fractures according to the anatomical level of the fractures, i.e. fractures of the anatomical neck, intertubercular fractures (which also include fractures along the epiphyseal line); subtubercular or surgical neck fractures (Figure 11.13). Such descriptions are of little assistance in the displaced proximal humeral fractures as two levels are often involved .

Classification based on the injurious mechanism were proposed by Watson-Jones (1943) and Dehne

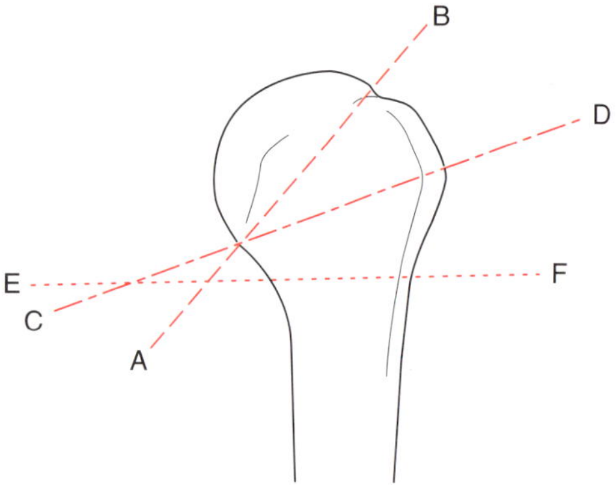

Figure 11.13. Kocher's classification of humeral head fractures (1896). Fractures of the anatomical neck (A–B); 2, intertubercular fractures (C–D); and 3, subtubercular or surgical neck fractures (E–F).

(1945). As the apex of angulation of these fractures is often directed anteriorly and seldom only in the coronal or scapular plane, description of these fractures as 'abduction' or 'adduction' fractures is misleading. Radiographic appraisal along these lines is often ambiguous (Neer, 1970a).

The Neer Classification

Codman (1934) proposed that fractures of the proximal humerus can be separated into four distinct fragments, occurring roughly along the lines of epiphyseal union (Figure 11.14). The four major fragments are the anatomical head, the greater tuberosity, the lesser tuberosity and the humeral shaft. He concluded that all fractures were some combination of these different fragments and that the musculotendinous cuff attaches to the more proximal fragments

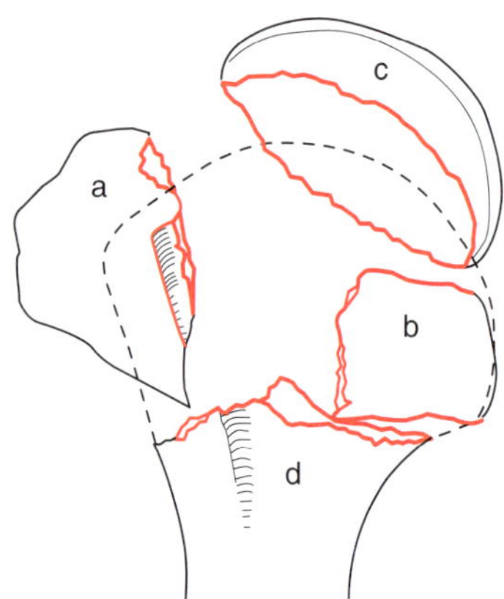

Figure 11.14. The four fragments in proximal fractures of the humerus according to Codman (1934). The four fragments are a, the greater tuberosity; b, lesser tuberosity; c, head and d, shaft of humerus.

and can hold the fractured fragments together. The degree of displacement of the fractured fragments are not included in this system.

Neer (1970a,b) based his four-part classification on Codman's ideas (Figure 11.15). The classification was reached following the analysis of 300 displaced proximal humeral fractures treated over a span of 14 years. His system takes into consideration the anatomy, biomechanical forces and, perhaps most importantly, the emphasis on displacement of the fracture fragments.

Radiographs consisting of anteroposterior, lateral and the Velpeau axillary views are recommended to identify the fracture fragments as well as their displacements. CT scans can be used to assess the amount of displacement of the tuberosities, splits of the humeral head, articular impression fractures and chronic fracture-dislocations. The Neer fracture system is a concept rather than a numerical classification and sets forth guidelines that are arbitrary and designed

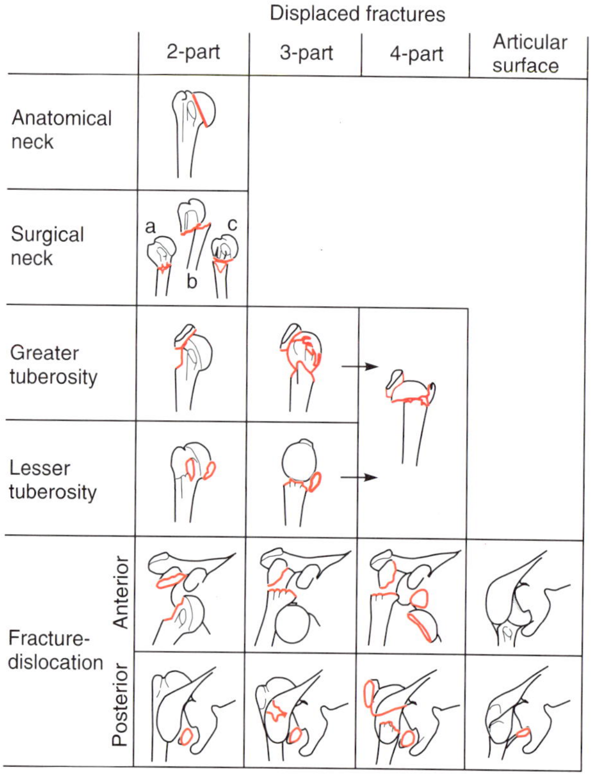

		Displaced fractures			
		2-part	3-part	4-part	Articular surface
Anatomical neck					
Surgical neck		a b c			
Greater tuberosity					
Lesser tuberosity					
Fracture-dislocation	Anterior				
	Posterior				

Figure 11.15. The Neer four-part classification of fractures of the proximal humerus (1970a). A fragment is considered displaced if there is more than 1 cm of separation or if the fragment is angulated by more than 45° from other fragments. Displaced fractures are thus either two-part, three-part or four-part. Fracture-dislocations are, similarly, either two-part, three-part or four-part. Impression fractures of the articular surface are associated with either an anterior or posterior dislocation.

to be helpful in recognizing displaced fractures. Emphasis is placed on the determination of the vascular supply to the humeral head, recognizing avascular necrosis as a common complication of such fractures.

The four parts are the same as Codman's description, namely the articular segment of the head of the humerus, the greater tuberosity, the lesser tuberosity and the shaft. Displacement is defined as separation of fragments by more than 10 mm and there is angulation of more than 45°. Hairline and fissure lines are not considered to be displaced. A fragment with several undisplaced components held in continuity by soft tissue is treated as single. A two-part fracture has a displaced fragment relative to the other three fragments. A three-part fracture has two fragments displaced in relation to each other but the head remains in contact with the glenoid fossa. In the four-part fracture, all four components are displaced and the head is out of contact with the glenoid and angulated laterally, anteriorly, posteriorly, inferiorly or superiorly.

A fracture-dislocation refers to a fracture when the head is displaced outside the joint space rather than being subluxated or rotated within the space. Fracture-dislocations are classified according to the direction (anteriorly or posteriorly in most cases) as well as the number of fracture fragments. Displaced fractures of the articular surface are grouped with fracture-dislocation.

Impression fractures of the articular surface are graded according to the percentage of the articular surface involved. A defect is classified as small if less than 20% of the articular surface is involved. If the impression involves more than 20% of the articular surface, redislocation tends to occur unless the articular fragment is stabilized. A large defect of more than 50% of the cartilage-covered head will result in an unstable and readily dislocatable head. Head-splitting fractures are recognized as a separate entity.

The AO/ASIF Classification

Jakob, Kristiansen and Ganz (1984) and the AO/ASIF group (Müller et al., 1990) published their classification based on a review of 730 fractures (Figure 11.16). As avascular necrosis is a well-documented complication of proximal humeral fractures, the vascularity of the fracture fragments plays a pivotal role in the prognosis of these fractures. This is emphasized in this classification. The alpha-numerical system provides an index of injury severity, and hence a type A fracture is less severe than a type B, which in turn, is less severe than the type C. In addition, each alphabetical group is subgrouped numerically in order of ascending degree of severity.

The type A fracture is extracapsular and involves two of the four primary fragments. There is no vascular isolation of the articular segments and the risk of avascular necrosis is minimal. The more severe type B fracture is intracapsular and three of the four primary segments are involved. There is partial isolation of the articular segment with a low risk of avascular necrosis. Type C fracture is the most severe, where there is total isolation of the articular segment with a high risk of avascular necrosis. This is an intracapsular fracture and all four primary segments are involved. As a general rule, subgroup 1 is comprised of undisplaced or minimally displaced fractures, subgroup 2 the displaced fractures and subgroup 3, the displaced fractures with additional complicating factors. This classification system was proposed to create a framework for more detailed prognostic and therapeutic guidelines.

Other Classifications

Duparc and Largier (1976) defined proximal humeral fractures as intra- or extra-articular according to the proposals of Malgaigne. This was based on a study of fracture-dislocations. Three forms of extra-articular fractures were described: subcapital (surgical neck) fractures, metadiaphyseal (spiral) fractures and 'vertical' fracture which involves the greater tuberosity. Intra-articular fractures may involve one or more humeral head fragments. The two groups differ therapeutically and in prognosis.

Habermeyer and Schweiberer (1989) modified the Neer classification by combining the number of displaced fragments with the anatomical level of fractures (Figure 11.17). The type A and type B fractures are extracapsular. Type A fractures are undisplaced, whilst the type B are displaced. Type C fractures are intracapsular with displaced fragments. Subgroups of these fractures are based on the number of fragments, as described along the lines described by Neer. Unlike the Neer classification, no special category was set aside for the fracture-dislocation; these are classified after the reduction of the dislocation.

Reproducibility of Classification of Proximal Humerus Fractures

Presently the Neer classification and the system introduced by the AO/ASIF group are the most widely accepted and their prognostic value has been proven

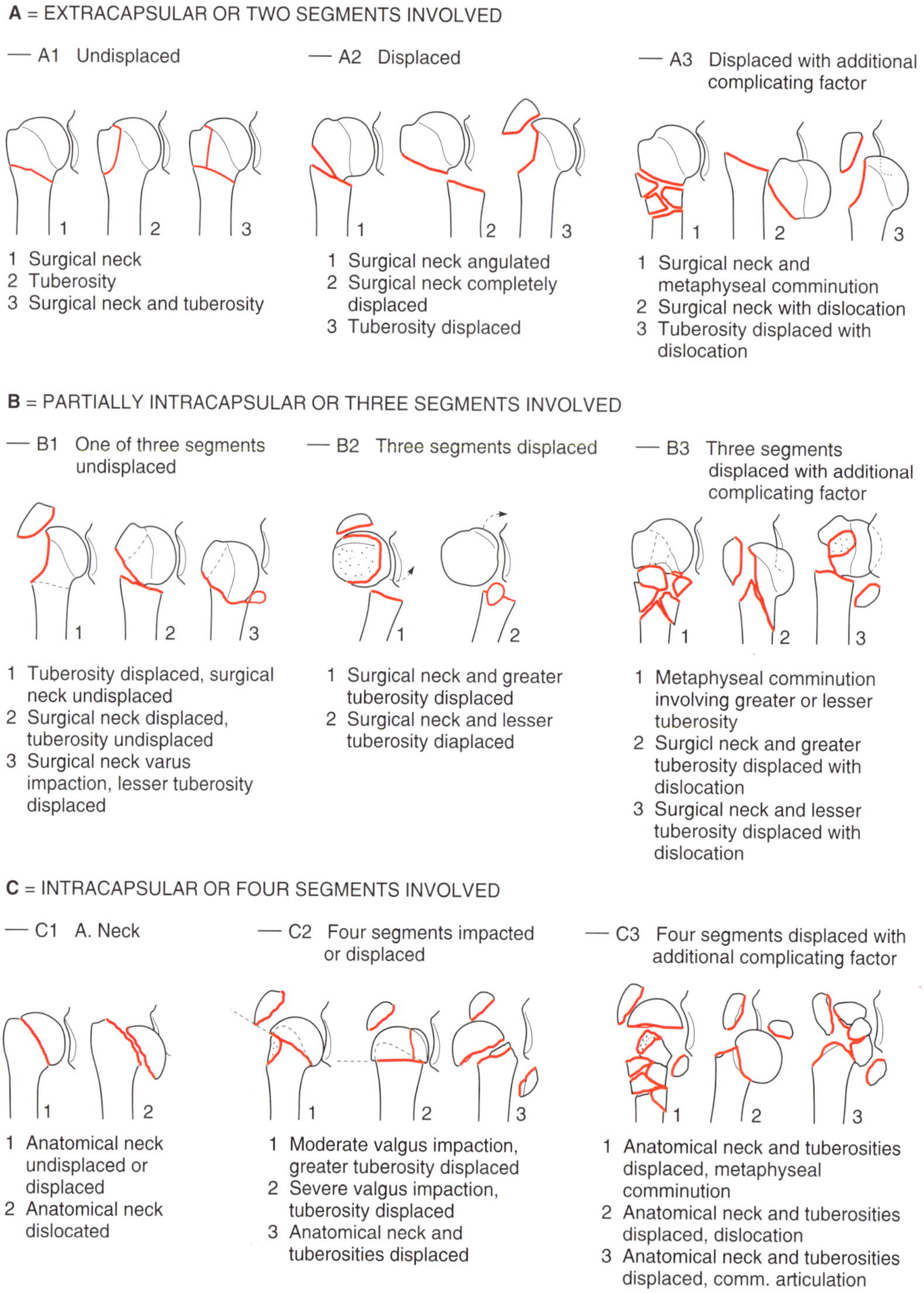

A = EXTRACAPSULAR OR TWO SEGMENTS INVOLVED

— A1 Undisplaced

1 Surgical neck
2 Tuberosity
3 Surgical neck and tuberosity

— A2 Displaced

1 Surgical neck angulated
2 Surgical neck completely displaced
3 Tuberosity displaced

— A3 Displaced with additional complicating factor

1 Surgical neck and metaphyseal comminution
2 Surgical neck with dislocation
3 Tuberosity displaced with dislocation

B = PARTIALLY INTRACAPSULAR OR THREE SEGMENTS INVOLVED

— B1 One of three segments undisplaced

1 Tuberosity displaced, surgical neck undisplaced
2 Surgical neck displaced, tuberosity undisplaced
3 Surgical neck varus impaction, lesser tuberosity displaced

— B2 Three segments displaced

1 Surgical neck and greater tuberosity displaced
2 Surgical neck and lesser tuberosity diaplaced

— B3 Three segments displaced with additional complicating factor

1 Metaphyseal comminution involving greater or lesser tuberosity
2 Surgicl neck and greater tuberosity displaced with dislocation
3 Surgical neck and lesser tuberosity displaced with dislocation

C = INTRACAPSULAR OR FOUR SEGMENTS INVOLVED

— C1 A. Neck

1 Anatomical neck undisplaced or displaced
2 Anatomical neck dislocated

— C2 Four segments impacted or displaced

1 Moderate valgus impaction, greater tuberosity displaced
2 Severe valgus impaction, tuberosity displaced
3 Anatomical neck and tuberosities displaced

— C3 Four segments displaced with additional complicating factor

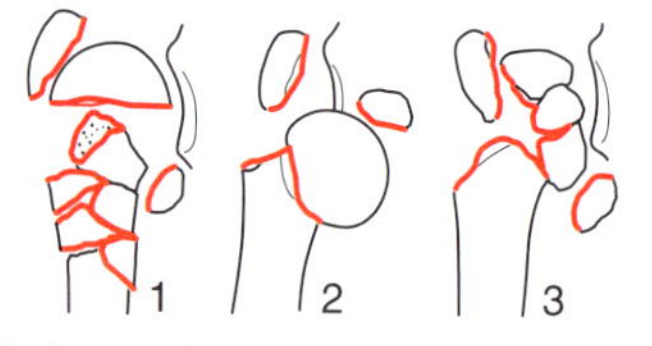

1 Anatomical neck and tuberosities displaced, metaphyseal comminution
2 Anatomical neck and tuberosities displaced, dislocation
3 Anatomical neck and tuberosities displaced, comm. articulation

Figure 11.16. The AO/ASIF classification of proximal humeral fractures (Jakob, Kristiansen and Ganz, 1984).

(Neer, 1970b; Mills and Horne, 1985; Kristiansen and Christensen, 1987; Szyszkowitz et al., 1993). Until recently, their reliability and reproducibility have rarely been questioned (Burstein, 1993).

Ackerman et al. (1986) reviewed 108 fractures and could not classify 13% of fractures into the Neer system and 24% into the AO/ASIF system. Kristiansen et al. (1988) evaluated the reliability of the Neer classification and reported a low degree of inter-observer agreement in the evaluation of a series of 100 proximal humerus fractures. Sidor et al. (1993) similarly assessed the inter-observer reliability and

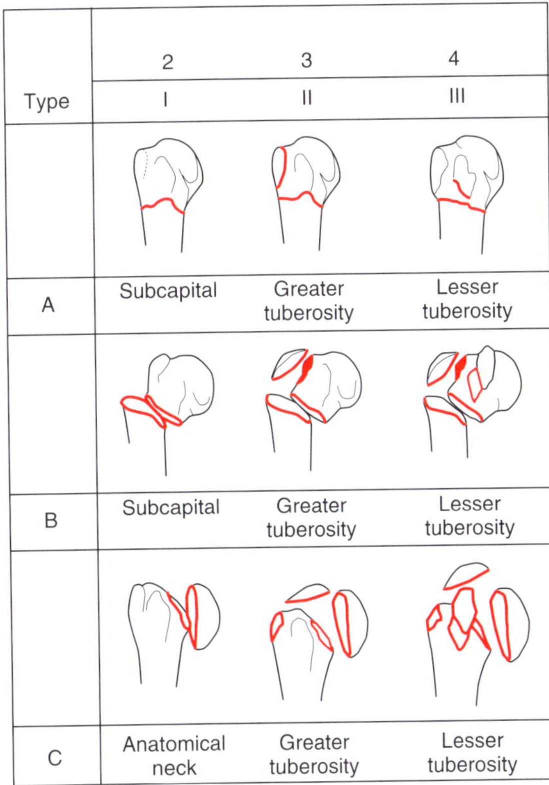

Figure 11.17. Modified classification of proximal humeral fractures by Habermeyer and Schweiberer (1989).

intra-observer reproducibility of the Neer classification. Fifty proximal humeral fractures were assessed using a complete trauma series. The inter-observer reliability was found to be moderate. The intra-observer reproducibility was found to be higher than the inter-observer reliability. The level of experience and expertise of the assessors were not found to be important factors, although intra-observer reproducibility improved with greater experience.

Siebenrock and Gerber (1993) compared the inter-observer reliability and intra-observer reproducibility of the Neer system with that of the AO/ASIF system. The AO/ASIF classification was found to be better than the Neer system for inter-observer reliability. Again, the intra-observer reproducibility were better than inter-observer reliability but both were considered moderate.

These studies imply that both the Neer and AO/ASIF systems are not sufficiently reproducible to allow meaningful comparison of similarly classified fractures in different studies. In addition, the fair to poor inter-observer reliability may well explain the controversy surrounding the recommendations for treatment and the wide variations seen in the result of treatment, particularly in the differences in incidence of avascular necrosis of the humeral head. Both classification systems are also thought to overestimate the importance of avascular necrosis, but underestimate the problems associated with malunion (Siebenrock and Gerber, 1992).

Both the Neer and the AO/ASIF systems are pathoanatomical classification and not merely radiographic. The limitations of the radiographic studies — the 'trauma series' which include an anteroposterior, scapular lateral and either a transthoracic, Velpeau axillary and rotational views — have been mentioned in the literature, particularly in determining the degree of displacement of the lesser tuberosity. Several authors have described their modifications to the original imaging studies to improve the radiographic assessment of proximal humeral fractures. Evaluation of complex humeral fractures with CT scan and even three-dimensional reconstruction may be necessary to provide even more information on these fractures. Often, the exact nature of these fractures are not apparent until exploration at surgery (Bigliani, Flatow and Pollock, 1994; Cowell, 1994).

The value of the Neer and AO/ASIF classification in allowing communication between individuals regarding the anatomical characteristics of fractures cannot be denied, but categorization of these fractures is difficult. It is likely that many of the reports of series using the various classifications should be treated with caution (Cowell, 1994).

Classification of Proximal Humeral Fractures in Children

Categorization of proximal humeral fractures in children is based on the location of the fractures, the amount of displacement and an assessment of stability.

Most proximal humeral fractures involve the growth plate and are well described by the Salter–Harris classification (1963) (see also Chapter 7) (Figure 11.18). The Salter–Harris type I fractures are the most common in neonates. In adolescents, the Salter–Harris type II make up 75% of all proximal humeral fractures, the other 25% are type I. The metaphyseal fragment of type II fractures are often in the posteromedial region of the thick periosteal sleeve. Types III to V fractures do occur but are rare.

Neer and Horowitz (1965) described a classification system based on the degree of fragment displacement. Grade I injuries are displaced less than 5 mm, grade II involving a displacement of up to a third of the width of the shaft, grade III by up to two-thirds of the width of the shaft and finally, grade IV where the

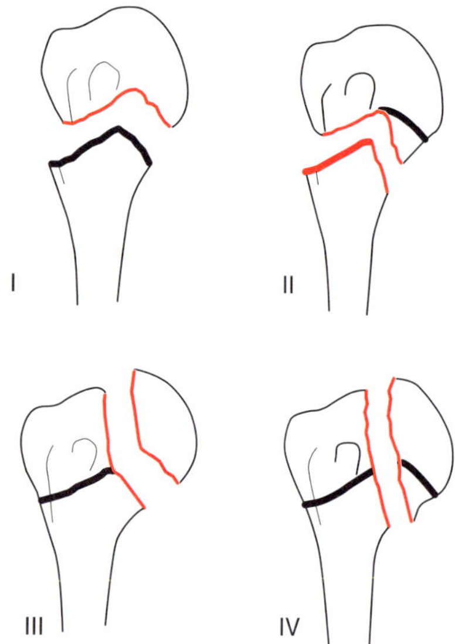

Figure 11.18. The Salter–Harris classification of epiphyseal proximal humeral fractures. Type I are more common in children under the age of 5 years. Type II injuries are the commonest, occurring in children more than 5 years of age. Types III and IV are rare.

displacement exceeds two-thirds of the shaft width. Grade IV fractures also include total displacement.

The combined assessment of stability together with the Neer and Horowitz displacement classification have been used by some authors to determine the method of treatment of such fractures. Closed reduction is usually necessary for the Neer and Horowitz grade III and IV (Dameron and Rockwood, 1984). There is tremendous potential for remodelling of immature proximal humeral fractures and this potential decreases with increasing age of the child. Age is thus a major factor to be considered in the management of such fractures. This variable, however, has not been included in any of the classification schemes of injuries in this region.

Group Discussion

The discussion group felt that classification systems of fracture of the scapula were predominantly descriptive. They had relatively little application in terms of directing treatment other than when they were broadly classified into those fractures involving the articular surface or displaced fractures of the neck of the glenoid. Although Ada and Miller (1991) had produced a complex classification system, when they actually looked at the outcome from scapula fractures they simplified the groups into those involving the blade, the glenoid neck and the articular surface.

Intra-articular fractures of the glenoid were of particular importance although it was felt that even these classification systems failed to address some of the important clinical issues, such as whether or not the fractures were associated with any major displacement of the articular surface or significant glenohumeral instability. This was particularly so for rim fractures either of the posterior or antero-inferior portions of the glenoid.

Classifications of coracoid fractures were felt to be relatively valueless in isolation and were much more important when associated with classifications of the acromioclavicular joint.

The acromioclavicular dislocation classification systems had some limited value in dictating treatment although it was felt that many clinicians relied more on clinical symptoms and function than the cosmetic appearance of the acromioclavicular dislocation. It was also noted that imaging of these injuries can sometimes be difficult and stress views are often required. The classification system of Bannister et al. (1992) had made some attempt at correlation with outcome and at reproducibility measurement. It was felt that this was the best buy in this area.

Classification systems for sternoclavicular dislocation were of limited value. None of them address the clinically relevant issues and therefore did not direct treatment or predict outcome.

The Allman (1967) and Craig (1991) classification systems of the clavicle had some limited value in predicting treatment and outcome. However, the most important area clinically are fractures of the distal clavicle addressed by Neer's classification. This classification system was felt to be very helpful in defining and dictating treatment. However, no reliability studies had been performed of any of these systems. The best system overall for the clavicle was felt to be the Craig system although the Neer and Rockwood (1984) system was much simpler when distal clavicle fractures alone are considered.

Classifications of clavicle fractures in children was also considered and a system has been described by Dameron and Rockwood (1984). This was felt to be of little relevance clinically and there was certainly no evidence to back-up its use in terms of predicting outcome or dictating treatment.

With respect to rotator cuff lesions it was noted that there were no particularly good classification systems for acute cuff ruptures. The classification systems that do exist are mostly for degenerative lesions and tears of degenerate rotator cuff. Of these classification systems, the one that might most appropriately be used for acute tears was that described by Post, Silver and Singh (1983).

Table 11.7 Summary of the group discussion of clavicle, scapular and humeral fractures

Site	*Classification system*	*Reliability tested?*	*Good for defining treatment?*	*Good for defining outcome?*	*Methods of examination*
Scapula	Desault (1805)	No	No	No	X-ray
	Zdravkovic and Damholt (1974)	No	No	No	X-ray
	Thompson et al. (1985)	No	No	No	X-ray
	Ada and Miller (1991)	No	Limited	Limited	X-ray
	Hardegger et al. (1984)	No	No	No	X-ray
	Ruedi and Schweiberer (1991)	No	Most useful	Most useful	X-ray
	Euler et al. (1992)	No	Limited	Limited	X-ray
Coracoid	Ogawa et al. (1990)	No	Possible	Possible	X-ray
	Eyres et al. (1995)	No	Possible	Possible	X-ray
Glenoid	Ideberg (1984)	No	Limited	Limited	X-ray
	Goss (1992)	No	Limited	Limited	X-ray
Clavicle	Allman (1967)	No	No	No	X-ray
	Craig (1991)	No	Yes	Yes	X-ray
	Thompson (1989)	No	Limited	Limited	X-ray
Distal clavicle	Neer and Rockwood (1984)	No	Yes	Yes	X-ray
	Buschle et al. (1983)	No	Yes, limited	Yes, limited	X-ray
Acromioclavicular joint	Tossy et al. (1963)	No	Limited	Limited	X-ray
	Allman (1967)				
	Post (1985)	No	Limited	Limited	X-ray
	Rockwood (1984)	No	Limited	Limited	X-ray
	Bannister et al. (1983, 1992)	No	Possible	Possible	X-ray
Clavicle in children	Salter and Harris (1963)	No	No	No	X-ray
Distal clavicle in children	Dameron and Rockwood (1984)	No	Limited	Limited	X-ray
Rotator cuff tears	Post et al. (1983)	No	Possible	Possible	US, MRI, Arthroscopy
Glenohumeral instability	None				
Proximal humerus	Neer (1970a,b)	Yes Challenged	Yes	Yes	X-ray
	Jakob et al. (1984) (AO/ASIF)	Yes Challenged	Yes Yes	Yes	X-ray
	Duparc and Largier (1976)	No	No	No	X-ray
	Habermeyer and Schweiberer (1989)	No	Possible	Possible	X-ray
	Kocher (1896)	No	Limited	Limited	X-ray
Proximal humerus in children	Salter and Harris (1963)	No	Limited	Limited	X-ray
	Neer and Horowitz (1965)	No	Possible	Possible	X-ray
	Schatzker, 1987	Yes	Yes	No	

The group also considered there was no particularly good classification system for acute glenohumeral instability. It was felt that the majority of subsequent problems arise because of recurrent instability and these classification systems are beyond the scope of this text.

The best set of classification systems in the shoulder region are those described for fractures of the proximal humerus. There are two main systems — the Neer (1970a,b) and the AO (Jakob, Kristiansen and Ganz, 1984). Both of these have definite value in defining treatment and outcome. They are also exceptional in that their reproducibility has been detailed. Unfortunately, these papers report some difficulties in reliably classifying these fractures. It is probable that additional imaging methods including CT scanning and MRI may help to further define the fractures. Nevertheless, the general principles introduced by the classification systems are helpful in defining both treatment and outcome (Table 11.7).

Acknowledgements

I wish to thank Mrs J. Liow and Dr T. Niemeyer for their help in the translation of manuscripts.

References

Ackerman, C., Lam, Q., Linder, P. et al. (1986) Zur problematik der Frakturklassifikation am proximalen Humerus. *Zeitschr. Unfallchir.*, **79**, 209–215.

Ada, J.R. and Miller, M.E. (1991) Scapular fractures: an analysis of 113 cases. *Clin. Orthop. Rel. Res.*, **269**, 174–180.

Allman, F.L. (1967) Fractures and ligamentous injuries of the clavicle. *J. Bone Joint Surg.*, **49A**, 774–784.

Bannister, G. (1983) The management of complete acromioclavicular dislocation. A randomised prospective controlled trial comparing early movement with coracoclavicular screw fixation. MCh, Orth. Thesis: University of Liverpool.

Bannister, G.C., Wallace, W.A., Stableforth, P.G. and Hutson, M.A. (1992) A classification of acute acromioclavicular dislocation: a clinical, radiological and anatomical study. *Injury*, **23**, 194–196.

Bigliani, L.U., Flatow, E.L. and Pollock, R.G. (1994) Correspondence. *J. Bone Joint Surg.*, **76A**, 790–791.

Bjerneld, H., Hovelius, L. and Thorling, J. (1983) Acromioclavicular separations treated conservatively: A 5-year follow-up study. *Acta Orthop. Scand.*, **54**, 743–745.

Böhler, L. (1929) Die Behandlung von Verrenkungsbruschen der Schulter. *Dtsch. Zeitschr. Chir.*, **219**, 238–245.

Bosworth, D.M. (1940) Analysis of 28 consecutive cases of incapacitating shoulder lesions radically explored and repaired. *J. Bone Joint Surg.*, **22**, 369.

Brunner, U., Habermeyer, P. and Schweiberer, L. (1992) Die Sonderstellung der lateralen Klavikulafraktur. *Orthopade*, **21**, 163–171.

Burstein, A.H. (1993) Editorial. Fracture classification systems: do they work and are they useful? *J. Bone Joint Surg.*, **75A**, 1743–1744.

Buschle, K.D., von Breitner, S. and Jager, M. (1983) Die laterale Claviculafraktur als Pendant zur Acromioclaviculargelenksprengung. In *Bandverletzungen am Schulter-, Knie- and Sprunggelenk* (R. Rahmanzadeh and M. Faensen, Hrsg.). Konstanz: Schnetztor.

Codman, E.A. (1934) *The Shoulder*. Boston: Thomas Todd.

Cofield, R.H. (1985) Current concepts review: Rotator cuff disease of the shoulder. *J. Bone Joint Surg.*, **67A**, 974–979.

Cowell, H.R. (1994) Editorial. Patient care and scientific freedom. *J. Bone Joint Surg.*, **76A**, 640–641.

Craig, E.V. (1991) Fractures of the clavicle. In *Fractures in Adults* (C.A. Rockwood, D.P. Green and R.W. Bucholz, eds), 3rd edn, vol. 1. Philadelphia: Lippincott.

Dameron, T.B. and Rockwood, C.A. Jr (1984) Fracture of the shaft of the clavicle. In *Fractures in Children*. (C.A. Rockwood Jr, K.E. Wilkins and R.E. King, eds). Philadelphia: Lippincott.

Dehne, E. (1945) Fractures of the upper end of the humerus. *Surg. Clin. North Am.*, **25**, 28–47.

Desault, P.J. (1805) *A Treatise on Fractures, Luxations and other Affections of the Bones*. Philadelphia: Fry and Kammerer.

Dias, J.J., Steingold, R.F., Richardson, R.A. et al. (1987) Five year review of conservatively treated acromioclavicular joint injuries. *J. Bone Joint Surg.*, **69B**, 719–722.

Duparc, J. and Largier, A. (1976) Les luxations-fractures de l'extremite superieure de l'humerus. *Rev. Chir. Orthop.*, **62**, 91–110.

Edwards, D.J., Kavanagh, T.G. and Flannery, (1992) Fractures of the distal clavicle: a case for fixation. *Injury*, **23**, 44–46.

Ellman, H., Hanker, G. and Bayer, M. (1986) Repair of the rotator cuff. End result study of factors influencing reconstruction. *J. Bone Joint Surg.*, **68A**, 1136–1144.

Euler, E., Habermeyer, P., Kohler, W. and Schweiberer, L. (1992) Scapular fractures: classification and differential therapy. *Orthopade*, **21**, 158–162.

Eyres, K.S., Brooks, A. and Stanley, D. (1995) Fractures of the coracoid process. *J. Bone Joint Surg.*, **77B**, 425–428.

Godsil, R.D. Jr and Linscheid, R.L. (1970) Intratendinous defects of the rotator cuff. *Clin. Orthop.*, **69**, 182–188.

Gore, D.R., Murray, M.P., Sepic, S.B. and Gardner, G.M. (1986) Shoulder muscle strength and range of motion following surgical repair of full thickness rotator cuff tears. *J. Bone Joint Surg.*, **68A**, 266–272.

Goss, T.P. (1992) Current concepts review: Fractures of the glenoid cavity. *J. Bone Joint Surg.*, **74A**, 299–305.

Habermeyer, P. and Schweiberer, L. (1989) Frakturen des proximalen humerus. *Orthopade*, **18**, 200–207.

Hardegger, F.H., Simpson, L.A. and Weber, B.G. (1984) The operative treatment of scapular fractures. *J. Bone Joint Surg.*, **66B**, 725–731.

Harryman, D.T. II, Mack, L.A., Wang, K.Y. et al. (1991) Repairs of the rotator cuff. *J. Bone Joint Surg.*, **73A**, 982–989.

Hawkins, R.J., Misamore, G.W. and Hobeika, P.E. (1985) Surgery for full thickness rotator cuff tears. *J. Bone Joint Surg.*, **67A**, 1349–1355.

Ideberg, R. (1984) Fractures of the scapula involving the glenoid fossa. In *The Surgery of the Shoulder* (J.E. Bateman and R.P. Welsh, eds). Philadelphia: Decker, pp. 63–66.

Imatani, R.J. (1975) Fractures of the scapula: a review of 53 fractures. *J. Trauma*, **15**, 473–478.

Jakob, R.P., Kristiansen, P. and Ganz, R. (1984) Proximale Humerusfrakturen – Klassifikation und aspekte der behandlung. In *Verletzungen und Erkrangkungen der Schulterregion* (G. Chapchal, ·ed.). Stuttgart: Georg Thieme.

Kavanagh, T.G., Sarkar, S.D. and Phillips, H. (1985) Complication of the displaced fractures (type 2) of the outer end of the clavicle. *J. Bone Joint Surg.*, **67B**, 492.

Kessel, L. and Watson, M. (1977) The refractory painful arc syndrome. *J. Bone Joint Surg.*, **59B**, 166–172.

Kocher, T. (1896) Die frakturen am oberen humerusende. In *Beitrage zur Kenntnis einiger praktisch wichtiger Frakturformen* (C. Sallman, Hrsg). Basel, Leipzig: Sallman, S7.

Kristiansen, B., Andersen, U.L.S., Olsen, C.A. and Varmarken, J.-A. (1988) The Neer classification of fractures of the proximal humerus. *Skeletal Radiol.*, **17**, 420–422.

Kristiansen, B. and Christensen, S.W. (1987) Proximal humeral fractures. *Acta Orthop. Scand.*, **58**, 124–127.

Mills, H.J. and Horne, G. (1985) Fractures of the proximal humerus in adults. *J. Trauma*, **25**, 801–805.

Müller, M.E., Nazarian, S., Koch, P. and Schatzker, J. (1990) *The Comprehensive Classification of Fractures of the Long Bone*. New York: Springer, pp. 54–63.

Neer, C.S. II (1968) Fractures of the distal third of the clavicle. *Clin. Orthop. Rel. Res.*, **58**, 43.

Neer, C.S. II (1970a) Displaced proximal humeral fractures. Part 1. Classification and evaluation. *J. Bone Joint Surg.*, **52A**, 1077–1089.

Neer, C.S. II (1970b) Displaced proximal humeral fractures. Part II. Treatment of the three-part and four-part displacement. *J. Bone Joint Surg.*, **52A**, 1090–1103.

Neer, C.S. II (1983) Impingement lesions. *Clin. Orthop. Rel. Res.*, **173**, 70–77.

Neer, C.S. II and Horowitz, B.S. (1965) Fractures of the proximal humeral epiphyseal plate. *Clin. Orthop. Rel. Res.*, **41**, 24–31.

Neer, C.S. II. and Rockwood, C.A. Jr (1984) Fractures and dislocations of the shoulder. In *Fractures in Adults* (C.A. Rockwood, Jr and D.P. Green, eds), 2nd edn. Lippincott: Philadelphia.

Ogawa, K., Toyama, Y., Ishige, S. and Matsui, K. (1990) Fractures of the coracoid process: its classification and pathomechanism. *J. Jap. Orthop. Assoc.*, **64**, 909–919.

Post, M. (1985) Current concepts in diagnosis and management of acromioclavicular dislocations. *Clin. Orthop. Rel. Res.*, **200**, 234–247.

Post, M. (1989) Current concepts in treatment of fractures of the clavicle. *Clin. Orthop. Rel. Res.*, **245**, 89–101.

Post, M., Silver, R. and Singh, M. (1983) Rotator cuff tear. *Clin. Orthop. Rel. Res.*, **173**, 78–91.

Recht, M.P. and Resnick, D. (1993) Magnetic resonance-imaging studies of the shoulder. *J. Bone Joint Surg.*, **75A**, 1244–1253.

Richards, R.R. (1993) Acromioclavicular joint injuries. *Instruct. Course Lect.*, **42**, 259–269.

Rockwood, C.A. Jr (1984) Injuries to the acromioclavicular joint. In *Fractures in Adults* (C.A. Rockwood Jr and D.P. Green, eds), 2nd edn, vol 1. Philadelphia: Lippincott, pp. 860–910.

Ruedi, T. and Schweiberer, L. (1991) Scapula, clavicle and humerus. In *Manual of Internal Fixation* (M.E. Müller, M. Allgower, R. Schneider, and H. Willenegger, eds). Berlin: Springer-Verlag, pp. 427–429.

Salter, R.B. and Harris, W.R. (1963) Injuries involving the epiphyseal plate. *J. Bone Joint Surg.*, **45A**, 587–622.

Sidor, M.L., Zuckerman, J.D., Lyon, T. et al. (1993) The Neer classification system for proximal humeral fractures. *J. Bone Joint Surg.*, **75A**, 1745–1750.

Siebenrock, K.A. and Gerber, C. (1992) Frakturklassifikation und problematik bei proximalen humerusfrakturen. *Orthopade*, **21**, 98–105.

Siebenrock, K.A. and Gerber, C. (1993) The reproducibility of classification of fractures of the proximal end of the humerus. *J. Bone Joint Surg.*, **75A**, 1751–1755.

Snyder, S.J. (1993) Evaluation and treatment of the rotator cuff. *Orthop. Clin. North Am.*, **24**, 173–192.

Szyszkowitz, R., Seggl, W., Schleifer, P. and Cundy, P.J. (1993) Proximal humeral fractures. Management techniques and expected results. *Clin. Orthop. Rel. Res.*, **292**, 13–25.

Thompson, D.A., Flynn, T.C., Miller, P.W. et al. (1985) The significance of scapular fractures. *J. Trauma*, **25**, 974–977.

Thompson, J.S. (1989) ORIF uniquely suited to displaced mid third clavicle fractures. *Orthopaedics Today*, **14**, 1.

Tossy, J.D., Mead, N.C. and Sigmond, H.M. (1963) Acromioclavicular separations: Useful and practical classification for treatment. *Clin. Orthop. Rel. Res.*, **28**, 111–119.

Urist, M.R. (1946) Complete dislocations of the acromioclavicular joint. *J. Bone Joint Surg.*, **28A**, 813–837.

Watson, M. (1985) Major rupture of the rotator cuff. The result of repair in 89 patients. *J. Bone Joint Surg.*, **67B**, 618–624.

Watson-Jones, R. (1943) *Fractures and Joint Injuries*, 3rd edn. Baltimore: Williams and Wilkins.

Wolfgang, G.L. (1974) Surgical repair of tears of the rotator cuff of the shoulder. *J. Bone Joint Surg.*, **56A**, 14–26.

Zanca, P. (1971) Shoulder pain: involvement of acromioclavicular joint: analysis of 1000 cases. *Am. J. Roentgenol.*, **112**, 493–506.

Zdravkovic, D. and Damholt, V.V. (1974) Comminuted and severely displaced fractures of the scapula. *Acta Orthop. Scand.*, **45**, 60–65.

12

The elbow, including humeral shaft, distal humerus and forearm

J. R. Williams

Introduction

The use of a classification system for trauma has to fulfil two aims. First, to help clinicians make management decisions to recommend to their patients. Secondly, to enable clinicians to predict the outcome of the trauma in the short and long term. These two aims may, of course, be linked. To achieve these aims in the busy clinical setting the classification system must be short, logical, comprehensive and easily communicated to others. These requirements are not always mutually compatible.

This chapter will include specific classification systems for injuries to the shaft of the humerus, the distal humerus, the coronoid process of the ulna, the olecranon, the radial head and neck, the forearm bones and, finally, elbow dislocations and instability. Generalized classification systems for traumatic injuries such as those for open fractures and growth plate injuries are found elsewhere in this volume (Chapters 2 and 7), as are overall scorings for injury severity such as the Mangled Extremity Severity Score (MESS) (see Chapter 1).

Children's fractures are discussed only where their classification is different from the equivalent adult fracture.

Shaft of Humerus

Fractures of the diaphyseal region or shaft of the humerus consist of 3% of all fractures (Ward, Savoie and Hughes, 1992). Approximately 7% of all diaphyseal fractures in the AO analysis of 2700 surgically treated diaphyseal fractures were of the humeral diaphysis (Müller et al., 1990). The ease of treatment of these fractures and their outcome depends on the position of the fracture and its 'personality'. Fractures of the distal third have a higher incidence of radial nerve injuries and are called by some the Holstein–Lewis fracture (Ward, Savoie and Hughes, 1992).

There is little in the literature specifically on the classification of humeral shaft fractures beyond that which pertains to all long bone shaft fractures. Ward et al. classify the fractures by anatomical location and fracture pattern. The humeral shaft is divided anatomically into three segments: that above the pectoralis major muscle insertion, that below the pectoralis major but above the deltoid muscle insertion and thirdly, that below the deltoid insertion. The fracture pattern description is a generic one using the terms transverse, oblique, spiral, segmental and comminuted. No data are produced to support these divisions and the classification system is not used any further in their work. This classification based on fracture position helps one understand how the fragments are positioned by the action of the muscles attached to them (Ward et al., 1992).

The most useful method for classifying fractures of the humeral shaft is that proposed for all long bones by the AO group (Müller et al., 1990). The humerus is bone number 1 in the AO system and the diaphysis is segment number 2. This is the middle segment of the bone that lies between the two squares that encompass the proximal and distal epiphyseal and metaphyseal regions (Figure 12.1a). In the AO study there were 200 diaphyseal humeral fractures falling into the 12-segment, 61% were type 12-A; simple fractures, 32% were type 12-B; wedge fractures and 7% were type 12-C; complex, multifragmentary fractures (Figure

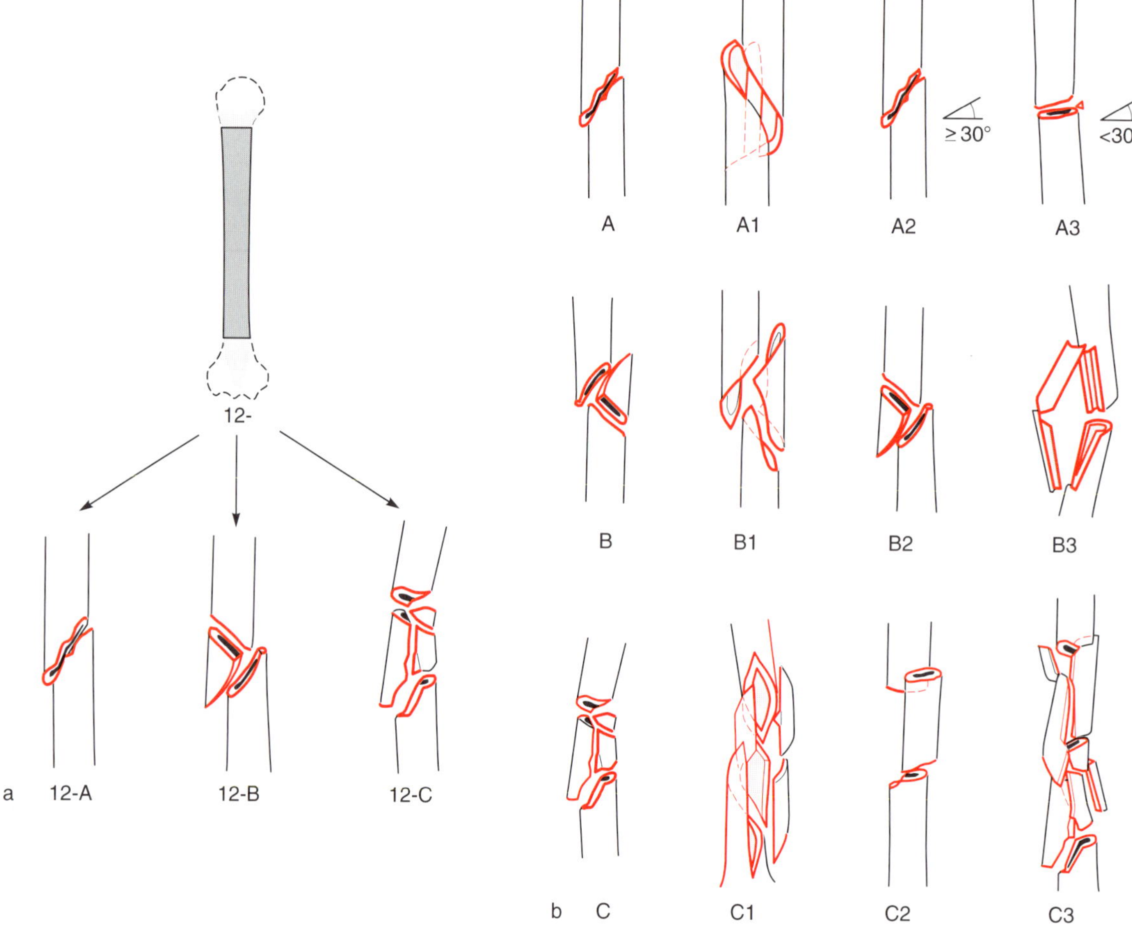

Figure 12.1. (a) The AO classification system showing the humeral diaphyseal segment 21- (Müller et al., 1990). (b) The AO classification system for humeral shaft fractures (Müller et al., 1990).

12.1b). Of the 12-C fractures segmental fractures (group 12-C2) were very rare (1%) with the spiral multifragmentary (group 12-C1) and the irregular multifragmentary (group 12-C3) making up 3% of the total each. The 1–3 groups of the A and B type of fractures are divided into .1, .2 and .3 sub-groups depending on their position in the diaphysis. The diaphysis is divided into three equal regions: proximal (.1), middle (.2) and distal (.3). The C type fractures are subgrouped (.1, .2, .3) according to the complexity of the fragmentation. This system is comprehensive and systematic for the humeral shaft and it is recommended as it is based on a large series of fractures. The classification is also consistent with that for other long bones such as the femur (Müller et al., 1990).

The Distal Humerus

The distal end of the humerus is one of the most complex skeletal areas to reconstruct surgically and more papers are devoted to this region than any other part covered in this chapter. There are, correspondingly, more classification systems, of which two have becoming widely accepted. Fractures of the distal humerus account for approximately 2% of all fractures and about one-third of all elbow fractures (Jupiter and Morrey, 1993). The manner of treatment has changed markedly over the past 30 years from strong advice against internal fixation (Riseborough and Radin, 1969), to excellent results with well executed and meticulous internal fixation (Holdsworth and Mossad, 1990; Jupiter, 1994). As internal fixation has become more widely accepted treatment for these fractures, classification systems that help in the understanding of operative reconstruction have gained favour (Jupiter, 1994).

The early classification systems describe fractures of the distal humerus by anatomical regions such as condyles and epicondyles, with condylar, transcondylar and bicondylar fracture patterns described. One of the earliest classifications in the English speaking literature was a posthumous publication by Henry Milch of New York (Milch, 1964). This has formed the basis for subsequent classifications and expounded some of the mechanistic thinking behind classifying distal humeral fracture patterns. Milch divided both medial and lateral unicondylar fractures into two types. Type 1 are those where the lateral wall of the trochlea remains intact and which can be treated either operatively or non-operatively. Type 2 fractures represent a fracture-dislocation where the lateral wall of the trochlea is violated and the fracture required internal fixation. Milch only considered two-part fractures, but more complex, so called T and Y, fractures, were described (Cassebaum, 1969). He divided the fractures of the distal humerus into four fragments. These consist of the remaining intact humeral shaft and three distally arranged fragments. Cassebaum's paper describes how to fix the various fragments into a stable construct and forms the basis for modern techniques of reconstruction (Jupiter et al., 1985).

Edward Riseborough and Eric Radin from Massachusetts General Hospital identified four fracture patterns in their series of 29 cases of intercondylar T fractures (Riseborough and Radin, 1969). Type I were undisplaced fractures. Type II fractures were where the trochlea and capitellar fragments were separated, but not rotated, in the frontal plane. Type III fractures were where there was both significant separation and rotation in the frontal plane. Type IV fractures had severe comminution of the articular surfaces and wide separation of the fragments in the frontal plane (Figure 12.2). The classification represented increasing severity of fracture and the results of the authors reflected this. It can be difficult to distinguish between type II and type III fractures on preoperative radiographs and so the Riseborough and Radin system has been simplified by the Orthopaedic Trauma Association (OTA). Fractures of the distal humerus are divided into either unicondylar or bicondylar (Figure 12.3). Type I bicondylar fractures are undisplaced. Type II bicondylar fractures are those with rotation and/or displacement of the condylar fragments. Type III bicondylar fractures have comminution of the articular surface (Henley, Bone and Parker, 1987). For unicondylar fractures there are six types with a seventh category for miscellaneous ones (Henley, 1987).

Shetty rejected the Riseborough and Radin classification as it did not include any recognition of comminution of the supracondylar ridges (Shetty, 1982). He defined type I fractures as T or Y intercondylar fractures with an intact tissue sleeve. Type II fractures had three separated fragments with disruption of the

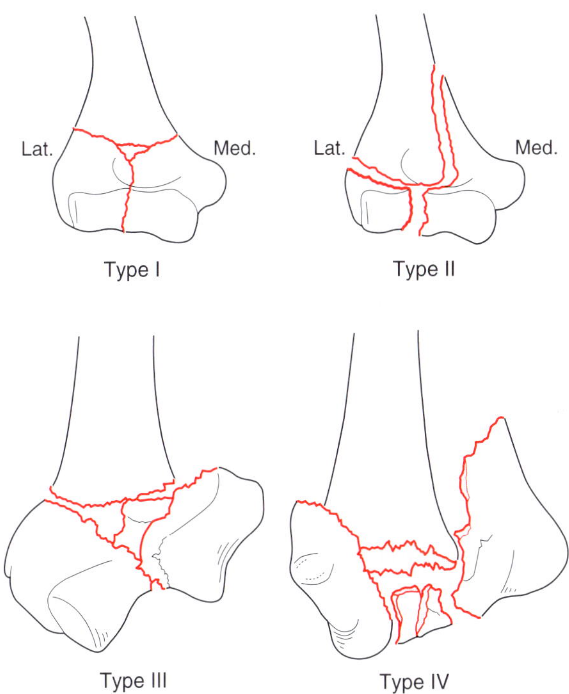

Figure 12.2. Riseborough and Radin's classification of distal humeral fractures (Riseborough and Radin, 1969).

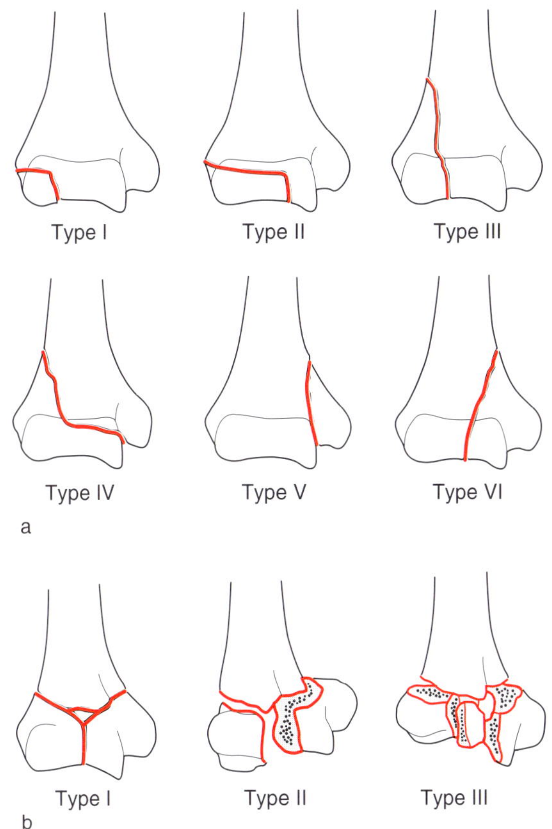

Figure 12.3. (a) OTA unicondylar, (b) bicondylar fracture classification for the distal humerus (Henley, 1987).

soft-tissue sleeve, but the fragments were reducible. Type III fractures were like type II, but, with an additional fragment from the shaft. The type III fractures are displaced but reducible. Type IV fractures represent those fractures with comminution that are not reducible. This classification system does not appear to have gained favour in the literature and does not represent modern surgical practice in the treatment of these fractures.

Horne and Pennal from Toronto (Horne, 1979) classified their group of 50 supracondylar distal humeral fractures into three distinct fracture types: unicondylar, supracondylar—transverse or oblique—and T, Y or M fractures. Their view was that the first two types should be treated by internal fixation but the T, Y and M fractures were best treated conservatively. The classification determined the treatment for each type.

Efforts to produce a unifying classification system for distal humeral fractures has lead to the AO classification (Müller et al., 1990), first developed in 1979. The AO classification divides the 13- segment fractures into type A: extra-articular, type B: partial intra-articular and type C: complete articular. The 1, 2 and 3 groups and .1, .2 and .3 subgroups define these further. This system is widely used (Letsch et al., 1989). Its limitations were pointed out by Schatzker particularly in the groups of the type A and B fractures. Although the AO system is anatomically rigorous it groups fractures together inappropriately in terms of their seriousness and difficulty to reconstruct, i.e. simple avulsion fractures of the epicondyles, group 13-A1, with difficult extra-articular supracondylar fractures, groups 13-A2 and 13-A3 (Schatzker, 1987a). Schatzker prefers to think of the distal humeral fractures as either 'simple' fractures—those of the lateral and medial epicondyles, of the lateral condyle and of the capitellum—or as 'complex' supracondylar fractures. The complex supracondylar fractures include 13-A2, 13-A3 and all the 13-C fractures in the AO classification. All these fractures share common problems of reconstruction and prognosis and should be considered together (Schatzker, 1987a). The subdivisions of the AO type C fracture are more satisfactory and many clinicians find they agree well with clinical practice (Figure 12.4) (Jupiter, 1994).

Jupiter has developed a different comprehensive system that is based on the surgical anatomy and the techniques used for reconstruction and subsequent prognosis of the fracture. The Jupiter system may prove to fulfil the requirements of Schatzker (Schatzker, 1987a; Jupiter, 1992). Although, Jupiter used the AO Müller classification in his 1985 series of

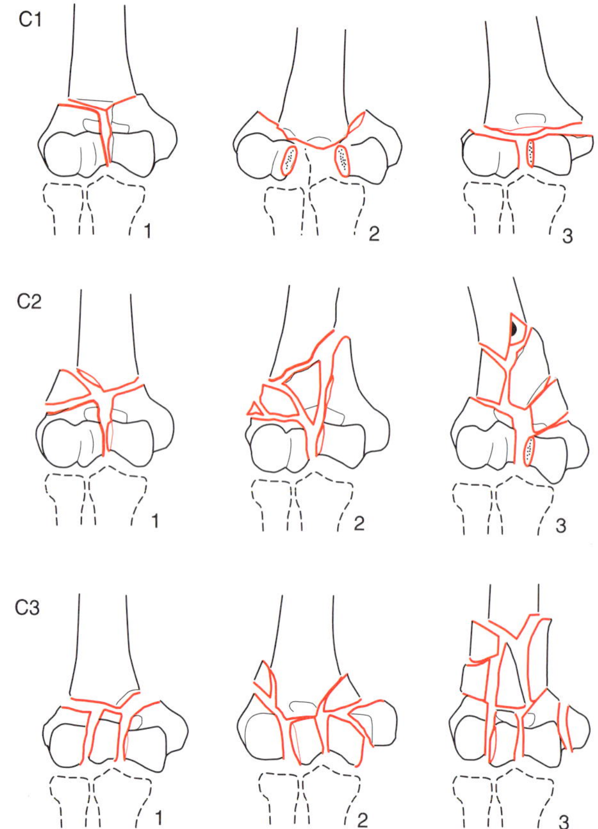

Figure 12.4. The AO classification of complex 13C type distal humeral fractures (Müller et al., 1990).

34 intercondylar fractures (Jupiter et al., 1985), it became clear that 'Historical classifications describing T or Y patterns of intercondylar fractures inadequately reflect the variety of fractures that are seen at operations' (Jupiter, 1994). The Jupiter system developed with David Mehne and Joel Matta divides the fractures into three basic categories: I intra-articular fractures, II extra-articular, intracapsular fractures, and III extracapsular fractures. Intra-articular fractures are divided into A: single column, similar to the classification of Milch, B: bicolumnar, C: capitellar and D: trochlea (Table 12.1 and Figure 12.5). The latter two groups may be part of a multiplanar distal humeral fracture which has a coronal plane split of the trochlea or capitellum in addition to the usual transverse and sagittal plane fractures of the T or Y pattern (Jupiter et al., 1993). The extra-articular, intracapsular fractures are all transcolumnar and are either high, and relatively easy to fix, or low and relatively difficult to fix. The extracapsular fractures represent fractures of the medial or lateral epicondyles. This system of Jupiter's is simpler than the AO system with only 19 possible categories as

Table 12.1 Jupiter, Mehne and Matto's classification of distal humeral fractures (Jupiter, 1994, with permission)

I. Intra-articular fractures
  A. Single-column
    1. Medial
      a. High
      b. Low
    2. Lateral
      a. High
      b. Low
  B. Bicolumnar
    1. T pattern
      a. High
      b. Low
    2. H pattern
    3. Lambda pattern
      a. Medial
      b. Lateral
  C. Capitellar
  D. Trochlear

II. Extra-articular intracapsular fractures
  A. Transcolumnar fractures
    1. High
      a. Extension
      b. Flexion
      c. Abduction
      d. Adduction
    2. Low
      a. Extension
      b. Flexion

III. Extracapsular fractures
  A. Medial epicondyle
  B. Lateral epicondyle

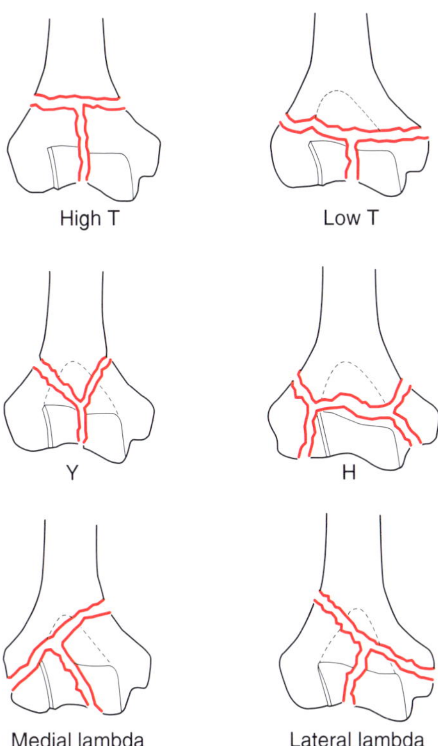

Figure 12.5. Jupiter, Mehne and Matta's classification of distal humeral fractures (Jupiter, 1994).

opposed to the 61 for the AO, if all the qualifications are used, or 27 if it is only taken to the subgroup level (.1, .2, .3). The concept of the two columns and the trochlea as the surgical anatomy basis for the system is inherently straight forward to understand (Jupiter and Mehne, 1992) and appears to allow fractures to be satisfactorily allocated (the capitellum is considered as the cartilage covered end of the lateral column). The system handles difficult fractures such as the divergent single column fracture described by Kuhn and colleagues well (Kuhn, Louis and Loder, 1995). This fracture pattern is not well handled by either the AO 13-B1.2 or 13-B2.2 subgroups. The Jupiter system also addresses the problem raised by Helfet and Schmeling, who used both the Riseborough and Radin and the AO systems, that neither system distinguishes high (above the olecranon fossa) from low (through the olecranon fossa) transcolumnar fractures (Helfet and Schmeling, 1993).

Isolated fractures of capitellum can either be included in an overall distal humerus classification system, such as the AO system (subgroups 13-B3.1(1-4) through 13-B3.3) (Müller et al., 1990), or they can be classified in their own right. There are two types of capitellar fracture: type I: those with a large fragment, the so-called Hahn–Steinthal fracture and type II: those which are only a thin osteochondral shell, the so called Kocher–Lorenz fracture (Alvarez et al., 1975; Hotchkiss and Green, 1991).

Fractures of the distal humerus are common in children and often present diagnostic difficulty due to the sequential appearance of the ossification centres in the capitellum, medial epicondyle, trochlea and lateral epicondyle. Fractures of the child's distal humerus can be grouped in three main classes (Wilkins, 1991). Class A: diaphyseal and metaphyseal supracondylar fractures, class B: physeal injuries and class C: T-condylar fractures. Supracondylar fractures are most commonly (97%) of the extension type but flexion fractures occasionally occur. Gartland classified extension fractures into three types: type I: undisplaced, type II: displaced, but, with an intact posterior cortex, and type III: totally displaced (Gartland, 1959). This system can also be used for the much rarer flexion fractures. Gartland's classification provides a good guide to treatment (Wilkins, 1990). Alternatively, supracondylar fractures can be classified under the system proposed by El-Ahwany (1973). This system does not include T or Y intercondylar fractures. Type 1 fractures are high transverse fractures above the olecranon fossa. Type 2 fractures are oblique fractures starting just above the capitellum and extending in a step-wise manner medially and proximally. Type 3 fractures are much steeper oblique fractures exiting high on the medial supracondylar ridge. Type 4 fractures start medially from the medial epicondyle and extend via the olecranon fossa to exit on the lateral supracondylar ridge. Type 5 fractures are low transverse fractures through the olecranon fossa. El-Ahwany suggests that this classification acts as a guide to treatment.

Physeal fractures in children involve one of five anatomical locations: the lateral condylar physis, the medial condylar physis, the total humeral physis, the medial epicondylar apophysis or the lateral condylar apophysis (Wilkins, 1991). Unicondylar fractures can be satisfactorily classified by the Milch classification as type I or II (Jakob et al., 1975; Wilkins, 1991). The amount of displacement has been classified as stage I: undisplaced, stage II: moderate displacement, and stage III: displaced and rotated (Wilkins, 1991). Total separation of the distal physis, are best classified by the general classification system of physeal injuries by Salter and Harris (see Chapter 7) and are usually type I or type II (DeLee et al., 1980). Apophyseal fractures commonly occur with dislocations and are taken on merit (Josefsson, Johnell and Wendeberg, 1987).

The Coronoid Process of the Ulna

Fractures of the coronoid process of the ulna may be present in 10–15% of elbow dislocations (Regan and Morrey, 1989). Increased elbow instability after reduction of a dislocation is associated with large coronoid fractures (Linscheid and O'Driscoll, 1993).

Only one article (Regan and Morrey, 1989) specifically classifies fractures of the coronoid process, although it is included in the AO 21-segment classification, but not separately (Müller et al., 1990). Regan and Morrey (1989) reviewed 35 fractures of the coronoid and divide the fractures into three groups dependent on the amount of the coronoid in the displaced fragment, based on an assessment of the initial radiographs. Type I: fractures avulsing just the tip of the coronoid, type II: those that involve less than 50% of the coronoid, either as a single fragment or multiple fragments, type III: those involving more than 50% of the coronoid (Figure 12.6). They subdivided these types into A: without elbow dislocation, and B: with elbow dislocation. Elbow dislocation was more common with increasing severity of the coronoid fracture. The more severe coronoid fractures were more commonly seen in multiply traumatized patients. Outcome was directly related to the type of fracture, with residual joint stiffness more common after type III injuries. The authors recommend early motion for type I and II fractures and open reduction and internal fixation for type III fractures, as these fractures predispose the patient to recurrent subluxation or dislocation (Jupiter, 1992).

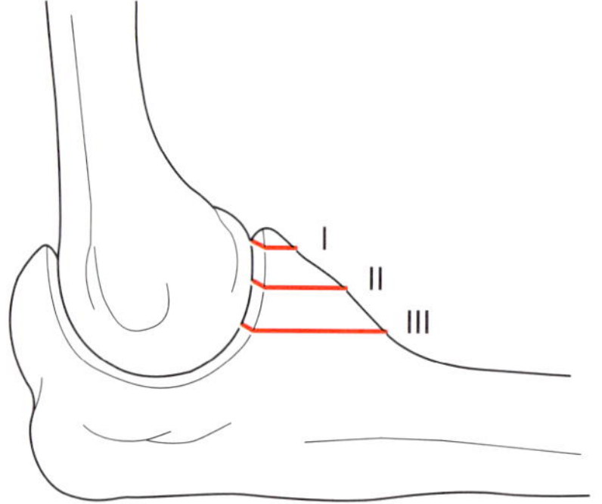

Figure 12.6. Regan and Morrey's classification of coronoid fractures (Regan and Morrey, 1989).

Radial Head and Neck

Fractures of the radial head are relatively common, comprising between 1.7% and 5.4% of all fractures (Morrey, 1993). They involve about 20% of all elbow trauma (Jupiter, 1992). The treatment of these injuries is still contentious, especially the role of radial head excision and, indeed, the importance of the radial head as a load-bearing structure in the elbow. Treatment options include aspiration of the joint and early movement, manipulation, open reduction and fixation and excision with or without prosthetic replacement.

All the classification systems for radial head fractures are based on the radiographic appearances of the fracture with the exception of the Scharplatz and Allgower (Scharplatz and Allgower, 1976) classification of elbow injuries which is based on the mechanism of injury (Jupiter, 1992). This latter classification is not as comprehensive as other earlier classifications and has not been widely adopted.

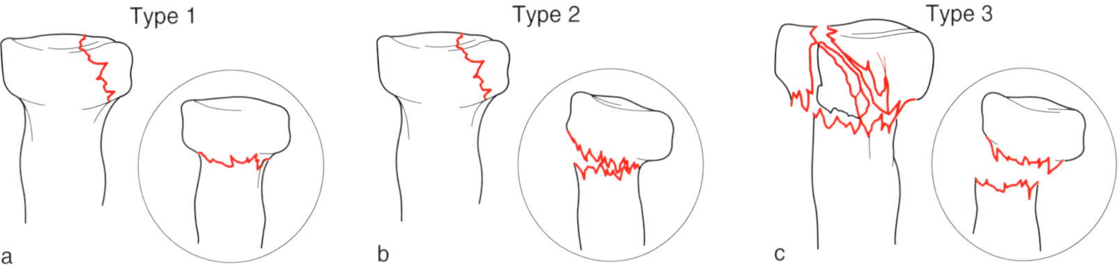

Figure 12.7. The 'modified' Mason classification of radial head fractures (Morrey, 1993).

The most commonly used classification for radial head fractures is that proposed by Mason (1954) based on his personal experience of 100 random cases, followed up for 12 months or more. He divided the fractures into three types. Type 1 (62%) are 'fissure fractures or marginal sector fractures without displacement', type 2 (20%) 'marginal sector fractures with displacement' and type 3 (18%) 'comminuted fractures involving the whole head'. No reference is made to the percentage of the articular surface that constitutes a 'marginal fracture'. Mason found an increasingly poor outcome with increasing severity of type and suggested 'If in doubt—resect'. A Mason type 4 fracture has subsequently been added by Johnston (1962) where there is also a dislocation of the elbow. Mason in his 1954 paper excluded radial neck fractures but these are now included in a 'modified' Mason classification: type 1: undisplaced, type 2: partial displaced, and type 3: totally displaced (Figure 12.7). In this modification type 2 is defined as involving more than 30% of the articular surface and displaced more than 2–3 mm (Morrey, 1993). Other classifications such as that of Bakalim place radial neck fractures into separate types, that is type I: undisplaced, type II: displaced (Bakalim, 1970). This makes it almost identical to the 'modified' Mason classification. Morrey has subdivided the 'modified' Mason classification into two main groups 'simple': 'modified' Mason types 1–3 and 'complex': fractures with associated soft-tissue injury, i.e. 'modified' Mason type 4 with medial or lateral collateral ligament injury and the Essex–Lopresti injury (Distal radio-ulnar joint (DRUJ) and interosseous membrane) (Morrey, 1993). This answers the criticism of the original Mason classification made by Davidson for not including associated soft-tissue injuries (Davidson, Moseley and Tullos, 1993).

McArthur uses a modification of Rockwood and Green's classification that is clearly based on Mason's and Johnson's ideas (McArthur, 1987). Type 1: undisplaced fractures, type 2A: two part marginal fractures with displacement, angulation or depression, type 2B: three part marginal fractures, type 3: comminuted fractures involving the entire radial head or fractures of the junction of the head and neck with extensive comminution and displacement, type 4: fractures of the radial head associated with dislocations of the elbow, fractures of the proximal ulna, fractures of the capitellum or fractures of the distal humerus. This has not been widely used, and although McArthur uses it as a basis for treatment it appears to offer little advantage over the 'modified' Mason classification and complicates the issues by including some fractures of the distal humerus.

Although there is some confusion whether Essex–Lopresti and Monteggia fractures should be included in the type 4 'modified' Mason classification this system has gained general acceptance and is related to treatment and outcome. Early work suggested that types 1 and 2 should be treated non-operatively and type 3 by excision of the radial head (Radin and Riseborough, 1966; Swanson, Jaeger and La Rochelle, 1981; Goldberg et al., 1986). With an increasing trend to open reduction and fixation of these fractures it is now suggested that type 1 fractures should be treated by aspiration of the haematoma and early motion only, type 4 fractures by open reduction and fixation. Type 2 fractures should be treated either as type 1 or as type 4 depending on the fracture's displacement and the attitude and ability of the surgical team. Type 3 fractures may be treated either by excision or fixation (Kanlic and Perry, 1992; Perry, Gibson and Kowalski, 1989). This increasing tendency to operate on radial head fractures has lead to a classification system from Schatzker that addresses the surgical issue of the fragments of the articular surface (Schatzker, 1987c). Type I: simple wedge fractures, type II: impacted fractures, type III: severely comminuted fractures, which can not be reconstructed. This classification has not taken over from the 'modified' Mason.

The one fracture involving the radial head that produces as much difficulty in classification as it does in management is the Essex–Lopresti injury, first described in 1951. The injury is a radial head fracture associated with longitudinal instability of the forearm bones and damage to the interosseus membrane and the ligaments of the distal radio-ulnar joint. This can be included in either the type 4 'modified' Mason or in injuries that involve ligamentous damage (Morrey, 1993). However, it may be better to specifically isolate this fracture as its rarity, behaviour, seriousness and difficulty of management make it very different from other radial head fractures. The Essex–Lopresti injury may, also, be classified with forearm injuries, but this is equally difficult and unsatisfactory.

The AO system classifies radial head fractures as part of the proximal radius and ulna segment 21- (see section on olecranon).

The Olecranon

The olecranon is the proximal articulating part of the ulna, proximal to the coronoid process but excluding the coronoid process itself. It includes the proximal epiphyseal and metaphyseal parts of the ulna. The subcutaneous position of the olecranon makes it vulnerable to trauma and a number of different fracture patterns are found with a variety of different treatment options depending on the pattern. The majority of these olecranon fractures are intra-articular except those of the very tip, although some of those in the centre of the trochlea notch may not involve articular cartilage.

Four main classification systems are available for these injuries, Colton, Schatzker, Mayo and AO (Colton, 1973; Schatzker, 1987b; Müller et al., 1990; Cabanela and Morrey, 1993). These are all remarkably similar. Colton's classification is a simple system that divides the fractures according to the anatomy and mechanism of the fracture. The fracture patterns are group 1: avulsion injuries of the olecranon tip, group 2: oblique fractures, group 3: fracture-dislocations (Monteggia), group 4: unclassified group. Colton describes five stages (a–e) of increasing severity in the oblique fractures of group 2 (Figure 12.8). He clearly shows how the injury pattern is caused and how treatment is based on the group into which the fracture falls. The classification, although having a clear anatomical basis has not been well understood as there are inaccurate interpretations in two standard textbooks (Jupiter, 1992; Cabanela and Morrey, 1993). Colton's classification is favoured by Rockwood and Green's textbook (Hotchkiss and Green, 1991).

Schatzker's classification system classifies olecranon fractures according to the mechanical influences and the requirements of internal fixation (Schatzker,

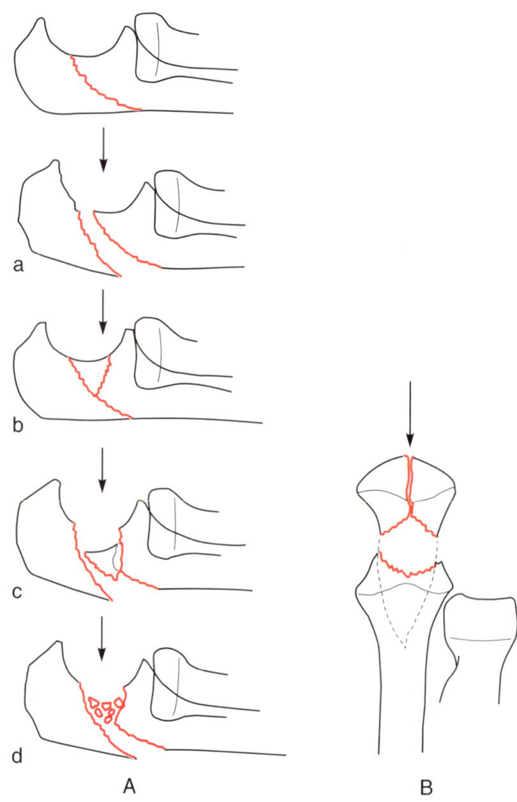

Figure 12.8. Colton's classification of group 2 fractures of the olecranon (Colton, 1973).

1987b). He divides the fractures, initially, into intra-articular and extra-articular; avulsion fractures of the tip of the olecranon. There are three types of intra-articular fractures. Type 1 fractures are transverse, either a simple fracture at the deepest part of the trochlea notch or a complex fracture with comminution or depression of the articular surface. Type 2 fractures are oblique fractures starting at the deepest point of the trochlea notch and running distally. Type 3 fractures are comminuted fractures with associated fractures; of which Schatzker defined three kinds: A: associated with a coronoid process fracture, B: oblique fractures that are very distal, and C: associated fractures or dislocations of the radial head. This latter group has obvious overlap with injuries classified under radial head fractures.

Morrey has proposed a classification system based on three variables; displacement, stability and comminution (Cabanela and Morrey, 1993). The treatment and prognosis are determined by the classification. Type I are undisplaced, stable injuries, type II: displaced, stable injuries and type III: displaced, unstable injuries. Type I comprise 5% of fractures and require only brief immobilization and have a good prognosis. Type II are the commonest making up 85% of fractures of the proximal ulna, they usually require fixation and have a good prognosis. Type III fractures make up the remaining 10% of fractures and require plate fixation, and in some cases hinged external fixation support. Type III have a poor prognosis. This system seems to have much to recommend it as it provides a simple guide to management and prognosis without being excessively complex.

The AO classification attempts to classify all fractures of the proximal radius and ulna under one heading, namely segment 21-. Type A fractures are extra-articular fractures of one or both bones, type B are intra-articular fractures of one bone and type C are intra-articular fractures of both bones (Müller et al., 1990). The classification system for the forearm, thus, differs from that for the other bones and this inconsistency makes it harder to follow. For type A and B fractures group 1 and 2 refer to the ulna and radius respectively and group 3 to both bones. For type C fractures group 1 are two simple fractures, one intra-articular of each bone, group 2 are simple intra-articular of one bone and multifragmentary intra-articular of the other and group 3 are intra-articular multifragmentary of both bones. Beyond this the sub-groupings and qualifications become too complex to describe other than pictorially. The AO system is comprehensive but not intuitive. It requires no fewer than 40 subgroups and qualifications thereof to describe fractures of the proximal radius and ulna.

Diaphyseal Forearm Bones (including Monteggia and Galeazzi fractures)

Forearm diaphyseal fractures are an important cause of unsatisfactory results in fracture management in adults and children, with high rates of loss of forearm rotation particularly with non-operative methods (Kellam and Jupiter, 1992). In adults operative fixation is recommended for most forearm diaphyseal fractures, particularly those of a single bone with a proximal or distal dislocation of the other bone (Tile, 1987), whereas, in children non-operative treatment is adequate in the majority of cases. Forearm diaphyseal fractures made up 18.5% of the AO series of 2700 long bone diaphyseal fractures (Müller et al., 1990).

A universally acceptable classification for forearm diaphyseal fractures has been slow to appear and has been complicated by a number of clearly understood and widely used eponymous fractures, i.e. Monteggia, Galeazzi, Night-stick and Essex–Lopresti. These have had to be incorporated into classification systems with variable levels of success and in most hospitals the eponymous names have remained, at least in verbal communications.

One of the earliest classifications of a diaphyseal forearm fracture was by Giovanni Battista Monteggia from Milan in 1814. The lesion he observed was 'a traumatic lesion distinguished by a fracture of the proximal third of the ulna and an anterior dislocation of the proximal epiphysis of the radius'. This definition was expanded by Malgaigne to include any ulnar fracture with a radial head dislocation (Bado, 1967). These fractures are relatively uncommon, comprising 1–2% of all forearm fractures (Kellam and Jupiter, 1992), and 7% of ulnar fractures (Morgan and Breen, 1994). It has been common parlance to describe the Monteggia fracture by the direction of the dislocation of the radial head, e.g. anterior, posterior or lateral. This manner of description originates from the classification system of Bado (1967) who coined the expression 'Monteggia lesion' to describe the group of fractures of the ulna with radial head dislocation. Bado only describes the direction of the radial head dislocation that is associated with an ulnar fracture at any level. Types I to III describe respectively anterior, posterior and lateral radial head dislocation and, also, the angulation of the ulnar fracture (Kellam and Jupiter, 1992). The discussion of angulation was not in the original Bado paper and is confusing due to the different way that angulation is described on either side of the Atlantic. The angulation of the ulnar fracture is best ignored as it is inherently obvious from the direction of the radial head dislocation. Bado expanded his classification to include a fourth type where there are fractures of both the ulnar and radius in their proximal thirds with an associated anterior radial head dislocation. Type I represent 55–78% of Monteggia fractures, type II 10–15%, type III 7–20% and type IV 5% (Morgan and Breen, 1994). The commonest point of fracture (60–70%) for the ulnar is at the junction of the proximal and middle thirds (Bado, 1967), although, to constitute a 'Monteggia lesion' it may be anywhere along the bone. Some workers have divided the classification further by including the position of the ulnar fracture in the system (Olney and Menelaus, 1989), but this is relatively unhelpful in such a small group of patients. The Bado classification worked well for children's fractures, as well as adult (Letts, Locht and Wiens, 1985; Wiley and Galey, 1985). In a study of 102 fractures from Melbourne Children's Hospital workers found a similar incidence of the different types to Bado; with types I and III being common, type II was rare and type IV was very rare (Olney and Menelaus, 1989). This paper found a higher number of Bado 'equivalent' lesion than reported else where. These are fractures with an atypical fracture pattern that are 'forced' into the Bado classification to avoid further subdivision. 'They possess similar characteristics, especially in their mechanism of action and in their treatment' (Bado, 1967). The most common Bado 'equivalent' is a fracture of the ulnar shaft with a radial neck fracture or separation of the proximal radial epiphysis—type I 'equivalent'. These had a high requirement for operative fixation unlike most of the other paediatric Monteggia fractures. Virtually all Bado 'equivalent' lesions are of the type I variety (Bado, 1967).

The AO classification of forearm fractures has been less successful in classifying the Monteggia fracture, by including all its variants under the 22-B1.3 subgroup. This classification is also not able to cope adequately with the Galeazzi fracture and includes all variants under the 22-B2.3 subgroup.

The Galeazzi fracture was described in 1934 by R. Galeazzi as a fracture of the middle and distal thirds of the radius associated with instability of the distal radio-ulnar joint (DRUJ). It is, like the Monteggia fracture, a rare injury of the forearm comprising 3–6% of all forearm fractures (Kellam and Jupiter, 1992). It has also been called the reverse Monteggia fracture, Piedmont or Darrah–Hughesston–Milch fracture. The Galeazzi fracture is mentioned by Bado, in his paper, as a type I 'equivalent' lesion (Bado, 1967). Galeazzi fractures can be divided into those that have a stable DRUJ after fixation of the radius fracture and those where the DRUJ remains unstable and further surgery is required to stabilize the DRUJ and repair the triangular fibrocartilage (TFC) if required. Separate categories can be defined where there is a fracture of the ulnar styloid that the TFC remains attached to, and fractures where the DRUJ is irreducible due to soft tissue interposition usually of extensor carpi ulnaris or extensor digiti minimi tendons (Morgan and Breen, 1994). This classification seems to have much to recommend it as it is simple and aids with the management and understanding of the injury. It agrees with the treatment algorithm presented by Kellam and Jupiter (Kellam and Jupiter, 1992).

Diaphyseal forearm fractures, without proximal or distal dislocations, are classified separately from Monteggia and Galeazzi fractures. In 1992, Kellam and Jupiter stated that, 'Currently, there is no universally accepted classification of diaphyseal fractures of the forearm'. This situation has changed little, although two systems are currently available that deal with these fractures in a comprehensive way, one from the AO group (Müller et al., 1990) and the other from the Orthopaedic Trauma Association (OTA) in the USA (Kellam and Jupiter, 1992).

The AO system tackles the forearm bones together as bone number 2 and the diaphyseal region as segment number 2, hence, all diaphyseal forearm fractures are 22 fractures. The type A, B and C fractures are the same as for other long bones describing simple, wedge and complex fracture patterns respectively (Figure 12.9). The incidence of these types in the AO series of 2700 diaphyseal fractures was 65%, 29% and 6%. For all three types the groups 1, 2 and 3 describe fractures of the ulna, radius and both bones respectively, rather than the more detailed pattern of the fracture as for the other long bones, e.g. humerus. In the AO series type A and type C fractures were fairly equally distributed between the three groups, but for the wedge fractures, 22-B, there were twice

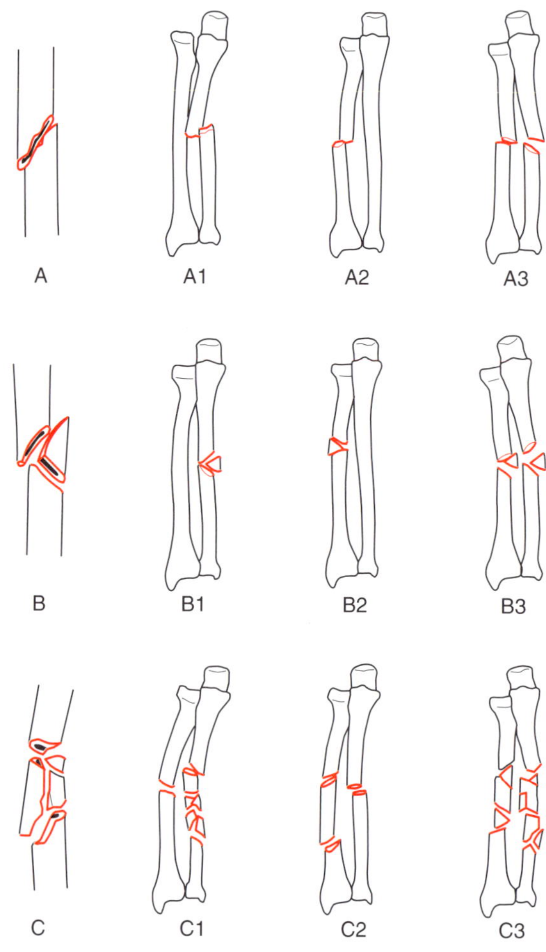

Figure 12.9. The AO classification of diaphyseal forearm fractures (Müller et al., 1990).

as many that affected both bones, 22-B3 (Müller et al., 1990). Describing the fracture pattern as transverse, oblique or with a proximal or distal dislocation is defined in the .1, .2 and .3 subgroups. Notwithstanding that the AO system is comprehensive it does not work as well for the twin bones of the forearm as it does for the isolated long bones, humerus and femur or the non-rotational twin bones the lower leg. As mentioned earlier, it does not solve the problems of the Monteggia and Galeazzi fractures. For these reasons it is not widely used.

The OTA system is somewhat simpler than the AO system as it deals with each forearm bone separately and in its own right. Although this has advantages it, like the AO system, can not cope with combined injuries such as the Monteggia and Galeazzi fracture. The OTA system uses seven groups from I to VII to cover transverse, oblique, spiral, wedge, comminuted, segmental and fractures with bone loss (Figure 12.10). Groups I–III correspond with AO type A fractures, group IV to the AO type B fractures and groups V–VII to the AO type C fractures. There are no specific reports of an association between severity and outcome, although, OTA groups IV to VII and AO type C fractures are regarded as complex and may benefit more from internal fixation and associated methods than their less complex counterparts (Morgan and Breen, 1994).

Children's forearm shaft fractures are usually classified by the type of the fracture; greenstick (incomplete), buckle or torus (compression) or complete, and the anatomical level; proximal, middle or distal third (King, 1991). This provides a good guide to their treatment and is in wide usage.

Elbow Dislocations and Instability

Acute elbow dislocation has an annual incidence of 6 per 100 000 population (Josefsson and Nilsson, 1986). The main incidence is from 5 to 30 years, with the mode in the late teens for men and women (Jupiter, 1992; O'Driscoll, 1994). Dislocation of the elbow joint is the commonest joint dislocation in the under 10 age group (Linscheid and O'Driscoll, 1993), and is the third commonest after glenohumeral and finger dislocations in adults (Jupiter, 1992). Elbow dislocations represent 20% of all dislocations (Jupiter, 1992)

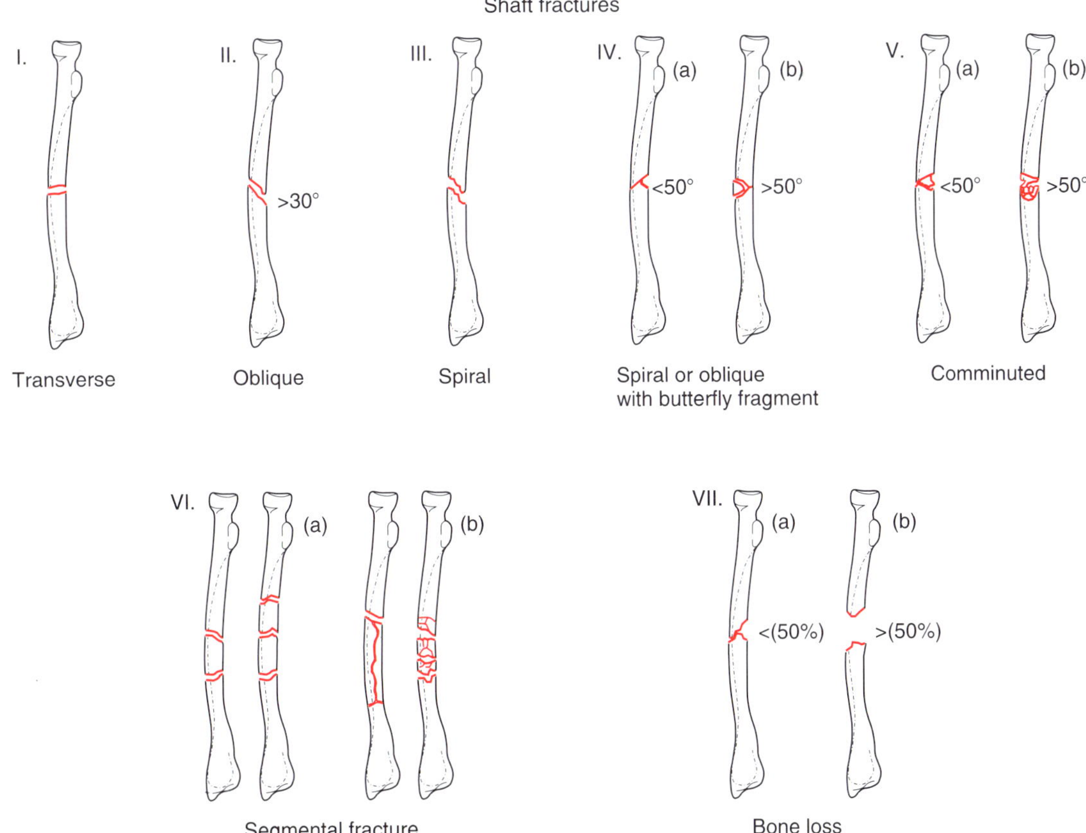

Figure 12.10. The OTA classification of radial forearm fractures (Kellam and Jupiter, 1992).

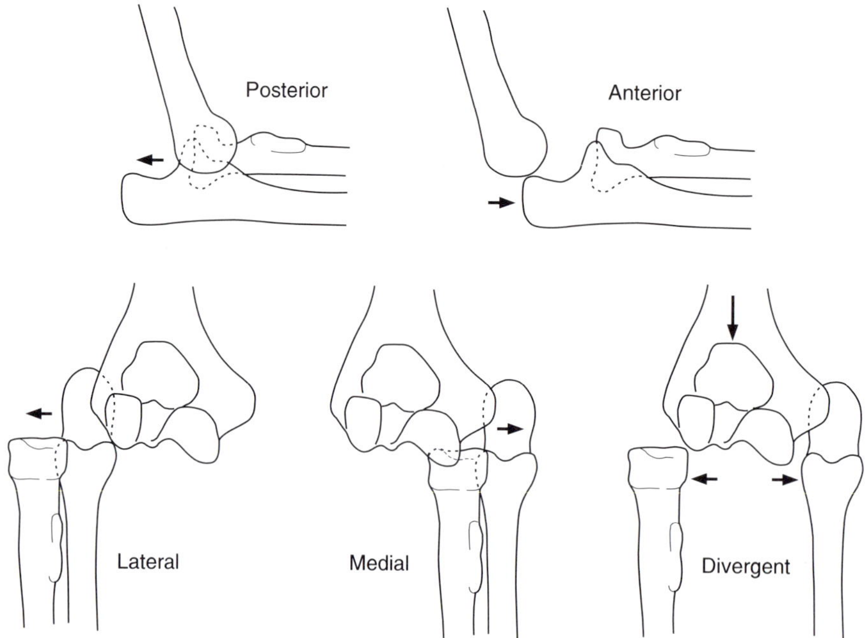

12.11. Classification of elbow dislocations (Jupiter, 1992).

and 60% of them occur in the non-dominant limb (Linscheid and O'Driscoll, 1993).

The overwhelming majority of elbow dislocations are of both bones. Those of the ulna only are so rare that no classification system includes it. Elbow dislocations are divided into simple: ligamentous injury only, and complex: those with associated bony fracture (Jupiter, 1992). The outcome from a simple dislocation is good (Josefsson, Johnell and Gentz, 1984). Those dislocations in which there is an associated fracture have a worse outcome (Linscheid and O'Driscoll, 1993).

Simple elbow dislocations are classified according to the direction of dislocation. This was originally based on the arthrographic findings of Johansson in 1962 who divided elbow dislocations into posterior where the dye leaked out anteriorly and lateral where the dye leaked out medially (Johansson, 1962). More recent work has shown that both lateral and medial collateral ligaments are torn in all elbow dislocations but lateral instability does not always result (Josefsson, Johnell and Wendeberg, 1987). Posterior dislocations are the commonest (80%) (Josefsson and Nilsson, 1986). The other three possibilities: anterior, medial, lateral and divergent are all rare, especially the last (Figure 12.11) (DeLee, 1981; Jupiter and Mehne, 1992; Linscheid and O'Driscoll, 1993). This directional classification is further defined in complex dislocations by a description of the associated fracture. The common associated fractures are of the radial head or neck, the coronoid, both of which have their own classification systems, and in children, avulsion injuries of the medial or lateral epicondyles that may be minimal or severe and locked into the joint (Linscheid and O'Driscoll, 1993).

Group Discussion

The group considered classification systems for injuries of the elbow. There are four classifications for fractures of the olecranon including those of Colton and the AO group. We were concerned that the AO system does not work well in general practice and felt that the best system was that proposed by Schatzker (1987b). None of these systems has any reliability testing.

We considered briefly classification of diaphyseal fractures of the forearm bones particularly the Monteggia system. There are various modifications of the Monteggia classification. Indeed the original Monteggia description was for a very limited problem of a fracture and anterior dislocation of the radius. The Bado modification (1967) is the most commonly used and was felt to be the best buy. For the radial head and neck, the classification system described by Mason (1954) was felt to be the most appropriate and the most usable. There are no good classification systems for Essex–Lopresti fractures. For fractures of

Table 12.2 Details of classification systems for injuries around the elbow

Site	*Author*	*Good for defining treatment*	*Good for defining outcome*	*Reliability tested?*	*Notes*
Humeral shaft	Ward et al., 1992	Limited	No	No	Applies to proximal shaft only
	AO (Müller et al., 1990)	No	No	No	Based on large series and consistent with other long bones
Distal humerus	Milch, 1964	Yes	Yes	No	More appropriate for children Unicondylar fractures only
	Cassebaum, 1969	No	No	No	Historical interest as introduced T & Y fractures
	Riseborough and Radin, 1969	Yes (but superseded)	Yes (but superseded)	No	Important paper
	Horne and Pennal, 1979	Yes	Yes	No	Meeting abstract only
	Shetty, 1982	No	No	No	Tried to account for comminution in Riseborough and Radin
	Schatzker, 1987	Yes	Yes	No	
	OTA (Henley, 1987)	No	No	No	
	AO (Müller et al., 1990)	Yes	Yes	No	Represents modern treatment methods
	Jupiter, 1994	Yes	Yes	No	Possible improvement on AO system
Coronoid process	Regan and Morrey, 1989	No	Yes	No	Helpful with dislocations
Radial head and neck	Mason, 1954	Yes	Yes	No	The origin and still the best, use 'modified' version with Johnston
	Schatzker, 1987	Yes	Yes	No	Helps with treatment but has not overtaken Mason
Olecranon	Colton, 1973	Yes	Yes	No	Misrepresented in some standard text, refer to original article
	Schatzker, 1987	Yes	Yes	No	Helps with internal fixation
	AO (Müller et al., 1990)	No	No	No	Difficult to apply
	Cabanela and Morrey, 1993	Yes	Yes	No	Takes account of stability
Diaphyseal forearm	Bado, 1967	Yes	Yes	Yes	Adults and children
	AO (Müller et al., 1990)	No	No	No	Does not work well considering both bones together
	OTA (Kellam and Jupiter, 1992)	No	No	No	Descriptive only
Elbow dislocations	Johansson, 1962	Yes	No	No	Original description of pathology
	Linscheid and O'Driscoll, 1993	No	No	No	Adds in associated fractures
Children supracondylar	Gartland, 1959	Yes	Yes	Yes	Mainly for extension fractures
	Milch, 1964	Yes	Yes	No	Better for children unicondylar than adults
	El-Ahwany, 1973	Yes	Yes	Yes	Best overall
Physeal fractures of distal humerus	Salter and Harris, 1963	Yes	Yes	Yes	Type V probably not found

the coronoid process, the description by Regan and Morrey (1989) appears to be the most useful as it relates to instability of the elbow which is the most practical issue. Of the fractures of the humeral shaft, the AO classification appears to be the most useful although extremely complicated, we would recommend this for clinical practice. Currently plain radiographs are used to describe the fractures and it may be that alternative forms of imaging may be used in the future. Of the descriptions of fractures of the condyles the original paper by Milch from 1964 is very useful in that it has a basic principle and it describes whether or not a fracture is likely to be stable or unstable. However, we felt that the classification systems of the AO group (Müller et al., 1990) and of Jupiter (1992) were the best buys for these fractures. The overall findings of the group are summarized in Table 12.2.

References

Alvarez, E., Patel, M.R., Nimberg, G. and Pearlman, H.S. (1975) Fracture of the capitulum humeri. *J. Bone Joint Surg.*, **57A**, 1093–1096.

Bado, J.L. (1967) The Monteggia lesion. *Clin. Orthop.*, **50**, 71–86.

Bakalim, G. (1970) Fractures of the radial head and their treatment. *Acta Orthop. Scand.,* **41**, 320–331.

Cabanela, M.E. and Morrey, B.F. (1993) Fractures of the proximal ulna and olecranon. In *The Elbow and its Disorders* (B.F. Morrey, ed.), 2nd edn. Philadelphia: Saunders, pp. 405–428.

Cassebaum, W.H. (1969) Open reduction of T & Y fractures of the lower end of the humerus. *J. Trauma*, **9**, 915–925.

Colton, C.L. (1973) Fractures of the olecranon in adults: classification and management. *Injury*, **5**, 121–129.

Davidson, P.A., Moseley, J.B. and Tullos, H.S. (1993) Radial head fracture. A potentially complex injury. *Clin. Orthop.*, **297**, 224–230.

DeLee, J.C. (1981) Transverse divergent dislocation of the elbow in a child. Case report. *J. Bone Joint Surg.*, **63A**, 322–323.

DeLee, J.C., Wilkins, K.E., Rogers, L.F. and Rockwood, C.E. (1980) Fracture-separation of the distal humeral epiphysis. *J. Bone Joint Surg.*, **62A**, 46–51.

El-Ahwany, M.D. (1973) Supracondylar fractures of the humerus in children with a note on the surgical correction of late cubitus varus. *Injury*, **6**, 45–56.

Essex-Lopresti, P. (1951) Fractures of the radial head with distal radio-ulnar dislocation. Report of two cases. *J. Bone Joint Surg.*, **33B**, 244–247.

Galeazzi, R. (1934) Uber ein besonderes Sydrom bei Verltzunger im Bereich der unter armknochen. *Arch. Orthop. Unfallchir.*, **35**, 557–562.

Gartland, J.J. (1959) Management of supracondylar fractures of the humerus in children. *Surg. Gynecol. Obstet.*, **109**, 145–154.

Goldberg, I., Peylan, J., Yosipovitch, Z. and Tiqva, P. (1986) Late results of excision of the radial head for an isolated closed fracture. *J. Bone Joint Surg.*, **68A**, 675–679.

Helfet, D.L. and Schmeling, G.J. (1993) Bicondylar intra-articular fractures of the distal humerus in adults. *Clin. Orthop.*, **292**, 26–36.

Henley, M.B. (1987) Intra-articular distal humeral fractures in adults. *Orthop. Clin. North Am.*, **18**, 11–23.

Henley, M.B., Bone, L.B. and Parker, B. (1987) Operative management of intra-articular fractures of the distal humerus. *J. Orthop. Trauma*, **1**, 24–35.

Holdsworth, B.J. and Mossad, M.M. (1990) Fractures of the adult distal humerus. Elbow function after internal fixation. *J. Bone Joint Surg.*, **72B**, 362–365.

Horne, G. (1979) Supracondylar fractures of the humerus in adults. *J. Bone Joint Surg.*, **61B**, 246.

Hotchkiss, R.H. and Green, D.P. (1991) Fractures and dislocations of the elbow. In *Rockwood and Green's Fractures in Adults* (C.A. Rockwood, D.P. Green and R.W. Bucholz, eds), Vol. 2, 3rd edn. Philadelphia: Lippincott, pp. 736–841.

Jakob, R., Fowles, J.V., Rang, M. and Kassab, M.T. (1975) Observations concerning fractures of the lateral humeral condyle in children. *J. Bone Joint Surg.*, **57B**, 430–436.

Johansson, O. (1962) Capsular and ligamentous injuries of the elbow joint. A clinical and arthrographic study. *Acta Chir. Scand.*, suppl. **287**, 1–159.

Johnston, G.W. (1962) A follow-up of one hundred cases of fracture of the head of the radius with a review of the literature. *Ulster Med. J.*, **31**, 51–56.

Josefsson, P.O., Gentz, C., Johnell, O. and Wendeberg, B. (1987) Surgical versus non-surgical treatment of ligamentous injuries following dislocation of the elbow joint. A prospective randomised study. *J. Bone Joint Surg.*, **69A**, 605–608.

Josefsson, P.O., Johnell, O. and Gentz, C. (1984) Long-term sequelae of simple dislocation of the elbow. *J. Bone Joint Surg.*, **66A**, 927–930.

Josefsson, P.O., Johnell, O. and Wendeberg, B. (1987) Ligamentous injuries in dislocations of the elbow joint. *Clin. Orthop.*, **221**, 221–225.

Josefsson, P.O. and Nilsson, B.E. (1986) Incidence of elbow dislocation. *Acta Orthop. Scand.*, **57**, 537–538.

Jupiter, J.B. (1992) Trauma to the adult elbow and fractures of the distal humerus. Part I. In *Skeletal Trauma. Fractures, Dislocations and Ligamentous Injuries* (B.D. Browner, J.B. Jupiter, A.M. Levine and P.G. Trafton, eds), Vol. 2. Philadelphia: Saunders, pp. 1125–1145.

Jupiter, J.B. (1994) Complex fractures of the distal part of the humerus and associated complications. *J. Bone Joint Surg.*, **76A**, 1252–1264.

Jupiter, J.B., Barnes, K.A., Goodman, L.J. and Saldana, A.E. (1993) Multiplane fractures of the distal humerus. *J. Orthop. Trauma*, **7**, 216–202.

Jupiter, J.B. and Mehne, D.K. (1992) Fractures of the distal humerus. *Orthopaedics*, **15**, 825–833.

Jupiter, J.B. and Morrey, B.F. (1993) Fractures of the distal humerus in the adult. In *The Elbow and its Disorders* (B.F. Morrey, ed.), 2nd edn. Philadelphia: Saunders, pp. 328–366.

Jupiter, J.B., Neff, U., Holzach, P. and Allgower, M. (1985) Intercondylar fractures of the humerus. An operative approach. *J. Bone Joint Surg.*, **67A**, 226–239.

Kanlic, E. and Perry, C.R. (1992) Indications and technique of open reduction and internal fixation of radial head fractures. *Orthopaedics*, **15**, 837–842.

Kellam, J.F. and Jupiter, J.B. (1992) Diaphyseal fractures of the forearm. In *Skeletal Trauma. Fractures, Dislocations and Ligamentous Injuries* (B.D. Browner, J.B. Jupiter, A.M. Levine and P.G. Trafton, eds), Vol. 2. Philadelphia: Saunders, pp. 1095–1124.

King, R.E. (1991) Fractures of the shafts of the radius and ulna. In *Rockwood and Green's Fractures in Children* (C.A. Rockwood, K.E. Wilkins and R.E. King, eds), Vol. 3, 3rd edn. Philadelphia: Lippincott, pp. 301–362.

Kuhn, J.E., Louis, D.S. and Loder, R.T. (1995) Divergent single-column fractures of the distal part of the humerus. *J. Bone Joint Surg.*, **77A**, 538–542.

Letsch, R., Schmit-Neuerburg, K.P., Sturmer, K.M. and Walz, M. (1989) Intraarticular fractures of the distal humerus. Surgical treatment and results. *Clin. Orthop.*, **241**, 238–244.

Letts, M., Locht, R. and Wiens, J. (1985) Monteggia fracture-dislocations in children. *J. Bone Joint Surg.*, **67B**, 724–727.

Linscheid, R.L. and O'Driscoll, S.W. (1993) Elbow dislocations. In *The Elbow and its Disorders* (B.F. Morrey, ed.), 2nd edn. Philadelphia: Saunders, pp. 441–452.

Mason, M.L. (1954) Some observations on fractures of the head of the radius with a review of one hundred cases. *Br. J. Surg.*, **42**, 123–132.

McArthur, R.A. (1987) Herbert screw fixation of fracture of the head of the radius. *Clin. Orthop.*, **224**, 79–87.

Milch, H. (1964) Fractures and fracture dislocations of the humeral condyles. *J. Trauma*, **4**, 592–607.

Morgan, W.J. and Breen, T.F. (1994) Complex fractures of the forearm. *Hand Clinics*, **10**, 375–390.

Morrey, B.F. (1993) Radial head fracture. In *The Elbow and its Disorders* (B.F. Morrey, ed.), 2nd edn. Philadelphia: Saunders, pp. 383–404.

Müller, M.E., Nazarian, S., Koch, P. and Schatzker, J. (1990) *The Comprehensive Classification of Fractures of Long Bones*. Berlin: Springer-Verlag.

O'Driscoll, S.W. (1994) Elbow instability. *Hand Clinics*, **10**, 405–415.

Olney, B.W. and Menelaus, M.B. (1989) Monteggia and equivalent lesions in childhood. *J. Pediatr. Orthop.*, **9**, 219–223.

Perry, C.R., Gibson, C.T. and Kowalski, M.F. (1989) Transcondylar fractures of the distal humerus. *J. Orthop. Trauma*, **3**, 98–106.

Radin, E.L. and Riseborough, E.J. (1966) Fractures of the radial head. A review of eighty-eight cases and analysis of the indications for excision of the radial head and non-operative treatment. *J. Bone Joint Surg.*, **48A**, 1055–1064.

Regan, W. and Morrey, B. (1989) Fractures of the coronoid process of the ulna. *J. Bone Joint Surg.*, **71A**, 1348–1354.

Riseborough, E.J. and Radin, E.L. (1969) Intercondylar T fractures of the humerus in the adult. A comparison of operative and non-operative treatment in twenty-nine cases. *J. Bone Joint Surg.*, **51A**, 130–141.

Scharplatz, D. and Allgower, M. (1976) Fracture-dislocations of the elbow. *Injury*, **7**, 143–159.

Schatzker, J. (1987a) Fractures of the distal end of the humerus. In *The Rationale of Operative Fracture Care* (J. Schatzker and M. Tile, eds). Berlin: Springer-Verlag, pp. 70–87.

Schatzker, J. (1987b) Fractures of the olecranon. In *The Rationale of Operative Fracture Care* (J. Schatzker and M. Tile, eds). Berlin: Springer-Verlag, pp. 89–95.

Schatzker, J. (1987c) Fractures of the radial head. In *The Rationale of Operative Fracture Care* (J. Schatzker and M. Tile, eds). Berlin: Springer-Verlag, pp. 97–101.

Shetty, S. (1982) Surgical treatment of T and Y fractures of the distal humerus. *Injury*, **14**, 345–348.

Swanson, A.B., Jaeger, S.H. and La Rochelle, D. (1981) Comminuted fractures of the radial head. The role of silicone-implant replacement arthroplasty. *J. Bone Joint Surg.*, **63A**, 1039–1049.

Tile, M. (1987) Fractures of the radius and ulna. In *The Rationale of Operative Fracture Care* (J. Schatzker and M. Tile, eds). Berlin: Springer-Verlag, pp. 103–129.

Ward, E.F., Savoie, F.H. and Hughes, J.H. (1992) Fractures of the diaphyseal humerus. In *Skeletal Trauma. Fractures, Dislocations and Ligamentous Injuries* (B.D. Browner, J.B. Jupiter, A.M. Levine and P.G. Trafton, eds), Vol. 2. Philadelphia: Saunders, pp. 1177–1200.

Wiley, J.J. and Galey, J.P. (1985) Monteggia injuries in children. *J. Bone Joint Surg.*, **67B**, 728–731.

Wilkins, K.E. (1990) The operative management of supracondylar fractures. *Orthop. Clin. North Am.*, **21**, 269–289.

Wilkins, K.E. (1991) Fractures and dislocations of the elbow. In *Rockwood and Green's Fractures in Children* (C.A. Rockwood, K.E. Wilkins and R.E. King, eds), Vol. 3, 3rd edn. Philadelphia: Lippincott, pp. 363–575.

13

The wrist

J. Chitnavis

Introduction

There are many classification systems to describe wrist injuries.

One reason for the abundance of descriptions is the complex anatomy of the wrist. Involved in the joint are the distal radius and ulna and the proximal carpal row of bones comprising scaphoid, lunate and triquetrum. The proximal row articulates closely with the capitate and hamate bones and, to a lesser degree, with the trapezoid and trapezium. All of the bones forming the wrist are linked by synovial joints and ligaments.

Another reason for the multiplicity of classifications of wrist injury is historical. The clinical description of several types of distal radial fracture dates from the late eighteenth century. Eponymous definitions of radial fractures including Colles', Smith's and Barton's types have remained in use since they were first described.

A third reason for the existence of a large number of classification systems for wrist injury is a result of recent interest in the operative management of complex wrist trauma. Fractures of the distal radius and scaphoid are the commonest and second commonest respectively of all fractures. A significant number of wrist injuries involve intra-articular multifragmentary fractures which, if managed conservatively, lead to considerable disability. With greater interest in the internal fixation of such injuries has come an increased awareness of the patterns of injuries involved. As a result, several new classifications of distal radial fractures, in particular, have been developed (Jupiter, 1991; Saffar, 1995).

The ideal classification of trauma to any joint should first allow a comprehensive description of common injuries to the bones, joints and soft tissues within that joint. The development of x-rays increased appreciation of the patterns of bony injuries to the wrist. Recently, newer techniques including MRI and arthroscopy have been applied to wrist injuries (Cobby and Watt, 1991; Johnston, Friedman and Kriegler, 1992; Metz and Gilula, 1993; Savoie, 1995; Whipple, 1995). Ideally, trauma classifications should anticipate the increased interest in associated soft-tissue and bony injuries that will result from these developments.

Secondly, in order to facilitate accurate communication of injury patterns, classifications must also be clear and concise. In order to compare outcomes from any specific injury in different studies, a system of coding injuries and then storing data on computer should be possible.

Thirdly, any classification should grade injury by severity. By considering the grade given, it should be possible to predict outcome and recommend appropriate intervention.

Unfortunately, none of the classification systems for wrist trauma achieves all of these goals (Swiontkowski, 1995; Pennig and Gausepohl, 1996). Several systems concisely and comprehensively describe injuries to only *one* bone in the wrist. However, close anatomical and functional relationships with adjacent bones and soft-tissues often result in associated injuries which cannot be encompassed by classifications limited to a single bone. Furthermore, there is little statistical evidence to support a direct correlation between severity by grade and eventual outcome for *any* classification. The current lack of validation of systems used to describe injuries to the wrist makes it difficult to prescribe optimum intervention based on classification alone.

This chapter will describe classification systems developed for:

1. Distal radial fractures.
2. Distal ulna fractures.
3. Radiocarpal dislocations.
4. Distal radio-ulnar dislocations.
5. Carpal fractures.
6. Carpal dislocations and fracture-dislocations.
7. Carpal instability.

Distal Radial Fractures

The many classifications of fractures involving the radial extremity could be subdivided as below:
1. Eponymous descriptions.
2. Classifications which consider extra- *and* intra-articular distal radial fractures.
3. Classifications for intra-articular distal radial fractures.

Eponymous Descriptions

The description of any fracture by the use of a single surname is not strictly a classification. The use of eponyms to describe injuries is imprecise and has created confusion in the past with regard to treatment and prognosis (Thompson and Grant, 1977; Jupiter, 1991; Szabo, 1993). Nevertheless, in the case of the distal radius, the eponyms used are generally better known than any classification and are therefore included here.

a. Colles' fracture

Abraham Colles wrote his classic description 'on the fracture of the carpal extremity of the radius' in 1814. Although Pouteau had written about distal radial fractures in 1783 (Mathoulin, Letrosne and Saffar, 1995), Colles described them more clearly, distinguished them from radiocarpal dislocations and suggested a means of reducing and stabilizing this fracture. He recognized a fracture 'about an inch and a half above' the end of the radius with 'the base of the metacarpus thrown backwards'. Colles was optimistic about prognosis and wrote that 'the limb will at some remote period again enjoy perfect freedom in all its motions, and be completely exempt from pain: the deformity, however, will remain undiminished through life'.

Despite radiographic evidence confirming that fractures fitting Colles' classic clinical description are often variably comminuted, displaced and intra-articular (Gartland and Werley, 1951; Bacorn and Kurtzke, 1953; Golden, 1963; Pool, 1973; Cooney, Dobyns and Linscheid, 1980a; Melone, 1984; Johnston, Friedman and Kriegler, 1992), they continue to be regarded and studied as a discrete entity (Smail, 1965; Cooney, Dobyns and Linscheid, 1980a; Jenkins et al., 1987; Villar et al., 1987).

With a few exceptions (Bacorn and Kurtzke, 1953; Mayer, 1940), most early studies predicted good outcomes for the majority with Colles' fracture. Unfortunately, studies suggesting a good prognosis were small, with short follow-up and no statistical support. (Smail, 1965; Cassebaum, 1950; Gartland and Werley, 1951).

More recently, critical analyses of factors affecting the prognosis of Colles' fractures have examined the effect of displacement and intra-articular extension into the radiocarpal and radio-ulnar joints. In a study of 235 Colles' fractures treated conservatively, Stewart, Innes and Burke (1985) were surprised to find no statistical difference in wrist function between those with comminuted intra-articular fractures and those with extra-articular fractures, six months following injury. At 3 year's follow-up of 90 consecutive Colles' fractures treated by closed reduction and cast immobilization, 30% of patients had wrist pain, 30% had residual deformity and 10% had mild functional disability (Millet and Rushton, 1995). Radial shortening following injury significantly reduced flexion, pronation and grip strength. Dorsal angulation at 1 week following injury significantly diminished the range of pronation and supination possible at 3 years post-fracture. However, neither dorsal nor radial angulation at 1 week following injury significantly reduced pronation or grip strength at 3 years follow-up (Villar et al. 1987). At an average of 5 years following injury, McQueen and Caspers (1988) reviewed 30 patients with Colles' fractures treated by closed reduction and plaster immobilization. None of the patients had radiocarpal extension of their fracture. The authors noted a significantly worse functional outcome in cases with malunited Colles' fractures compared to controls in whom fractures had united with minimal displacement.

b. Smith's fractures

In 1847, Robert Smith (see Palmer, 1993) described distal radial fractures just proximal to the radiocarpal joint 'in which the lower fragment and the carpus were displaced forwards in relation to the forearm'. One of Colles' successors, Smith, recognized the injury he had described as 'a reversed Colles' fracture'. In 1957, Thomas differentiated Smith's fractures into three types. Unfortunately, Types 1 and 3 are not very easily distinguished by his description. Moreoever, Thomas' Type 2 is, in fact, a volarly-displaced radiocarpal fracture-dislocation, which had already been described by Barton.

c. Barton's fracture

In 1838, John Barton (see Palmer, 1993) wrote about displacement of the wrist 'consequent to a fracture through the articular surface of the carpal extremity of the wrist'. He stated that the distal radial fragment was 'usually' from the dorsal side of the radius but that 'rarely' there was a fracture involving the palmar side of the radius. Barton appreciated that the fractures he described were unstable and required immobilization in traction. Confusingly, American literature has considered only the *dorsally*-displaced variant as a 'Barton's fracture' whereas British and European writers have regarded the *volarly*-displaced fracture-dislocation as 'Barton's fracture' (Thomas' Type 2). This uncertainty regarding a basic difference in fracture configuration highlights the inadequacies of eponymous terminology. Thompson and Grant (1977) suggested the use of 'anterior' and 'posterior' fracture-dislocations of the radiocarpal joint to clarify the picture.

A study comparing the outcomes of 12 anterior and eight posterior fracture-dislocations of the wrist treated similarly, found no significant differences in the range of wrist flexion, extension, pain or degenerative change between the two groups (Pattee and Thompson, 1988).

Classifications which Consider Extra- and Intra-Articular Distal Radial Fractures

a. Gartland and Werley (1951)

Type 1 Simple Colles' fracture with no involvement of the radial articular surface.

Type 2 Comminuted Colles' fracture with fractures of the radial articular surface in which the fragments are undisplaced.

Type 3 Comminuted Colles' fracture with fractures of the radial articular surface in which the fragments are displaced.

This arrangement of distal radial fractures was intended for Colles' fractures only and cannot be used to evaluate undisplaced distal radial fractures nor fractures with volar displacement of the distal fragment. The authors used their scheme to classify 60 Colles' fractures treated by closed reduction (Gartland and Werley, 1951). They developed a scoring system to assess the results. Pain, residual deformity, wrist function and complications including arthritis, nerve injury and finger stiffness were evaluated. In general, outcomes appear to correlate quite well with the classification, although no statistical validation is provided. Their results acknowledge the tendency of increased dorsal tilt of the distal fragment and radial shortening to be associated with poor prognosis. However, their classification does not address these issues. Nevertheless, an independent assessment of the Gartland and Werley classification has suggested that it is predictive of anatomical outcome (Solgaard, 1984) and it is acknowledged as having formed the basis of subsequent attempts to classify distal radial fractures (Toh and Jupiter, 1994). A study of 30 patients with fractures of the distal radius treated by closed reduction and percutaneous 'K' wire fixation showed that the Gartland and Werley scheme was predictive of flexion and extension at union but not of rotation nor of the degree of dorsiflexion (Clancey, 1984).

b. Lidstrom (1959)

Type 1 Fissure fractures and fractures with no appreciable displacement.

Type 2 Fractures with posterior displacement:
- A. Fractures with dorsal angulation, not involving the joint surface.
- B. Fractures with dorsal angulation, involving the joint surface but without comminution of the articular surface.
- C. Fractures with complete displacement, not involving the joint surface.
- D. Fractures with complete displacement, involving the joint but without comminution of the articular surface.
- E. Fractures with complete displacement and comminution of the joint surface.

Type 3 Fractures with volar displacement.

Lidstrom used this classification in a study of 515 adults with displaced (Types 2 and 3 only) distal radial fractures treated by closed reduction and plaster immobilization. Outcomes were assessed at an average follow-up of nearly four years and based on subjective and objective criteria including pain, deformity and function. Lidstrom's classification correlated quite well with outcome in his series of patients and also in a more recent independent survey (Solgaard, 1984). However, in a survey of 115 distal radial fractures treated by closed reduction and cast immobilization, Porter and Stockley (1985) did not detect a significant correlation between Lidstrom fracture type and wrist function based on factors including grip strength. Type 3 fractures are not subdivided in this classification by comminution and intra-articular involvement. Although Lidstrom's survey suggested that distal radio-ulnar joint involvement increased risks of impaired function, injury to this joint is not appreciated by his classification.

c. Older, Stabler and Cassebaum

Type I Colles' fracture: non-displaced
1. Loss of some volar angulation and up to 5° of dorsal angulation.
2. No significant radial shortening—2 mm or more above the distal ulna.

Type II Colles' fracture: displaced with minimal comminution
1. Comminution of dorsal radius.
2. Shortening—usually not below the distal ulna but occasionally up to 3 mm below it.
3. Minimal comminution of the dorsal radius.

Type III Colles' fracture: displaced with comminution of the dorsal radius
1. Comminution of dorsal radius.
2. Shortening—usually below the distal ulna.
3. Comminution of distal radial fragment—usually not marked and often characterized by large pieces.

Type IV Colles' fractures: displaced with severe comminution of distal radius
1. Comminution of dorsal radius marked.
2. Comminution of distal radial fragment—shattered.
3. Shortening—usually 2–8 mm below the distal ulna.
4. Poor volar cortex in some cases.

The authors studied 100 Colles' fractures, 95 of which were treated by closed reduction and plaster immobilization whilst five cases required fixation using pins and plaster. The anatomical and functional outcome at four months, as assessed by their own scoring system, correlated well with fracture severity. Older, Stabler and Cassebaum recommended reduction of all type 2 fractures where there was any radial shortening or dorsal tilt in excess of 10°. They suggested reduction of all type 3 fractures except stable injuries in the elderly. Finally, they advised reduction and internal fixation where necessary for all type 4 fractures.

Although the broad categories of this classification are clear, communication of the subtypes is hampered by complexity and imprecise definitions. The authors concluded that distal radio-ulnar joint involvement was a major determinant of outcome in 10% of cases, yet this factor is excluded from the classification.

Several fracture criteria considered by this classification, and in particular radial length and dorsal angulation, are clearly important in determining outcome, as recognized by several independent studies (Gartland and Werley, 1951; Altissimi et al., 1986; Villar et al., 1987; Fernandez, 1982).

In fact, one study of the validity of classification systems in determining the anatomical result after distal radial fractures has favoured the Older system above four other classifications (Solgaard, 1984).

d. Frykman (1967)

Type I Extra-articular fractures without fracture of the distal ulna.

Type II Extra-articular fractures accompanied by fracture of the distal ulna.

Type III Intra-articular fractures involving the radiocarpal joint but without fracture of the distal ulna.

Type IV Intra-articular fractures involving the radiocarpal joint and accompanied by fracture of the distal ulna.

Type V Intra-articular fractures involving the distal raido-ulnar joint but without fracture of the distal ulna.

Type VI Intra-articular fractures involving the distal radio-ulnar joint and accompanied by fracture of the distal ulna.

Type VII Intra-articular fractures involving both the radiocarpal and the radio-ulnar joint but without fracture of the distal ulna.

Type VIII Intra-articular fractures involving both the radiocarpal and the distal radio-ulnar joint and accompanied by fracture in the distal ulna.

Frykman conducted a detailed analysis of a series of 430 fractures; 295 of these required closed reduction and plaster immobilization, 135 were immobilized in plaster without reduction and seven fractures were internally fixed. He assessed the results at an average of 2 years and 7 months following injury. The assessment was based on subjective criteria as used by Lidstrom but with the addition of weakness as a symptom. Objective criteria in evaluating outcome included range of movement and grip strength as well as the appearance of the fracture on radiography. Overall, results from Frykman's analysis do not appear to parallel the severity of the fracture. Extra-articular injuries (types I, II) fare better than intra-articular injuries but there is little variation in outcome from intra-articular fractures (types III–VIII). Statistical analyses of the data are presented to show that the type of fracture has a significant influence on outcome but it is not clear from the study what exactly is meant by this. Of interest, Frykman noted a significant effect of radial shortening with poor functional outcome, yet his classification does not evaluate this parameter.

Frykman's classification has been used in more studies of the end results of fractures of the distal radius than any other classification (Toh and Jupiter, 1994). This is surprising since independent assessors have questioned its validity. Altissimi et al. (1986) found no correlation between Frykman's classification and outcome from 297 distal radial fractures treated conservatively. Clancey (1984) also did not find any relation between Frykman fracture type and wrist anatomy or range of movement at union. As with Lidstrom's classification, Porter and Stockley (1985) also failed to relate functional outcome with the Frykman classification in their series of radial fractures. These investigators also found difficulty in interpreting involvement of the distal radio-ulnar joint on radiographs of distal radial fractures. The Frykman's classification does not consider the degree of comminution of intra-articular fractures, nor does it describe volarly-displaced fractures. Moreoever, it does not suggest treatment options (Linscheid, 1995).

Nevertheless, the Frykman classification was the first to consider the importance of the distal radio-ulnar joint in wrist fractures. Several commentators have since appreciated the importance of this joint and some have included it in their own classifications of wrist injury (Scheck, 1962; Melone, 1984; Bruckner, Lichtman and Alexander, 1992; Kauer, 1992).

e. Fernandez (1993)

Fernandez classified distal radial fractures by mechanism of injury (Table 13.1).

- Type I — Bending fractures where the thin metaphyseal cortex fails to tensile stresses and the opposite cortex undergoes a certain degree of comminution (extra-articular Colles' or Smith's fractures).
- Type II — Shearing fractures of the joint surface (Barton's and reversed Barton's fractures, radial styloid fractures).
- Type III — Compression fractures of the joint surface, with impaction of subchondral and metaphyseal cancellous bone (intra-articular comminuted fractures of the distal radius).
- Type IV — Avulsion fractures of ligamentous attachments (ulnar and radial styloid).
- Type V — Combinations of bending, compression, shearing or avulsion mechanisms (high velocity injuries).

The advantage of this scheme is that it is intended to make it easier to select treatments. The reduction of choice is one that reverses the original force causing the fracture. For example, bending type fractures are treated by placing tension on the side of the concavity of the angulation. Also, compression type fractures are treated by distracting the distal fragments and attached ligaments using pins or external fixators.

Support exists in the literature for classifications based on mechanism of injury (Linscheid, 1995).

However, there are difficulties. First, the mechanism of injury might not be apparent in every case (Johnson, 1980). Secondly, although the initial subdivision into fracture types is easy to communicate, the details of subclassification are unwieldy. Thirdly, several parameters that are important in determining outcome, such as radial shortening and distal radio-ulnar joint involvement are not considered. So far, validation is not forthcoming either from the author or from independent reviewers.

f. The AO classification of distal radius/ulna (Müller et al., 1990)

The classification of the Association for the Study of Internal Fixation (AO/ASIF) is the most detailed of all systems describing the patterns of distal radial fractures (Figure 13.1 and Table 13.2). It represents an alpha-numeric grading designed to enable computer storage and retrieval of data. A similar format covers all long bones, thus helping with recollection of the basic types. This classification was based on increasing severity of injury where 'severity' represents the difficulty of treatment as well as probable complications and prognosis. In the words of Maurice M. Müller, a founder of the AO Group, 'a classification is useful only if it considers the severity of the bone lesion and serves as a basis for treatment and for evaluation of the results'.

An alpha-numeric grade for soft-tissue injuries to skin, muscles, tendons, nerves and blood vessels has also been developed (Müller et al., 1991).

Problems with the AO Classification include its complexity and, so far, its lack of validation. Its development was not accompanied by a comprehensive assessment of outcome and, furthermore, published independent studies of outcome are awaited.

Table 13.1 Classification of Fernandez (1993)

Type	*Stability*	*Displacement*	*Fragments*	*Associated lesions*	*Treatment*
I. Bending of metaphysis	Stable Unstable	Non-displaced Dorsal (Colles) Volar (Smith) Proximal Combined	Always 2 main Metaphyseal comminution	Uncommon	Conservative (if stable) Percutaneous pins External fixation Bone graft, exceptionally
II. Shearing of joint surface	Unstable	Dorsal Radial Volar Proximal Combined	2 part 3 part Comminuted	Common	Open reduction Screw plate fixation
III. Compression of joint surface	Stable Unstable	Non-displaced Dorsal Volar Proximal Combined	2 part 3 part 4 part Comminuted	Frequent	Conservative Closed/open reduction Percutaneous pins, combined external and internal fixation/bone graft
IV. Avulsion fractures	Unstable	Dorsal Radial Volar Proximal Combined	2 part (radial/ulnar styloid) 3 part (volar, dorsal, margin) Comminuted	Very frequent	Closed or open reduction Pin or screw fixation Tension wiring
V. Combination of mechanisms	Unstable	Dorsal Radial Volar Proximal Combined	Comminuted (frequently intra-articular, open)	Always present	Combined method

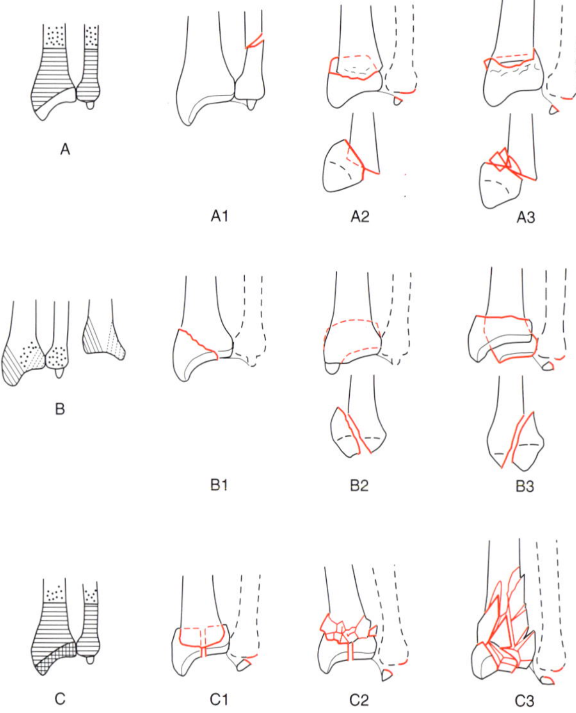

Figure 13.1. The AO/ASIF classification of distal radial fractures (Müller et al., 1990).

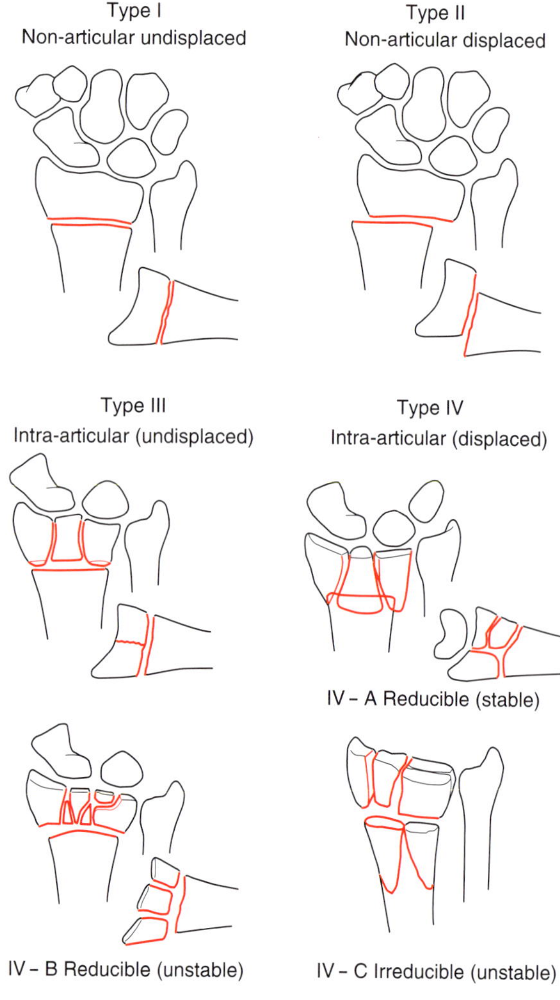

Figure 13.2. The 'Universal Classification' of distal radial fractures (Cooney, 1993).

g. 'Universal Classification' (Cooney, 1993) (Figure 13.2)

Type I — Non-articular, non-displaced
Type II — Non-articular, displaced
- IIA Reducible and stable
- IIB Reducible and unstable
- IIC Irreducible

Type III — Articular, non-displaced
Type IV — Articular, displaced
- IVA Reducible (ligamentotaxis), stable
- IVB Reducible, unstable
- IVC Irreducible
- IVD Complex

This system is simple to communicate. However, it does not distinguish between radiocarpal and distal radio-ulnar joints. Its foundation rests on intra-articular involvement, the displacement of fragments and whether the fracture is reducible and stable. By considering fracture reducibility and stability, this classification provides a guide for treatment. Cooney (1993) advised cast immobilization for types I, IIA and III; percutaneous pins for IIB and IVA; open reduction for IIC, IVC and IVD including percutaneous pins in the case of IVD. However, factors associated with poor outcome such as radial shortening are not inherent and studies evaluating this scheme are currently lacking.

h. McMurtry and Jupiter (1992) (Figure. 13.3)

McMurtry and Jupiter divided the fractures into three types:

1. Anatomic type

Extra-articular

Do not involve radiocarpal or distal radio-ulnar joint—characteristically occur in distal 3–4 cm of radius.

If displaced, injury to distal radio-ulnar joint is present unless ulna has fractured proximal to this joint.

Intra-articular

Any fracture extending into radiocarpal or radio-ulnar joint and displaced by more than 1–2 mm.

Subdivided by 'parts' where a 'part' is a fragment

Table 13.2 The Classification of the Association for the Study of Internal Fixation (AO/ASIF) of distal radial fractures (Müller et al., 1991)

(Alpha-numeric code 23-)		
Type A	(Extra-articular)	
	A1	Extra-articular fracture of ulna, radius intact
	.1	Styloid process
	.2	Metaphyseal simple
	.3	Metaphyseal multifragmentary
	A2	Extra-articular fracture of radius, simple, impacted
	.1	Without tilt
	.2	With dorsal tilt (Colles type)
	.3	With volar tilt (Smith type)
	A3	Extra-articular fracture of radius, multifragmentary
	.1	Impacted with axial shortening
	.2	With a wedge
	.3	Complex
Type B	(Partial articular fracture)	
	B1	Partial articular fracture of the radius, sagittal
	.1	Lateral simple
	.2	Lateral multifragmentary
	.3	Medial
	B2	Partial articular fracture, radius, dorsal rim (Barton)
	.1	Simple
	.2	With lateral sagittal fracture
	.3	With dorsal dislocation of the carpus
	B3	Partial articular fracture, radius, volar rim (reverse Barton)
	.1	Simple with small fragment
	.2	Simple with a large fragment
	.3	Multifragmentary
Type C	(Complete articular fracture)	
	C1	Complete articular fracture of the radius, articular simple, metaphyseal simple
	.1	Postero-medial fragment
	.2	Sagittal articular fracture line
	.3	Frontal articular fracture line
	C2	Complete articular fracture, radius, articular simple, metaphyseal multifragmentary
	.1	Sagittal articular fracture line
	.2	Frontal articular fracture line
	.3	Extending into the diaphysis
	C3	Complete articular fracture of the radius, multifragmentary
	.1	Metaphyseal simple
	.2	Metaphyseal multifragmentary
	.3	Extending into the diaphysis

of bone of sufficient size to be functionally significant and capable of being manipulated and/or internally fixed.

Two-part intra-articular fracture

Typically associated with radiocarpal subluxation. Opposite portion of radiocarpal joint intact and in continuity with remainder of radius.

- Dorsal Barton's
- Volar Barton's
- Radial styloid/Chauffeur's
- Dorso-ulnar impacted/'die punch' fracture

Three-part intra-articular fracture

Typically involves lunate and scaphoid facets of the distal radius and consists of separate fragments displaced both from each other and from the proximal radius. Lunate facet critical as it articulates with radiocarpal and distal radio-ulnar joint.

Four-part intra-articular fracture

Same as three-part fracture except that lunate facet is fractured into dorsal and volar fragments.

All displaced intra-articular fractures extending into the lunate facet in the coronal plane are associated with a fracture involving the distal radio-ulnar joint.

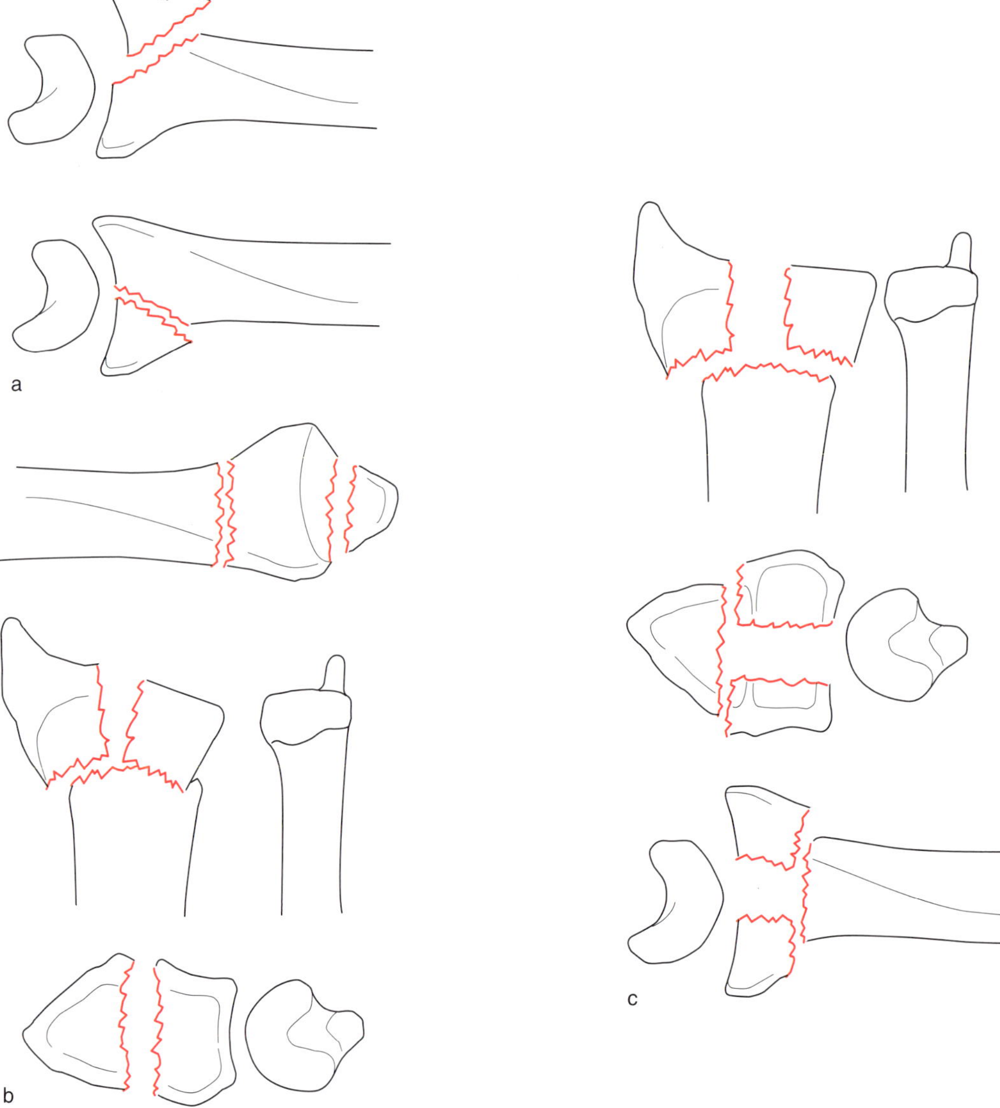

Figure 13.3. McMurtry and Jupiter (1992) classification of distal radial fractures. (a) two-part fracture; (b) three-part fracture; (c) four-part fracture.

Intra-articular fracture with five or more parts

Highly comminuted and varied distal radius fractures.

Extent of disruption can preclude direct manipulated or fixation.

2. Local factors

Comminution

Increases with energy of injury and patient age.

Extent of cortical comminution important in predicting the intrinsic stability of fracture reduction.

Bone quality

Osteopenia directly related to fracture's tendency to shorten and the ability of the fracture to be internally fixed.

Energy of injury

High energy fractures involve greater displacement, comminution and instability.

Displacement

Increased displacement increases risk of soft-tissue stripping and instability.

Risk of swelling, neuro-vascular compromise and open fractures rises with increased displacement.

3. General patient considerations

Loading expectations of hand and wrist determined by age, occupation, handedness and lifestyle as well as patient's psychological outlook, associated medical conditions and compliance.

McMurtry and Jupiter have designed a classification which can be used to describe most distal radial fractures. However, this system does not include an option to describe associated injuries to the ulna, carpal bones or soft tissues. The degree of radial shortening and radial deviation in sagittal or coronal planes cannot be communicated. Whilst the inclusion of local factors and general patient considerations helps in conveying the 'personality' of the fracture and should suggest appropriate intervention, there is a chance that these criteria might overload the classification. Thus far, this classification has not been validated and does not appear to have been widely used in the literature.

i. Hooper, Dias and Gregg (1996)

Undisplaced distal radial fractures
All those without a clinical deformity.
(Undisplaced intra-articular extensions are disregarded.)
Displaced fractures
All those with a clinical deformity.

Type 1 Metaphyseal displacement.
(Any intra-articular extension is essentially undisplaced.)

Type 2 Metaphyseal and intra-articular displacement.
(Intra-articular fracture taken arbitrarily as > 2 mm gap or step at fractures site.)
Integrity of scapholunate ligament must be established.

Type 3 Gross comminution of metaphysis involving a large area of the dorsum only or also involving some of the volar cortex and resulting in an unstable fracture.

General point
Bone quality. Osteoporosis precludes most attempts at internal fixation.

This is a practical classification in which a fracture is considered to be displaced depending on whether a reduction was needed. Whilst a comprehensive description of injury is not possible, prognosis is predicted by consideration of the degree of intra-articular congruity. Knirk and Jupiter (1986) provided some evidence that an articular step-off of more than 2 mm at the radiocarpal joint is predictive of traumatic osteoarthritis at an average of approximately 6 years post-injury in young adults. By describing the integrity of the scapholunate ligament, this classification also hints at associated carpal injuries which have been linked with a poor outcome from distal radial fractures (Pattee and Thompson, 1988). As yet, there are no studies to validate this classification.

Other classifications of extra- and intra-articular fractures not described here include Sarmiento et al. (1975); Porter and Stockley (1985); Linscheid (1995); Saito (1995) and Mathoulin, Letrosne and Saffar (1995).

j. Childhood fractures

A variety of different fractures are possible in childhood and include greenstick fractures, compression fractures, complete fractures and epiphyseal injuries. No widely recognized classification encompasses all of these injuries. The best known is the Salter and Harris classification of epiphyseal fractures, described in detail in Chapter 7.

The distal radial epiphysis is particularly important, not only as it is more commonly injured than all other epiphyses combined but also because it contributes much more to radial length than the proximal radial epiphysis (Salter and Harris, 1963).

Classifications which Consider only Intra-articular Distal Radial Fractures

a. Melone, 1984 and 1993 (Figure 13.4)

Four fragments comprising radial shaft, radial styloid and two fragments representing a 'medial complex' formed from a dorsal medial fragment and a palmar medial fragment.

Type 1 Minimally comminuted fractures. Stable after closed reduction. Undisplaced or with variable displacement of medial complex as a unit.

Type 2 (Die punch) Comminuted fractures. Moderate or severe displacement of medial complex as a unit. Comminution of both posterior and anterior cortices of radius is typical.
Associated features include separation of medial from styloid fragments; radial shortening greater than 5–10 mm and considerable angulation, usually exceeding 20°. Unstable after closed reduction.

Type 2A and Type 2B (1993)
Type 2B fractures have greater comminution and displacement that is refractory to closed methods of reduction.

Type 3 (Spike fractures) Comminuted fractures. Spike from radial shaft projects into flexor compartment and causes nerve or tendon injury.
Displacement of medial complex as a unit.
Unstable.

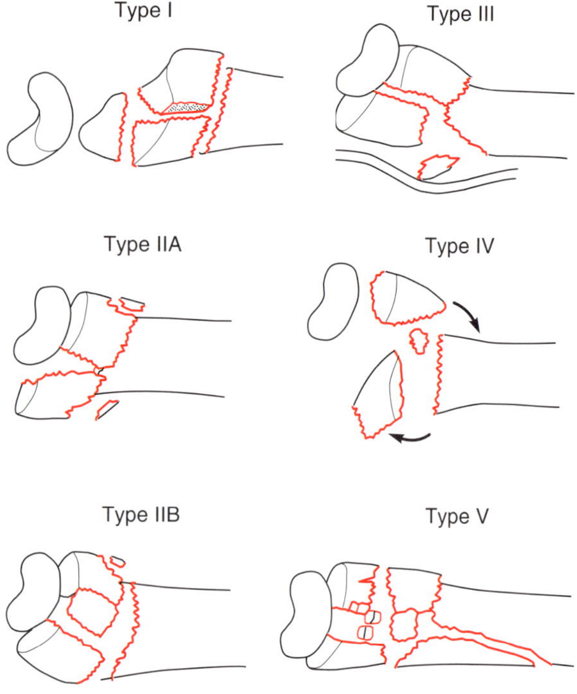

Figure 13.4. The 'Melone' classification (1984, 1993).

Type 4 (Split fractures) Medial complex compressed by lunate resulting in wide separation or rotation of dorsal and palmar medial fragments.
Unstable fractures.

Type 5 (1993)
Type 5 fractures result from a severe force comprising both compression and crush that cause extensive comminution, often extending from the articular surface to the diaphysis.

This classification is centred around the medial complex, a fragment which engages both radiocarpal and distal radio-ulnar joints and has ligamentous attachment to the ulnar styloid and the proximal carpal row. It was designed to help classify complex intra-articular fractures and thereby assist in open reduction and internal fixation of such injuries in young adults. Although the classification is reported to be based on 300 fractures, the selection and distribution of cases between different types are not described.

The basic four-part fracture pattern is easy to appreciate but even when undisplaced and multifragmentary (type 1), it is described as 'non-comminuted'. Melone acknowledges that it is often difficult to distinguish some fracture types by radiographic appearance alone and that classification is sometimes only possible whilst reducing the fracture. Melone has added to his classification since its first description in 1984. Unlike the other types of fracture in his classification, types 2B and 5 (1993) are not based on fracture configuration but on ease of reduction and mechanism of injury. Inconsistency in methods to differentiate fracture types adds confusion and impairs communication of the 1993 classification. However, his scheme for describing intra-articular fractures of the distal radius has clinical relevance. Type 1 fractures are treated by cast immobilization whilst type 2 fractures may require external fixation coupled with percutaneous wires. Type 2B fractures require open reduction with external fixation and bone graft or, if volarly displaced, with plate and screw fixation. Type 3 fractures are treated by closed or open reduction and screw or percutaneous wire internal fixation, usually with soft-tissue repair. Type 4 fractures demand open reduction and careful restoration of the articular surface and soft-tissue integrity. Finally, type 5 injuries require provisional stabilization with external fixation, prior to definitive reconstructive surgery.

This classification awaits validation.

b. Mayo classification (Missakian et al., 1992) (Figure 13.5)

Type I Undisplaced intra-articular fracture of radiocarpal joint.

Type II Displaced, intra-articular fracture of the radioscaphoid joint involving a significant portion of the articular surface of the distal radius (more than a radial styloid fracture). Associated dorsal angulation and shortening are necessary components.
The intra-articular step-off usually exceeds 2 mm; a scaphoid fracture or scapholunate ligament tear is not uncommon.

Type III Displaced intra-articular fracture of the radiolunate joint that presents as a 'die-punch' fracture of the lunate fossa of the distal radius. A displaced fracture component into the distal radio-ulnar joint (sigmoid notch) is common. Open reduction and internal fixation are often necessary.

Type IV Displaced intra-articular fracture involving both radioscaphoid and radiolunate joint surfaces and usually includes a fracture component into the sigmoid fossa of the distal radio-ulnar joint. Usually comminuted, with severe displacement and often a radial shaft component.

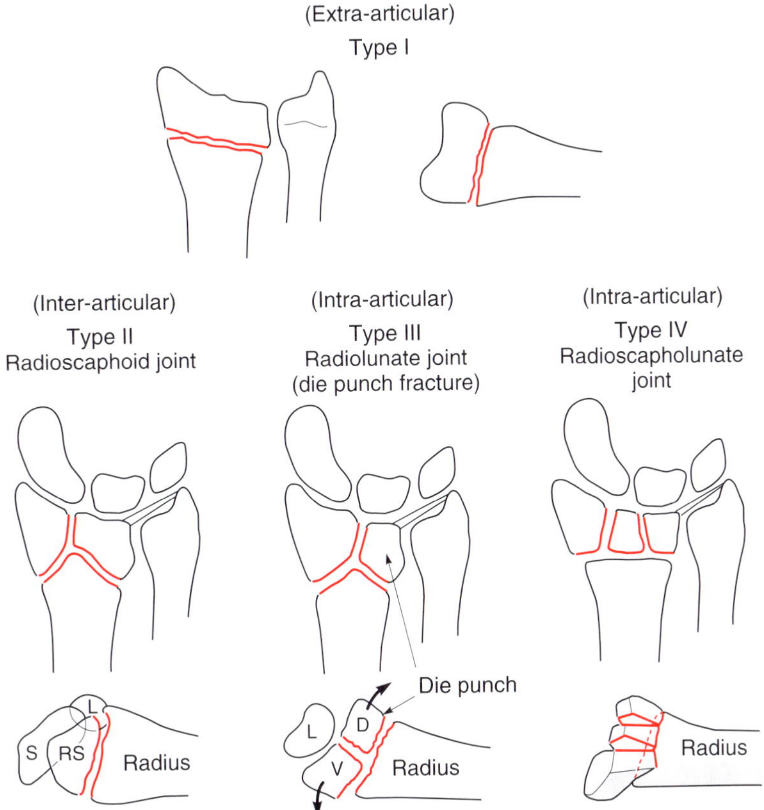

Figure 13.5. Mayo Classification of distal radial fractures (Missakian et al., 1992). S, scaphoid; L, lunate; V, volar; D, dorsal; RS, radio-scaphoid.

This classification of intra-articular fractures is comprehensive and easy to convey. It is the first classification described so far which considers associated carpal injury. Although several authors have described associated injuries to the carpus following distal radial fractures (Taleisnik and Watson, 1984; Pattee and Thompson, 1988; Bickerstaff and Bell, 1989; Bradway, Amadio and Cooney, 1989; Linscheid, 1995), it is remarkable that so few classifications of distal radial fracture have considered these related injuries.

The authors recognize that they might be attempting to classify 'a myriad of different types of fracture patterns that defy a systematic analysis' (Bassett, 1987). However, they feel that their classification is of therapeutic significance and suggest that type II fractures require plate fixation and type III Kirschner wires whilst type IV merit a combination of external and internal fixation. A few studies (Knirk and Jupiter, 1986; Bradway, Amadio and Cooney, 1989) have provided evidence, based on a small number of distal radial intra-articular fractures, that the onset of traumatic arthritis at the wrist is related to the degree of articular congruity at union. Although the authors recognize this, articular incongruity is not specifically considered in their classification. Like the Melone system, the Mayo classification has yet to be evaluated.

An arrangement of intra-articular fracture patterns similar to those above has been described by Mathoulin, Letrosne and Saffar (1995).

Distal Ulnar Fractures

Unlike fractures of the distal radius, distal ulna fractures have received little attention in classification systems. This is surprising since fractures of these two bones often coexist. There is a close anatomical link between the distal ulna and radius not only via the inferior radio-ulnar joint but also through the attachments of the triangular fibrocartilage which is anchored to the base of the styloid process. In addition, the ulnar styloid process also serves as attachment for the ulnocarpal ligament which helps stabilize the carpus (Sokolow, 1995).

a. Dymond (1984)

Type I < 50% displacement of fracture. Stable.
Type II > 50% displacement of fracture. Unstable.

Dymond's classification of distal ulnar fractures was validated by a small cadaveric and clinical study. Soft-tissue disruption to the periosteum and interosseous membrane was limited in cases of distal ulnar fracture showing less than 50% overlap at the fracture site. Fractures with greater displacement at the fracture site, had more soft-tissue damage and were less stable on pronating and supinating the forearm. Dymond's limited clinical study suggested that type I fractures could be treated satisfactorily in a below elbow cast and type II fractures in an above elbow cast.

b. AO classification, Müller et al. (1990) (Figure 13.6)

(Alpha-numeric code 23-)
Type A (Extra-articular)
A1 Extra-articular fracture of ulna, radius intact.
.1 Styloid process
.2 Metaphyseal simple
.3 Metaphyseal multifragmentary

A scheme for describing ulnar fractures is also available in the AO classification in conjunction with distal radius fractures 23-A2 to 23-C3:

1. Radio-ulnar dislocation (fracture of styloid process)
2. Simple fracture of ulnar neck
3. Multifragmentary fracture of ulnar neck
4. Fracture of ulnar head
5. Fracture of ulnar head and neck
6. Fracture of ulna proximal to the neck

This is a wide-ranging description of possibilities. However, fractures of the base of the ulnar styloid process, which detach the triangular fibrocartilage and ulnocarpal ligament are believed to be more significant injuries than those involving the tip of the ulnar styloid. The AO scheme does not specifically distinguish between these variants. Whether the prognosis from ulnar fractures is directly related to this grading system remains uncertain. Furthermore, it is not clear how this scheme will assist in deciding treatment options.

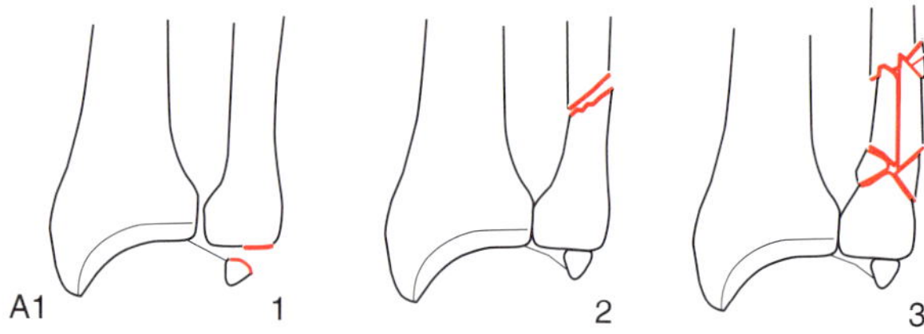

Figure 13.6. The AO/ASIF classification of distal ulnar fractures (Müller et al., 1990).

c. Frykman (1967)

Types I, IV and VIII already described under distal radial fractures involve distal ulnar fractures in association with distal radial fractures.

Radiocarpal Dislocations

Prior to the work of Pouteau in 1783 (see Dumontier, Lenoble and Saffar, 1995) and Colles in 1814, which established the existence of distal radial fractures, all wrist injuries resulting in deformity were thought to be radiocarpal dislocations.

a. Dumontier, Lenoble and Saffer (1995)

Type I Isolated dislocation or dislocation with ulnar styloid fracture. In type I, radiocarpal ligaments are torn and there is a risk of anterior or ulnar translation.
Isolated dislocations are very rare with only a handful described in the literature. Associated injuries to the ulna and carpus have been noted.
Type II Radiocarpal dislocations with fracture of the radial styloid. In type II the radiocarpal ligament is intact and there is less risk of translation.

Dumontier's classification was based on a review of only 13 cases. He recommends open reduction and internal fixation in both types with, in addition, ligament suture in type I where outcome is considered to be less favourable.

b. Moneim, Bolger and Omer (1985)

Type I Ligamentous disruption between distal radius and carpus as one unit.
Radial and ulnar styloid fractures and injury to median and ulnar nerves noted.
Type II Severe injury.
Ligamentous injury involves several areas in radiocarpal joint and carpus in association

with radial and ulnar styloid fractures. Tendency to volar dislocation.

Moneim's classification was based on only four type I and three type II cases. As described above, difficulty in distinguishing between the two types might be anticipated. Closed reduction for type I where possible and open reduction and repair for type II was suggested by the author.

c. Fracture subluxation of the radiocarpal joint

These have been described under Barton's fractures. Certain authorities regard fracture subluxations at the radiocarpal joint differently from fracture-dislocations at that site (Dumontier, Lenoble and Saffar, 1995).

Cooney et al. (1987) have classified radiocarpal fracture-dislocations under carpal dislocations (see below).

Dislocation of the Distal Radio-ulnar Joint

Dislocation of this joint can occur in conjunction with radial fractures or, very rarely, in isolation.

Isolated dislocation

A series of 10 cases by Dameron (1972) comprised an equal number of volar (anterior) and dorsal (posterior) dislocations of the distal ulna. Outcomes appeared similar. No formal classification exists for these injuries.

Dislocation in Association with Radial Fracture

a. Galleazi fracture

A Galleazi fracture typically involves a fracture of the distal radial diaphysis with distal radio-ulnar joint dislocation. Extension of the fracture does occasionally occur into the distal radius (Herzberg, 1995). Dislocation of the inferior radio-ulnar joint implies disruption of the triangular fibrocartilage and accurate reduction of this joint is necessary for a good outcome from Galleazi fractures (Mikic, 1975).

b. Essex–Lopresti fractures

Like Galleazi fractures, Essex–Lopresti fractures involve dislocation of the distal radio-ulnar joint, in this case in association with radial *head* fractures. Once again, outcomes depend considerably on the reduction of the dislocated distal radio-ulnar joint (Herzberg, 1995).

c. AO/ASIF other classifications

The AO Classification of ulna fractures (described above) is the only widely used classification of wrist injuries that considers dislocation at the distal radio-ulnar joint but only in association with ulnar styloid fractures.

Carpal Fractures

After the radius, the scaphoid is the most commonly fractured of all bones (Ruby, 1992).

The scaphoid is also, by far, the most vulnerable of all carpal bones to fracture, accounting for more than 70% of all fractures in this region (Teisen and Hjarbaek, 1988). Isolated fractures to several bones in the carpus are so rare that only short descriptions of fracture patterns exist, rather than a comprehensive scheme of classification. Indeed, the only carpal bone to have a validated fracture classification is the scaphoid.

a. Scaphoid Fractures

1. Russe (1960)
   - I. Horizontal oblique fracture line
     - Distal third
     - Middle third
     - Proximal third
   - II. Transverse fracture line
   - III. Vertical oblique fracture line

Russe's system for classifying scaphoid fractures was based on experience of 220 fractures all assessed by a standardized set of radiographs. He felt that the horizontal oblique and transverse patterns were associated with compressive forces between the fracture fragments whilst fractures with a vertical oblique configuration involved interfragmentary shear. Russe also recognized that most blood vessels entering the scaphoid do so distal to its waist. Accordingly, he forecast prolonged times for union in horizontal oblique fractures in the proximal thirds and all vertical oblique types but few details of case selection or follow-up accompany his classification. Fractures with a horizontal oblique pattern in the middle or distal thirds and all transverse fractures were immobilized for 6 weeks. However, all other patterns were immobilized for 10–12 weeks. More recently, the poor outcome forecast for proximal third fractures has

received support both from vascular studies (Gelberman and Menon, 1980) and a large study demonstrating significantly longer times to radiological union in these fractures (Leslie and Dickson, 1981). The merits of standardized scaphoid views have been incorporated in other studies (Leslie and Dickson, 1981; Herbert and Fisher, 1984).

2. Weber (1980)
   1. Non-displaced. Dorsal and volar soft-tissue attachments intact.
      Fracture line open less than 1 mm.
   2. Angulated. Either dorsal or volar soft tissues disrupted.
      Angulation of fracture, usually with lunate rotated dorsally (dorsal intercalated segment instability).
   3. Displaced. Dorsal and volar soft tissues torn.
      Anteroposterior view of scaphoid shows a step-off at fracture site.
      Avascular necrosis of proximal fragment likely.

Weber's scheme was based on a study of 36 acute scaphoid waist fractures of all three types above. No details of the radiograph series used to classify these fractures were provided. Although progressively fewer fractures united as the degree of displacement increased, there were not enough cases to record a statistically significant relationship. Weber's treatment protocol, based on his study, suggested immobilization for non-displaced fractures and angulated fractures in increasing degrees of radial deviation. He suggested early operative stabilization of displaced fractures.

3. Cooney, Dobyns and Linscheid (1980b)
   1. Undisplaced, stable. No displacement on anteroposterior, lateral or oblique views.
   2. Displaced, unstable. >1 mm displacement on anteroposterior or oblique views or >15° lunocapitate angulation or >45° scapholunate angulation on lateral views.

This is a simple classification complicated only by the need to measure intercarpal angles. It does not differentiate fractures by their site despite evidence that fracture location affects prognosis (Leslie and Dickson, 1981).

In a study of 45 acute fractures, including 32 undisplaced and 13 displaced types, non-union occurred in two (6%) of undisplaced and six (46%) of displaced varieties. Before drawing conclusions from these findings it is important to realize that there were several differences in the mode and duration of immobilization between displaced and undisplaced fractures. Based on the results of this study, Cooney advocated using cast immobilization to treat displaced and undisplaced fractures providing closed reduction was possible, with open reduction and internal fixation as a salvage procedure.

4. Herbert and Fisher (1984) (Figure 13.7)

Type A Acute stable fractures
- A1 fractures of the tubercle
- A2 undisplaced 'crack' fracture of waist

Type B Acute unstable fractures
- B1 oblique fractures of distal third
- B2 displaced or mobile fractures of waist
- B3 proximal pole fractures
- B4 fracture dislocations of carpus
- B5 comminuted fractures

Type C Delayed union

Type D Established non-union
- D1 fibrous non-union
- D2 sclerotic non-union (pseudarthrosis)

Herbert's alpha-numeric classification encompasses a wide variety of acute and established fractures which can easily be distinguished without the use of precise measurements. Furthermore, it is relatively easy to remember and convey and is coded for installation into a computer database. In a study of 158 fractures

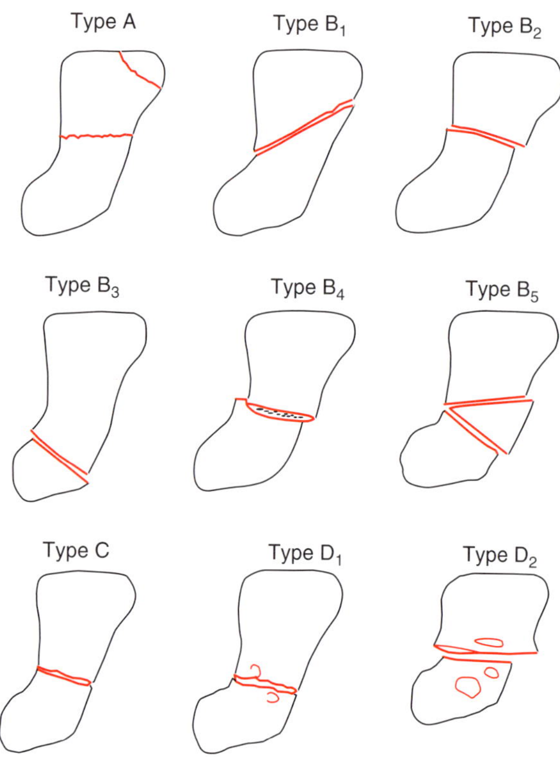

Figure 13.7. Classification of scaphoid fractures (Herbert and Fisher, 1984).

of types B, C and D, all treated by open reduction and internal fixation using a screw-fixation system developed by Herbert, outcome was graded radiographically using specified views. There was an increasing trend towards non-union with increased grade of fracture. Despite the fact that statistical validation was not provided, this classification has found favour with leading authorities. As a guide to treatment, the author's recommendation is to use his screw fixation for all fractures, including type A fractures which have not united radiologically by 6 weeks.

A classification of distal scaphoid fractures was suggested by Prosser, Brenkel and Irvine in 1988 but this was primarily descriptive and was not based on prognosis or treatment options.

b. Lunate Fractures

Lunate fractures account for approximately 1% of all carpal fractures in most series (Teisen and Hjarbaek, 1988). Although there have been several classifications proposed for Kienböck's disease, which is generally considered to be a sequel to lunate trauma, only one system has classified fresh lunate fractures.

1. Fresh lunate fractures (Teisen and Hjarbaek, 1988)

Group I	Fracture of volar pole, possibly affecting the volar nutrient artery.
Group II	Chip fracture which does not affect the main blood supply.
Group III	Fracture of the dorsal pole of the lunate, possibly affecting the dorsal nutrient artery.
Group IV	Sagittal fracture through the body of the lunate.
Group V	Transverse fracture through the body of the lunate.

Based on only 17 cases of lunate fractures, none of which had any evidence of Kienböck's disease, the authors subdivided fractures based on site of fracture in relation to blood supply (Gelberman et al., 1980). Repeat radiograph in 11 patients between 4 and 31 years following fracture failed to demonstrate any sign of Kienböck's disease in any fracture type. This classification has not therefore distinguished fractures types by prognosis.

2. Kienböck's disease

a. Lichtman et al. (1977)

Stage I	X-rays normal but rarely linear/compression fracture of lunate. Usually preceding injury to wrist. Wrist pain slowly settles.
Stage II	Definite density change in lunate relative to other carpal bones. Size, shape, anatomical relationships of carpus not significantly altered. Loss of height of lunate on radial side apparent late in stage II. Recurrent pain, swelling and wrist tenderness.
Stage III	Complete collapse of lunate with proximal migration of the capitate and disruption of carpal architecture. Ribbon-like elongation of lunate on lateral view. Clinically identical to stage II but with marked wrist stiffness.
Stage IV	Features of stage I but with generalized degenerative changes in carpus. Clinical findings of degenerative arthritis of wrist with loss of wrist motion.

Lichtman et al. used this classification to assess the outcomes from conservative treatment in 22 patients and silicone replacement arthroplasty in 14 patients, the latter with an average follow-up of 27 months. All patients with stage I disease were treated by cast immobilization. Of these, 19 had unsatisfactory results. Following replacement arthroplasty, 8 out of 9 patients with stage II disease but only 6 out of 10 patients with stage III disease had satisfactory outcomes. The only patient with stage IV disease to receive an implant had a poor result. Based on these findings, Lichtman et al. recommended silicone replacement of the lunate before the lunate collapses (stages I and II).

Taleisnik (1985) simplified Lichtman's four stages of lunatomalacia and suggested a treatment plan for each stage. However, his modified classification was not based on a study of outcomes.

A short ulna has long been associated with Kienböck's disease (Taleisnik, 1985) but this parameter is not encompassed by current classifications of Kienböck's disease. Other important factors in this disease which are not part of these classifications include age and carpal anatomy (Waldram and Kleinert, 1991).

c. Fractures of Carpal Bones other than Scaphoid and Lunate

As many of these fractures are rare, studies are often limited to small series and case reports. The following account summarizes available descriptions and provides references.

Triquetrum

These fractures have been subdivided into dorsal chip fractures and fractures involving the body of the triquetrum. It is believed that whilst the latter have a favourable prognosis when treated conservatively, the former often fail to heal and might require excision of the fragment with ligament repair (Bartone and Grieco, 1956; Bryan and Dobyns, 1980; Taleisnik, 1985).

Pisiform

These have been subdivided into linear, comminuted and simple chip fractures (Vasilas, Grieco and Bartone, 1960).

Hamate

Milch (1934) subdivided fractures into those involving the body and those involving the hook of the hamate. In general, fractures of the body unite after immobilization but fractures of the hook are often missed until symptomatic non-union is present. Excision might then be necessary (Hooper, Dias and Gregg, 1996).

Capitate

There are no widely recognized classifications for isolated capitate fractures but a description of scapho-capitate fractures has been presented by Vance, Gelberman and Evans (1980). Based on a series of 21 cases, they established six patterns of injury all essentially involving transverse fractures of the scaphoid and capitate associated with a 180° rotation of the proximal capitate fragment. Their intention was to highlight scaphocapitate injuries. The results of treatment based on at least 1 year's follow-up concluded that open reduction and internal fixation was their preferred treatment in all cases of scaphocapitate fracture dislocations.

Trapezoid

Isolated trapezoid fractures are extremely rare (Hooper, Dias and Gregg, 1996). There is currently no classification of these injuries.

Trapezium

In general, fractures have been subdivided into those involving the body, usually in association with a trapeziometacarpal dislocation (Cordrey and Ferrer-Torrells, 1960; Hooper, Dias and Gregg, 1996) and those involving the volar ridge. Palmer (1981) subdivided volar ridge fractures into type 1, involving the base of the trapezial ridge and type 2, involving the tip of the trapezial ridge. Of only three cases in his series, cast immobilization enabled union of one type 1 fracture but open reduction and internal fixation was necessary for both type 2 fractures.

Carpal Dislocations and Fracture-dislocations

These are complex and varied injuries which are poorly understood (Dobyns and Linscheid, 1980).

The classifications that exist for these injuries reflect the spectrum of possible carpal injuries. In general, although these classifications have increased awareness of different injuries to the carpus, there have been few studies to support any of the proposed descriptions of injury.

1. Green and O'Brien (1980)

- Type I — Dorsal perilunate/volar lunate dislocation
- Type II — Dorsal trans-scaphoid perilunate dislocation
- Type III — Volar perilunate/dorsal lunate dislocation
- Type IV — A. Transradial styloid perilunate dislocation
  - B. Scapho-capitate syndrome
  - C. Transquetral fracture-dislocation
  - D. Miscellaneous
- Type V — Isolated rotary scaphoid subluxation
  - A. Acute subluxation
  - B. Recurrent subluxation
- Type VI — Total dislocation of the scaphoid

This classification is based on the premise that lunate and perilunate dislocations are different stages of the same injury. Displacements might be dorsal or volar, and in the opinion of Green and O'Brien, both are indications for closed reduction and, if necessary, correction of residual instability of the scaphoid or of the lunate.

Types II–IV fractures necessitate accurate closed or open reduction and immobilization, by internal fixation if necessary. The authors had no personal experience with type V injuries.

Green and O'Brien's classification is comprehensive and concise. However, it was not based on any published study and, despite general recommendations regarding treatment which apply equally to all six types, it does not suggest the anticipated outcomes from these injuries.

2. *Mayo classification (Cooney et al., 1987)*

Type I	Dorsal perilunate/palmar lunate dislocation A. Rotary subluxation of the scaphoid B. Late post-trauma carpal instability
Type II	Transcarpal perilunate fracture dislocations A. Transradial styloid perilunate B. Trans-scaphoid perilunate C. Scaphocapitate syndrome D. Transquetral perilunate
Type III	Radiocarpal fracture-dislocations
Type IV	Axial loading fracture-dislocations
Type V	Complete carpal bone dislocations (scaphoid)

Cooney et al. (1987) differentiated pure ligamentous dislocations (type I) from fracture-dislocations (type II) in order to focus attention on the latter group of injuries. They studied 30 different type II injuries and followed them up for an average of 3 years after treatment. Nine patients were treated conservatively whilst 21 had operative treatment. Outcomes, though favourable, did not vary significantly between different subtypes of the classification. Furthermore, recommended treatments followed general principles and did not differ between subtypes.

3. *Mayfield (1980)*

Stage I	Scaphoid dislocation or instability Scapholunate and radioscaphoid ligament injury.
Stage II	Capitate dislocation.
Stage III	Triquetral dislocation Radiotriquetral ligament injury
Stage IV	Lunate dislocation Radiocarpal, radiotriquetral and dorsal radiocarpal ligament injury.

Mayfield's classification considers a spectrum of injuries resulting from dorsiflexion, ulnar deviation and intercarpal supination. By loading 32 cadaveric wrists to 'failure' using two different machines, 13 perilunate dislocations were produced. The cadaveric wrists were dissected out and examined for bony and ligamentous damage and carpal instability. Different degrees of perilunate injury were evident beginning with scapholunate disruption and progressing through scaphocapitate and scaphotriquetral ligament injury.

Although this classification provides a valuable insight into mechanisms behind perilunate dislocation, it does not encompass fracture-dislocations. Johnson has suggested that perilunate injuries involving radial, scaphoid, capitate and triquetral fractures could be considered by a scheme similar to that of Mayfield. Furthermore, Mayfield's classification does not advise specifically on treatment options.

Carpal Instability

Instability of the carpus has been defined as the failure of carpal bones to maintain their anatomical alignment either in the passive state (static instability) or when undergoing stress or movement (dynamic instability) (Leslie, 1994). Instability can result from ligamentous or bony disruption or a combination of the two.

Central to an understanding of carpal instability is an appreciation of two common theories of carpal biomechanics. The 'Row Theory' regards carpal function as dependent on a proximal row comprising scaphoid, lunate and triquetrum and a distal row consisting of the trapezium, trapezoid, capitate and hamate bones. This arrangement creates two joints, the first between the radius and the proximal row (radiocarpal joint) and the second between the proximal and distal rows (midcarpal joint) (Ruby, 1992). The 'Columnar Theory', however, regards carpal structure as arranged with a central column comprising lunate and capitate proximally and the trapezium, trapezoid and hamate distally; a lateral column comprising the scaphoid and a medial column consisting of the triquetrum (Taleisnik, 1985).

In general, the existing classifications of carpal instability distinguish different patterns of injury without predicting outcome or suggesting specific treatment options.

a. *Linscheid et al. (1972)*

Dorsiflexed Intercalated Segment Instability (DISI):

'On lateral x-ray of the wrist, the lunate is dorsiflexed and displaced ventrally with respect to the longitudinal axes of the radius and the capitate, which are no longer co-linear.

Usually some rotational displacement of the longitudinal axis of the scaphoid, so that it is almost perpendicular to the radius.'

Volar-flexed (Palmar-flexed) Intercalated Segment Instability (VISI):

'Palmar flexion of the lunate relative to the longitudinal axes of the radius and the capitate.

There may be dorsal subluxation of the lunate on the radius.

There is no increase in scapholunate gap.'

This was the first widely recognized classification of carpal instability. Linscheid et al. described the characteristic radiographic appearance of a small series of patients with DISI and VISI without forecasting outcomes or providing therapeutic options. Two further types of instability were described by them in 1975. The first, 'ulnar translocation', was characterized by an increased space between radial styloid and scaphoid, whilst the second, 'dorsal carpal subluxation', involved dorsal displacement of the entire carpus relative to the articular surface of the radius (Taleisnik, 1980).

b. Taleisnik (1980)

Lateral Carpal Instability
- Scaphoid-trapezium-trapezoid dissociation
- Scaphoid-capitate diastasis
- Post-traumatic scaphoid-lunate dissociation

Central Carpal Instability
- Analogous to a dynamic dorsal intercalated instability

Medial Carpal Instability
- Analogous to a volar intercalated instability pattern

Proximal Carpal Instability
- Ulnar translocation
- Dorsal carpal translocation
- Volar subluxation of carpus.

Taleisnik's classification is based on the columnar theory of carpal structure. It describes ligamentous disruption between the bones comprising the central, lateral and medial columns. In addition, it includes ligamentous instability at the radiocarpal joint and within the central column. Like Linscheid's classification, it was not supported by a clinical study and does not offer prognosis or preferred options for treatment.

Both of these descriptions of carpal instability exclude fractures of the carpus and they do not describe injuries in which the cleavage plane transgresses the described 'rows and columns', such as capitate-hamate diastasis (Ruby, 1992). Recently, others have suggested modified arrangements describing carpal instability which address some of these limitations (Ruby, 1992; Leslie, 1994).

c. Mayfield (1980)

The Mayfield classification of carpal dislocation as described above applies equally to carpal instability.

d. Palmer (1989)

(Traumatic Triangular Fibrocartilage Complex Abnormalities)

A Central perforation

B Ulnar avulsion
- With distal ulnar fracture
- Without distal ulnar fracture

C Distal avulsion

D Radial avulsion
- With sigmoid notch fracture
- Without sigmoid notch fracture

Palmer developed this classification to highlight the importance of the triangular fibrocartilage complex (TFCC) and to establish a framework from which appropriate treatment options could be developed, based on injury patterns.

This classification is described under carpal instability because type C injuries, involving avulsion of the TFCC from its attachment either to the lunate (ulnolunate ligament) or to the triquetrum (ulnotriquetral ligament), results in ulnar carpal instability. However, as the TFCC also stabilizes the distal raido-ulnar joint, Palmer's classification is of equal relevance to distal radio-ulnar instability.

Thus far this classification remains purely descriptive and awaits the results of outcome studies.

e. Mayo Classification (Linscheid and Dobyns, 1989)

Dobyns developed a descriptive yet comprehensive classification of a variety of complex injuries (Table 13.3). His system distinguishes instability into 'dissociative' types associated with torn perilunar ligaments (e.g. CID, carpal instability dissociative) and 'non-dissociative' types in which the perilunar ligaments are not torn (e.g. CIND, carpal instability non-dissociative). The abbreviations VISI and DISI are explained under Linscheid's 1972 classification described above. This scheme also encompasses carpal instability resulting from carpal and distal radial fractures.

Summary

This account of classifications of wrist trauma has identified limitations with most of the commonly-used schemes to describe injuries to this joint complex.

Table 13.3 Mayo classification of carpal injuries (Linscheid and Dobyns, 1989)

Instability	*Dislocation*	*Fracture-dislocation*
Radiocarpal	Radiocarpal	Radiocarpal
CIND type	Dorsal	Dorsal Barton's
VISI	Volar	Volar Barton's
DISI	Ulnar	Radial styloid (+carpal translation)
Ulnar translation		Lunate fossa (+carpal translation)
Partial/residual perilunate	Perilunate	Transosseous perilunate
CID type	Various stages	Trans-scaphoid perilunate
Scapholunate dissociation (+DISI)		Transradiostyloid perilunate
Lunotriquetral dissociation (+VISI)		Other combinations
Midcarpal	Midcarpal	Midcarpal
CIND type		Malunited Colles' fracture
Triquetrohamate (VISI > DISI)		
Scaphotrapeziotrapezoidal (VISI > DISI)		
Capitolunate (DISI > VISI)		
Diffuse laxity (DISI > VISI)		

A minority of classifications have been supported by clinical studies in which the classification of fracture severity has correlated with outcomes resulting from a given treatment. In such cases, it is possible to predict the prognosis for any injury from the classification employed. However, such classifications have been limited to injuries involving a single unit within the wrist, for example the distal radius or scaphoid. With more complex injuries, there are insufficient data to validate the classifications used. Perhaps as a consequence of these deficiencies, few systems of classifying wrist trauma have enjoyed widespread support.

Group Discussion

Most classifications of wrist injuries remain, essentially, descriptions of patterns of injury. The discussion group found this section one of the most challenging, particularly with respect to injuries of the distal radius. It was felt important that the readers of this chapter should refer themselves to the original descriptions of Colles (1814). Barton (see Thompson and Grant, 1977) and Smith (see Palmer, 1993) to be certain that they understood precisely what these eponymous descriptions meant. There are a large number of classification systems of distal radial fractures. There was no consensus as to which provided the best buys. Note was made that Older's classification from 1965 provided a good basis for descriptive purposes. The AO classification is also valuable but has some limitations. The most recent additions by Jupiter (1991) and Hooper, Dias and Gregg (1996) also have some value. The classification system of Frykman (1967) was felt not to be particularly useful and, indeed, a number of papers have been written describing its limitations.

For classification systems of the scaphoid the original description of Russe (1960) was felt to provide a good basis for describing these fractures. However, no mention of displacement or angulation is present in this classification which was felt to be an omission. There are classification systems for the individual carpal bones including the lunate but the group did not have any strong recommendations about these. The very complex area of carpal dislocations and carpal instability was considered. This area is obviously evolving rapidly with new concepts of management and treatment of these injuries and it was felt difficult to be certain as to which classification system was the best. The group's findings are summarized in Table 13.4.

Table 13.4 A summary of the results of the discussion group

Site	*Author/year*	*Good for defining outcome*	*Good for defining treatment*	*Reliability tested*	*Notes*
Distal radius fracture	Gartland and Werley, 1951	Defines anatomical outcome	Limited	Yes (Solgaard, 1984)	Important early classification See Solgaard, 1984
Distal radius fracture	Lidstrom, 1959	Yes	Limited	No	Emphasizes dorsal angulation
Distal radius fracture	Older et al., 1965	Defines anatomical outcome better than most	Unknown	Yes (Solgaard, 1984)	Subtypes imprecise
Distal radius fracture	Frykman, 1967	No	No	No	Most quoted. Emphasizes distal radio-ulnar joint involvement
Distal radius fracture	Fernández, 1993	No	Designed to define treatment	No	Based on mechanism of injury
Distal radius fracture	AO Group Müller et al., 1990	No	No	No	Most comprehensive. Likely to grow in importance
Distal radius fracture	'Universal', Cooney, 1993	No	Designed to guide treatment	No	Simple but does not distinguish radiocarpal from radio-ulnar joint involvement
Distal radius fracture	McMurtry and Jupiter, 1992	No	No	No	Comprehensive. Describes 'personality' of fracture. Potentially complex
Distal radius fracture	Hooper, Dias, Gregg, 1996	Limited	Designed to guide treatment	No	Clinically relevant and simple
Distal radius fracture	Melone, 1984, 1993	No	Designed to guide treatment of intra-articular fractures	No	Only for intra-articular fractures. Imprecise subtypes
Distal radius fracture	'Mayo', Missakian et al., 1992	No	Designed to guide treatment	No	Considers associated carpal injury
Distal ulna fracture	Dymond, 1984	No	No	Limited	Basic
Distal ulna fracture	AO Group, Müller et al., 1990	No	No	No	Comprehensive

Table 13.4 ***(Continued)***

Site	*Author/year*	*Good for defining outcome*	*Good for defining treatment*	*Reliability tested*	*Notes*
Distal ulna fracture	Frykman, 1967	No	No	No	Ulna fractures incorporated in distal radius classification (see above)
Radiocarpal joint dislocation	Dumontier et al., 1995	No	No	No	Based on small series
Radiocarpal joint dislocation	Moneim et al., 1985	No	No	No	Based on small series
Scaphoid fracture	Russe, 1960	Yes	Yes, limited	Yes, Leslie and Dickson, 1981; Gelberman and Menon, 1980	Important early classification
Scaphoid fracture	Weber, 1980	Yes	Defined treatment with classification	No	Waist fractures only
Scaphoid fracture	Cooney et al., 1980b	Yes	Limited	No	Measurement of intercarpal angles necessary
Scaphoid fracture	Herbert and Fisher, 1984	Yes	Limited	No	Comprehensive with alpha-numeric classification
Lunate fracture	Teisen and Hjarbaek, 1988	No	No	No	Based on small series
Carpal dislocations and Fracture-dislocations	Green and O'Brien, 1980	No	No	No	Descriptive only
Carpal dislocations and fracture-dislocations	'Mayo', Cooney et al. 1987	No	No	No	Descriptive only
Carpal dislocations and fracture-dislocations	Mayfield, 1980	No	No	No	Descriptive only
Carpal instability	Linscheid et al., 1972	No	No	No	Linscheid et al. paper key early work. Descriptive only
Carpal instability	Taleisnik, 1980	No	No	No	Descriptive only
Carpal instability	Mayfield, 1980	No	No	No	Same classification used for carpal dislocation (see above)

References

Altissimi, M., Anteucci, R., Fiacca, C. and Mancini, G. (1986) Long term results of conservative treatment of fractures of the distal radius. *Clin. Orthop. Rel. Res.*, **206**, 202–211.

Bacorn, R. and Kurtzke, J. (1953) Colles' fracture. A study of two thousand cases from the New York State Workmen's Compensation Board. *J. Bone Joint Surg.*, **35A**(3), 643–658.

Bartone, N. and Grieco, R. (1956) Fractures of the triquetrum. *J. Bone Joint Surg.*, **38A**, 353–356.

Bassett, R. (1987) Displaced intra-articular fractures of the distal radius. *Clin. Orthop. Rel. Res.*, **214**, 148–152.

Bickerstaff, D. and Bell, M. (1989) Carpal malalignment in Colles' fractures. *J. Hand Surg.*, **14B**(2), 155–160.

Bradway, J., Amadio, P. and Cooney, W. (1989) Open reduction and internal fixation of displaced, comminuted intra-articular fractures of the distal end of the radius. *J. Bone Joint Surg.*, **71A**(6), 839–847.

Bruckner, J., Lichtman, D. and Alexander, A. (1992) Complex dislocations of the distal radio-ulnar joint. Recognition and management. *Clin. Orthop. Rel. Res.*, **275**, 90–103.

Bryan, R. and Dobyns, J. (1980) Fractures of the carpal bones other than lunate and navicular. *Clin. Orthop. Rel. Res.*, **149**, 107–111.

Cassebaum, W. (1950) Colles' fracture. A study of the end results. *JAMA*, **143**, 963–965.

Clancey, G. (1984) Percutaneous Kirschner wire fixation of Colles' fractures. A prospective study of 30 cases. *J. Bone Joint Surg.*, **66A**(7), 1008–1014.

Cobby, M. and Watt, I. (1991) Investigating the traumatised joint. *Curr. Orthop.*, **5**(2), 125–134.

Colles, A. (1814) On the fracture of the carpal extremity of the radius. *Edinb. Med. Surg. J.*, **10**, 182–186.

Cooney, W. (1993) Fractures of the distal radius. A modern treatment-based classification. *Orthop. Clin. North Am.*, **24**(2), 211–216.

Cooney, W., Bussey, R., Dobyns, J. and Linscheid, R. (1987) Difficult wrist fractures. Perilunate fracture-dislocations of the wrist. *Clin. Orthop. Rel. Res.*, **214**, 136–147.

Cooney, W., Dobyns, J. and Linscheid, R. (1980a) Complications of Colles' fractures. *J. Bone Joint Surg.*, **62A**(4), 613–619.

Cooney, W., Dobyns, J. and Linscheid, R. (1980b) Fractures of the scaphoid: a rational approach to management. *Clin. Orthop. Rel. Res.*, **149**, 90–97.

Cordrey, L. and Ferrer-Torrells, M. (1960) Management of fractures of the greater multangular. *J. Bone Joint Surg.*, **42A**, 1111–1118.

Dameron, T. (1972) Traumatic dislocation of the distal radio-ulnar joint. *Clin. Orthop. Rel. Res.*, **83**, 55–63.

Dobyns, J. and Linscheid, R. (1980) Carpal bone injuries. *Clin. Orthop. Rel. Res.*, **149**, 2–3.

Dumontier, C., Lenoble, E. and Saffar, P. (1995) Radiocarpal dislocations. In *Fractures of the Distal Radius* (P.C. Saffar and W.P. Cooney, eds). London: Martin Dunitz.

Dymond, I. (1984) The treatment of isolated fractures of the distal ulna. *J. Bone Joint Surg.*, **66B**(3), 408–410.

Fernandez, D.L. (1982) Correction of post-traumatic wrist deformity in adults by osteotomy, bone grafting and internal fixation. *J. Bone Joint Surg.*, **64A**(8), 1164–1178.

Fernandez, D.L. (1993) Fractures of the distal radius: operative treatment. Instructional Course Lectures; American Academy of Orthopedic Surgeons. **42**, 73–88.

Frykman, G. (1967) Fracture of the distal radius including sequelae—shoulder-hand-finger syndrome, disturbance in the distal radio-ulnar joint and impairment of nerve function. *Acta Orthop. Scand.*, Suppl. 108.

Gartland, J. and Werley, C. (1951) Evaluation of healed Colles' fractures. *J. Bone Joint Surg.*, **33A**(4), 895–907.

Gelberman, R. and Menon, J. (1980) The vascularity of the scaphoid bone. *J. Hand Surg.*, **5**(5), 508–513.

Gelberman, R., Bauman, T., Menon, J. and Akeson, W. (1980) The vascularity of the lunate bone and Kienbock's disease. *J. Hand Surg.*, **5**(3), 272–278.

Golden, G. (1963) Treatment and prognosis of Colles' fracture. *Lancet*, **1**(March), 511–515.

Green, D. and O'Brien, E. (1980) Classification and management of carpal dislocations. *Clin. Orthop. Rel. Res.*, **149**, 55–72.

Herbert, T. and Fisher, W. (1984) Management of the fractured scaphoid using a new bone screw. *J. Bone Joint Surg.*, **66B**(1), 114–123.

Herzberg, G. (1995) Galleazi and Essex-Lopresti fractures. In *Fractures of the Distal Radius* (P.C. Saffar and W.P. Cooney, eds). London: Martin Dunitz, pp. 264–266.

Hooper, G., Dias, J. and Gregg, P. (1996) The wrist. In *Fractures and Dislocations* (P.S. Gregg, J.G. Stevens and P.H. Worlock, eds). Oxford: Blackwell Science, pp. 447–493.

Jenkins, N., Jones, D., Johnson, S. and Mintowt-Czyz, W. (1987) External fixation of Colles' fractures. An anatomical study. *J. Bone Joint Surg.*, **69B**(2), 207–211.

Johnson, R. (1980) The acutely injured wrist and its residuals. *Clin. Orthop. Rel. Res.*, **149**, 33–44.

Johnston, G., Friedman, L. and Kriegler, J. (1992) Computerized tomographic evaluation of acute distal radial fractures. *J. Hand Surg.*, **17A**(4), 738–744.

Jupiter, J. (1991) Current Concepts Review. Fractures of the distal end of the radius. *J. Bone Joint Surg.*, **73A**(3), 461–469.

Kauer, J. (1992) The distal radio-ulnar joint. Anatomic and functional considerations. *Clin. Orthop. Rel. Res.*, **275**, 37–45.

Knirk, J. and Jupiter, J. (1986) Intra-articular fractures of the distal end of the radius in young adults. *J. Bone Joint Surg.*, **68A**(5), 647–659.

Leslie, I. (1994) Carpal instability. *Curr. Orthop.*, **8(1)**, 14–22.

Leslie, I. and Dickson, R. (1981) The fractured carpal scaphoid. Natural history and factors affecting outcome. *J. Bone Joint Surg.*, **63B**(2), 225–230.

Lichtman, D., Mack, G., MacDonald, R. et al. (1977) Kienbock's disease: The role of silicone replacement arthroplasty. *J. Bone Joint Surg.*, **59A**(7), 899–907.

Lidstrom, A. (1959) Fractures of the distal end of the radius. A clinical and statistical study of end results. *Acta Orthop. Scand.*, Suppl. 41.

Linscheid, R. (1995) Classification of distal radial fractures by mechanism of injury. In *Fractures of the Distal Radius* (P.C. Saffar and W.P. Cooney, eds). London: Martin Dunitz, pp. 41–49.

Linscheid, R. and Dobyns, J. (1989) Carpal instability. *Curr. Orthop.*, **3**, 106–114.

Linscheid, R., Dobyns, J., Beabout, J. and Bryan, R. (1972) Traumatic instability of the wrist. Diagnosis, classification and pathomechanics. *J. Bone Joint Surg.*, **54A**(8), 1612–1632.

Mathoulin, C., Letrosne, E. and Saffar, P. (1995) Classification of intra-articular fractures of the distal radius. In *Fractures of the Distal Radius* (P.C. Saffar and W.P. Cooney, eds). London: Martin Dunitz, pp. 126–130.

Mayer, J. (1940) Colles' fractures. *Br. J. Surg.*, **27**, 629–642.

Mayfield, J. (1980) Mechanism of carpal injuries. *Clin. Orthop. Rel. Res.*, **149**, 45–54.

McMurtry, R. and Jupiter, J. (1992) Fractures of the distal radius. In *Skeletal Trauma: fractures, dislocations and ligament injuries* (B.J. Browner, J.B. Jupiter, A.M. Levine and P.G. Trafton, eds). Philadelphia: Saunders, pp. 1063–1094.

McQueen, M. and Caspers, J. (1988) Colles' fracture: Does the anatomical result affect the final function? *J. Bone Joint Surg.*, **70B**(4), 649–651.

Melone, C. (1984) Articular fractures of the distal radius. *Orthop. Clin. North Am.*, **15**(2), 217–236.

Melone, C. (1993) Distal radial fractures: patterns of articular fragmentation. *Orthop. Clin. North Am.*, **24**(2), 239–253.

Metz, V. and Gilula, L. (1993) Imaging techniques for distal radius fractures and related injuries. *Orthop. Clin. North Am.*, **24**(2), 217–228.

Mikic, Z. (1975) Galleazi fracture dislocations. *J. Bone Joint Surg.*, **57A**(8), 1071–1080.

Milch, H. (1934) Fractures of the hamate bone. *J. Bone Joint Surg.*, **16A**, 459–462.

Millet, P. and Rushton, N. (1995) Early mobilization in the treatment of Colles' fracture: a 3-year prospective study. *Injury*, **26**(10), 671–675.

Missakian, M., Cooney, W., Amadio, P. and Glidewell, H. (1992) Open reduction and internal fixation for distal radius fractures. *J. Hand Surg.*, **17A**(4), 745–755.

Moneim, M., Bolger, J. and Omer, G. (1985) Radio-carpal dislocation—classification and rationale for management. *Clin. Orthop. Rel. Res.*, **192**, 199–209.

Müller, M., Nazarian, S., Koch, P. and Schatzker, J. (1990) *The Comprehensive Classification of Fractures of Long Bones*. Berlin: Springer-Verlag.

Müller, M., Allgower, M., Schneider, R. and Willenegger, H. (1991) *Manual of Internal Fixation. Techniques Recommended by the AO-ASIF Group*, 3rd edn. Berlin: Springer-Verlag.

Older, T., Stabler, E. and Cassebaum, W. (1965) Colles' fracture: evaluation and selection of therapy. *J. Trauma*, **5**(4), 469–476.

Palmer, A. (1981) Trapezial ridge fractures. *J. Hand Surg.*, **6**, 561–564.

Palmer, A. (1989) Triangular fibrocartilage complex lesions: a classification. *J. Hand Surg.*, **14A**(4), 594–606.

Palmer, A. (1993) Fractures of the distal radius. In *Operative Hand Surgery* (D. Green, ed.). Edinburgh: Churchill Livingstone, pp. 929–971.

Pattee, G. and Thompson, G. (1988) Anterior and posterior marginal fracture dislocations of the distal radius. An analysis of the results of treatment. *Clin. Orthop. Rel. Res.*, **231**, 183–195.

Pennig, D. and Gausepohl, T. (1996) External fixation of the wrist. *Injury*, **27**(1), 1–15.

Pool, C. (1973) Colles' fracture. *J. Bone Joint Surg.*, **55B**(3), 540–544.

Porter, M. and Stockley, I. (1985) Fractures of the distal radius. Intermediate and end results in relation to radiologic parameters. *Clin. Orthop. Rel. Res.*, **220**, 241–252.

Prosser, A., Brenkel, I. and Irvine, H. (1988) Articular fractures of the distal scaphoid. *J. Hand Surg.*, **13B**(1), 87–91.

Ruby, L. (1992) Fractures and dislocations of the distal radius. In *Skeletal Trauma: fractures, dislocations and ligament injuries* (B.J. Browner, J.B. Jupiter, A.M. Levine and P.G. Trafton, eds). Philadelphia: Saunders, pp. 1025–1062.

Russe, O. (1960) Fractures of the carpal navicular. Diagnosis, Non-operative treatment and operative treatment. *J. Bone Joint Surg.*, **42A**, 759–768.

Saffar, P. (1995) Current trends in treatment and classification of distal radial fractures. In *Fractures of the Distal Radius* (P.C. Saffar and W.P. Cooney, eds). London: Martin Dunitz, pp. 12–18.

Saito, H. (1995) Classification and treatment of intra-articular fractures of the distal radius. In *Fractures of the Distal Radius* (P.C. Saffar and W.P. Cooney, eds). London: Martin Dunitz, pp. 131–142.

Salter, R. and Harris, W. (1963) Injuries involving the epiphyseal plate. *J. Bone Joint Surg.*, **45A**(3), 582–622.

Sarmiento, A., Pratt, G., Berry, N. and Sinclair, W. (1975) Colles' fractures. Functional bracing in supination. *J. Bone Joint Surg.*, **57A**(3), 311–317.

Savoie, F. (1995) The role of arthroscopy in the diagnosis and management of cartilaginous lesions of the wrist. *Hand Clin.*, **11**(1), 1–5.

Scheck, M. (1962) Long-term follow up of treatment of comminuted fractures of the distal end of the radius by transfixation with Kirschner wires and cast. *J. Bone Joint Surg.*, **44A**(2), 337–351.

Smail, G. (1965) Long-term follow-up of Colles' fracture. *J. Bone Joint Surg.*, **47B**(1), 80–85.

Sokolow, C. (1995) Fractures of the ulnar styloid. In *Fractures of the Distal Radius* (P.C. Saffar and W.P. Cooney, eds). London: Martin Dunitz, pp. 297–300.

Solgaard, S. (1984) Classification of distal radius fractures. *Acta Orthop. Scand.*, **56**, 249–252.

Stewart, H., Innes, A. and Burke, F. (1985) Factors affecting the functional outcome of Colles' fracture: an anatomical and functional study. *Injury*, **16**, 289–295.

Swiontowski, M. (1995) Outcomes measurement in orthopaedic trauma surgery. *Injury*, **26**(10), 653–657.

Szabo, R. (1993) Extra-articular fractures of the distal radius. *Orth. Clin. North Am.*, **24**(2), 229–237.

Taleisnik, J. (1980) Post-traumatic carpal instability. *Clin. Orthop. Rel. Res.*, **149**, 73–82.

Taleisnik, J. and Watson, H. (1984) Mid-carpal instability caused by malunited fractures of the distal radius. *J. Hand Surg.*, **9A**(3), 350–357.

Taleisnik, J. (1985) *The Wrist*. New York: Churchill Livingstone.

Teisen, H. and Hjarbaek, J. (1988) Classification of fresh fractures of the lunate. *J. Hand Surg.*, **13B**(4), 458–462.

Thomas, F. (1957) Reduction of Smith's fracture. *J. Bone Joint Surg*, **39B**(3), 463–470.

Thompson, G. and Grant, T. (1977) Barton's fractures – Reverse Barton's fractures. Confusing eponyms. *Clin. Orthop. Rel. Res.*, **122**, 210–221.

Toh, C. and Jupiter, J. (1984) Distal radial fractures. *Curr. Orthopaedics*. **8**, 3–13.

Vance, R., Gelberman, R. and Evans, E. (1980) Scaphocapitate fractures. Patterns of dislocation, mechanisms of injury and preliminary results of treatment. *J. Bone Joint Surg.*, **62A**(2), 271–276.

Vasilas, A., Grieco, R. and Bartone, N. (1960) Roentgen aspects of injuries to the pisiform bone and pisotriquetral joint. *J. Bone Joint Surg.*, **42A**, 1317–1328.

Villar, R., Marsh, D., Rushton, N. and Greatorex, R. (1987) Three years after Colles' fracture. A prospective review. *J. Bone Joint Surg.*, **69B**(4), 635–638.

Waldram, M. and Kleinert, M. (1991) Avascular necrosis of the carpal bones. *Curr. Orthopaedics*, **5**(1), 4–12.

Weber, E. (1980) Biomechanical implications of scaphoid waist fractures. *Clin. Orthop. Rel. Res.*, **149**, 83–89.

Whipple, T. (1995) The role of arthroscopy in the treatment of intra-articular wrist fractures. *Hand Clin.*, **11**(1), 13–18.

14

Hand fractures and tendon injuries

N. Blewitt

Introduction

Hand fractures can be complicated by deformity from no treatment, stiffness from over treatment, and both deformity and stiffness from poor treatment.

Swanson, 1970

Any classification is only useful if it indicates the severity of injury, acts as a guide to treatment and helps predict prognosis. Reproducible classification systems are also useful for the comparison of results of new techniques used in the treatment of hand injuries reported from different centres.

The aim of this chapter is to present a logical account of available classifications of hand injuries. Clinical factors concerning the patient's individual circumstances, the surgeon's individual experience and the rehabilitation service available locally will, in addition to the specific injury pattern, determine the choice of treatment and the subsequent plan of therapy towards full recovery.

Nomenclature

The Bone and Joint Injury Committee produced an agreed system of nomenclature following hand injuries (Dobyns, Linscheid and Cooney, 1983). Dobyns' fracture nomenclature applies to the bones, joints and ligaments of the hand (Table 14.1). The specific pattern of fractures, dislocations and sprains is outlined in Table 14.2. Associated soft-tissue damage and relevant terminology is addressed in Table 14.3.

Fracture Stability

When a fracture is classified the degree of displacement and soft-tissue injury will contribute to the fracture stability. The functionally stable fracture has been defined by Pun et al. (1989) based on a prospective study of 284 digital hand fractures, as the ability of the patient actively to move the adjacent joints by more than 30% of the expected normal range and with such movement the alignment of the fracture stays within the acceptable criteria. The acceptable alignment as determined by radiographs are:

- 10° angulation in both planes (sagittal and coronal).
- Except in the metaphyseal region where 20° sagittal plane alignment is accepted.
- 45° of sagittal angulation is acceptable in the neck of the fifth metacarpal.
- Up to 50% bony overlap (product of percentage apposition at fracture side on AP and lateral radiograph).
- No rotational deformity.

Significant soft-tissue injury:

- Tendon laceration.
- Large skin defect requiring reconstruction.

Specific Location and Classifications

Fractures of the Distal Phalanx

There are two useful classifications for fractures of the distal phalanx.

1. Kaplan (1940) classification (Figure 14.1)
   Type 1 Longitudinal split
   Type 2 Comminuted tuft
   Type 3 Transverse fracture
2. Dobyns, Beckenbaugh and Bryan (1982) classification of distal phalangeal fractures.

This is an anatomical classification based upon the site of fracture, tuft, shaft or base, it is then subdivided into an open or closed fracture (Table 14.4).

Table 14.1 Agreed nomenclature following hand injuries (Dobyns, Linscheid and Cooney, 1983)

Bone		*Joint*	*Ligament*
Distal phalanx		Distal interphalangeal DIP	DIP ligaments
	Tuft	Base of distal phalanx	Dorsal capsule: palmar plate
	Diaphysis	Dorsal, palmar	Radial collateral (RCL), accessory RCL
	Metaphysis	Radial, ulnar	Ulnar collateral (UCL), accessory UCL
	Epiphysis	Condyles of middle phalanx	
	Physis	Radial, ulnar	
		Dorsal, palmar	
Middle phalanx		Proximal interphalangeal joint PIP	PIP ligaments
	Distal metaphysis	Head of middle phalanx	Dorsal capsule: palmar plate
	Diaphysis	Dorsal, palmar	RCL, accessory RCL
	Proximal metaphysis	Radial, ulnar	UCL, accessory UCL
	Epiphysis	Condyles of proximal phalanx	
	Physis	Radial, ulnar dorsal, palmar	
Proximal phalanx		Metacarpal phalangeal MP	MP ligaments
	Distal metaphysis	Head of proximal phalanx	Dorsal capsule: palmar plate
	Diaphysis	Dorsal, palmar	RCL, accessory RCL
	Proximal metaphysis	Radial, ulnar	UCL, accessory UCL
	Epiphysis	Condyles of metacarpal head	
	Physis	Dorsal, palmar radial, ulnar	
Metacarpal		Carpometacarpal CMC	CMC ligaments
	Distal metaphysis	Head of metacarpal	Dorsal CMC
	Diaphysis	Dorsal, palmar	Volar CMC
	Proximal metaphysis	Radial, ulnar	Intermetacarpal
	Epiphysis	Distal articular facet of carpal	Thumb
	Physis	Dorsal, palmar radial, ulnar	Anterior oblique CMC
			Posterior oblique CMC
			Dorsoradial
			First intermetacarpal

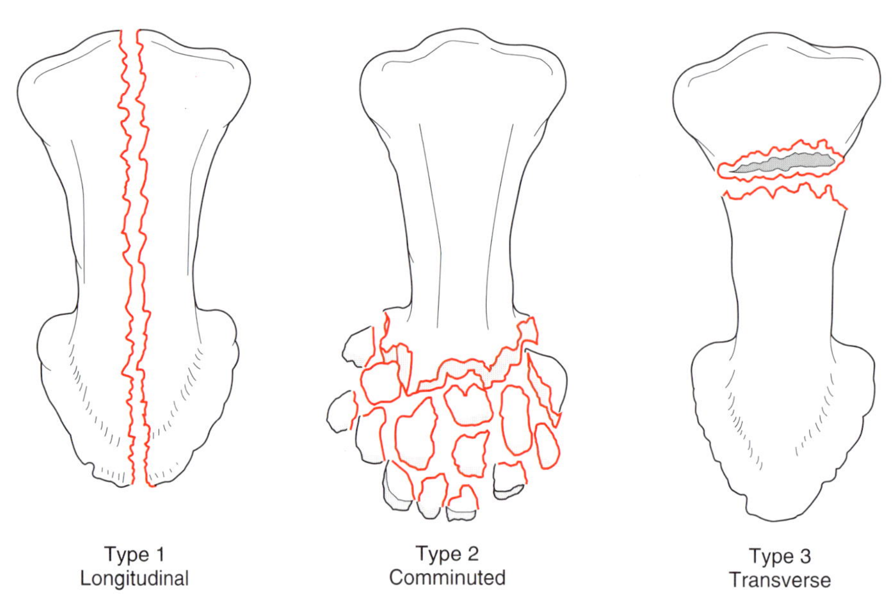

Figure 14.1. The Kaplan (1940) classification of fractures of the distal phalanx.

Table 14.2 Fractures, dislocations and sprains

Fractures	
Simple	Comminuted
Open	Closed
Cartilaginous	Osteocartilaginous
Displaced	Undisplaced
Articular	Non-articular
Compressed	Distracted
Stable	Unstable
Reducible	Irreducible
Oblique short	Oblique long spiral
Transverse	Longitudinal
Angulated dorsal	Angulated palmar
Angulated ulnar	Angulated radial
Rotated supination	Rotated pronated
Shortened	Lengthened
Dislocations	
Open	Closed
Partial subluxation; instability	Complete
Reducible	Irreducible
Stable	Unstable
Simple	Complex with fracture/STI
Displaced	Undisplaced
Sprains	
Stable	Unstable
Simple	Complex STI
Diffuse	Specific indicate ligament

Table 14.3 Soft-tissue structures

	Yes	*No*	*Anatomical segment*
Artery			
Nerve			
Vein			
Tendon			
Muscle			
Sheath			
Retinaculum			
Fascial layer			
Fibrofatty			
Skin			
Nail			

Table 14.4 Dobyns, Beckenbaugh and Bryan (1982) classification of distal phalangeal fractures

Site	*Closed*	*Open*
Tuft	Stable	? Nail bed injury
Shaft	Non-displaced	Nail bed injury
Base–Physeal	Unstable	non-union Eponychium injured angulated malrotated

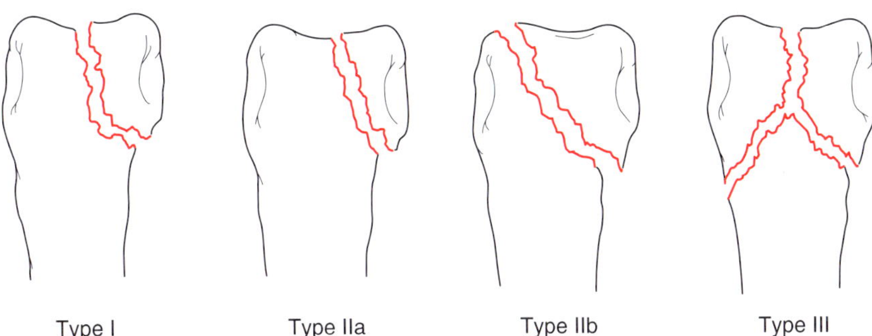

Figure 14.2. London's (1971) classification of interphalangeal joint fracture.

Interphalangeal Joint Condylar Fractures

London's (1971) classification (Figure 14.2) addresses condylar fractures of the interphalangeal joints. To identify whether the fracture is unicondylar or bicondylar and to determine the degree of metaphyseal extension present, each stage has therapeutic and prognostic implications.

Intra-articular Fractures of the Base of the Phalanx

These fractures have been simply divided into three groups (Hastings and Carroll, 1988) (Figure 14.3).

The collateral ligament avulsion fracture: a comminuted compression fracture involving the articular surface and a vertical intra-articular fracture extending upwards into the diaphysis.

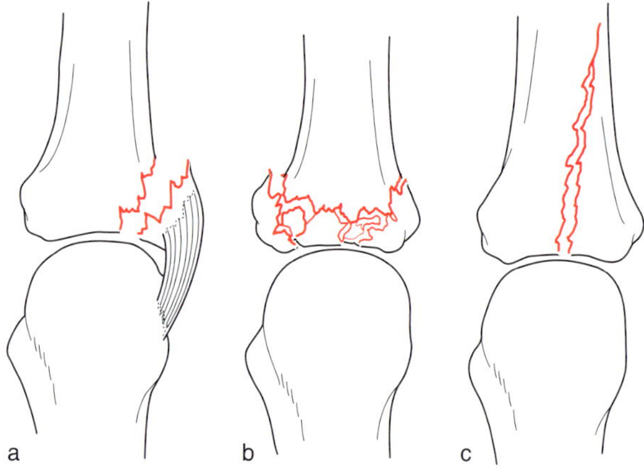

Figure 14.3. The Hastings and Carroll (1988) classification of fractures of the base of the phalanx.

Avulsion fractures from the base of the proximal phalanx can occur in a variety of patterns. These have been further subdivided into six types (Jupiter and Belsky, 1992). Each type reflects the degree of comminution, articular involvement and the site of bony avulsion (Figure 14.4).

The dorsal fracture-dislocation of the metacarpophalangeal joint with an avulsion fracture from the base of the proximal phalanx warrants special consideration regarding joint stability. Such stability depends upon the proportion of collateral ligament remaining attached to the proximal phalanx fragment (Figure 14.5). When 50% of the palmar articular base of the middle phalanx is involved in the fracture, any attachment of the true collateral ligament is lost and the joint is unstable (Dobyns, Beckenbaugh and Bryan, 1982; Hastings and Carroll, 1988). See later section on PIPJ dislocation.

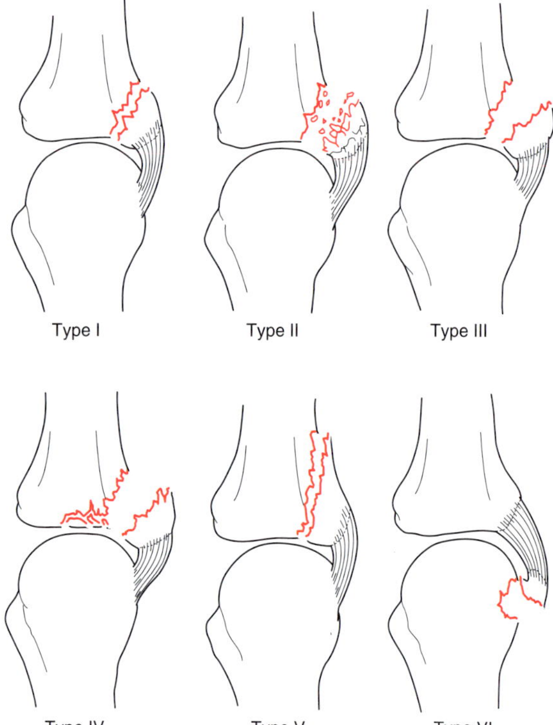

Figure 14.4. Avulsion fracture of the base of the phalanx (from Jupiter and Belsky, 1992).

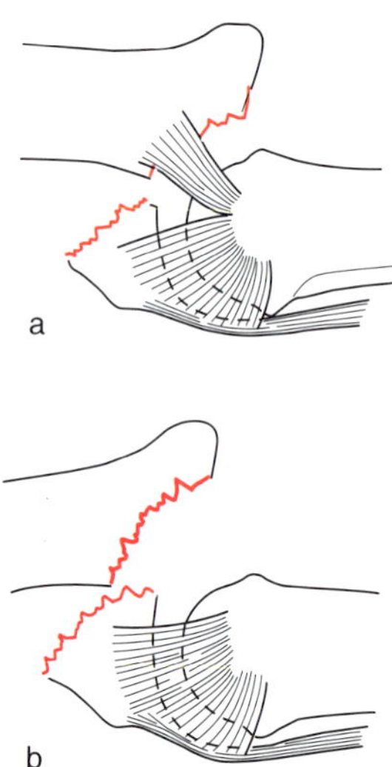

Figure 14.5. Dorsal fracture-dislocation of the metacarpophalangeal joint (from Jupiter and Belsky, 1992).

Phalangeal Shaft Fractures

Fractures of the phalanges are classified according to the general nomenclature previously outlined and detailed in Tables 14.1–14.3.

Metacarpal Fractures

Metacarpal Head Fractures

Fractures of the metacarpal head were classified by McElfresh and Dobyns into:

Type 1 Oblique sagittal fractures of shaft entering joint.

Type 2 Vertical coronal fractures extending into the joint.

Type 3 Horizontal transverse fractures extending into the joint.

Type 4 Comminuted fractures (the most common type).

Metacarpal Neck Fractures

Metacarpal neck fractures are classified as stable or unstable according to the previously outlined criteria and also on the direction of displacement.

Metacarpal Shaft Fractures

Metacarpal shaft fractures are classified according to the agreed system of nomenclature outlined in Tables 14.1–14.3.

Metacarpal Base Fractures

Metacarpal base fractures of the thumb have been extensively investigated in the literature and controversy still surrounds optimal treatment. However, there is agreement on the classification of these injuries. There are four types of metacarpal base fracture of the thumb (Figure 14.6):

1. Extra-articular epibasal, either transverse or oblique.
2. Bennett (1982) fracture-subluxation. The fracture fragment is held in place by the strong oblique retinacular ligament; the metacarpal subluxes radially.
3. Rolando (1910) subdivided fractures into 'Y' and 'T' patterns.
4. A comminuted fracture.

Metacarpal Base Fractures—Little Finger

There are similarities in the metacarpal base fracture of the little finger and the thumb. Again four types of fractures are recognized (Figure 14.7) (Dobyns, Beckenbaugh and Bryan, 1982; Jupiter and Belsky, 1992):

1. Extra-articular epibasal.
2. Two-part intra-articular fracture.
3. Three-part intra-articular fracture.
4. Comminuted fracture with impaction.

Ligamentous Injuries and Dislocation

Injuries to the ligaments of the thumb and fingers have been studied by Moberg and Stener (1954). Essentially their classification depends upon the anatomical structures affected. Their detailed anatomical and physiological studies provide a complete classification for such injuries and document the direction and magnitude of force required to produce each subtype. In general, injuries can affect the radial collat-

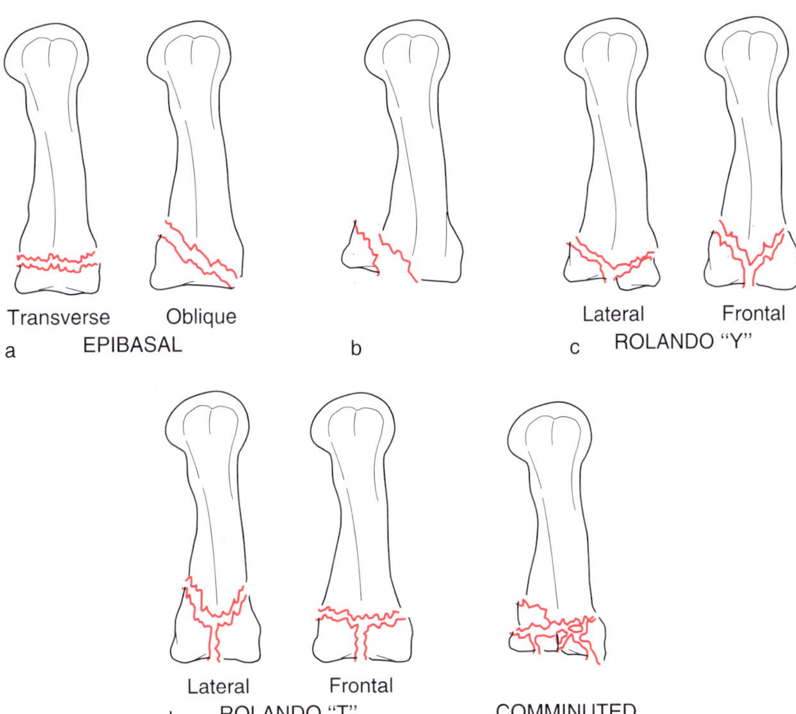

Figure 14.6. Thumb metacarpal base fracture (from Jupiter and Belsky, 1992).

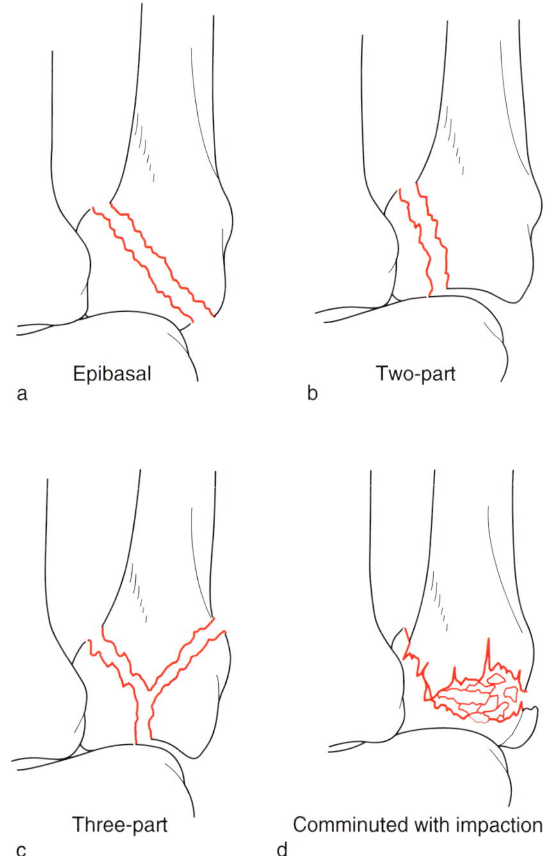

Figure 14.7. Little finger metacarpal base fracture (from Jupiter and Belsky, 1992).

eral ligament, the ulnar collateral ligament or the volar ligaments or indeed all three ligaments can be affected together. Dislocation is accompanied by at least two structures affected. The site of rupture is also documented according to whether a distal, a proximal or an insubstance rupture occurs. They also noted the presence of bony avulsion fractures within this spectrum of injury.

Distal Interphalangeal Joint Dislocations

Dislocation of this joint is uncommon but when it occurs is usually dorsal in direction; however, palmar or lateral dislocation has been reported.

Proximal Interphalangeal Joint (PIPJ) Dislocation

Dislocation of this joint is commonly encountered. The direction of dislocation may be dorsal, palmar or lateral and can further be classified as reducible or irreducible (complex) with interposition of soft tissue (Figure 14.8).

The system of classification has been formalized by Dray and Eaton (1993) into three major types for the PIPJ:

Type I (Hyperextension)
- Volar plate avulsion
- Longitudinal split in collateral ligaments

Type II (Dorsal dislocation)
- Avulsion volar plate
- Major bilateral split in collateral ligament system

Type III (Fracture dislocation)
- Fracture volar base middle phalanx
  - A Stable fracture-dislocation. Small triangle < 40% volar articular arc. Dorsal part of collateral ligaments remain attached to middle phalanx
  - B Unstable fracture-dislocation. Fracture impaction large segmental articular surface of middle phalanx > 40%. Loss of attachment of volar plate and collateral ligaments.

Metacarpophalangeal Joint (MCPJ) Dislocation

Dislocation of the metacarpophalangeal joints are usually in a dorsal direction. They are classified as simple or complex. In complex dislocations there is interposition of soft-tissue. This tissue comprises the palmar plate and sometimes tendon trapped in the joint preventing successful closed reduction. A com-

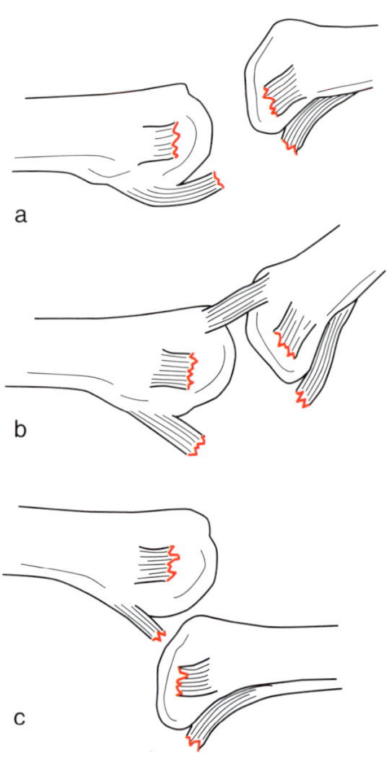

Figure 14.8. Proximal interphalangeal joint dislocation (from Jupiter and Belsky, 1992).

plex metacarpophalangeal joint dislocation can be suspected by the presence of a skin dimple in the palmar surface over the metacarpal head, if specifically sought. Open reduction is required. Prediction of complex dislocation of the thumb MCPJ can be assisted by a plain radiograph. If in a dorsal dislocation the sesamoids are in normal proximity to the base of the proximal phalanx the metacarpal head will have buttonholed through the palmar tendons which will block attempted closed reduction.

Thumb Metacarpophalangeal Joint Ulnar Collateral Ligament Rupture

A forced abduction injury to the thumb metacarpophalangeal joint may result in a complete rupture of the ulnar collateral ligament (Figure 14.9). Often the ligament ruptures at its site of insertion into the base of the proximal phalanx and during injury retracts to lie superficial to the adductor aponeurosis. In the absence of ligament tissue apposition, healing would not be possible and future instability accompanying pinch grip of thumb to index would be predictable. This was described fully by Stener (1962) and is referred to as the 'Stener lesion'.

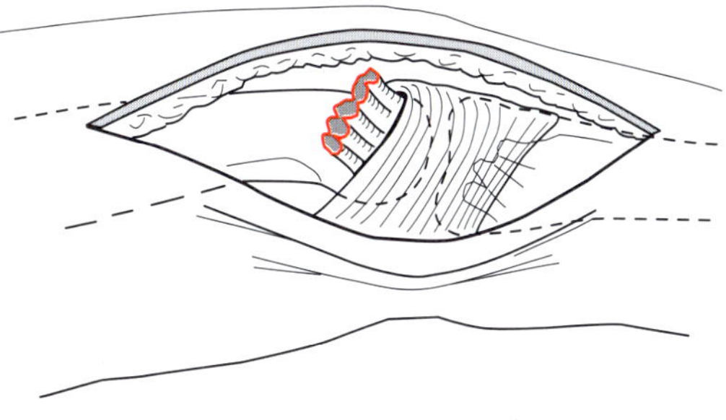

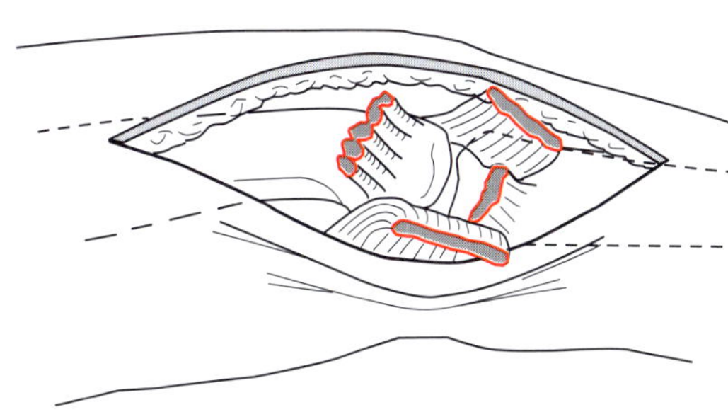

Figure 14.9. Thumb metacarpophalangeal joint dislocation. Ulnar collateral ligament rupture (Stener, 1962).

Carpometacarpal Joint Dislocation

Carpometacarpal joint dislocations of the fingers are usually accompanied by a fracture of the base of the metacarpals. Carpometacarpal joint dislocation of the thumb represents a rotational dislocation of the joint around the oblique palmer ligament and can be treated by closed reduction and percutaneous pinning.

Tendon Injuries

Flexor Tendons

Flexor tendon injuries are classified according to the tendon injured and its anatomical zone. Verdan's zones (Verdan and Michon, 1961) for flexor tendon site of injury are universally accepted although often modified (Milford, 1980) (Figure 14.10) (Leddy, 1993).

- Zone I — Distal to the FDS insertion.
- Zone II — Within the fibrous flexor sheath.
- Zone III — Proximal to the fibrous flexor sheath to the carpal tunnel.
- Zone IV — Within the carpal tunnel.
- Zone V — Proximal to the carpal tunnel.

Zone two (often called 'No man's land' in the past) is the area where much controversy still abounds. Zone two is a complex zone with genuine anatomical changes occurring from proximal to distal. Reflecting this a subdivision of the Bunnell zone II has been proposed by Tang (1994) (Figure 14.11).

- Zone 2A — Tendons within extending from the proximal C2 to distal A4 pulley.
- Zone 2B — Extending from the proximal C1 pulley to distal A3 pulley.
- Zone 2C — Within the A2 pulley.
- Zone 2D — Extending from the proximal border of the A2 pulley to the mouth of the fibrous flexor sheath.

Flexor Digitorum Profundus Avulsion

Bony avulsion fractures from the insertion of FDP have been classified by Leddy and Packer (1979) into three types (Figure 14.12).

- Type 1. FDP rupture and retraction into palm.
- Type 2. FDP rupture and retraction to proximal interphalangeal joint level with or without a bone flake.
- Type 3. Large bony avulsion from flexor digitorum profundus insertion.

This classification system has been validated by others (Dray and Eaton, 1993).

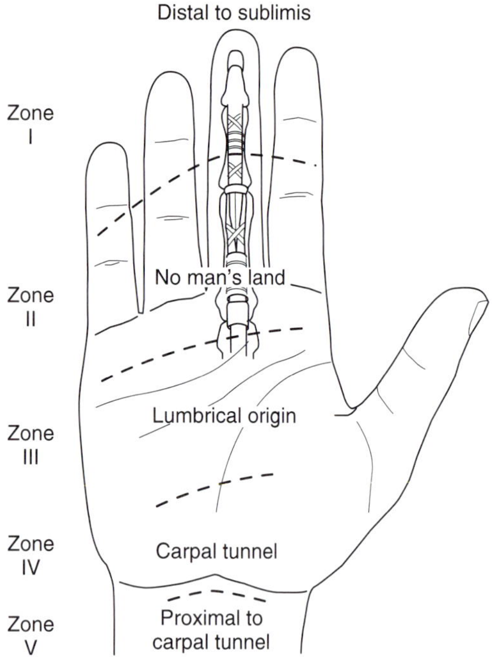

Figure 14.10. Flexor tendon injuries: zones of the hand (Milford, 1980).

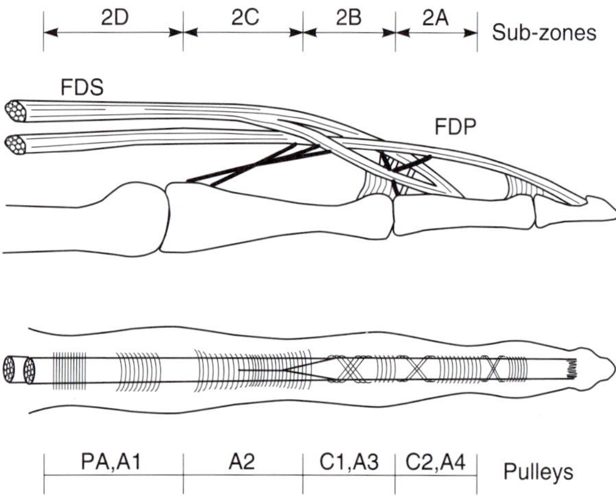

Figure 14.11. Flexor tendon injuries. Subdivision of zone II (Tang, 1994).

Mallet Deformity

There are two main classifications of the mallet finger which are complementary. Watson Jones (1956) describes three types (Figure 14.13):

Type 1. Stretching of the tendon but continuity is maintained.

Type 2. Complete disruption through the tendon structure substance.

Type 3. Flake of bone avulsion at tendon insertion.

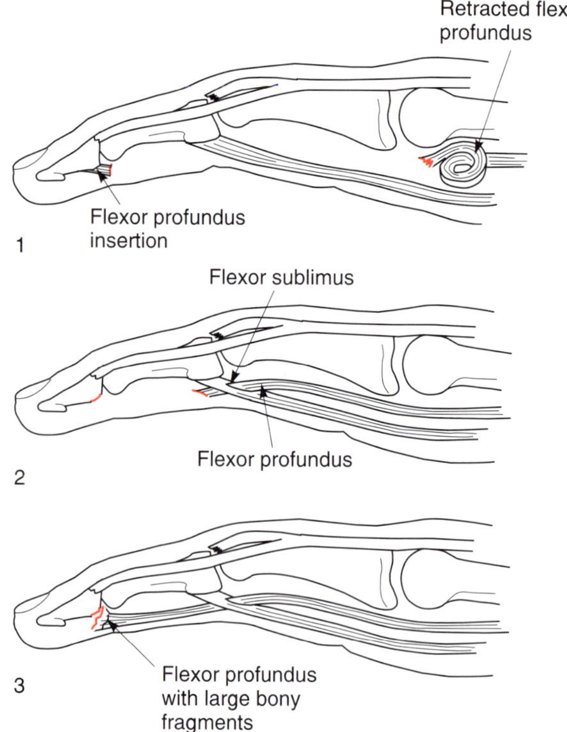

Figure 14.12. Flexor tendon injuries. Flexor digitorum profundus avulsion (Leddy and Packer, 1979).

The Watson Jones Type 3 mallet fingers have been further subdivided by Wehbe and Schneider (1984) into three more types (Figure 14.14):

Type 1. Bony injury without subluxation of the distal interphalangeal joint.

Type 2. Fracture with subluxation of distal interphalangeal joint.

Type 3. Epiphyseal and physeal injuries

Subtype A: Fragment less than one third articular surface.

B: Fragment between one third and two thirds articular surface.

C: Fragment greater than two thirds articular surface.

In addition to these classifications for mallet fingers Doyle (1993) introduced another related to the presence and degree of associated soft-tissue defect.

Type I: Closed or blunt trauma with loss of tendon continuity with or without a small avulsion fracture.

Type II: Laceration at or proximal to the DIP joint with loss of tendon continuity.

Type III: Deep abrasion with loss of skin, subcutaneous cover, and tendon substance.

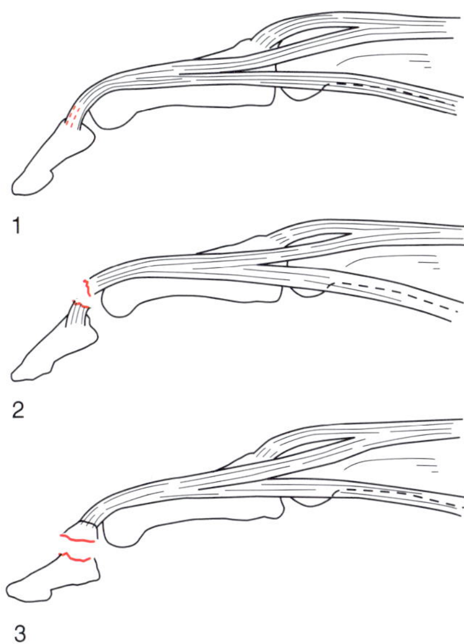

Figure 14.13. Extensor tendon injuries. Mallet deformity (Watson-Jones, 1956).

Type IV: (A) Trans-epiphyseal plate fracture in children;
(B) Hyper-flexion injury with fracture of articular surface of 20% to 50%;
(C) Hyper-extension injury with fracture of the articular surface usually greater than 50% and with early or late subluxation of distal phalanx.

Boutonnière Deformity

Following an extensive review of boutonnière deformities, the Souter (1967) classification emerged.

Group 1: Uncomplicated central slip rupture.
Group 2: Complicated by tissue loss over proximal interphalangeal joint.
Group 3: Complicated by fracture of proximal phalanx.
Group 4: A: Complicated by intra-articular fracture of PIPJ.
B: Complicated by gross disruption of PIPJ.
Group 5: Complicated by multiple severe injuries to the hand.

Subdivided according to open or closed and detailing individual tissue loss.

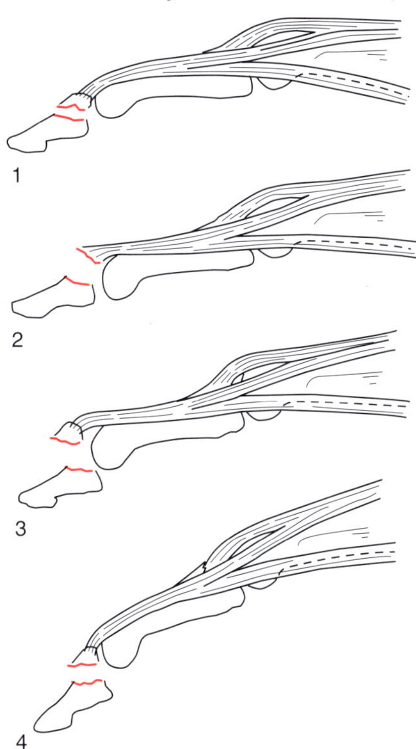

Figure 14.14. Extensor tendon injuries. Bony mallet deformity (Wehbe and Schneider, 1984).

Extensor Tendon Injury

Like flexor tendon laceration, the anatomical level of extensor tendon injury is used for classification. This system is initially attributed to Klinert and Verdan (1983) and is illustrated in Figure 14.15 (Milford, 1980). A comprehensive modification has been described utilising nine zones extending from the finger tip to the elbow (Doyle, 1993) (Figure 14.16).

Soft-Tissue Injuries to the Hand

Soft-tissue injuries of the hand are very common. In addition, soft-tissue injury of varying degree always accompanies bony fracture and relates to the violence and mechanism of injury. Soft-tissue injuries are classified depending upon the anatomical structure affected according to the agreed nomenclature (see Table 14.3), although specific injuries can be addressed.

Crush Injuries

Hand injuries caused by wringers have been classified by Matev (1967) into three types:

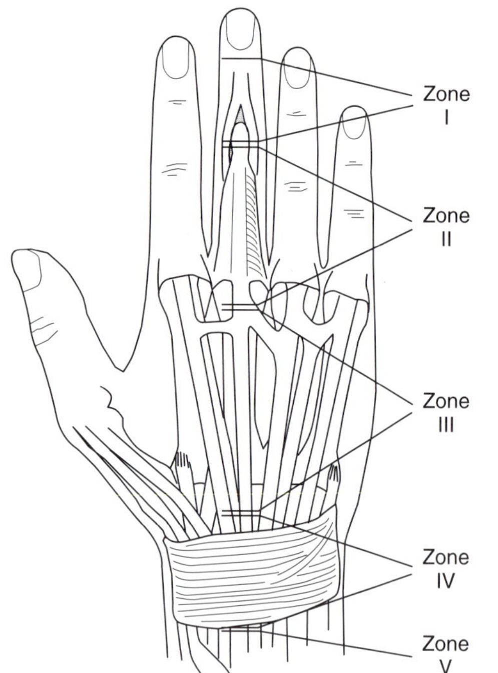

Figure 14.15. Extensor tendon injury zones (Klinert and Verdan, 1983).

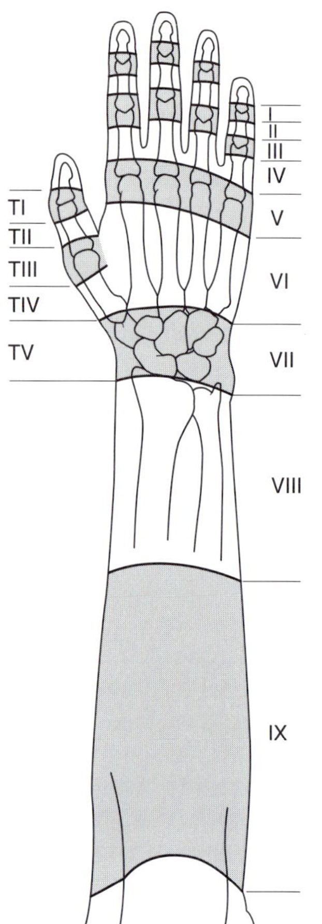

Figure 14.16. Zones of extensor tendon injury (Doyle, 1993).

Type 1 Partial or total denudation of the hand often caused by smooth rapidly moving rollers.

Type 2 Lacerated wounds with wide flaying of surrounding skin often accompany fractures caused by smooth rollers with a 2–3 cm gap between them. Often a small wound with severe deep tissue injury.

Type 3 Numerous lacerations with avulsions often caused by rapidly moving planes with sharp cutting edges. Often severe mutilation results.

Extravasation Injuries in the Hand

Classification of extravasation injuries in the upper extremities (Loth and Eversmann, 1991).

- Vesicants Likely to cause necrosis.
- Irritants Local inflammation but unlikely to cause necrosis.
- Innocuous Rapidly resolving without apparent tissue damage.

Fingertip Injury

Allen's classification 1980 of fingertip injuries comprises four groups depending on level of injury and degree of tissue loss as outlined in Figure 14.17. The letter 'c' is appended for crush injuries and 'a' for clean amputations. It does not concern injuries specifically of the nail or address injury to the extensor mechanism.

Traumatic Amputations

The classification of traumatic amputation of the hand is based on anatomical zone systems. Such systems of classification are useful not only for diagnosis but also for the comparison of results of replantation between different units. Three such systems are in current use (Daniel and Terzis, 1977; Biemer, 1980; Tamai, 1982) (Figure 14.18). The system described by Tamai is primarily concerned with digital amputation level I injuries occur through the distal phalanx, level II injuries through the DIPJ and level III injuries occur distal to the finger PIPJ or thumb MCPJ. Level IV injuries occur between the base of the proximal and middle phalanx while level V injuries are through the metacarpal. The systems of Daniel and Biemer are quite similar with level I, II and III corresponding to the digital segments and level IV through

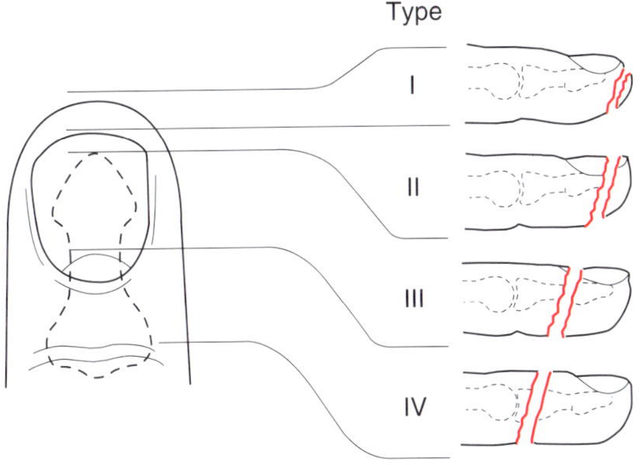

Figure 14.17. Allen's classification of finger tip injuries (Allen, 1980).

the metacarpals and level V injuries more proximal. The Daniel level VI injury is proximal to the wrist joint.

Discussion

The group considered the classification systems for hand fractures and tendon injuries. Overall this region of the body has been well researched.

Two general systems are recommended by the group, that of Dobyns, Linscheid and Cooney (1983) for fractures and that of Pun et al. from 1989 for classifications of stability. For classifications of the phalangeal fractures the classic work is that of London (1971) and is recommended for classifications of both flexors and extensor tendons. There are a number of other classification systems for less common injuries. For mallet finger the classification of Watson-Jones (1956) is most useful and for flexor digitorium profundus avulsions the classification of Leddy (1993). For boutonnière abnormalities the classifications of Souter (1967) appears to be the best buy. A summary of the discussion groups' findings is given in Table 14.5.

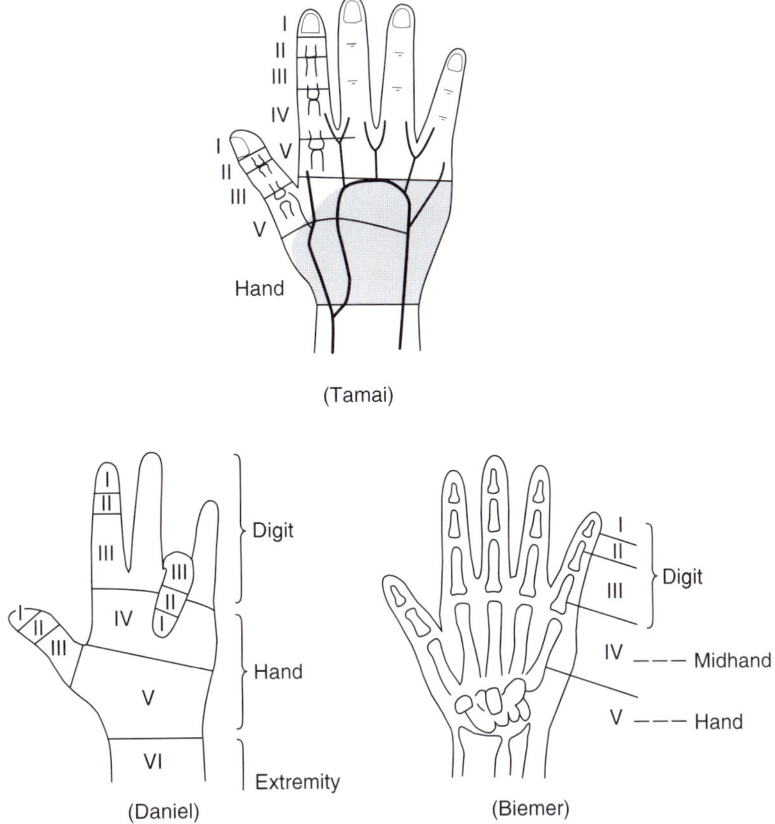

Figure 14.18. Three classifications of traumatic amputations of the hand (Tamai, 1982)

Table 14.5 Discussion groups' summary

Site	*Author*	*Good for defining treatment*	*Good for defining outcome*	*Reliability tested*	*Methods, e.g. exam X-ray CT, MRI etc*
General	Dobyns et al., 1983	Yes	Yes	No	X-ray
	Pun et al., 1989	Yes	Yes	No	Stability
Distal phalanx	Kaplan, 1940	No	No	No	X-ray
	Dobyns et al., 1982	Yes	Yes	No	
Interphalangeal joint condylar fracture	London, 1971	Yes	Yes	No	X-ray
Base fracture	Hastings and Carroll, 1988	No	No	No	X-ray
	Jupiter and Belsky, 1992	No	No	No	Exam
FDP avulsion	Leddy and Packer, 1979	Yes	Yes	No	Exam
Mallet finger	Watson-Jones, 1956	Yes	Yes	No	Exam
Mallet finger	Wehbe and Schneider, 1984	Yes	Yes	No	Exam
Boutonnière	Souter, 1967	Yes	Yes	No	Exam
Extensor tendon	Milford, 1980	Yes	Yes	No	Exam
Soft-tissue injury	Matev, 1967	Yes	Yes	No	Exam
Ring avulsion injury	Urbaniak et al., 1981	Yes	Yes	No	Exam
Metacarpal head fracture	McElfresh and Dobyns, 1983	Yes	Yes	No	X-ray
Metacarpal base fracture	Jupiter and Belsky, 1992	No	No	No	X-ray

References

Allen, M.J. (1980) Conservative treatment of fingertip injuries. *Hand*, **12**, 257.

Bennet, E.H. (1882) Fractures of the metacarpal bones. *Dublin J. Med. Sci.*, **73**, 72.

Biemer, E. (1980) Definition and classification in replantation surgery. *Br. J. Plast. Surg.*, **33**, 164.

Daniel, R.K. and Terzis, J.K. (1977) *Reconstructive Microsurgery*. Little, Brown: Boston, p. 128.

Dobyns, J.H., Beckenbaugh, R.D. and Bryan, R.S. (1982) In *Hand Surgery* (J.E. Flynn, ed.), 3rd edn. Williams & Wilkins: Baltimore, pp. 111–180.

Dobyns, J.H., Linscheid, R.L. and Cooney, W.P. (1983) Fractures and dislocations of the wrist. Then and now. *J. Hand Surg.*, **8A**, 690–692.

Doyle, J.R. (1993) Extensor Tendons—Acute Injuries. In *Operative Hand Surgery* (D.P. Green, ed.), 3rd edn. Edinburgh: Churchill Livingstone, pp. 1931–1934.

Dray, G.J. and Eaton, R.G. (1993) Dislocation and ligament injuries in the digits. In *Operative Hand Surgery* (D.P. Green, ed.), 3rd edn. Edinburgh: Churchill Livingstone, pp. 767–798.

Hastings, H. and Carroll, C. (1988) Treatment of closed articular fractures of the metacarpophalangeal and proximal interphalangeal joints. *Hand Clin.*, **4**, 503–528.

Jupiter, J.B. and Belsky, M.R. (1992) Fractures and dislocations of the hand. In *Skeletal Trauma* (B.D. Browner, J.B. Jupiter, A.M. Levine and P.G. Trafton, eds). Philadelphia: Saunder.

Kaplan, L. (1940) The treatment of fractures and dislocations of the hand: technique of unpadded casts for carpal metacarpal and phalangeal fractures. *Surg. Clin. North Am.*, **20**, 1695–1720.

Klinert, H.E. and Verdan, C. (1983) Report of the committee on tendon injuries. *J. Hand Surg.*, **8**, 794–798.

Leddy, J.P. (1993) Flexor Tendon Injuries. In *Operative Hand Surgery* (D.P. Green, ed.), 3rd edn. Edinburgh: Churchill Livingstone, pp. 1823–1824.

Leddy, L.P. and Packer, J.W. (1979) Avulsion of the profundus insertion in athletes. *J. Hand Surg.*, **4**, 461–464.

London, P.S. (1971) Sprains and fractures involving the interphalangeal joints. *Hand*, **3**, 155–158.

Loth, T.S. and Eversmann, W.W. (1991) Extravasation injuries in the upper extremity. *Clin. Orthop.*, **272**, 248–254.

McElfresh, E.C. and Dobyns, J.H. (1983) Intra articular metacarpal head fractures. *J. Hand Surg.*, **8**, 383–393.

Matev, I. (1967) Wringer injuries of the hand. *J. Bone Joint Surg.*, **49B**, 722–730.

Milford, L.W. (1980) The Hand. Flexor Tendon Injuries. (Extensor tendon injuries). In *Campbell's Operative Orthopaedics* (A.S. Edmonson and A.H. Crenshaw (eds), VIth edn. St Louis: Mosby, pp. 148–153.

Moberg, E. and Stener, B. (1954) Injuries to the ligaments of the thumb and fingers. *Acta Chir. Scand.*, **106**, 166–186.

Pun, W.K., Chow, S.P., So, Y.C. et al. (1989) A prospective study on 284 digital fractures of the hand. *J. Hand Surg.*, **14A**, 474–481.

Rolando, S. (1910) Fracture de la base du premier metacarpien et principalement sur une varieté non encore écrite. *Presse Med.*, **33**, 303–304.

Souter, W.A. (1967) The boutonnière deformity. *J. Bone Joint Surg.*, **49B**, 4, 710–721.

Stener, B. (1962) Displacement of the ruptured ulnar collateral ligament of the metacarpophalangeal joint of the thumb. *J. Bone Joint Surg.*, **44B**, 869–879.

Swanson, A.B. (1970) Fractures involving the digits of the hand. *Orthop. Clin. North Am.*, **1**, 261–274.2.

Tamai, S. (1982) 20 Years Experience of Limb Replantation: Review of 293 upper extremity replants. *J. Surg.*, **7**, 549–556.

Tang, J.B. (1994) Flexor tendon repair in zone 2C. *J. Hand Surg.*, **19B**, 1, 72–75.

Urbaniak, J.R., Evans, J.P. and Bright, D.S. (1981) Microvascular management of ring avulsion injuries. *J. Hand Surg.*, **6A**, 25–30.

Verdan, C.E. and Michon, J. (1962) Le Traitment des plaies des tendons fléchisseurs des doigts. *Rev. Chir. Orthop.*, **47**, 285.

Watson-Jones, R. (1956) *Fractures and Joint Injuries*. Edinburgh: Livingstone, pp. 45–646.

Wehbe, M.A. and Schneider, L.H. (1984) Mallet fractures. *J. Bone Joint Surg.*, **66A**, 658–669.

15

Proximal femoral fractures

M. J. Parker

Introduction

The term 'proximal femoral fractures' is synonymous with the more general term of 'hip fracture'. It refers to any fracture of the femur from the proximal articular surface to a level below the lower border of the lesser trochanter. The distance below the lesser trochanter varies between classifications from 3 to 10 cm. Fractures within this region can be subdivided into three main groups, femoral head, intracapsular femoral neck fractures, and extracapsular (Figure 15.1).

These three groups are illustrated in Figure 15.1. Points to note are:

1. The division between femoral head and intracapsular femoral fractures is clearly defined by the articular margin.
2. The division between intracapsular and extracapsular fractures is defined by the anatomical position of the capsular attachment to the femur. Anteriorly the capsule inserts into the intertrochanteric line, thereby enclosing the whole of the femoral neck. Posteriorly, the attachment is half way along the femoral neck. A small proportion of fractures will traverse the capsular boundaries and can be considered as both extracapsular and intracapsular.
3. The term 'extracapsular' encompasses both trochanteric and subtrochanteric fractures. The distal border of a subtrochanteric fractures is ambiguous, varying markedly with the different classification systems (see Figure 15.14).

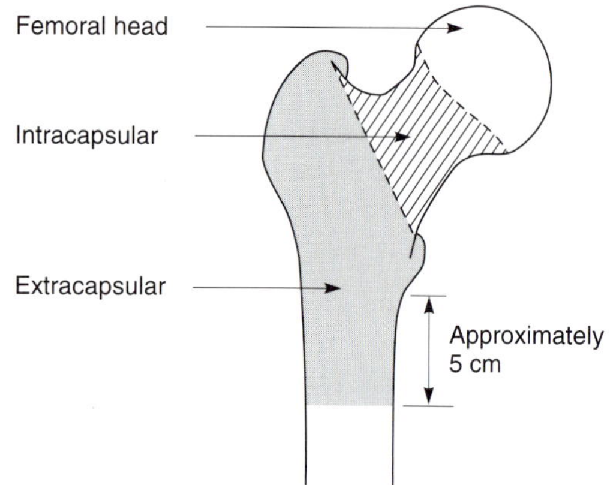

Figure 15.1. Primary division of the proximal femur into three main areas.

The description of each classification given in this chapter is, as far as possible, in the words of the original author.

Femoral Head Fractures

These fractures involve the articular surface of the proximal femur. These are rare and only three methods of classification have been described.

1. Pipkin Classification

Described by Garrett Pipkin of Kansas City in 1957 (Figure 15.2).

Type I	Dislocation of the hip with fracture of the femoral head caudad to the fovea capitis femoris.
Type II	Dislocation of the hip with fracture of the femoral head cephalad to the fovea capitis femoris.
Type III	Type I or type II injury associated with fracture of the femoral neck.

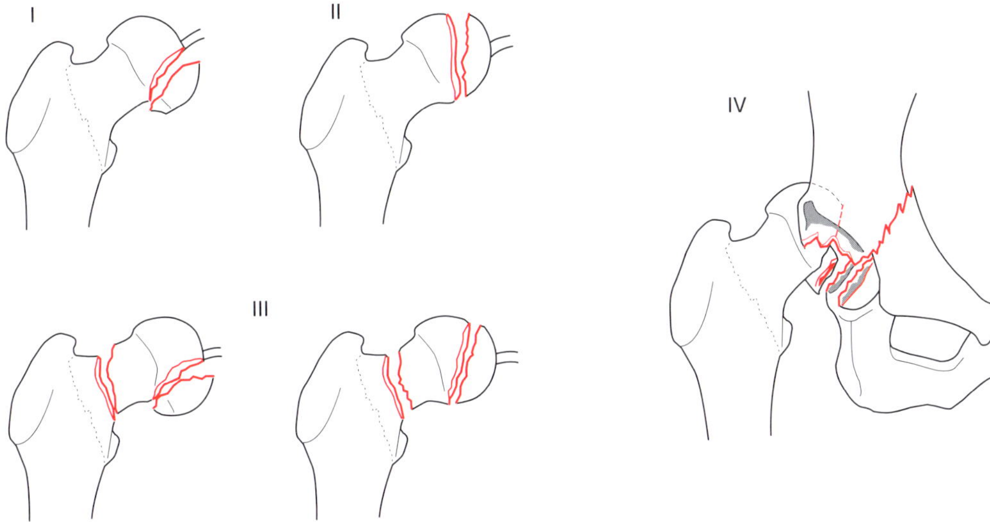

Figure 15.2. The Pipkin (1957) classification of femoral head fractures.

Type IV Type I or type II injury associated with fracture of the acetabular rim.

2. Brumback Classification

Described at the meeting of the American Hip Society in 1986 by Brumback and colleagues (Swiontkowski, 1991). There are five main types.

- Type 1 A posterior hip dislocation with femoral head fracture involving the inferomedial, non-weight bearing portion of femoral head.
- Type 1A With minimum or no fracture of the acetabular rim and stable hip joint after reduction.
- Type 1B With significant acetabular fracture and hip joint instability.
- Type 2 A posterior hip dislocation with femoral head fracture involving the superomedial, weight bearing portion of femoral head.
- Type 2A With minimum or no fracture of the acetabular rim and stable hip joint after reduction.
- Type 2B With significant acetabular fracture and hip joint instability.
- Type 3 Dislocation of the hip (unspecified direction) with associated femoral neck fracture.
- Type 3A Without fracture of the femoral head.
- Type 3B With femoral head fracture.
- Type 4 Anterior dislocation of the hip with fracture of the femoral head.
- Type 4A Indentation type: depression of the superolateral weight-bearing surface of the femoral head.
- Type 4B Transchondral type: osteocartilaginous shear fracture of the weight-bearing surface of the femoral head.
- Type 5 Central dislocation of the hip with fracture of the femoral head.

3. AO Classification

Figure 15.3 gives the AO classification of femoral head fractures (Müller et al., 1990).

31-C1 Femoral head fracture, split.
- .1 Avulsion of the ligamentum teres.
- .2 With rupture of the ligamentum teres.
- .3 Large fragment.

31-C2 Femoral head fracture, with depression.
- .1 Posterior and superior.
- .2 Anterior and superior.
- .3 Split-depression.

31-C3 Femoral head fracture, with neck fracture.
- .1 Split and transcervical neck fracture.
- .2 Split and subcapital neck fracture.
- .3 Depression and neck fracture.

Note, however, that in the AO classification in the Manual of Internal Fixation (Müller et al., 1991), C3.1 and C3.2 are changed.

Inter-observer Reliability

There have been no studies to compare inter-observer reliability for any of these classifications.

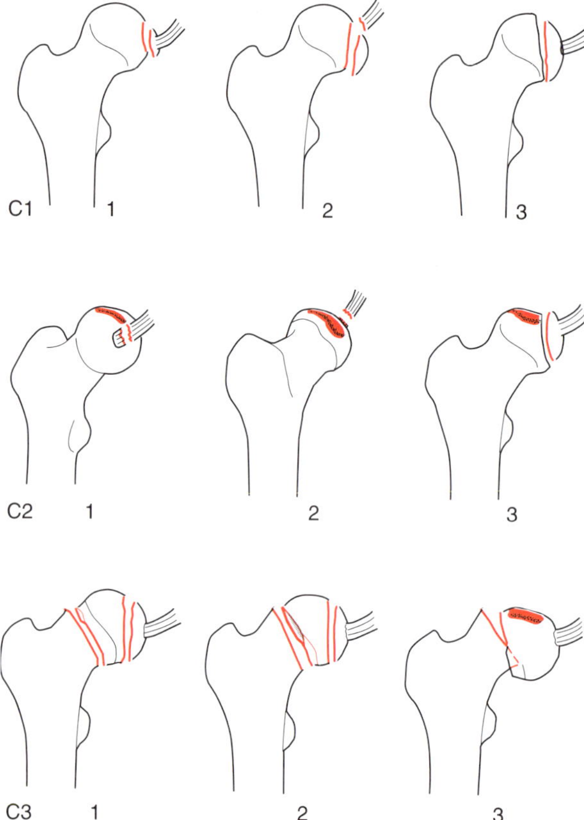

Figure 15.3. The AO/ASIF classification of femoral head fractures (Müller, 1990).

Published Reports Using the Different Classification Systems

All the reports of treatment and outcome of femoral head and neck fractures have used the Pipkin classification, with none using the AO or Brumback system. Pipkin based his classification on a series of 23 patients with femoral head fractures. The number of patients within each of the four groups (seven type 1, nine type 2, four type 3, three type 4) were too small to be able to define any preferred method of treatment for each group. A number of case series of femoral head fractures have since been published using the Pipkin classification but again reported number of patients have been small.

A summary of published reports of femoral head fractures found only 265 cases (Swiontkowski, 1991), of which 170 were classified by the Pipkin system. For these patients there was a wide variety of treatment regimes with different methods of assessment making it difficult to determine the merit for the classification system.

Factors not defined within the Pipkin classification system are probably more important in determining treatment. These factors include the size of the femoral head fragment, the degree of displacement of the femoral head fragment, the ability to perform a closed reduction, the presence of other injuries and the age and general physical health of the patient.

In summary, the currently available classification systems are of unproven value in determining the choice of treatment.

Value of Classification System in Prediction Outcome

Assessment of outcome for different types of femoral head fractures is difficult as the injury is generally overshadowed by other associated injures such as acetabular fractures. Swiontkowski (1991) summarizing the literature concluded that Pipkin types III and IV fractures carried the worst prognosis. For type III fractures the prognosis was determined by the degree of displacement of the femoral neck fracture.

Summary

Femoral head fractures are rare and the treatment and prognosis following these injuries are determined more by their associated injuries than the femoral head fracture. The only classification system which has been used in clinical practice is the Pipkin system but unfortunately not all fractures can be described by this method. The more recently described indentation or depression fracture of the femoral head, associated with anterior dislocation of the hip described by DeLee and colleagues (1980) cannot be classified by the Pipkin system. This led to the development of the fracture classification system described by Brumback.

Intracapsular Femoral Neck Fractures

This refers to fractures traversing the femoral neck. Figure 15.4 refers to some of the terms that have been used to describe these fractures. Points to note are:

1. The term 'subcapital fracture', if used precisely, refers to a fracture immediately distal to the articular surface. However, nowadays it is generally used to describe all intracapsular fractures.
2. The terms 'cervical, transcervical and midcervical', if used precisely, refer to a fracture in the mid part of the femoral neck. They are frequently used interchangeably to describe all intracapsular fractures.
3. The separation of fractures into subcapital and cervical is probably unhelpful. In clinical practice

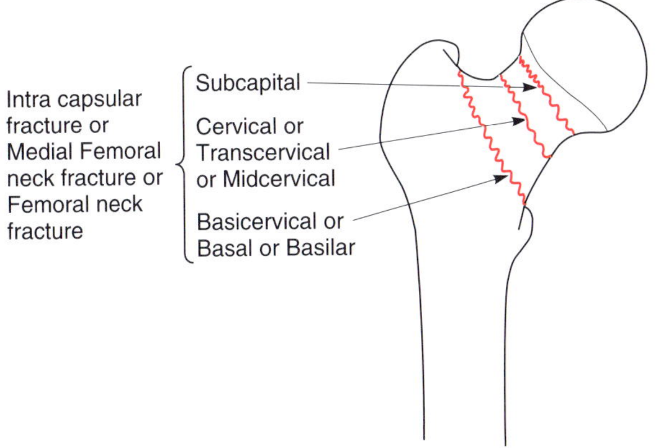

Figure 15.4. Different descriptive terms used to describe intracapsular fractures.

most intracapsular fractures start from a fixed point on the superior aspect of the femoral neck but cross the calcar at a distance of between 0 and 2.5 cm from the articular surface (Figure 15.5). The classical cervical fracture where the fracture line only involves the middle of the femoral neck is rare.

4. Basicervical, basal or basilar fractures may be regarded as either intracapsular or extracapsular as the fracture line may cross the capsular attachments. They are considered as intracapsular fractures in some fracture classification systems and extracapsular in others.
5. The terms 'femoral neck fracture' and 'medial femoral neck fractures' are of historical use. Medial femoral neck fractures were divided into abduction (impacted) and adduction (displaced) fractures.

Reported Classification Systems

1. Waldenström classification—1924

In the early 1900's impaction of the fracture was seen as one of the most important signs of stability and thereby uneventful fracture healing. Waldenström in 1924, divided intracapsular fracture into impacted or abduction fractures and displaced or adduction fractures.

2. Pauwels classification—1935

This classification was based on the theory of greater shearing forces at the fracture site for those fractures with a more vertical fracture line. The classification is determined by measuring the fracture angle on the anteroposterior (AP) radiograph (Figure 15.6). Pauwels divided fractures into three groups of:

Type I—Pauwels angle less than 30°
Type II—Pauwels angle 30 to 70°
Type III—Pauwels angle more than 70°

Since there are few fractures greater than 70° some papers have changed this angle to 50° or 60°. Other have used the shearing angle, this is the angle of the fracture to the femoral shaft (Brown and Abrami, 1964).

3. Garden classification—1961

This divided intracapsular fractures into four groups dependent on the trabeculae of the femoral head as seen on the AP radiograph (Figure 15.7).

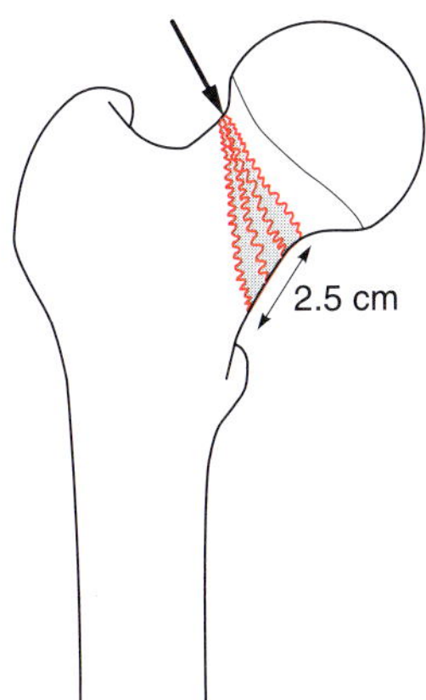

Figure 15.5. A cervical fracture as illustrated in Figure 15.4 is rare, as the fracture line invariably originates from a point just distal to the articular surface superiorly, to cross the calcar at more variable point.

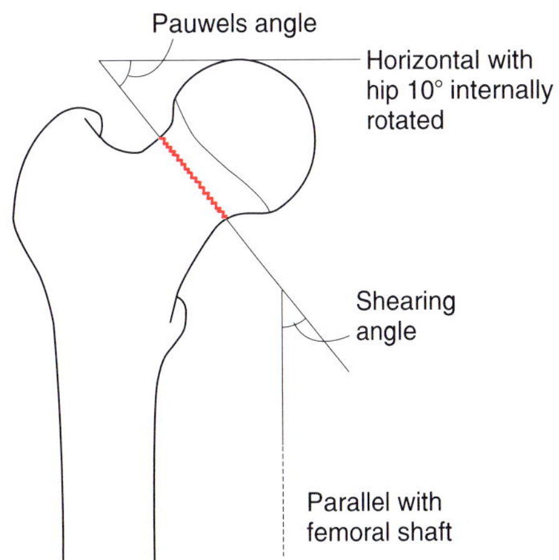

Figure 15.6. Pauwels (1935) angle and shearing angle (Brown and Abrami, 1964). The shearing angle is 90° – Pauwels angle.

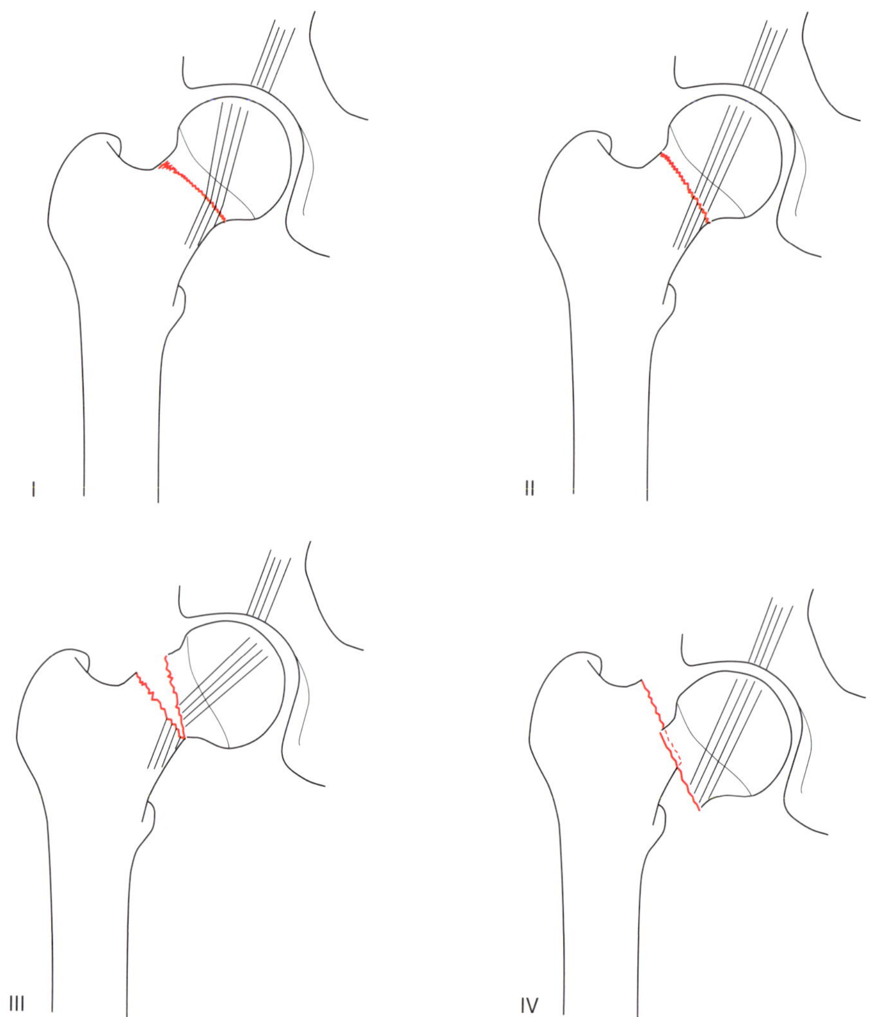

Figure 15.7. The Garden (1961) classification.

Stage I Incomplete subcapital fracture. The 'abducted' or 'impacted' injury in which the fracture of the inferior cortical buttress is greenstick in nature and a minimal degree of lateral rotation of the distal upon the proximal fragment creates the radiological illusion of impaction. The medial lamellae of the internal weight-bearing system in the distal fragment lie in abduction as compared with those in the capital fragment, which itself is adducted. If unprotected, this fracture may become complete at any time.

Stage II Complete subcapital fracture without displacement. The inferior cortical buttress has been broken but no tilting of the capital fragment has taken place. As in the stage I fracture, the closely opposed fragments in this complete fracture may succumb to lateral rotation forces and show the classical displacement of 'subcapital separation'.

Stage III Complete subcapital fracture with partial displacement. The two fragments retain their posterior retinacular attachment and crushing of the posterior cervical cortex has not yet taken place. Lateral rotation of the distal fragment, therefore, tilts the capital fragment into abduction and medial rotation as shown radiologically by the direction of the medial weight-bearing lamellae in the femoral head. If the tendency for the limb to rotate laterally is not resisted by external or internal fixation, stripping of the retinacular attachments and crushing of the thin posterior cervical cortex will allow the full displacement of stage IV to occur.

Stage IV Complete subcapital fracture with full displacement. This stage is reached when the retinacular hinge is detached from the posterior surface of the neck and collapse of the posterior cortical shell has taken place. The fragments are then divorced from each other and the capital fragment at once returns to a more normal position in the acetabulum. Its medial weight-bearing lamellae are then seen radiologically to lie in alignment with their fellows in the pelvis.

Some subsequent descriptions of the Garden classification, particularly in the American journals, have used an alternative description of the Garden classification in which stage III and IV fractures are differentiated by the degree of displacement of the fracture, rather than the appearance of the trabeculae (Figure 15.8).

4. Classifications related to the level of the fracture

Early descriptive terms of intracapsular fracture used the terms subcapital, cervical and basal to describe the level of the fracture (see Figure 15.4). Banks (1962) described four types of fracture (Figure 15.9). An alternative method of determining the level of the fracture using the epiphyseal scar was proposed by Tamai and colleagues in 1983. It divided fractures into three types (Figure 15.10).

A more precise method of determining the level of the fracture is shown in Figure 15.11 (Brown and Abrami, 1964).

5. Tronzo—1984

This divided intracapsular fractures into impacted, stable (displaced but fracture surfaces not separated) and unstable (complete separation of the lateral view). The stable and unstable groups are then subdivided into three groups dependent upon the radiographic appearance of the fracture.

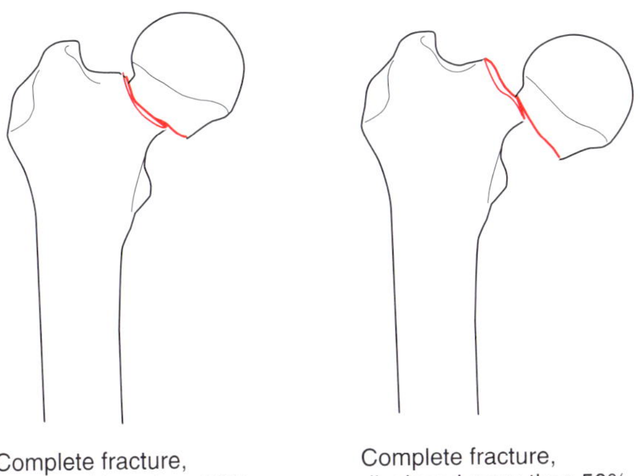

Figure 15.8. Modified description of Garden classification for stage III and IV fractures. Stage III. Complete fracture. Less than 50% displacement of the fracture surfaces. Stage IV. Complete fracture. More than 50% displacement of the fracture surfaces (see text).

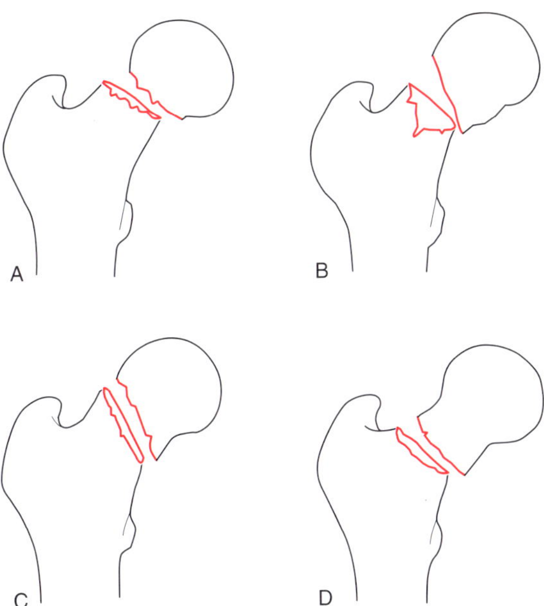

Figure 15.9. Banks' (1962) classification. Type A, classical subcapital fracture; Type B, wedge subcapital fracture; Type C, inferior-neck beak fracture; Type D, mid-neck fracture.

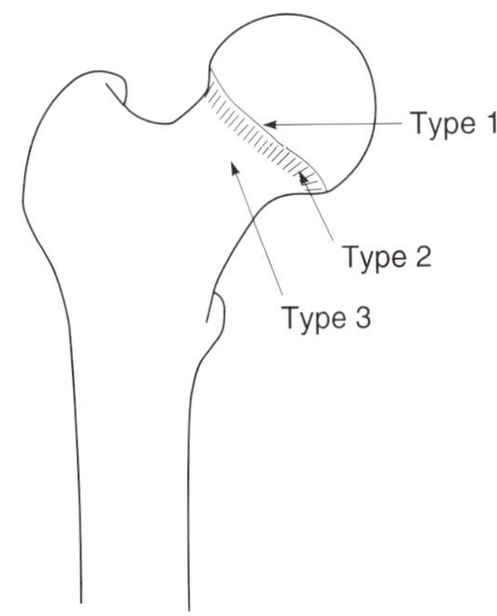

Figure 15.10. Tamai classification related to the epiphyseal scar. Type 1, fracture along the epiphyseal scar; Type 2, fracture line in subepiphyseal area; Type 3, fracture line in subcapital area.

6. AO classification (Müller et al.)—1990

This newer classification system uses a mixture of terminology from earlier articles. It includes the degree of fracture displacement, the level of the fracture and the angle of the fracture line (Figure 15.12).

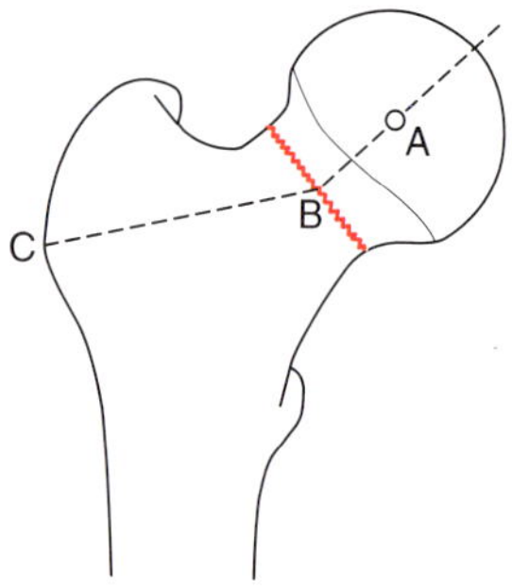

Figure 15.11. Brown and Abrami (1964) method of accurately determining the level of the fracture. Point A is the centre of the femoral head. The line AB bisects the fracture line. The ratio of CB to AB determines the level of the fracture.

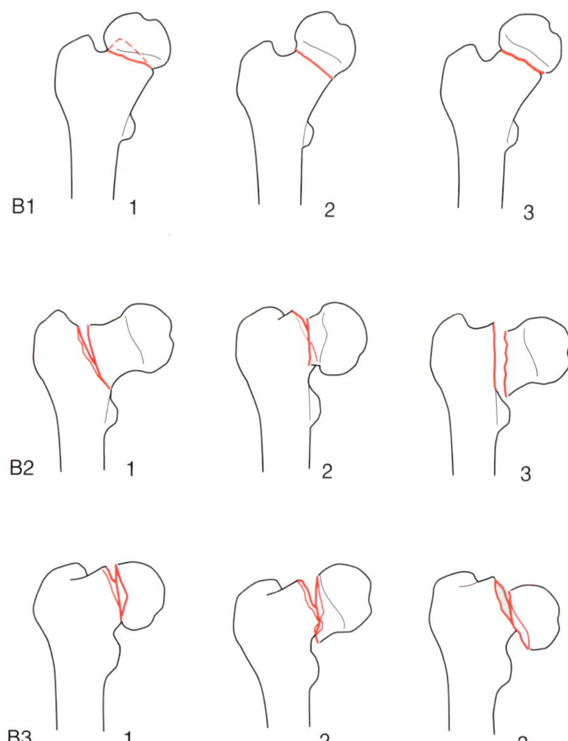

Figure 15.12. AO/ASIF classification of femoral neck fractures (Müller, 1990)

B1 Neck fracture, subcapital, with slight displacement.
- .1 impacted in valgus 15°
  - 1. Posterior tilt $< 15^\circ$ 2. Posterior tilt $> 15^\circ$
- .2 impacted in valgus $< 15^\circ$
  - 1. Posterior tilt $< 15^\circ$ 2. Posterior tilt $> 15^\circ$
- .3 non-impacted

B2 Neck fracture, transcervical.
- .1 basicervical
- .2 midcervical adduction
- .3 midcervical, shear

B3 Neck fracture, subcapital, non-impacted, displaced.
- .1 moderate displacement in varus and external rotation
- .2 moderate displacement with vertical translation and external rotation
- .3 marked displacement
  - 1. in varus 2. with translation

Interobserver Reliability

Two studies have reported acceptable levels of intra- and interobserver variation for simple division into displaced and undisplaced fractures (Blundell et al., 1988, Thomsen et al., 1996). However Frandsen et al. (1988) reported disagreement in 33% of fractures studied.

Two studies of the interobserver variation for the Garden classification (Frandsen et al., 1988, Thomsen et al., 1996), found a high degree of interobserver variation. Thomsen et al. (1996) reported a mean kappa value for interobserver variation of 0.39. In the study of Frandsen and colleagues (1998), there was agreement between the 8 observers for only 22% of the 100 fractures studied. A study of the Pauwels and Garden classification (McCabe et al., 1994) found an unacceptably high degree of interobservation variation for both classification systems. For the Garden system there was complete agreement by all four observers in only 17% of cases.

The AO classification system has been found to have very poor intra- and interobserver reliability, with kappa values between 0.25 and 0.36 for interobserver variation and 0.33 and 0.61 for intraobserver variation (Blundell et al., 1988). Another study of the AO system found that only seven out of 18 observers were able to correctly classify an intracapsular fracture (Johnstone et al., 1993).

Clinical Value of Different Classification Systems

For intracapsular fractures the choice of treatment and outcome are intimately linked. Any classification system of intracapsular fractures aims to predict the risk of non-union or avascular necrosis for the fracture whether treated operatively or conservatively. For intracapsular fractures the term 'non-union' includes those fractures which develop early displacement or re-displacement after reduction and internal fixation.

1. Waldenström classification

This is the original concept of classifying intracapsular fractures into those which are essentially undisplaced and those which are displaced. Undisplaced fracture includes those which are impacted or abduction fractures and can be equated to Garden stages I and II fractures. Displaced fractures refers to Garden stages III and IV fractures.

In excess of 100 different papers have used this simple classification. All have consistently reported a significant increase in the incidence of non-union for displaced fractures (Table 15.1).

2. Pauwels classification of intracapsular fractures

Pauwels originally measured the angle on the AP radiograph after internal fixation of the fracture with the leg in 10° of internal rotation. Clinical studies have since used this angle measured on the preoperative AP radiograph. Errors will occur with this method as, in the acute situation the X-ray is frequently taken in some degree of external rotation, which alters the Pauwels angle. A summary of studies of the Pauwels classification is give in Table 15.2.

Other studies have used the shearing angle. Most have reported no relationship between the fracture angle and the incidence of non-union or avascular necrosis (Hargadon and Pearson, 1963; Boyd and Salvatore, 1964; Brown and Abrami, 1964; Graham, 1968; Barnes et al., 1976; Svenningsen et al., 1984; Otremski et al., 1990; Raaymakers and Marti, 1991; Parker, 1994; Parker and Dynam, 1998). A few studies have found that undisplaced fracture with a more vertical fracture line have a higher risk of non-union (Eklund and Eriksson, 1964; Crawford, 1969; Bunata, Fahey and Drennan, 1973).

3. Garden classification of intracapsular fractures

The Garden classification uses the AP radiograph to divide intracapsular fractures into four stages. Stages 1 and 2 fractures are essentially undisplaced whilst stages 3 and 4 are displaced. Although stage 1 fractures are impacted and essentially undisplaced, some displacement may have occurred on the lateral radiograph. This is not considered within the Garden classification. An analysis of the degree of displacement of intracapsular fractures related to the Garden stage (Eliasson, Hansson and Kärrholm, 1988), found that there was no significant difference in the degree of

Table 15.1 Summation of the papers which have reported the incidence of healing complications for intracapsular fractures (from Parker and Pryor, 1993)

	Non-union	*Avascular necrosis*
Undisplaced intracapsular fracture treated conservatively	15%	10%
Undisplaced intracapsular fracture treated operatively	5%	10%
Displaced intrcapsular fracture treated by reduction and	internal fixation	21%

Table 15.2 Summary of trials relating the incidence of non-union or avascular necrosis to the Pauwels angle (percentages)

		Grade 1	*Grade 2*	*Grade 3*
Non-union				
Pugh	1955	0/2	2/13	2/14
Flatmark and Lone	1962	0/3	2/32	2/28
Cassebaum and Nugent	1963	2/10	11/31	8/29
Crawford	1969	0/20	4/47	0/6
Fielding et al.	1974	0/13	8/78	2/7
Parker and Dynam	1998	4/10	44/174	44/151
TOTAL		6/58(10%)	76/375(20%)	58/235(25%)
Avascular necrosis				
Crawford	1969	7/20	4/47	0.6
Massie	1973	8/27	53/191	20/49
TOTAL		15/47(32%)	57/238(24%)	20/55(36%)

displacement between Garden stages I and II fractures. Neither was there a significant difference between stages 3 and 4.

The Garden classification has been used in numerous publications as a predictor of non-union and avascular necrosis. One study of Garden stages I and II fractures treated conservatively, reported a 11% non-union rate for stage I fractures in comparison to 40% for stage II fractures (Jensen and Høgh, 1983). Tables 15.3 and 15.4 give a summary of publications that have used the Garden system after internal fixation of intracapsular fracture.

The merit of the Garden classification has been questioned because the difference between stages III

Table 15.3 Publications that have used the Garden stage as a predictor for non-union (percentage)

Author	*Year*	*Garden I*	*Garden II*	*Garden III*	*Garden IV*
Garden	1961	0/9	0/11	2/30	13/30
Brown and Abrami	1964			2/23	44/144
Fielding et al.	1974			10/80	2/20
Smyth and Shar	1974			7/37	7/41
Barnes et al.	1976	3/236	1/19	126/408	158/443
Graham	1968			7/46	19/66
Bracey	1977			16/36	8/20
Brown and Court-Brown	1979	0/8	3/27	5/35	34/130
Frandsen	1979			10/68	27/63
Kofoed and Alberts	1980			2/63	14/49
Frandsen and Andersen	1981			28/133	49/116
Høgh et al.	1982			8/49	11/44
Frandsen et al.	1984			10/73	31/83
Svenningsen et al.	1984	0/60	0/22	16/151	9/81
Cobb and Gibson	1986	0/24	0/19	0/17	3/11
Skinner and Powles	1986			8/73	10/34
Stappaerts and Broos	1987			9/39	18/46
Parker	1994	12/165	3/30	27/168	15/83
TOTAL		15/502 (3%)	7/128 (3%)	293/1529 (19%)	472/1504 (31%)

Table 15.4 Publications that have used the Garden stage as a predictor for avascular necrosis (percentage)

Author	*Year*	*Garden I*	*Garden II*	*Garden III*	*Garden IV*
Brown and Abrami	1964			4/23	24/144
Smyth and Shar	1974			11/37	10/41
Fielding et al.	1974			7/80	10/20
Barnes et al.	1976	34/236	1/19	72/408	76/443
Graham	1968			13/46	18/66
Bracey	1977			3/36	2/20
Frandsen	1979			12/68	7/63
Kofoed and Alberts	1980			5/63	7/49
Frandsen and Andersen	1981			21/133	15/116
Høgh et al.	1982			6/49	8/44
Nordkild and Sonne-Holm	1984	0/8	0/9	8/56	5/14
Cobb and Gibson	1986	1/24	2/19	3/17	6/11
Skinner and Powles	1986			15/73	15/34
Stappaerts and Broos	1987			13/39	6/46
TOTAL		35/268 (13%)	3/47 (6%)	193/1128 (17%)	209/1111 (19%)

and IV fractures is nominal and Garden stage II fractures are rare. It has been argued that it is only necessary to have a division into displaced and undisplaced fractures (Parker, 1993a; Eliasson et al., 1988).

4. Classifications related to the level of the fracture

Theoretical studies have suggested that the more proximal fracture will have an increased risk of the fracture healing complications. The method described by Tamai relates the epiphyseal scar to the site of the fracture. It was implied that only the type 3 distal fractures had a viable femoral head. The classification system has not been used or studied in clinical practice.

The subdivision of the fracture level using the terms 'subcapital' and 'cervical' has not been reported to be of clinical value. Basal fractures are regarded by many as being more akin to an extracapsular fracture; there have been inadequate reports of their outcome to determine if a separate subclassification is justifiable.

Regarding those studies which looked at the fracture level using the method shown in Figure 15.11, most found that the fracture level had no effect on the risk of fracture healing complications (Svenningsen et al., 1984; Unger and Shuster, 1986; Alho et al., 1991; Parker, 1994). Two studies (Graham, 1968; Barnes et al., 1976), found an increased risk of non-union for those fractures which were more proximally based, whilst one study reported the opposite (Fielding, Wilson and Ratzan, 1974). Brown and Abrami (1964) reported an increased risk of avascular necrosis for more proximally based fractures.

5. AO classification of femoral neck fractures

This more complex system utilizes features from a number of different classification systems to divide fractures into three broad groups, undisplaced/minimally displaced, transcervical/basicervical and displaced subcapital fractures. Each of these groups is then subdivided into three groups.

Two studies of the AO classification system for hip fractures have found it unnecessarily complicated and that it did not play a useful role in planning management (Newey, Ricketts and Roberts, 1993; Blundell et al., 1998).

Summary

Many other factors not seen on the radiograph will also influence the risk of non-union and avascular necrosis. Such factors include the patient's age, time interval between injury and surgery and the presence of metabolic bone disease or rheumatoid arthritis. No method of classification dependent solely on the radiographic appearance can, therefore, be reliably used to predict fracture healing.

Extracapsular Fractures

Terminology used to describe extracapsular fractures is confusing. Anatomical boundaries are not clearly defined and many of the anatomical descriptive terms are ambiguous. Figure 15.13 gives the terms that will be used for this chapter. Points to note from Figures 15.13 are:

1. A basal fracture is a two-part fracture just proximal to the lesser and greater trochanters. As it traverses the capsular boundaries it is considered as both an intracapsular and an extracapsular fracture by different classification systems.
2. A trochanteric fracture refers to those cases in which the principal fracture line traverses the femur within the area shown.
3. A proximal trochanteric fracture refers to a fracture line which runs in an infero-medial direction between the greater and lesser trochanters.
4. A fracture in the distal trochanteric area refers to one in which the fracture line is predominately within the area of bone at the level of the lesser trochanter. Such fractures lines may be transverse, oblique running infero-medially or oblique and running infero-laterally, the so-called reverse fracture line.
5. The proximal and distal borders of a subtrochanteric fracture are ambiguous and many of the classification systems for subtrochanteric fractures fail

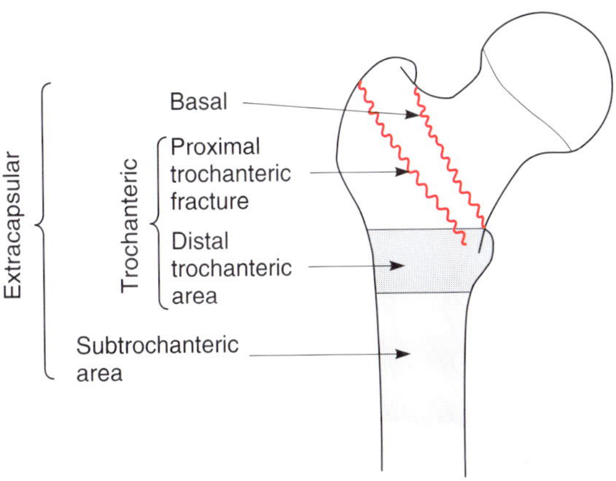

Figure 15.13. Terminology used to describe the different anatomical areas in this chapter.

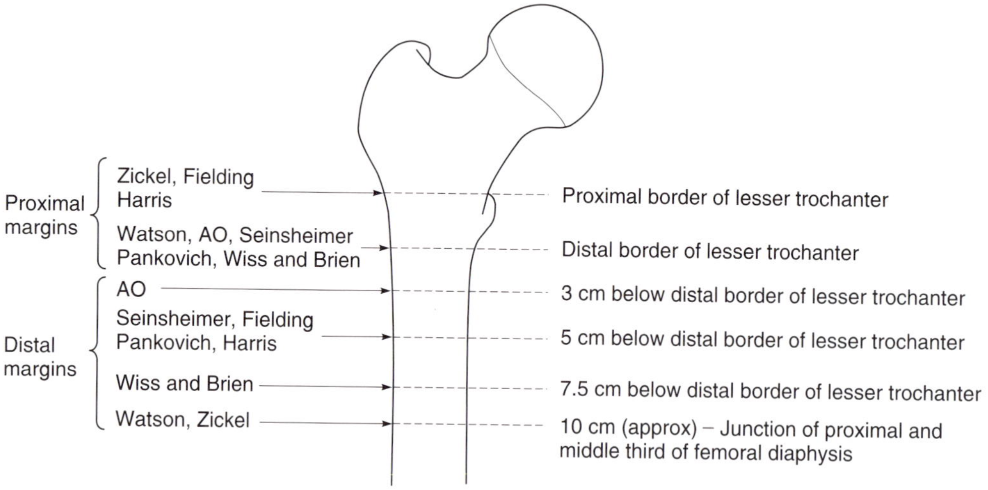

Proximal margins
Zickel, Fielding Harris — Proximal border of lesser trochanter
Watson, AO, Seinsheimer Pankovich, Wiss and Brien — Distal border of lesser trochanter
Distal margins
AO — 3 cm below distal border of lesser trochanter
Seinsheimer, Fielding Pankovich, Harris — 5 cm below distal border of lesser trochanter
Wiss and Brien — 7.5 cm below distal border of lesser trochanter
Watson, Zickel — 10 cm (approx) – Junction of proximal and middle third of femoral diaphysis

Figure 15.14. Different proximal and distal borders used to define a subtrochanteric fracture.

to define the anatomical boundaries. The proximal and distal borders of a subtrochanteric fracture for those classification systems which do give a definition are shown in Figure 15.14.

6. The term 'pertrochanteric' is poorly defined and ambiguous. It is sometimes used for all types of trochanteric fractures and in a few classification systems for fractures running obliquely between the lesser and greater trochanter (Figure 15.15).
7. Likewise, the exact meaning of the term 'intertrochanteric fracture' is inadequately defined. The AO classification uses the term 'intertrochanteric' exclusively for fracture in the distal trochanteric area. Surgeons may use the term loosely to include fractures in the region of the 'intertrochanteric' line.

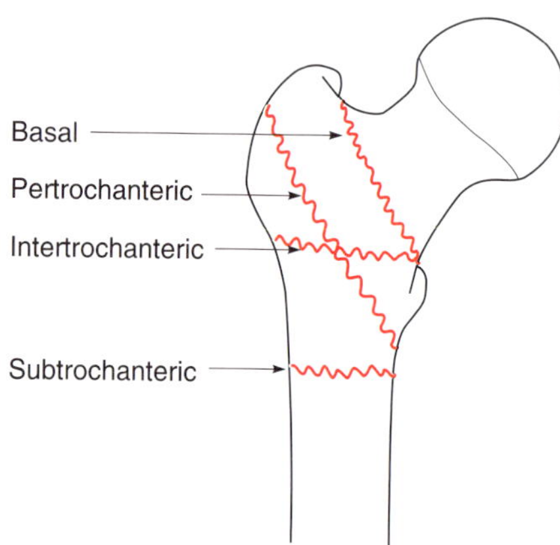

Figure 15.15. Classification of Moore (1939).

Reported Classification Systems

Classification methods for these fractures are based on either morphology, the degree of comminution or the extent of displacement of the fracture.

A large number of different systems have been described for extracapsular fractures (Table 15.5).

Moore—1939

This early descriptive article on fractures of the proximal femur used the four descriptive terms shown in Figure 15.15. The exact meaning of each of the terms used was not, however, specified.

Briggs and Keats—1946

This early classification system divided trochanteric fractures into five groups:

Type 1 Basal fractures.
Type 2 Two-part trochanteric fractures.
Type 3 Three-part involving the lesser trochanter.
Type 4 Comminuted pertrochanteric.
Type 5 Pertrochanteric with fracture of the lesser trochanter and may extend into the subtrochanteric region.

Boyd and Griffin—1949

This classification system considered both the anatomical site and stability of the fracture after reduction (Figure 15.16):

Table 15.5 Published method of classifying extracapsular femoral fractures and the area of bone to which they apply. PT = Proximal trochanteric fracture, DT = fractures in the distal trochanteric area, ST = subtrochanteric fracture

Study	*Year*	*PT*	*DT*	*ST*	*Number of subdivisions*
Moore	1939	+	+	+	4
Briggs and Keats	1946	+	+	–	5
Boyd and Griffin	1949	+	+	–	4
Evans	1949	+	+	–	5
Murray and Frew	1949	+	+	+	5
Böhler	1957	+	+	–	4
Watson et al.	1964	–	–	+	–
Massie	1964	+	+	–	12
Fielding and Magliato	1966	–	+	+	3
May and Chacha	1968	+	–	–	2
Ender	1970	+	+	–	3
Tronzo	1973	+	+	–	6
Cech and Sosna	1974	–	–	+	5
Jensen and Michaelsen	1975	+	–	–	5
Zickel	1976	–	+	+	6
Seinsheimer	1978	–	–	+	8
Waddell	1979	–	+	+	3
Kyle et al.	1979	+	+	–	4
Pankovich et al.	1979	+	+	+	6
Harris	1980	+	+	+	6
Wilson et al.	1980	+	–	–	3
Malkawi	1982	–	–	+	5
Tronzo	1984	+	+	–	10
AO (Müller)	1990	+	+	+	27
Wiss and Brien	1992	–	–	+	6

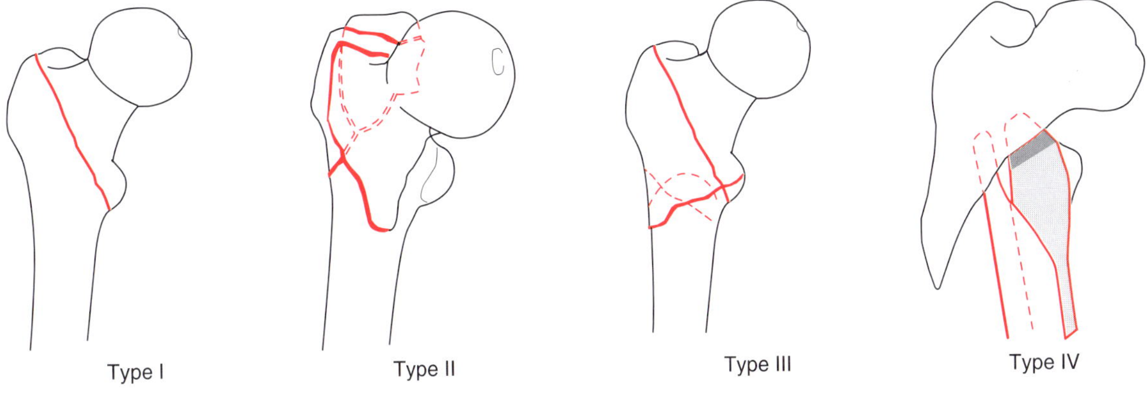

Figure 15.16. Boyd and Griffin (1949) classification of proximal femoral fractures.

Type I	Linear fracture between the trochanters. Reduction is simple and maintained.
Type II	Comminuted trochanteric fracture.
Type III	Fracture line in the region of, or below the lesser trochanter. May be comminuted.
Type IV	Comminuted fracture from the trochanteric region extending into the shaft of the femur.

Evans—1949

Evans proposed a classification of trochanteric fractures, based on the stability of the fracture as determined by the pre and post-reduction radiographs (Figure 15.17).

Type 1

STABLE	Undisplaced fracture
	Displaced but when reduced, overlap of the inner cortical buttress makes the fracture stable
UNSTABLE	Displaced and the medial cortical buttress is not restored by reduction of the fracture
	Displaced and comminuted fracture in which the medial cortical buttress is not restored by reduction of the fracture

Type 2

UNSTABLE	Reversed obliquity fracture line

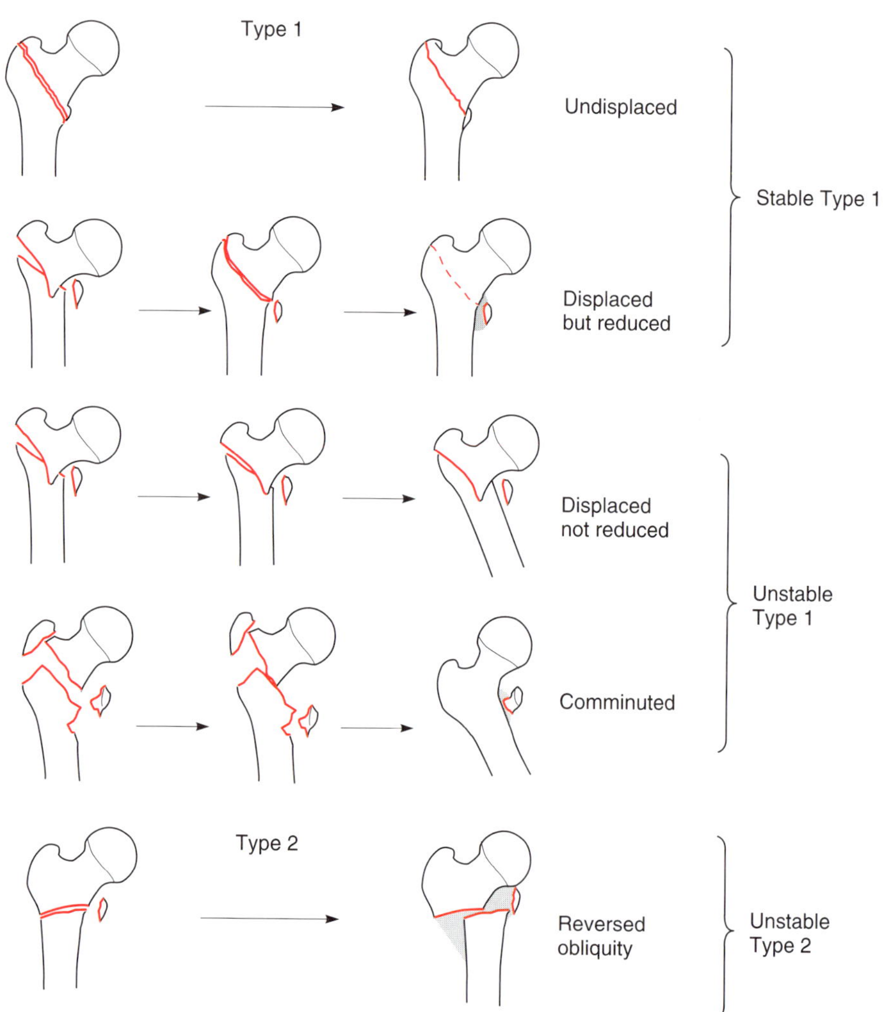

Figure 15.17. Evans' (1949) classification. Type 1 'stable' fractures when reduced. Type 1 'unstable' fractures when reduced. Type 2 are all 'unstable' when reduced.

Murray and Frew—1949

Extracapsular fractures were divided into five groups, based on their anatomical characteristics (Figure 15.18):

Type 1 Basal
Type 2 Intertrochanteric
Type 3 Comminuted intertrochanteric
Type 4 Oblique subtrochanteric
Type 5 Transverse subtrochanteric

Böhler—1957

This system divided trochanteric fracture into four groups:

1. Basal.
2. Fracture line between the trochanters with external rotation of the leg and separation of the fragments.
3. Base of neck driven deeply into the spongy mass of the trochanters.
4. Fracture line in region of the trochanters often with comminution.

Watson, Campbell and Wade—1964

This system used three numbers to describe subtrochanteric fractures. The first number referred to the distance in centimetres from the lesser trochanter to the start of the fracture line, the second was the length of the fracture and the third the number of fragments.

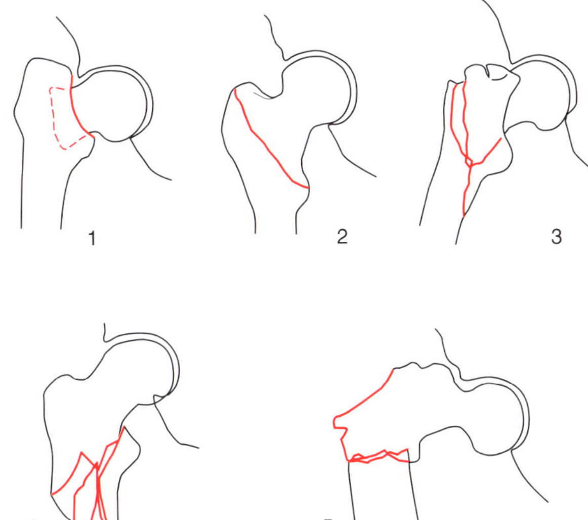

Figure 15.18. Classification of Murray and Frew (1949) based on radiographic tracings.

Massie—1964

This divided extracapsular fractures into three groups of stable undisplaced, displaced but stable, and displaced with instability. Each of these three groups was further divided into one of four subgroups:

Trochanteric—fracture between trochanters or basal
Intertrochanteric transverse
Intertrochanteric, spiral or oblique
Comminuted intertrochanteric.

Fielding and Magliato—1966

This classification is only for fractures at the level of and for two inches distal to the lesser trochanter (Figure 15.19). As many fractures will traverse these boundaries the level for which the major part of the fracture is found is used.

May and Chacha—1968

This classification divided trochanteric fractures into two groups depending on the appearance of the greater trochanter and femoral neck on the AP and lateral radiographs. For type I fractures the proximal fragment consists of the head and neck alone. For type II fractures the head, neck and major part of the greater trochanter constitutes the proximal fragment.

Ender—1970

This classification divided trochanteric fractures into three groups:

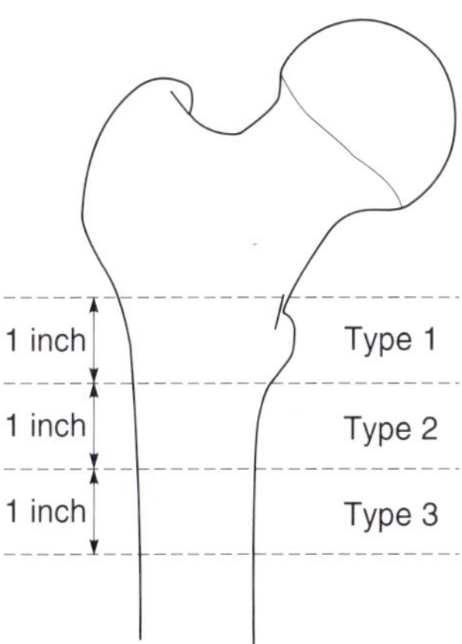

Figure 15.19. Fielding and Magliato (1966) classification of subtrochanteric fractures based on the proximal part of the lesser trochanter. Type 1 at the level of the lesser trochanter; Type 2 2.5–5 cm below; and Type 3 5–7.5 cm below the lesser trochanter.

Type 1 Eversion fracture with a posteromedial fragment.

Type 2 Impaction fracture with inversion and adduction of the neck fragment and varus collapse of the fracture.

Type 3 Diatrochanteric fracture. Fracture line extending subtrochanterically or a reversed fracture line.

Tronzo—1973

This was a reinterpretation and modification of the Boyd and Griffin classification system.

Type 1 Incomplete fracture only involving greater trochanter.

Type 2 Un-comminuted fracture with or without slight displacement. Intact posterior wall and relatively small lesser trochanter fragment.

Type 3 Comminuted posterior wall with telescoping of the neck spike into the femoral shaft. Larger lesser trochanter fragment.

Type 4 As for type 3 but the greater trochanter is also in a separate fragment.

Type 5 Comminuted posterior wall without telescoping of the two major fragment. Neck spike displaced outside the femoral shaft. Larger lesser trochanter fragment.

Type 6 Reversed fracture line with medial displacement of the shaft.

A later classification system by the same author (Tronzo, 1984) divides trochanteric fractures into stable and unstable with five subgroups in each group dependant on the radiographic appearance of the fracture.

Cech and Sosna—1974

This applied to subtrochanteric fractures.

Type 1 Subtrochanteric area with extension into the trochanteric region.

Type 2 Proper subtrochanteric. These fractures are subdivided into simple fractures which are two-part transverse or oblique and into comminuted fractures.

Type 3 Pathological subtrochanteric fractures.

Type 4 Subtrochanteric fractures in children.

Jensen and Michaelsen—1975

This is a modification of the Evans classification. It is based on the initial radiographs of the fracture (Figure 15.20). The description of this classification is only readily applied to fractures in the proximal trochanteric region, and is difficult to apply to fractures at the level of the lesser trochanter. Reversed fracture lines (Evans type 2), fractures are re-classified as type 3 fractures. It is not intended for subtrochanteric fractures. Type 1 and 2 were considered as stable and 3, 4 and 5 as unstable.

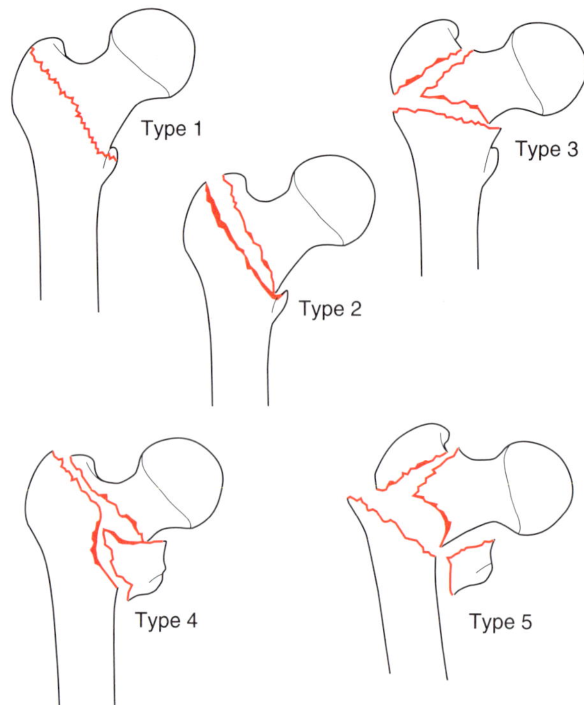

Figure 15.20. Jensen and Michaelsen (1975) classification of trochanteric fractures.

Type 1 Undisplaced two-fragment fracture.

Type 2 Displaced two-fragment fracture.

Type 3 Three-fragment fracture, where the greater trochanter constitutes the third part, giving loss of posterolateral support.

Type 4 Three-fragment fracture where the lesser trochanter constitutes the third part, giving loss of medial support.

Type 5 Four-fragment fracture involving both the lesser and greater trochanter, with loss of medial and posterolateral support.

Zickel—1976

This classification defined a subtrochanteric fracture as any fracture within the area of femur from the proximal border of the lesser trochanter, to the junction of the upper and middle third of the femoral diaphysis. Three different types of fracture were

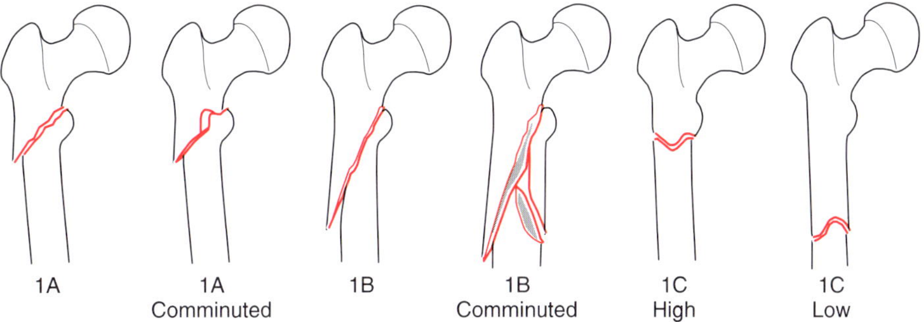

Figure 15.21. The Zickel (1976) classification of proximal femoral fractures. Type 1A short oblique fracture without/with comminution; Type 1B long oblique fracture without/with comminution; Type 1C high and low transverse fractures.

described, each of which could be subdivided into two types (Figure 15.21).

Type 1A Short oblique fracture and short oblique fracture with comminution.

Type 1B Long oblique fracture and long oblique fracture with comminution.

Type 1C High and low transverse fractures.

Seinsheimer—1978

Although this is primarily a classification system for subtrochanteric fractures, the definition of fractures to be included was an fracture with part of the fracture line within the area of bone from the inferior aspect of the lesser trochanter to a point 5 cm distal to this. Therefore some femoral shaft fracture and trochanteric fractures with subtrochanteric extensions will be included. Figure 15.22 illustrates the classification.

Type I Non-displaced fractures. Any subtrochanteric fracture with displacement of less than 2 mm.

Type IIA A two-part transverse femoral fracture.

Type IIB A two-part spiral fracture with lesser trochanter attached to the proximal fragment.

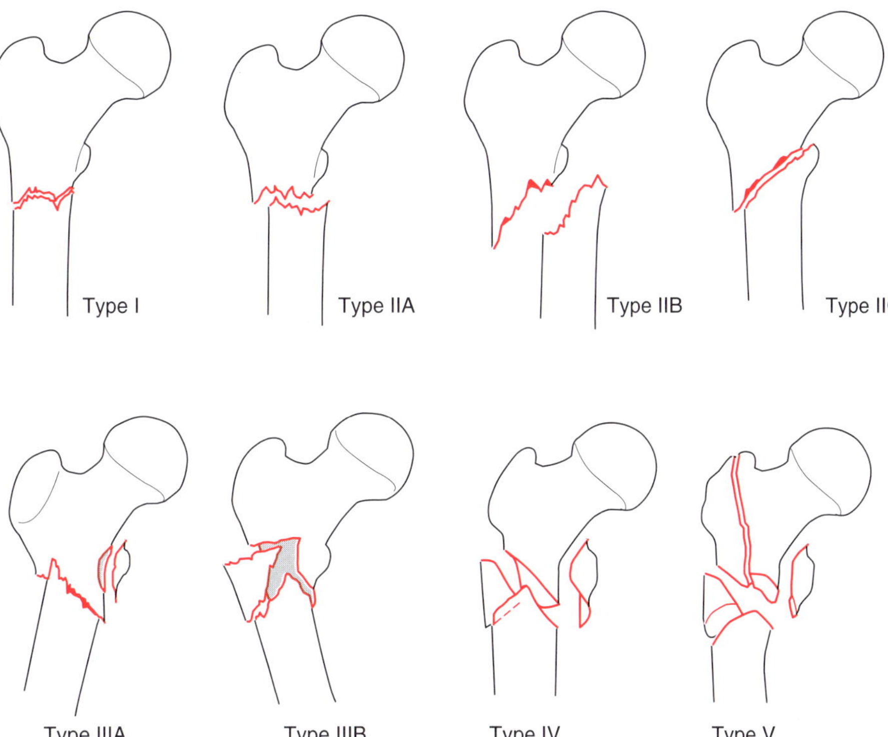

Figure 15.22. Seinsheimer classification.

Type IIC — A two-part spiral fracture. Oblique fracture line, with lesser trochanter attached to the distal fragment.

Type IIIA — A three-part spiral fracture in which the lesser trochanter was part of the third fragment which had an interior spike of the cortex of varying lengths.

Type IIIB — A three-part spiral fracture of the proximal one-third of the femur with the third part a butterfly fragment.

Type IV — A comminuted fracture with four or more fragments.

Type V — Subtrochanteric and intertrochanteric fractures. Any subtrochanteric fracture with extension through the greater trochanter.

Waddell—1979

A simple classification of subtrochanteric fractures into three groups.

Type 1 Transverse or short oblique, minimal comminution.

Type 2 Long. Minimal or no comminution.

Type 3 Comminuted.

Kyle, Gustilo and Premer—1979

This is a modification of the Evans classification, dividing trochanteric fractures into four groups (Figure 15.23).

Type I — Stable, undisplaced intertrochanteric line fracture.

Type II — Stable, displaced intertrochanteric line fracture. Varus deformity. Fracture of lesser trochanter.

Type III — Unstable, displaced fracture of greater trochanter. Posterior medial comminution. Varus deformity.

Type IV — Unstable, displaced comminuted inter-subtrochanteric fracture. Fracture of greater trochanter. Posteromedial comminution with subtrochanteric component.

Pankovich et al.—1979 and 1980

This is a classification system for all types of extracapsular fractures which were primarily divided into intertrochanteric and subtrochanteric fractures.

Intertrochanteric

1. Stable intertrochanteric (two-part fracture).
2. Unstable intertrochanteric (four-part fracture).

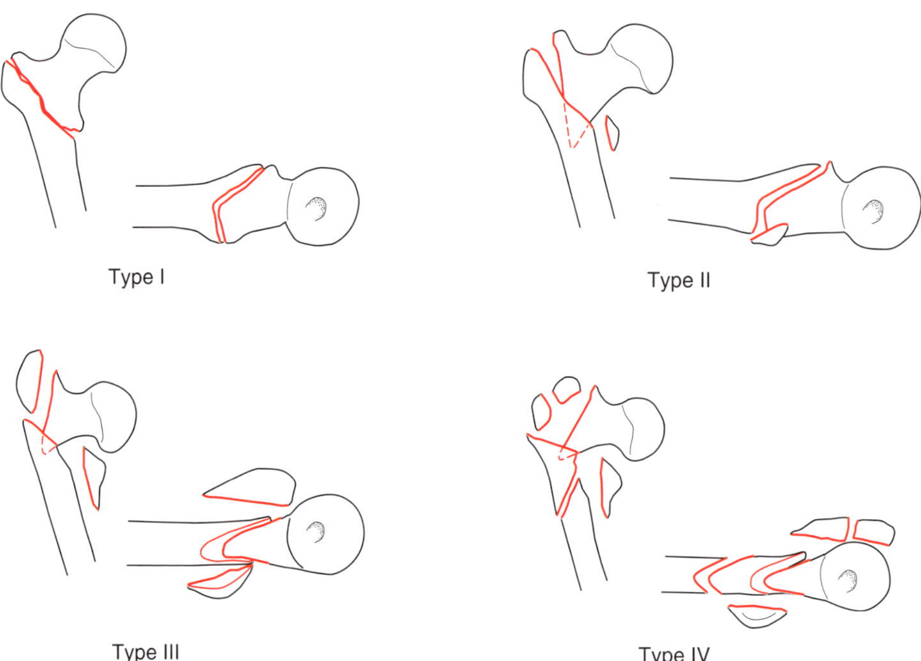

Figure 15.23. Classification of Kyle et al. (1979).

Subtrochanteric

1. Simple and short oblique, stable post-operatively.
2. Long oblique or spiral, potentially unstable.
3. Single butterfly fragment or communition on only one side, usually stable.
4. Comminution of the entire circumference of the cortex. Potentially very unstable.

Harris—1980

This clearly-described classification divided extracapsular into three main groups (Figure 15.24): transtrochanteric, stable subtrochanteric and unstable subtrochanteric:

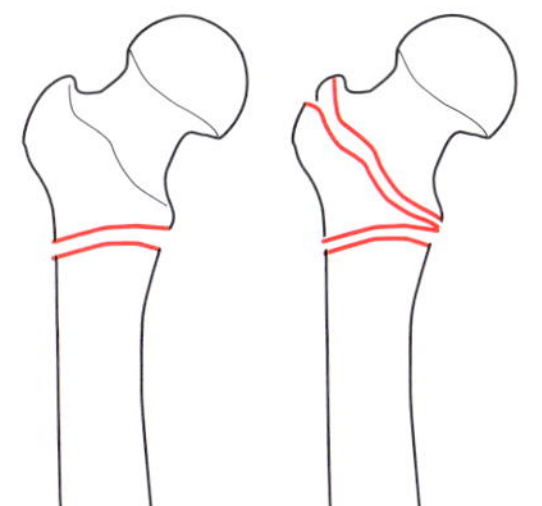

Transtrochanteric

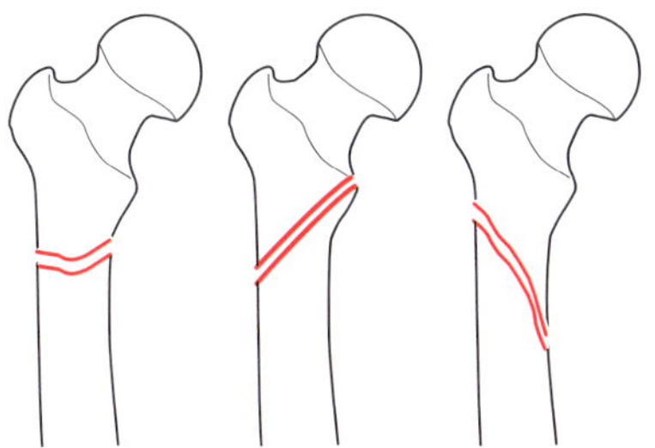

Stable subtrochanteric

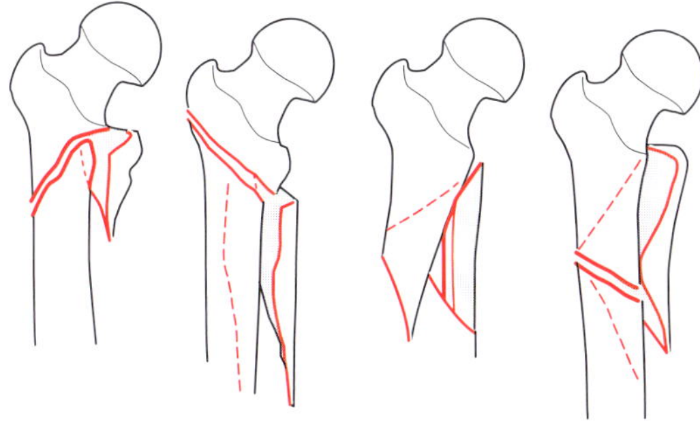

Unstable subtrochanteric

Figure 15.24. Harris (1980) classification of subtrochanteric fractures.

1. Transtrochanteric: horizontal or slightly oblique at the level of lesser trochanter, with or without extension into the intertrochanteric region.
2. Subtrochanteric fractures, horizontal or short oblique fractures at a level between lower margins of lesser trochanter and 5 cm distally. The major fracture line measures between 90 and 45° to the axis of the femoral shalf, and there is no significance posterior and medial communication.
3. Unstable subtrochanteric fractures: horizontal or short oblique fractures with posteromedial comminution or long oblique fractures. The major fracture line of long oblique fractures is angulated less than 45° to the axis of the shaft (usually approximately 30°) and these fractures are most often associated with significant comminution.

Wilson et al.—1980

This classification divided trochanteric fractures into three groups.

1. Paratrochanteric—usually stable. Occurs through the base of the neck in the region of the intertrochanteric line.
2. Pertrochanteric—fracture between the greater and lesser trochanters, usually with comminution. Loss of the medial buttress and subsequent instability with varying deformation.
3. Pertrochanteric with subtrochanteric extension.

Malkawi—1982

A classification system for subtrochanteric fractures (Figure 15.25).

Type I Non-comminuted, subdivided into
- I-A Transverse fractures
- I-B Oblique fractures

Type II Comminuted, subdivided into
- II-A With medial cortical comminution
- II-B With medial cortical comminution and greater trochanter comminution
- II-C With lateral cortical comminution

AO classification (Müller)—1990

Fractures in the trochanteric region down to the lower border of the lesser trochanter are divided into nine subgroups (Figure 15.26).

A1 Trochanteric area fracture, pertrochanteric simple.

.1 Along the intertrochanteric line
1. non-impacted 2. impacted

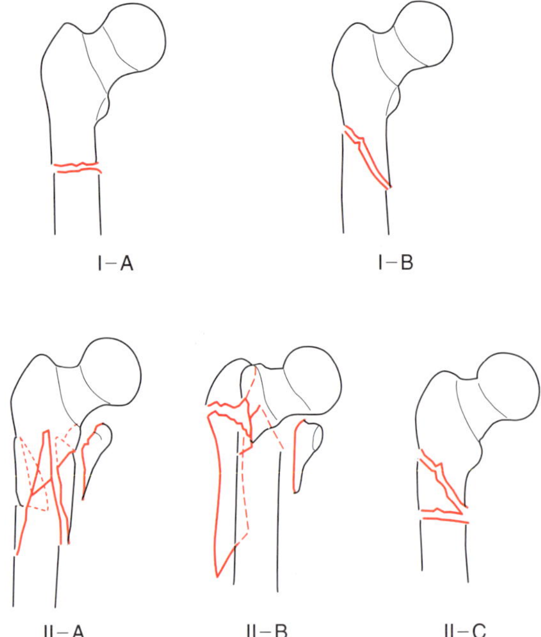

Figure 15.25. Malkawi (1982) classification of subtrochanteric fractures.

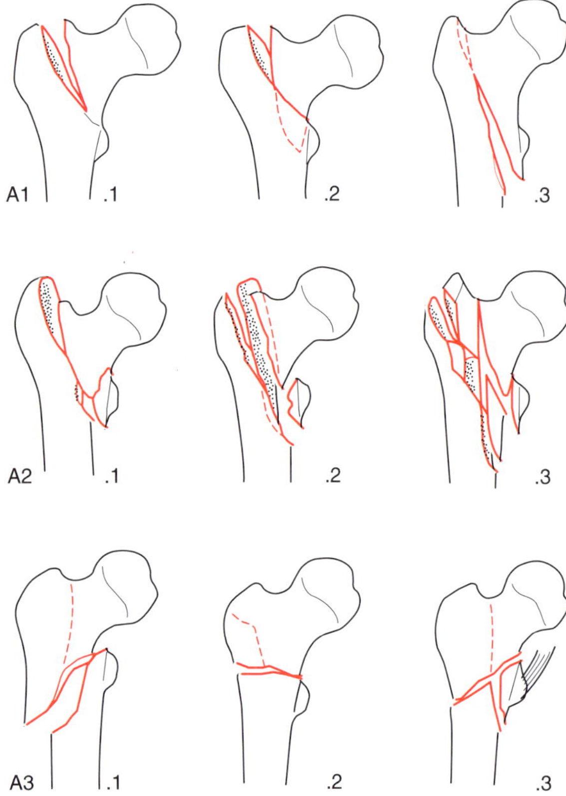

Figure 15.26. AO/ASIF classification of trochanteric fractures (Müller, 1990).

.2 through the greater trochanter
1. high variety 2. low variety
.3 Below the lesser trochanter

A2 Trochanteric area fracture, pertrochanteric multifragmentary
.1 With one intermediate fragment
.2 With several intermediate fragments
.3 Extending more than 1 cm below the lesser trochanter

A3 Trochanteric area fracture, intertrochanteric
.1 Simple, oblique
.2 Simple, transverse
.3 Multifragmentary

1. Extending to the greater trochanter
2. Extending to the neck of femur

Basal fractures are excluded, and classified within the AO classification of femoral neck fractures (see Figure 15.12). Femoral diaphyseal fractures in the subtrochanteric zone are classified as shown in Figure 15.27.

A1.1 Simple fracture spiral
A2.1 Simple fracture oblique ($\geq 30°$)
A3.1 Simple fracture transverse ($< 30°$)

B1.1 Wedge fracture, spiral wedge
B2.1 Wedge fracture, bending wedge
B3.1 Wedge fracture, fragmented wedge

C1. Complex fracture, spiral
.1 With two intermediate fragments
.2 With three intermediate fragments
.3 With more that three intermediate fragments

C2. Complex fracture, segmental
.1 With one intermediate segmental fragment
.2 With one intermediate segmental fragment and additional wedge fragment(s)
.3 With two intermediate segmental fragments

C3.1 Complex fracture, irregular
.1 With two or three intermediate fragments
.2 With limited shattering (< 5 cm)
.3 With extensive shattering (≥ 5 cm)

The subtrochanteric zone is defined as the 3 cm of bone distal to the lower border of the lesser trochanter. This area of bone is smaller than that for other classification systems (see Figure 15.14). The subdivisions are based on the anatomical site and number of fragments and does not consider the degree of displacement.

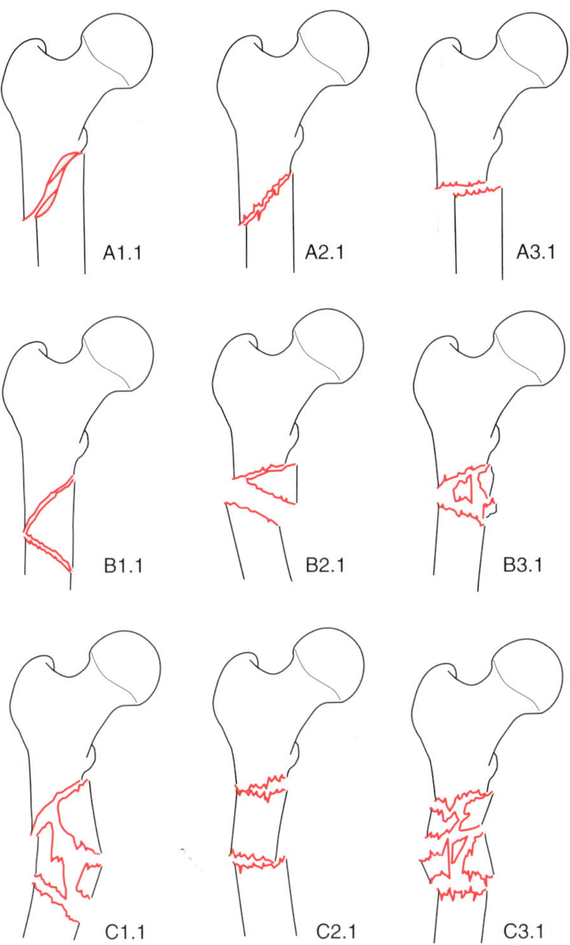

Figure 15.27. AO/ASIF classification of femoral diaphysis fractures, as applied to subtrochanteric fractures (Müller, 1990).

Wiss and Brien—1992

The subtrochanteric area was defined as the 7.5 cm of femur distal to the lesser trochanter. Fractures in which the major part of the fracture line was within this area, were then classified according to the method described by Winquist, Hansen and Clawson (1984). The primary division was into three groups of transverse, oblique or comminuted. Comminuted fractures can then be further divided into four grades of comminution.

Interobserver Reliability

In a study of the Jensen and Michaelsen classification system (Andersen, Jorgensen and Hededam, 1990), 50 trochanteric fracture were assessed by six observers. Only nine fractures were classified identically by all observers. There was agreement in assessing fracture stability (grade 1 or 2 versus 3, 4 or 5), in only 57% of cases. A later study (Gehrchen, Nielsen and Olesen, 1993), found that four observers agreed on fracture classification in only 23 out of 53 radiographs. When assessing stability only there was agreement for 34 out of 52 radiographs. The same observers when re-presented with the same radiographs six weeks later gave a different classification in a quarter of cases. In both these studies. A study of the Seinsheimer classification concluded that it has no value in clinical practice due to the high level of intra-observer variation.

A study of the AO classification system using an example of a trochanteric fracture with a subtrochanteric extension, found disagreement from the correct classification for 15 of the 18 observers (Johnstone, Radford and Parnell, 1993).

Clinical Value of Different Classification Systems

Whilst numerous fracture classifications have been described, there have been very few clinical studies that have found them to be of value in determining the choice of treatment, risk of fracture healing complications or prognosis. Many authors have used a simplified classification such as the simple division into displaced or undisplaced fractures (Hafner, 1951; Wade, 1959). Another method is to divide fractures into those which are two part and those with more than two parts, that is comminuted. This method has been used by a number of authors (Dimon and Hughston, 1967; Laros and Moore, 1974; Kyle, Gustilo and Premer, 1979; Jensen, 1980; Pankovich and Tarabishy, 1980). All these reports found an increase in complications of surgical fixation for the comminuted fractures.

The majority of papers on trochanteric fractures simply divide the fracture into two groups, stable and unstable. Many have used a simplification of the Evans, Jensen, or Kyle classification system (Kyle, Gustilo and Premer, 1979; Jensen, 1980; Wolfgang, Bryant and O'Neill, 1982; Herrlin et al., 1988) and others have used their own definitions of what constitutes a stable or unstable fracture (Dimon and Hughston, 1967; Dimon, 1973; Harrington and Johnson, 1973; Sarmiento, 1973; Laros and Moore, 1974; Simpson, Varty and Dodd, 1989). All these studies have consistently reported an increased incidence of complications for unstable fracture.

There have been few assessments of the value of the different classification methods in prediction outcome, particularly with reference to fixation failure after surgery. Jensen, in a study comparing classification methods for trochanteric fractures (Jensen, 1980), found the Ender classification to be of no use and the method described by Tronzo to have limited

Table 15.6 Incidence of fixation failure related to Jensen classification (percentage)

Study	*Year*	*1*	*2*	*3*	*4*	*5*
Jensen	1980	1/25	0/9	7/95	1/28	5/77
Larsson et al.	1990	0/87	1/121	5/190	2/55	7/154
Davis et al.	1990		0/35	7/49	1/24	12/89
Parker	1993b	0/60	5/139	3/106	8/172	12/186
TOTAL		1/172 (0.6%)	6/304 (2.0%)	22/440 (5.0%)	12/279 (4.3%)	36/506 (7.1%)

benefit. He reported that his modification of the Evans system (Jensen and Michaelsen, 1975) was more reliable.

The Jensen and Michaelsen classification has become one of the most frequently used systems. Types 1 and 2 fractures are considered as stable fractures, whilst types 3 to 5 are considered to be unstable fractures. However, this classification system is difficult to use for fractures at the level of the lesser trochanter such as reversed fracture lines which were originally given a separate category in the Evans classification. For trochanteric fractures treated by internal fixation, four studies have reported on results related to the Jensen classification (Table 15.6).

Because subtrochanteric fractures are uncommon there are no large series of results of treatment. Most of the classification systems have only been used in one or two papers. The most frequently used have been the Seinsheimer, Zickel and AO systems. No classification has been shown to be of value in determining the method of treatment or risk of fracture healing complications. The only exception to this was with the Seinsheimer classification where three studies have found type III fractures to have a higher risk of fixation failure (Seinsheimer, 1978; Bajaj et al., 1988; Lechner et al., 1990).

Summary of Extracapsular Fractures

The primary requirement before any detailed classification system can be used is to define the anatomical boundaries for fracture to be included. Furthermore difficulties are found in classifying fractures which cross boundaries. For example, a trochanteric fractures with a subtrochanteric extension is classified as subtrochanteric with the Seinsheimer system, but as trochanteric fractures by most other methods. Few classification systems give a clear description of which fractures are included within their system and which are excluded.

Proximal Femoral Fractures in Children

Reported Classification Systems

Colonna—1929

This was proposed by Delbet and popularized by Colonna. Fractures are divided into four groups.

Type I	Epiphyseal
Type II	Transcervical
Type III	Basal
Type IV	Trochanteric

Parker and Pryor—1993

This is the same classification as described by Colonna but with the addition of type V, for subtrochanteric fractures. Each of the five groups are subdivided into displaced and undisplaced fracture.

Clinical Value of the Classification System

Table 15.7 is a review of reported results of treatment based on the classification system described above (Parker and Pryor, 1993).

Recommendations

Prior to classifying a fracture, it is essential to define the anatomical area related to that fracture. Whilst

Table 15.7 Reported results related to classification of hip fractures in children (percentage) (Parker and Pryor, 1993)

	Non-union	*Avascular necrosis*
Type I	3/25 (12%)	18/27 (67%)
Type II	9/91 (10%)	50/129 (39%)
Type III	3/52 (6%)	31/100 (67%)
Type IV	1/33 (3%)	5/33 (15%)
Type V	0/38 (0%)	0/38 (0%)

this is clear for those fractures of the femoral head, there is an ambiguous division between intracapsular and extracapsular fractures. Furthermore there is no consensus of what constitutes a subtrochanteric fracture. Patently the first aim of any discussion of a classification system for proximal femoral fractures must be to first define the different anatomical boundaries.

Those divisions detailed in Figure 15.13 are to be recommended with the following expositions:

1. A basal fracture should be considered as an extracapsular fracture. Any fractures proximal to a basal fracture should be considered as an intracapsular fracture.
2. The division between trochanteric and subtrochanteric fractures is at the lower border of the lesser trochanter.
3. The distal border of the subtrochanteric area is 5 cm distal to the lower border of the lesser trochanter.
4. For those fractures which traverse these anatomical boundaries the area in which the predominate part of the fracture line traversing the femur is found should be used.

Femoral Head Fractures

The group recommends that the Pipkin (1957) system is the least cumbersome method of classifying these injuries. An additional category for indentation or depression fractures should be considered. The increased complexity of the Brumback (1986) and AO systems have not been justified, so far, by any improvements in treatment or in assessing prognosis. The AO system has been changed since its original description, thereby negating its clinical value.

Intracapsular Fractures

The group came to the conclusions that only the simplest classification is sufficiently reliable. This is the Waldenström (1924). This system splits fractures into displaced and undisplaced. Undisplaced fractures includes impacted fractures and those which are 'minimally' displaced. This simple division is of proven value in defining management and outcome. Based on current available evidence there appears to be little justification to use any further subdivision of intracapsular fractures, than into those which are essentially undisplaced and those which are displaced. Whilst there may be some differences in the incidence of non-union related to the level of the fracture, clinical evidence to date is insufficient to justify using this in routine clinical practice. In specific situations the pragmatic surgeon may consider using the Pauwels classification when considering whether to treat an undisplaced intracapsular fracture conservatively. Likewise, Garden stage 2 fractures appear to be more unstable than impacted stage 1 fractures and, therefore, unsuitable for conservative treatment. Table 15.4 indicates that the differences between Garden stage 3 and 4 fractures is strongly dependent on observer variation and this makes the Garden classification too imprecise for routine clinical use. Further clinical research is, therefore, required to justify the routine use of these more complex systems.

Proximal Extracapsular Fractures

The multiplicity of classification systems reflects the fact that there is no ideal system for these common fractures. Surgeons will manage these fractures depending on the availability of implants and technical expertise. This makes generalizing a classification system difficult because of the wide variety and difference in implant systems available. The Jensen and Michaelsen (1975) system has had several studies assessing its reliability and has also been used in prognosis. The Jensen classification can also be used in a simplified form, reducing the fractures into two-part fractures (stable) and more than two-part fractures (comminuted or unstable). Surgeons may prefer to consider basal fractures as a separate subgroup with division into undisplaced and displaced two-part fractures.

Classification of fractures in the distal trochanteric area is difficult with the Jensen and Michaelsen system. Surgeons may prefer to consider these fractures separately, particularly those of the reverse oblique type. The AO system does identify this but the complexity and doubtful reliability of this system precludes its routine use. Subtrochanteric fractures should be seen as a separate group but, to date, none of the available classification systems has been shown to assist in sub-dividing these fractures in any useful way.

Figure 15.28 summarizes our classification of hip fractures in adults.

Proximal Femoral Fractures in Children

The Parker and Pryor (1993) system appears to be useful in defining complications. It is likely that the anatomical site will influence treatment although it is not possible to make any specific recommendations.

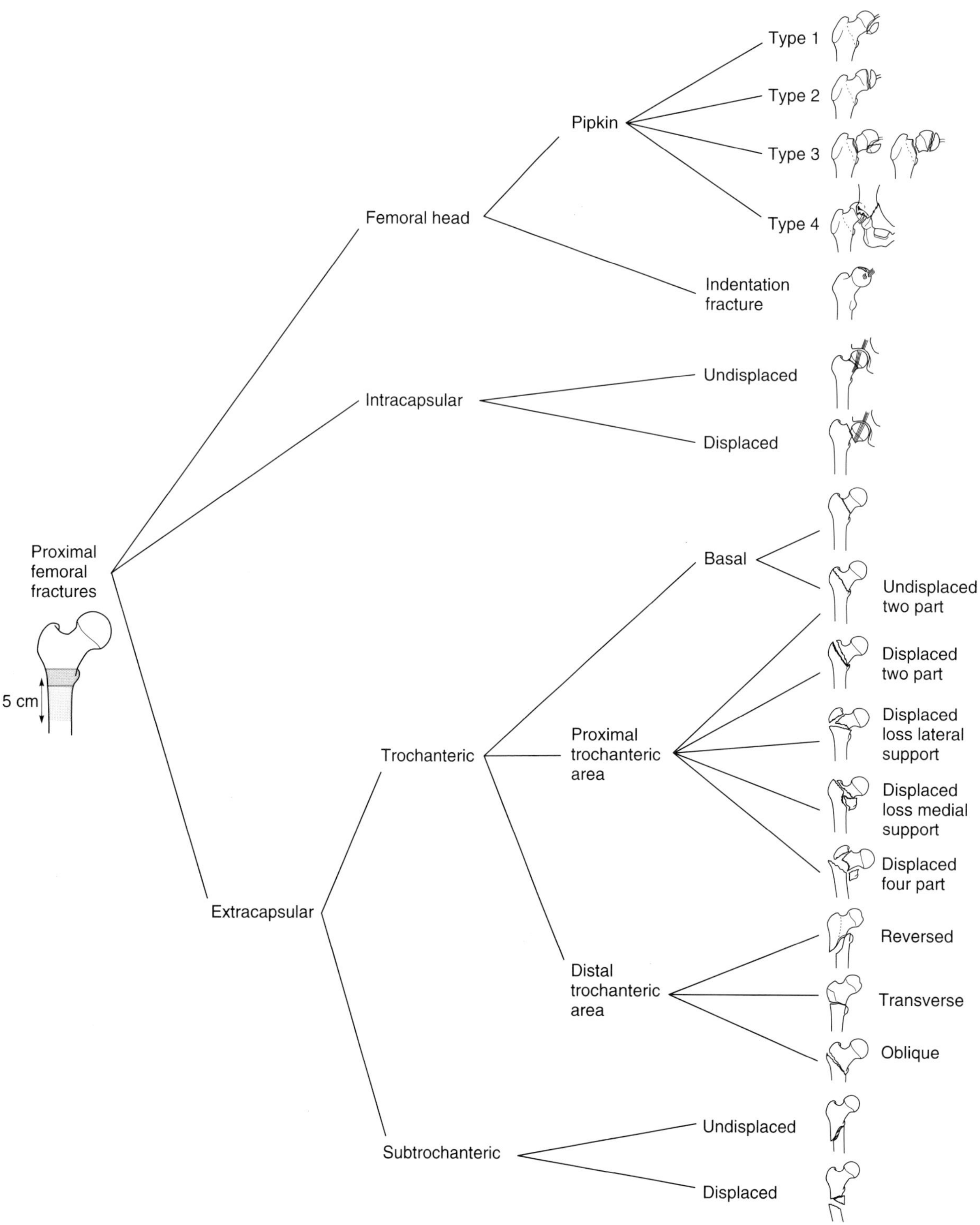

Figure 15.28. Summary of classifications of the proximal femoral fractures in adults.

References

Alho, A., Benterud, J.G., Ronningen, H. and Hoiseth, A. (1991) Radiographic prediction of early failure in femoral neck fracture. *Acta Orthop. Scand.*, **62**, 422–426.

Andersen, E., Jorgensen, L.G. and Hededam, L.T. (1990) Evans' classification of trochanteric fractures: an assessment of the interobserver and intraobserver reliability. *Injury*, **21**, 377–378.

Bajaj, H.N., Rao, P.S., Kumar, B. and Chacko, V. (1988) Subtrochanteric fractures of the femur—an analysis of the results of operative and non-operative management. *Injury*, **19**, 169–171.

Banks, H.H. (1962) Factors influencing the result in fractures of the femoral neck. *J. Bone Joint Surg.*, **99A**, 937–964.

Barnes, R., Brown, J.T., Garden, R.S. and Nicoll, E.A. (1976) Subcapital fractures of the femur: a prospective review. *J. Bone Joint Surg.*, **58B**, 2–24.

Blundell, C.M., Parker, M.J., Pryor, G.A., Hopkinson-Wooley, J. and Bonsall, S.S. (1998) An assessment of the AO classification of intracapsular fractures of the proximal femur. *J. Bone Joint Surg.*, **80B**, 679–683.

Böhler, L. (1957) *The Treatment of Fractures*, Vol. 11, 5th edn. New York: Grune & Stratton.

Boyd, H.B. and Griffin, L.L. (1949) Classification and treatment of trochanteric fractures. *Arch. Surg.*, **58**, 853–866.

Boyd, H.B. and Salvatore, J.E. (1964) Acute fracture of the femoral neck: Internal fixation or prosthesis? *J. Bone Joint Surg.*, **46A**, 1066–1068.

Bracey, D.J. (1977) A comparison of internal fixation and prosthetic replacement in the treatment of displaced subcapital fractures. *Injury*, **9**, 1–4.

Briggs, H. and Keats, S. (1946) Management of intertrochanteric fractures of the femur by skeletal traction with the beaded Kirschner wire. *Am. J. Surg.*, **71**, 788–796.

Brown, J.T. and Abrami, G. (1964) Transcervical femoral fracture: a review of 195 patients treated by sliding nail-plate fixation. *J. Bone Joint Surg.*, **46B**, 648–663.

Brown, T.I.S. and Court-Brown, C. (1979) Failure of sliding nail-plate fixation in subcapital fractures of the femoral neck. *J. Bone Joint Surg.*, **61B**, 342–346.

Brumback, R.J., Kenzora, J.E., Levitt, L.E. et al. (1986) *Proceedings of the American Hip Society*. St. Louis: Mosby, pp. 181–206.

Bunata, R.E., Fahey, J.J. and Drennan, D.B. (1973) Factors influencing stability and necrosis of impacted femoral neck fractures. *JAMA*, **223**, 41–44.

Cassebaum, W.H. and Nugent, G. (1963) Predictability of bony union in displaced intracapsular fractures of the hip. *J. Trauma*, **3**, 421–424.

Cech, O. and Sosna, A. (1974) Principles of the surgical treatment of subtrochanteric fractures. *Orthop. Clin. North Am.*, **5**, 651–662.

Cobb, A.G. and Gibson, P.H. (1986) Screw fixation of subcapital fractures of the femur—a better method of treatment. *Injury*, **17**, 259–264.

Colonna, P.C. (1929) Fractures of the neck of the femur in childhood. *Am. J. Surg.*, **6**, 793–796.

Crawford, H.B. (1969) Impacted femoral neck fractures. *Clin. Orthop.*, **66**, 90–93.

Davis, T.R.C., Sher, J.L., Horsman, A. et al. (1990) Intertrochanteric femoral fractures: mechanical failures after internal fixation. *J. Bone Joint Surg.*, **72B**, 26–31.

DeLee, J.C., Evans, J.A. and Thomas, J. (1980) Anterior dislocation of the hip and associated femoral-head fractures. *J. Bone Joint Surg.*, **62A**, 960–964.

Dimon, J.H. (1973) The unstable intertrochanteric fracture. *Clin. Orthop.*, **92**, 100–107.

Dimon, J.H. and Hughston, J.C. (1967) Unstable intertrochanteric fractures of the hip. *J. Bone Joint Surg.*, **49A**, 440–450.

Ender, J. (1970) Probleme beim frischen per- und subtrochanteren oberschenkelbruch. *Hefte Unfallheilk*, **106**, 2–11.

Eklund, J. and Eriksson, F. (1964) Fractures of the femoral neck: with special regard to the treatment and prognosis of stable abduction fractures. *Acta Chir. Scand.*, **127**, 315–337.

Eliasson, P., Hansson, L.I. and Kärrholm, J. (1988) Displacement in femoral neck fractures: a numerical analysis of 200 fractures. *Acta Orthop. Scand.*, **59**, 361–364.

Evans, E.M. (1949) The treatment of trochanteric fractures of the femur. *J. Bone Joint Surg.*, **31B**, 190–203.

Fielding, J.W. and Magliato, H.J. (1966) Subtrochanteric fractures. *Surg. Gynecol. Obstet.*, **122**, 555–560.

Fielding, J.W., Wilson, S.A. and Ratzan, S. (1974) A continuing end-result study of displaced intracapsular fractures of the neck of the femur treated with the Pugh nail. *J. Bone Joint Surg.*, **56A**, 1464–1472.

Flatmark, A.L. and Lone, T. (1962) The prognosis of abduction fractures of the neck of the femur. *J. Bone Joint Surg.*, **44B**, 324–327.

Frandsen, P.A. (1979) Osteosynthesis of displaced fractures of the femoral neck: a comparison between Smith–Petersen osteosynthesis and sliding-nail-plate osteosynthesis — a radiological study. *Acta Orthop. Scand.*, **50**, 443–449.

Frandsen, P.A. and Andersen, P.E. (1981) Treatment of displaced fractures of the femoral neck: Smith–Petersen osteosynthesis verses sliding-nail-plate osteosynthesis. *Acta Orthop. Scand.*, **52**, 547–552.

Frandsen, P.A., Andersen, P.E., Christoffersen, H. and Thomsen, P.B. (1984) Osteosynthesis of femoral neck fracture: the sliding-screw-plate with or without compression. *Acta Orthop. Scand.*, **55**, 620–623.

Frandsen, P.A., Andersen, E., Madsen, F. and Skjodt, T. (1988) Garden's classification of femoral neck fractures: an assessment of inter-observer variation. *J. Bone Joint Surg.*, **70B**, 588–590.

Garden, R.S. (1961) Low-angle fixation in fractures of the femoral neck. *J. Bone Joint Surg.*, **43B**, 647–663.

Gehrchen, P.M., Nielsen, J.Ø. and Olesen, B. (1997) Poor reproducibility of Evans' classification of the trochanteric fracture: assessment of 4 observers in 52 cases. *Acta Orthop. Scand.*, **64**, 71–72.

Gehrchen, P.M., Nielsen, J.Ø., Olesen, B. and Andresen, B.K. (1997) Seinsheimer's classification of subtrochanteric fractures: poor reproducibility of 4 observers' evaluation of 50 cases. *Acta Orthop. Scand.*, **68**, 524–526.

Graham, J. (1968) Early or delayed weight-bearing after internal fixation of transcervical fracture of the femur: a clinical trial. *J. Bone Joint Surg.*, **50B**, 562–569.

Hafner, R.H.V. (1951) Trochanteric fractures of the femur: a review of eighty cases with a description of the 'low-nail' method of fixation. *J. Bone Joint Surg.*, **33B**, 513–516.

Hargadon, E.J. and Pearson, J.R. (1963) Treatment of intracapsular fractures of the femoral neck with the Charnley compression screw. *J. Bone Joint Surg.*, **45B**, 305–311.

Harrington, K.D. and Johnston, J.O. (1973) The management of comminuted unstable intertrochanteric fractures. *J. Bone Joint Surg.*, **55A**, 1367–1376.

Harris, L.J. (1980) Closed retrograde intramedullary nailing of peritrochanteric fractures of the femur with a new nail. *J. Bone Joint Surg.*, **62A**, 1185–1193.

Herrlin, K., Stromberg, T., Lingren, L. et al. (1988) Trochanteric fractures: classification and mechanical stability in McLaughlin, Ender and Richard osteosynthesis. *Acta Radiol.*, **29**, 189–196.

Høgh, J., Jensen, J. and Lauritzen, J. (1982) Dislocated femoral neck fractures: a follow-up study of 98 cases treated by multiple AO (ASIF) cancellous bone screws. *Acta Orthop. Scand.*, **53**, 245–249.

Jensen, J.S. (1980) Classification of trochanteric fractures. *Acta Orthop. Scand.*, **51**, 803–810.

Jensen, J.S. and Michaelsen, M. (1975) Trochanteric femoral fractures treated with McLaughlin osteosynthesis. *Acta Orthop. Scand.*, **46**, 795–803.

Jensen, J. and Høgh, J. (1983) Fractures of the femoral neck: a follow-up study after non-operative treatment of Garden's stage 1 and 2 fractures. *Injury*, **14**, 339–342.

Johnstone, D.J., Radford, W.J.P. and Parnell, E.J. (1993) Interobserver variation using the AO/ASIF classification of long bone fractures. *Injury*, **24**, 163–165.

Kofoed, H. and Alberts, A. (1980) Femoral neck fractures: 165 cases treated by multiple percutaneous pins. *Acta Orthop. Scand.*, **51**, 127–136.

Kyle, R.F., Gustilo, R.B. and Premer, R.F. (1979) Analysis of six hundred and twenty-two intertrochanteric hip fractures. *J. Bone Joint Surg.*, **61A**, 216–221.

Laros, G.S. and Moore, J.F. (1974) Complications of fixation in intertrochanteric fractures. *Clin. Orthop.*, **101**, 110–119.

Larsson, S., Friberg, S. and Hansson, L-I. (1990) Trochanteric fractures: mobility, complications, and mortality in 607 cases treated with the sliding-screw technique. *Clin. Orthop.*, **260**, 232–241.

Lechner, J.D., Rao, J.P., Stashak, G. and Adibe, S.O. (1990) Subtrochanteric fractures: a retrospective analysis. *Clin. Orthop.*, **259**, 140–145.

McCabe, J., O'Farrell, D., O'Byrne, J. and O'Brien, T. (1994) Observer variation in the radiographic classification of femoral neck fractures. *J. Bone Joint Surg.*, **76B** (Suppl. II), 120.

Malkawi, H. (1982) Bone grafting in subtrochanteric fractures. *Clin. Orthop.*, **168**, 69–72.

Massie, W.K. (1973) Treatment of femoral neck fractures emphasizing long term follow-up observations on aseptic necrosis. *Clin. Orthop.*, **92**, 16–59.

Massie, W.K. (1964) Fractures of the hip. *J. Bone Joint Surg.*, **46A**, 658–690.

May, J.M.B. and Chacha, P.B. (1968) Displacements of trochanteric fractures and their influence on reduction. *J. Bone Joint Surg.*, **50B**, 318–323.

Moore, J.R. (1939) Fractures of the upper end of the femur including fracture dislocations at the hip joint. *Am. J. Surg.*, **44**, 117–134.

Müller, M.E., Nazarian, S., Koch, P. and Schatzker, J. (1990) *The AO Classification of Fracture of Long Bones.* Berlin: Springer-Verlag.

Müller, M.E., Allgöwer, M., Schneider, A. and Willenegger, H. (1991) *Manual of Internal Fixation. Techniques Recommended by the AO-ASIF Group.* Berlin: Springer-Verlag.

Murray, R.C. and Frew, J.F.M. (1949) Trochanteric fractures of the femur: A plea for conservative treatment. *J. Bone Joint Surg.*, **31B**, 204–219.

Newey, M.L., Ricketts, D. and Roberts, L. (1993) The AO classification of long bone fractures: an early study of its use in clinical practice. *Injury*, **24**, 309–312.

Nordkild, P. and Sonne-Holm, S. (1984) Sliding screw-plate for fixation of femoral neck fracture. *Acta Orthop. Scand.*, **55**, 616–619.

Otremski, I., Katz, A., Dekel, S. et al. (1990) Natural history of impacted subcapital femoral fractures and its relevance to treatment options. *Injury*, **21**, 379–381.

Pankovich, A.M., Goldflies, M.L. and Pearson, R.L. (1979) Closed Ender nailing of femoral-shaft fractures. *J. Bone Joint Surg.*, **61A**, 222–232.

Pankovich, A.M. and Tarabishy, I.E. (1980) Ender nailing of intertrochanteric and subtrochanteric fractures of the femur. *J. Bone Joint Surg.*, **62A**, 635–645.

Parker, M.J. (1993a) Garden grading of intracapsular fractures: meaningful or misleading? *Injury*, **24**, 241–242.

Parker, M.J. (1993b) Valgus reduction of trochanteric fractures. *Injury*, **24**, 313–316.

Parker, M.J. (1994) Prediction of fracture union after internal fixation of intracapsular femoral neck fractures. *Injury*, **25**(Suppl. 2), 3–6.

Parker, M.J. and Dynam, E. (1998) Is Pauwel's classification still valid? *Injury*, **29**, 521–523.

Parker, M.J. and Pryor, G.A. (1993) *Hip Fracture Management*. Oxford: Blackwell Science.

Pauwels, F. (1935) *Der Schenkelhalsbruch. Ein mechanisches Problem.* Stuttgart: F. Enke.

Pipkin, G. (1957) Treatment of grade IV fracture-dislocation of the hip; a review. *J. Bone Joint Surg.*, **39A**, 1027–1042.

Pugh, W.L. (1955) A self-adjusting nail-plate for fractures about the hip joint. *J. Bone Joint Surg.*, **37A**, 1085–1093.

Raaymakers, E.L.F.B. and Marti, R.K. (1991) Non-operative treatment of impacted femoral neck fractures: a prospective study of 170 cases. *J. Bone Joint Surg.*, **73B**, 950–954.

Sarmiento, A. (1973) Unstable intertrochanteric fractures of the femur. *Clin. Orthop.*, **92**, 77–85.

Seinsheimer, F. (1978) Subtrochanteric fractures of the femur. *J. Bone Joint Surg.*, **60A**, 300–306.

Simpson, A.H.R.W., Varty, K. and Dodd, C.A.F. (1989) Sliding hip screws: modes of failure. *Injury*, **20**, 227–231.

Skinner, P.W. and Powles, D. (1986) Compression screw fixation for displaced subcapital fractures of the femur: success or failure? *J. Bone Joint Surg.*, **68B**, 79–87.

Smyth, E.H.J. and Shah, V.M. (1974) The significance of good reduction and fixation in displaced subcapital fractures of the femur. *Injury*, **5**, 197–209.

Stappaerts, K.H. and Broos, P.L.O. (1987) Internal fixation of femoral neck fractures: a follow-up study of 118 cases. *Acta Chir. Belg.*, **87**, 247–251.

Svenningsen, S., Benum, P., Nesse, O. and Furset, O.I. (1984) Internal fixation of femoral neck fractures: compression screw compared with nail plate fixation. *Acta Orthop. Scand.*, **55**, 423–429.

Swiontkowski, M.F. (1991) Femoral head fractures. *Curr. Orthop.*, **5**, 99–105.

Tamai, K., Azuma, H. and Kako, K. (1983) A new anatomic classification of capital fragments in femoral neck fractures with the epiphyseal scar as a guide. *Clin. Orthop.*, **179**, 147–151.

Thomsen, N.O.B., Jensen, C.M., Skovgaard, N., Pedersen, M.S., Pallesen, P. et al. (1996) Observer variation in the radiological classification of fractures of the neck of the femur using Garden's system. *Int. Orthop.*, **20**, 326–329.

Tronzo, R.G. (1973) *Surgery of the Hip Joint*. Philadelphia: Lea & Febiger.

Tronzo, R.G. (1984) *Surgery of the Hip Joint*. New York: Springer-Verlag.

Unger, A.S. and Shuster, H.F. (1986) Predicting the healing of displaced subcapital hip fractures via postoperative roentgenographic factors. *Orthop. Rev.*, **15**, 575–580.

Waddell, J.P. (1979) Subtrochanteric fractures of the femur: a review of 130 patients. *J. Trauma*, **19**, 582–591.

Wade, P.A., Campbell, R.D. and Kerin, R.J. (1959) Management of intertrochanteric fractures of the femur. *Am. J. Surg.*, **97**, 634–643.

Waldenström, J. (1924) Fractures récentes du col fémoral—traitement opératoire ou orthopédique. *J. Chir.*, **24**, 129.

Watson, H.K., Campbell, R.D. and Wade, P.A. (1964) Classification, treatment, and complications of the adult subtrochanteric fracture. *J. Trauma*, **4**, 457–480.

Wilson, H.J., Rubin, B.D., Helbig, F.E. et al. (1980). Treatment of intertrochanteric fractures with the Jewett nail: experience with 1,015 cases. *Clin. Orthop.*, **148**, 186–191.

Winquist, R.A., Hansen, S.T. and Clawson, D.K. (1984) Closed intramedullary nailing of femoral fractures. *J. Bone Joint Surg.*, **66A**, 529–539.

Wiss, D.A. and Brien, W.W. (1992) Subtrochanteric fractures of the femur: results of treatment by interlocking nailing. *Clin. Orthop.*, **283**, 231–236.

Wolfgang, G.L., Bryant, M.H. and O'Neill, J.P. (1982) Treatment of intertrochanteric fracture of the femur using sliding screw plate fixation. *Clin. Orthop.*, **163**, 148–158.

Zickel, R.E. (1976) An intramedullary fixation device for the proximal part of the femur. *J. Bone Joint Surg.*, **58A**, 866–872.

16

Femoral shaft and distal femoral fracture classification

D. E. Porter

Introduction

In their classic paper on juxta-epiphyseal injuries, Salter and Harris (1963) described the attributes of any useful classification system as:

- simple
- understandable
- easy to remember
- based on scientific and clinical data
- correlating with either prognosis or treatment or both.

Such exacting criteria should be the benchmark by which we judge the classification systems available today. Unfortunately, although numerous studies on treatment of femoral fractures exist, a search of the orthopaedic literature yields a thin harvest of research which correlates actual fracture type with outcome.

There are many factors other than fracture site and configuration which affect outcome and best treatment regime. They include bone quality, the presence of polytrauma and important other local fractures. These do not appear in any graded classification system, but should be considered separately as adjuncts to the standard classifications.

Anatomy

The femur is the longest and strongest bone in the body. Distal to the trochanters its shaft is fairly uniform in calibre, convex anteriorly, with a smooth surface except along the posteriorly situated linea aspera. The linea aspera is especially prominent in the middle third of the shaft, dividing below to form the supracondylar ridges. Above, the linea aspera continues medially as the spiral and pectineal lines, and laterally blends with the gluteal tuberosity.

The lower extremity of the femur is broadened for articulation with the tibia. Two condyles, separated by a deep intercondylar fossa, occupy the inferior and posterior aspects of the articular surface. The condylar surfaces are indistinctly separated anteriorly from the laterally supported patellar surface. The lateral condyle is broader, but shorter than the medial condyle. Between shaft and condyles are situated the epicondylar prominences. The superior margin of the medial epicondyle is thickened to form the adductor tubercle (Woodburne and Burkel, 1994).

The anatomical point at which femoral shaft becomes distal femur is empirical. In the literature, 'distal femur' is the lower third or quarter of the femur, between 7.5 and 15 cm proximal to the knee joint. In this chapter, the femoral shaft is defined as the diaphysis which extends distal to the lesser trochanter. The distal femur is the distal femoral metaphysis and epiphysis as defined by the AO nomenclature (Müller et al., 1991).

Femoral Shaft Fracture Classifications

Two modern, general classification systems of femoral shaft fractures have been described. These are the AO classification (Müller et al., 1991) and the OTA (Orthopaedic Trauma Association) classification (Gustilo, 1990). Perhaps surprisingly, corroborative studies which specifically relate AO or

OTA fracture grade to treatment regimes or prognosis are uncommon. Other important specific classifications such as the Gustilo open fracture classification (Gustilo, Mendoza and Williams, 1984), the Winquist classification of comminution (Winquist and Hansen, 1980), and classifications of peri-prosthetic fractures will be discussed, as will several other factors which affect treatment and outcome (see also Chapter 2).

The AO Classification (Müller et al., 1991) (Figure 16.1)

1. Description by fracture SITE (proximal, middle or distal third of diaphysis).
2. Description of fracture TYPE:

Grade A1 Simple fracture, spiral
A2 Simple fracture, oblique
A3 Simple fracture, transverse
B1 Wedge fracture, spiral wedge
B2 Wedge fracture, bending wedge
B3 Wedge fracture, fragmented wedge
C1 Complex fracture, spiral
C2 Complex fracture, segmental
C3 Complex fracture, irregular

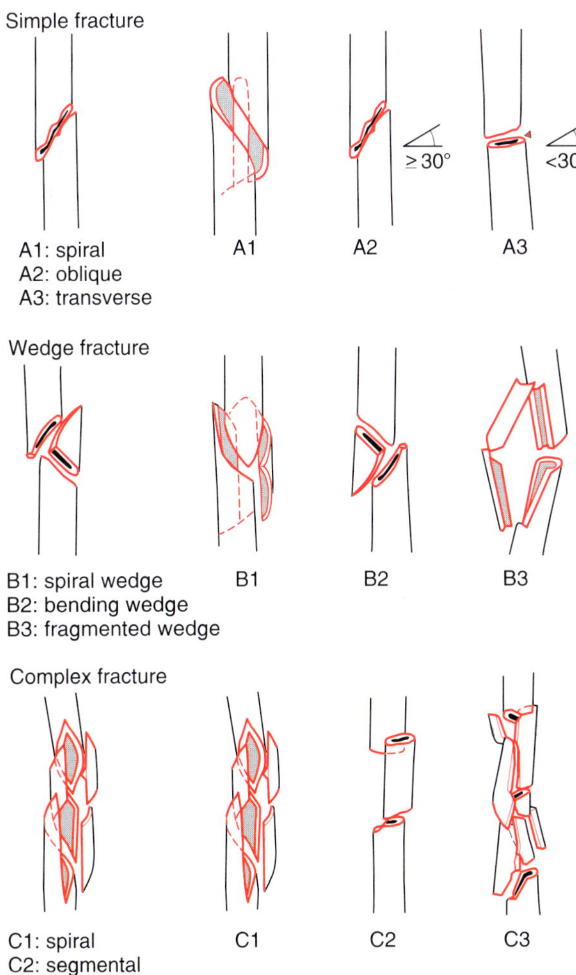

Figure 16.1. AO/ASIF classification of femoral shaft fractures (Müller et al., 1991).

The OTA Classification (Gustilo, 1990) (Figure 16.2)

1. Description of fracture SITE.
2. Description of fracture TYPE:

Grade I Transverse (< 30°)
II Oblique (> 30°)
III Spiral (without butterfly fragment)
IV Spiral (with butterfly fragment).
V Comminuted
VI Segmental
VII Bone loss

Fracture site appears to have little impact on either management or outcome. Fractures at or below the level of the lesser trochanter can be successfully treated in the same manner as more distal diaphyseal fractures, usually by means of an intramedullary nail, with similar results (Johnson, 1988). Distal third diaphyseal (infra-isthmal) fractures may be treated likewise with no adverse effect (Browner et al., 1984; Richards et al., 1984). Winquist, Hansen and Clawson (1984) found that simple proximal third femoral shaft fractures required a proximal locking bolt to prevent malrotation, whereas midshaft fractures did not. In very distal diaphyseal fractures, it appears that two distal locking bolts, rather than one, are required to prevent malalignment; Grover and Wiss (1995) in a trial of 182 complex femoral fractures randomized to nailing with either one or two distal bolts found that the only case of malalignment occurred in a distal segmental fracture which had only one locking bolt.

The underlying assumption of these two classification systems is that as fracture grade increases, so does fracture instability. This has implications for treatment and outcome. Simple fracture patterns vary in their inherent stability (AO grade A, OTA grades I–III). Spiral fractures, for example, always require interlocking nails whereas oblique fractures require them when close to either end of the shaft (Winquist et al.,1984). Increasing fracture comminution (AO grades B–C, OTA grades IV–VII) is indicative of increasing instability. This is reflected in studies which reveal worst outcomes in comminuted fractures treated conservatively, and the best in those reduced anatomically and fixed rigidly (Grover and Wiss, 1995). Johnson, Johnston and Parker (1984)

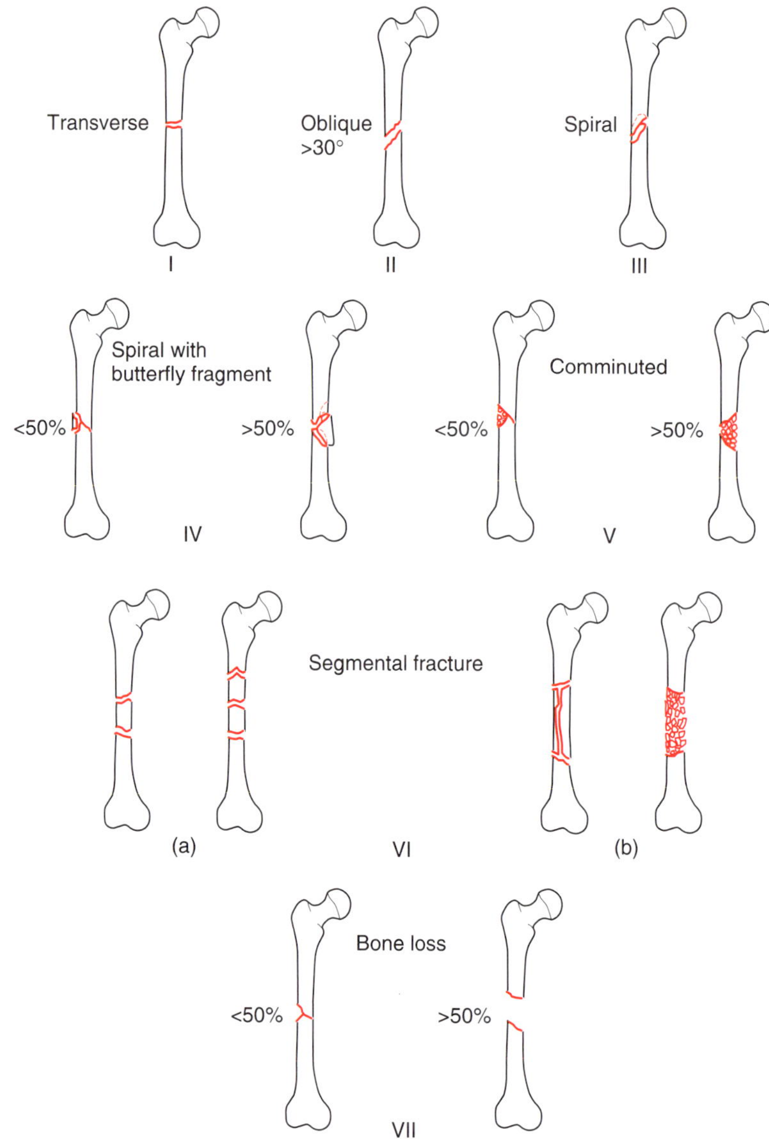

Figure 16.2. OTA classification of femoral shaft fractures (Gustilo, 1990).

showed that in 79 comminuted shaft fractures, 66% of those treated by roller traction had significant complications, compared with only 4% of those fixed with an interlocking femoral nail. However, internal fixation of comminuted fractures is technically demanding; in one study, 11 out of 40 comminuted fractures had either shortening of over 1 cm, or angulatory/rotatory deformity of over 5° (Sojbjerg, Eiskjaer and Møller-Larsen, 1990). Segmental fractures often imply violent injury, with dispersed areas of damage and some devascularization of segments. Outcome is thereby compromised even if optimal reduction and fixation is achieved (Hirsch, Cavandias and Nachemson, 1955).

Winquist's classification of comminution omits the simple fracture patterns. Although incomplete, it is widely quoted in the literature, perhaps because comminuted fractures of the femoral shaft provide one of the most difficult technical challenges to the surgeon.

The Winquist and Hansen Classification of Comminution (Winquist and Hansen 1980) (Figure 16.3)

Grade O: No comminution
Grade I: Insignificant butterfly fragment
Grade II: Large butterfly fragment (> 50% cortical contact)

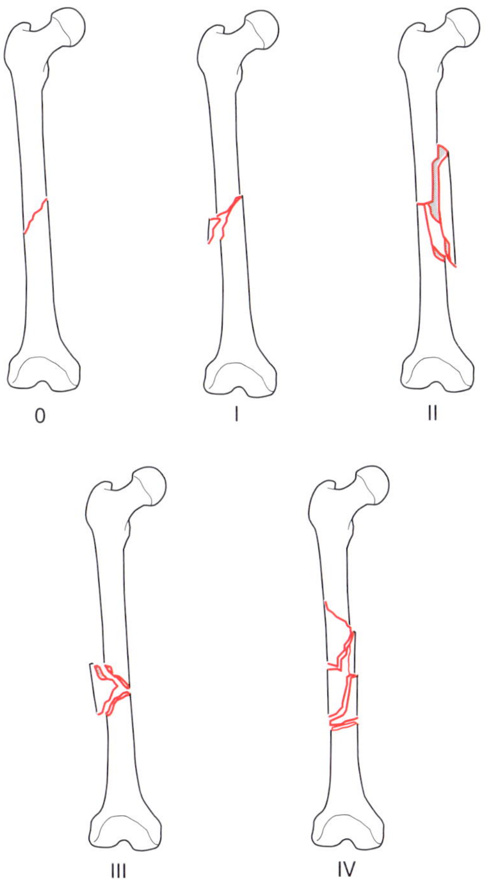

Figure 16.3. The Winquist and Hansen (1980) classification of comminution. 0, no comminution; I, insignificant butterfly fragment; II, butterfly with > 50% circumferential contact; III, butterfly with < 50% circumferential contact; IV, segmental comminution.

Grade III: Large butterfly fragment (< 50% cortical contact)

Grade IV: Segmental comminution (no major contact between main shaft fragments to prevent shortening)

In a later publication, Winquist, Hansen and Clawson (1984) analysed 520 femoral shaft fractures treated by intramedullary nailing. They determined that reliable results without shortening or malrotation could be achieved if attention was paid to the degree of comminution. Fractures classified grade I could be successfully managed without interlocking. Grade II usually required locking bolts, and grades III and IV always required them to avoid shortening. These treatment recommendations have been confirmed in another large series (Johnson, Johnston and Parker, 1984), and even in open gunshot injuries (Brien et al., 1995). Following these treatment guidelines, Brumback et al. (1988a) found that shortening of over 2 cm still occurred in 10% of dynamically locked nails. This appeared to be either due to mis-grading of fracture comminution, or due to intra-operative fracture extension, for example converting a Winquist grade I to a grade III. They recommended that intra-operative fracture extensions should always be statically locked. In a further publication, this group prospectively analysed outcome in 100 shaft fractures, all statically locked. Results indicated that union was not significantly delayed (Brumback et al., 1988b). Primary static locking has now become standard practice, even in relatively stable fracture configurations. Errors of technique can often be identified in re-fractures of the femoral diaphysis. In a retrospective analysis of 148 shaft fractures, most of the 19 re-fractures occurred after plate fixation. Failure of medial cortical buttressing or grafting, and inadequate external plate compression were identified as factors which increased the risk of re-fracture after fixation in comminuted, devascularized femoral fractures (Breederveld, Patka and Van-Mourik, 1985).

Open Fractures

The Gustilo classification of open fractures of long bones has prognostic significance in femoral fractures (Anderson and Gustilo, 1980; Gustilo, Mendoza and Williams, 1984). In the same way that Winquist's classification of comminution gives prognostic information for shortening and malrotation, Gustilo's classification provides information on the likelihood of complications, including infection. In one study, bone infection and subsequent osteomyelitis was noted in eight out of 63 open femoral fractures (O'Brien et al., 1991). Infections are least frequent in type I open femoral fractures. In 22 consecutive open shaft fractures, Rittmann et al. (1979) found that types II and III fractures contributed 85% of all infections. Lhowe and Hansen (1988) analysed 67 open femoral fractures (over one-third were type I injuries) in which the only two infections occurred in types II and III fractures. A separate study of 86 open femoral fractures (of which 43 were grade III), found that the only two infections occurred in the 17 type IIIB fractures. All open fractures require thorough local irrigation and débridement within 8 hours of injury to achieve these low infection rates (Brumback et al., 1989). Surgery within 6 hours of injury is currently considered acceptable practice. Complications other than infection also significantly increase from grade IIIA to IIIC injuries (Brumback et al., 1989). After débridement, primary intramedullary nailing achieves good results across all Gustilo grades (Kaltenecker, Wruhs and Quaicoe, 1990; O'Brien et al., 1991). In 18 open femoral shaft fractures treated by intramedullary nailing, excellent knee flexion (>135°) was achieved and no infections occurred, compared with one infection and only <100° of knee flexion in seven similar fractures treated by other methods of internal fixation (Chapman and Mahoney, 1979). Other than the Gustilo classification, no open femoral fracture classification is in common use.

Although intramedullary nailing is the treatment most widely used for open femoral fractures, other methods of fracture fixation or immobilization have been reported. External fixators have been advocated, particularly in type IIIC open fractures and in severely contaminated wounds (Dabezies et al., 1984: Browner et al., 1992). Soft-tissue infection rates, however, mostly in pin-sites, have been reported in up to 30%. The incidence of infection after careful plate fixation of open femoral fractures is as low as 3% in some studies, in contrast to the 15% bone re-fracture or plate breakage rate (Magerl et al., 1979). Skeletal traction of open fractures is not well tolerated over long periods, and provides the least satisfactory degree of soft-tissue and bony stability of all treatment options (Browner et al., 1992).

Special Considerations

Gunshot Injuries

A special type of open femoral fracture is that associated with gunshot injuries. Recent studies suggest that outcome is not significantly different from other open injuries provided appropriate wound care is undertaken. Brien et al. (1995) reviewed 27 gunshot-induced comminuted femoral fractures, and found that 93% were Winquist grades III–V. As might be predicted from other studies on open fracture management, in the absence of severe haemodynamic instability, immediate closed interlocking nailing in all fractures was recommended for best results. Only Winquist type I and II fractures could be managed with dynamically locked nails. All fractures healed without infection. In 14 gunshot injuries treated within 8 hours of injury, Nicholas and McCoy (1995) found that static nailing produced results approximating to those expected in closed injuries with 13 fractures proceeding to bony union within 22 months. Bergman et al. (1993) reviewed their experience in 65 femoral fractures treated with immediate reamed intramedullary nailing. They had no infections, and morbidity was decreased compared with studies of delayed femoral nailing in gunshot injuries. They advocated delay, however, in injuries causing major devitalization of soft tissue and bone.

Vascular injuries are common in gunshot-induced femoral fractures. Femoral vascular injury is nine times more likely to require repair in the presence of a fracture than if no fracture occurred (Payne, Gabriel and Massoud, 1995). A classification of gunshot injuries by zone of thigh penetration has shown some correlation with incidence of vascular compromise. A missile entering the femur between isthmus and distal metaphysis is the most likely zone to produce a significant vascular injury (Payne et al., 1995). In femoral fractures in general, associated vascular injuries carry a high morbidity. Adequate and immediate fasciotomies are often required. Cargile, Hunt and Purdue (1992) observed that 6 out of 27 fractures associated with vascular injuries eventually required amputation. Although the ratio of femoral venous to arterial injuries is only 1:3, it is the sequelae of venous injuries which result in most chronic morbidity (Cargile et al., 1992).

Multiply-injured Patients

In addition to associated local injuries, the presence of life-threatening conditions can directly affect both optimal fracture management and outcome. Grading of a multiple-injury constellation is done through the Injury Severity Score (ISS). The three most severely injured body regions are graded by the Abbreviated Injury Scale (AIS) from 1 to 5 (minor to critical), and the squares of these three values are summated to produce an ISS for that individual. A closed femoral shaft fracture scores '3' by the AIS scale, whereas that associated with a femoral artery laceration scores '4'. An ISS of 18–20 implies a multiply-injured patient (Baker, O'Neill and Haddon, 1974; Copes et al., 1988) (see Chapter 1).

In their study of 520 femoral fractures, Winquist, Hansen and Clawson (1984) revealed that adequate fluid resuscitation and early primary nailing in the multiply-injured patient resulted in no increase in fat embolism compared to those patients in which nailing was delayed beyond 7 days. Many more recent studies reveal that immediate stabilization of a femoral shaft fracture in a multiply-injured patient results in a better overall prognosis. This effect may be related to the biological effect of fracture stabilization (ten Duis et al., 1988). In a prospective trial of 178 patients with femoral fracture randomized to fixation either within 24 hours or delayed beyond 48 hours, Bone et al. (1989) found that early fixation reduced the incidence of ARDS and pneumonia, and shortened hospital stay. This effect was most marked in those with an ISS >18. Johnson, Cadambi and Seibert (1985) also found that the greater the ISS score, the greater the benefit from early fixation. An ISS > 40 resulted in a fivefold increase in ARDS if delayed rather than immediate fixation was employed. Further evidence in favour of early fixation was revealed in a study of 33 patients with femoral fractures and an ISS >36 with a significantly reduced incidence of pulmonary shunting and pneumonia, compared with those undergoing delayed stabilization (Behrman et al., 1990). Although Nutz and Katholnigg (1994) suggested that the worst outcome resulted in fractures fixed during the 'intermediate' time interval of days 2–7 post-injury, an epidemiological study of patients with ISS >15 in contrast revealed that fixation during days 2–4 post-injury resulted in the lowest mortality (Fakhry et al., 1994).

The ideal method of fixation is a source of contention. Brumback et al. (1989) considered that in severe grade IIIB open femoral fractures, an external fixator was quicker and less life-threatening than a primary intramedullary nail. Nutz and Katholnigg (1994) also found that in 40 multiply-injured patients, primary intramedullary nailing resulted in higher pulmonary complications than external fixation of the femoral fracture. After removal of external fixation, either secondary intramedullary nailing or plate fixation produced similar acceptable outcomes (Siebert et al., 1995), although concerns exist that infection rates might be higher than after a single procedure. Pape et al. (1993) considered that an ISS >30, together with evidence of pulmonary contusion, consistently predicted patients at high risk of ARDS. These authors advocated temporary external fixation, followed by definitive stabilization 3–4 days post-injury.

Fatigue Fractures

Fatigue fractures of the femur are a significant problem in new military recruits. First recognized as a potential femoral shaft injury during the Second World War, undisplaced fractures were treated with plaster casts and physiotherapy (Branch, 1944). More recently, the asymptomatic nature of early fractures has been shown to result in under-reporting. In a prospective trial, Milgrom et al. (1985) discovered six symptomatic and 13 asymptomatic previously undiagnosed femoral stress fractures out of a cohort of 300 soldiers investigated by scintigraphic bone scan. They found that pre-displacement stress fractures could be successfully treated by rest. A radiographic lucent line or lytic defect, however, indicated the need for prophylactic fixation to prevent displacement.

Fractures in the Elderly

Bone quality in elderly patients is often poor. Young cortical bone has a typically dense cortex, whereas old osteoporotic bone is brittle with a wide medullary canal. Low energy impacts and falls can cause significant comminution in osteoporotic bone. Moran, Gibson and Cross (1990) found that half of all femoral shaft fractures were Winquist grade III or IV. Using standard fixation with interlocking nails, peri-operative mortality was 17% and overall complication rate 54%. Many technical problems were encountered, for example, secondary to the restricting effects of hip joint arthritis. A study of 250 cases of femoral shaft fractures does confirm that internal fixation provides the a better outcome than conservative treatment (Taylor, Bannerjee and Alpar, 1994). As in younger individuals, early internal fixation of femoral shaft fractures in the elderly multiply-injured patient confers a benefit through reduced pulmonary insult. (Behrman et al., 1990).

Pathological Fractures

Pathological fractures due to tumour or metabolic bone disease can often be diagnosed and treated prior to displacement since pain occurs in microscopic infraction. Metastatic disease is by far the commonest cause, with the breast as the most frequent primary site. Internal fixation of femoral shaft fractures can be augmented with cement spacing. In 61 fractures treated in this way, 1-year survival was 50% (Haberman, 1982). There are no comparative studies of survival in conservatively managed patients, but the reduced level of pain, and better ambulation is thought to confer a qualitative advantage. The complication rate is about 10%, and pain relief is achieved in 90% (Dijkstra et al., 1994).

The femur is the commonest site for Paget's disease of bone, and its sequelae include bony deformation which complicates internal fixation. Soft bone in metabolic bone disease often compromises plate and screw fracture fixation. Osteomalacia produces Looser's compression fracture zones distal to the lesser trochanter. In a study of osteomalacia-induced proximal femoral shaft fractures, two out of four patients treated with plate and hip screw fixation failed to heal, possibly due to soft bone. The other two were treated by intramedullary nailing and achieved union without complication (Chalmers, 1970).

Other specific injuries which bear on treatment and prognosis should be identified. Fractures of the ipsilateral femoral neck are problematic. A theoretical compromise of femoral head blood supply may result from intramedullary nailing, particularly if the proximal femur is extensively reamed. In 33 such injuries, Wu and Shih (1991) found that union of both shaft and neck fractures was always achieved with intramedullary nailing and supplementary neck pinning. No cases of avascular necrosis of the femoral head were identified, but average follow-up was only 2 years. Retrograde femoral nailing has been advocated in ipsilateral shaft and neck fractures to avoid vascular damage. Eight patients treated in this way resulted in two diaphyseal non-unions. These poor results, however, may have been due to initial injury severity (Patterson et al., 1995). Fracture-dislocations of the hip are also at high risk of avascular necrosis of the femoral head. It has been advocated that associated ipsilateral femoral shaft fractures should undergo primary external fixation after immediate hip reduction. Delayed reamed intramedullary nailing should then be performed. In 16 such patients followed up for 2–8 years, all obtained a satisfactory outcome (Wu, Shih and Chen, 1993). Femoral shaft fractures associated with unstable pelvic fractures can be treated by early intramedullary nailing of the femur (90% survival and 95% of survivors proceeding to union), and either internal or external pelvic fixation. Twenty such patients were followed up for over 2 years with satisfactory knee and hip function (Wu and Shih, 1993).

Ipsilateral Tibial Shaft Fractures

Ipsilateral femoral and tibial shaft fractures result in the 'floating knee'. Karlstrom and Olerud (1977) found that in 32 patients, rigid fixation of both long bones resulted in 28 with good or excellent outcome. This compared with none of the patients who underwent any other treatment modality. McAndrew and Pontarelli (1988) showed that cast bracing and traction in these injuries often resulted in tibial malunion, poor ankle movement, and a higher rate of pneumonia and thrombo-embolic complications. Furthermore, ipsilateral femoral and tibial fractures frequently result in knee ligament injuries, so that after rigid fixation knee stability should be also examined under anaesthetic (McAndrew and Pontarelli, 1988).

Acute Spinal Cord Lesions

Acute spinal cord lesions and femoral fracture are not infrequently associated injuries in falls and road traffic accidents. Garland, Rieser and Singer (1985) analysed their results in 27 patients. *Five* of 11 patients managed conservatively developed a non-union, resulting in delayed operative fixation. The remaining 16 patients underwent primary internal fixation. Operative stabilization within 6 weeks was shown to confer the greatest benefit, with no non-unions, and least episodes of orthopaedic and medical complications. Decubitus ulceration was more common when patients with complete cord lesions were treated conservatively, presumably due to the compromising effect of unstable fractures on nursing care.

Periprosthetic Fractures

The advent of hip joint arthroplasty has produced a large cohort of elderly individuals liable to fracture around their femoral prosthesis. These fractures require special assessment, and a number of separate classification systems have been advocated. The first published classification system consisted of three types, depending on fracture site in relation to the prosthesis.

The Johansson classification of periprosthetic femoral fractures (Johansson et al., 1981)

Type I Fracture entirely proximal to prosthesis tip

Type II Fracture commences in proximal femur and extends distal to prosthesis tip

Type III Fracture entirely distal to prosthesis

This classification has been shown to have prognostic value. The authors analysed data from 35 patient and found that fractures around an intact prosthesis (type I) could be treated either conservatively or by revision, but outcome was poor in approximately 50%. In contrast, conservative treatment of fractures at the stem tip (type II) always produced poor results, compared with 100% experiencing a satisfactory outcome after long stem revision and internal fixation. Of four patients fractured below the prosthesis (type III) a satisfactory result was achieved in all, whether managed in traction or by internal fixation.

Bethea et al. (1982) modified Johansson's classification by dividing fractures entirely proximal to the prosthesis tip into two further groups, although they ignored fractures entirely distal to the prosthesis. This became the model for most subsequent classification systems.

The Bethea classification of periprosthetic fractures (Bethea et al., 1982)

Type A Fracture at stem tip

Type B Spiral fracture around stem
Type C Comminuted fracture around stem

The authors recommended revision to a long stem prosthesis in fractures at the stem tip (type A), which were at high risk of non-union. Spiral fractures around the stem (type B) could be treated either conservatively or by surgery, but stem loosening was a significant complication in those conservatively managed. Comminuted fractures around the stem (type C) were fixed immediately to allow early mobilization, often involving a long stem prosthetic revision. All type B and C fractures proceeded to union.

A subsequent review of 29 intra- and postoperative periprosthetic femoral fractures utilized Bethea's classification. The authors advocated early surgical stabilization for all fractures. A good outcome was obtained in revision arthroplasty (eight out of eight achieved union), and in patients treated with plating and bone graft (nine out of 11 achieved union). Worse results were obtained in those treated with wire loop fixation (one out of three achieved union), and in traction in which two patients died from sepsis (Su et al., 1988). In a large study of 75 periprosthetic fractures of the proximal femoral shaft, Cooke and Newman (1988) used a classification system which differed only slightly from that of Bethea. They found that transverse fractures at the stem tip required either internal fixation or long stem revision with 70% obtaining a good or excellent outcome, compared with only a 60% union rate in those treated conservatively. Oblique or comminuted fractures around the stem were also best managed by revision, with 70% achieving good function compared with only 20% of those treated conservatively. Eighty per cent of fractures distal to the stem tip were treated consevatively, all with satisfactory results.

Mont and Maar (1994) performed a large retrospective analysis of the published data on 195 periprosthetic femoral fractures which allowed classification into six types.

The Mont and Maar classification of periprosthetic fractures (Mont and Maar, 1994)

Type 1 Intertrochanteric fracture
Type 2 Proximal femoral fracture
Type 3 Fracture spanning the prosthesis tip
Type 4 Fracture distal to prosthesis tip
Type 5 Comminuted blowout fracture
Type 6 Supracondylar fracture

A statistical analysis of treatment and outcome in each type was performed. For fractures types around the prosthesis and at its tip, circlage wiring or long stem revision produced significantly better results than all other treatment modalities ($P < 0.02$). Fractures distal to the prosthesis tip treated with either traction or long stem revision achieved successful results in about 75%, and both treatments were superior to circlage wiring or plate fixation ($P < 0.01$).

Distal Femoral Fracture Classifications

Distal femoral fractures comprise 4–7% of all femoral fractures. They are often the result of either high-energy trauma in the young, or falls in the elderly with associated osteoporosis. Early studies of these fractures reported significant problems and poor outcome with both closed and open treatment methods, although later reports reveal a considerable improvement in outcome when principles of rigid fixation combined with biological principles of fracture care are applied.

Important general classifications of distal femoral fractures include those of Neer, Sensheimer and the AO group.These will be discussed in detail, together with others in the literature. Whether the fracture is open or closed is important, and other factors relating to bone biology and associated injuries will receive attention as adjuncts to rigid classification systems since they impose heavily on the management decision-making process.

Early incomplete classifications differentiated supracondylar (extra-articular) fractures from intercondylar (intra-articular) 'T' or 'Y' fractures. The importance of this distinction lies in the importance (and difficulty) of achieving accurate reduction and fixation of the intra-articular component of the fracture. 'T' and 'Y' fractures thus exist as an important component in most current classifications The exception to this rule is the first complete modern classification system, devised by Neer.

The Neer Classification (Neer, Grantham and Shelton, 1967) (Figure 16.4)

Grade I Minimally displace fracture
Grade IIA Displacement of condyle(s) medially
Grade IIB Displacement of condyle(s) laterally
Grade III Concomitant metaphyseal and distal diaphyseal fractures (comminuted)

Type I injuries arise in low energy impacts and falls. Type IIA are more commonly in the elderly. Type IIB tend to occur in younger patients sustaining a blow from the lateral side. Type III fractures usually result from high energy impacts.

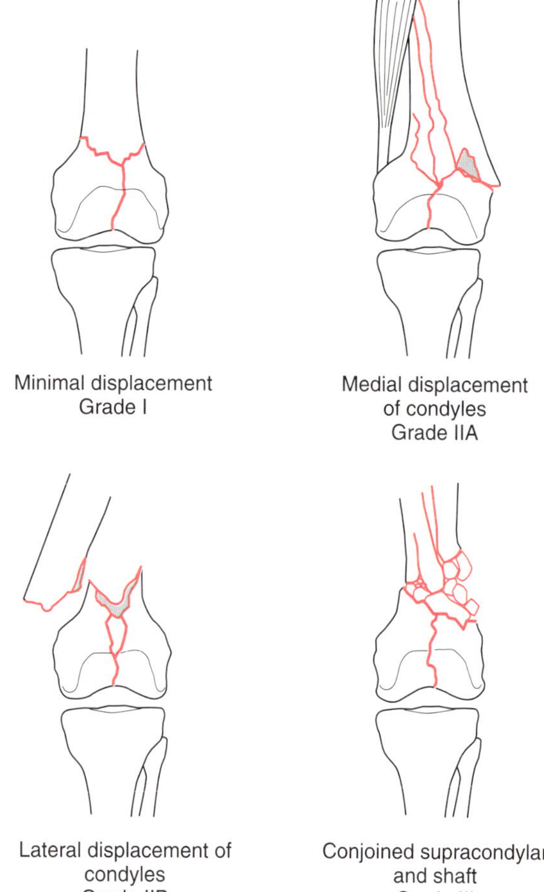

Figure 16.4. The Neer classification of distal femoral fractures (Neer, Grantham and Shelton, 1967).

Neer retrospectively analysed data from 110 unselected patients. Each grade contributed approximately one-quarter to the total. Half were treated with blade plate or other forms of fixation, and half in traction. Internal fixation resulted in a poor outcome (reflected in function, motion, work and anatomy) in 50%, compared with only 10% in the traction group. Furthermore, 48% of patients were dissatisfied with operative treatment compared with only 10% managed conservatively. Overall, recovery was slowest in type IIA and III fractures. By present-day standards, the outcome defined as 'satisfactory' in Neer's study is clearly suboptimal. Recent publications which have more stringent outcome standards are therefore of greater relevance.

In Neer's publication, all fractures were illustrated as intra-articular. Clearly, however, a large proportion of distal femoral fractures are extra-articular. In fact it is possible to utilize Neer's classification for either intra-articular fractures alone, or both intra and extra-articular fractures by using supplementary grading. Chiron and Casey (1974) utilized a classification in which Neer's system was included solely for intra-articular 'T' and 'Y' fractures. In contrast to Neer's findings, two-thirds of type IIA and III fractures achieved a good or excellent outcome although AO fixation techniques were used in all cases. Type I fractures also did well, although only one-third of type IIA fractures obtained a good or excellent result. Olerud (1972) also reviewed 15 solely intra-articular fractures fixed with an angled blade plate. All grade I and II fractures, and 75% of grade III fractures healed with good or excellent results. In contrast, Mize, Bucholz and Grogan (1982) classified 30 both intra- and extra-articular fractures by Neer's system. They fixed all by AO technique and achieved good or excellent results in 80%, although they failed to stratify outcome according to grade.

An inclusive fracture classification for the distal 9 cm of femur was published by Sensheimer. His was a mechanistic classification, which differentiated between low and high energy impacts, old and young bone.

The Sensheimer Classification (Seinsheimer, 1980) (Figure 16.5)

Type I	Non-displaced fracture (< 2 mm displacement)
Type IIA	Fracture only supracondylar (two-part)
Type IIB	Fracture only supracondylar (comminuted)
Type IIIA	Fracture involves intercondylar notch (medial condyle separate)
Type IIIB	Fracture involves intercondylar notch (lateral condyle separate)
Type IIIC	Fracture involves intercondylar notch (both condyles separate)
Type IVA	Comminuted fracture involves medial condylar surface
Type IVB	Comminuted fracture involves lateral condylar surface
Type IVC	Comminuted fracture involves both condylar surfaces, or one condyle and supracondylar region

Seinsheimer retrospectively analysed 47 patients with distal femoral fractures for mechanism of injury, bone quality, pre-existing medical problems, management and outcome. Type I and IIA fractures (non-displaced or simple supracondylar), and type IIIA and IIIB fractures (simple intercondylar notch) made up 25% of the total, and all occurred in osteoporotic patients with pre-existing systemic disease.

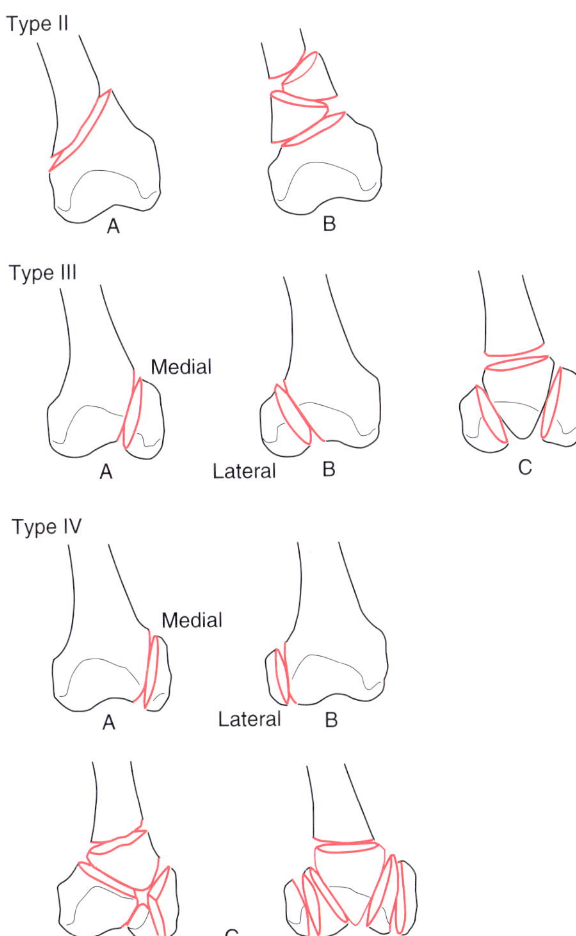

Figure 16.5. The Seinsheimer (1980) classification, the Type I undisplaced fracture is omitted.

Type IIB fractures (comminuted supracondylar), and type IIIC fractures (bicondylar into the notch) contributed 65% to the total, and either occurred after high energy trauma in mainly young males, or after simple falls in elderly females. Type IV fractures (comminuted intercondylar) made up 10% of the total, and all occurred in the young after high energy impacts. In general, treatment was traction for the elderly, and internal fixation for the young, especially in type III and IV fractures. Outcome was considered to be unrelated to method of treatment, but those treated by immobilization gained less knee flexion than those treated operatively. If conservative treatment was employed, a cast brace allowed better knee flexion to be obtained than a rigid cast. As with Neer's classification, Sensheimer's grading system is now seen to be of mainly historical interest.

Subsequently, the AO technique of rigid fixation with condylar lag-screw and side-plate was reviewed by Giles et al. (1982) in their management in 14 Seinsheimer grade I/II fractures, and 10 grade III/IV fractures. All achieved union without infection, with the only two cases of shortening and loss of reduction occuring in grade III fractures.

The classification of Healy and Brooker (1983) is dissimilar to other classifications, and divides distal femoral fractures into four types.

The Healy Classification (Healy and Brooker, 1983)

Simple, extra-articular
Simple, intra-articular
Complex, extra-articular
Complex, intra-articular

Complex fractures were defined as either angulated, comminuted, or displaced over 2 mm. The authors compared outcome in 98 fractures. Half were treated by AO techniques and over 90% of these achieved fair or good results regardless of fracture pattern. In contrast, outcome in the half treated conservatively depended heavily on fracture type, with 100% of simple extra-articular fractures achieving a fair or good result, compared to 35% of complex intra-articular fractures (Healy and Brooker, 1983). Pritchett (1984) used a virtually identical classification system in an analysis of 23 fractures fixed with a supracondylar lag-screw and AO plate. Only one of 14 extra-articular fracture fixations failed, compared with two of nine patients with intra-articular fractures.

The AO group has documented data on thousands of distal femoral fractures. They have updated their mechanistic, inclusive classification system to reflect both severity of injury and outcome. An extensive body of trauma surgeons use the AO classification as a basis for decision making in distal femoral fractures (Mize, 1989; Mast, Jakob and Ganz, 1989).

The AO Classification (Müller, Nazarian and Koch, 1987) (Figure 16.6)

Type A1 Extra-articular, simple two-part supracondylar fracture
Type A2 Extra-articular, metaphyseal wedge fracture
Type A3 Extra-articular, comminuted supracondylar fracture
Type B1 Unicondylar, lateral condylar, sagittal fracture
Type B2 Unicondylar, medial condylar, sagittal fracture
Type B3 Unicondylar, coronal fracture
Type C1 Bicondylar, non-comminuted supracondylar 'T' or 'Y' fracture

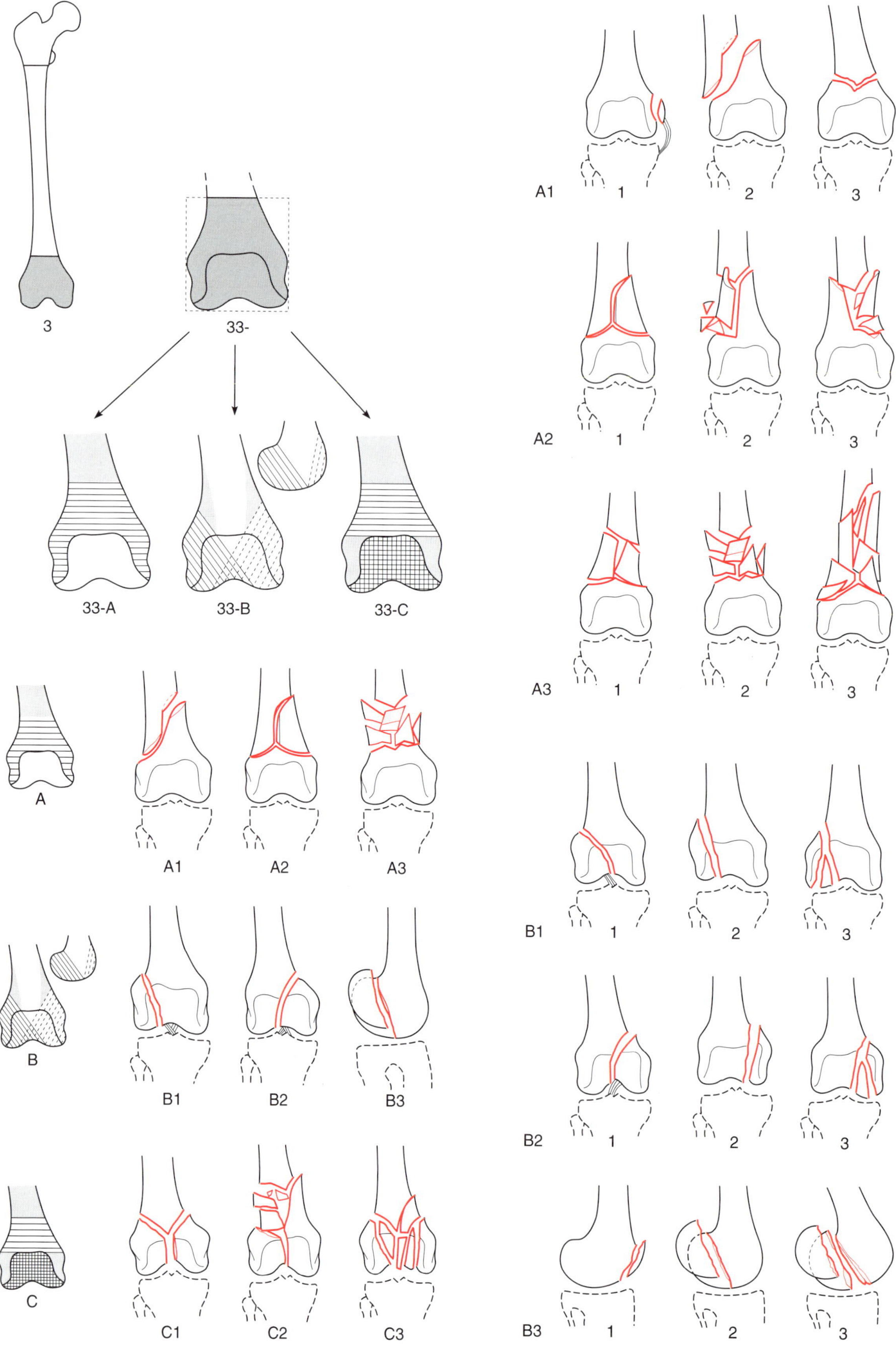

Figure 16.6. The AO/ASIF classification of distal femoral fractures (Müller, Nazarian and Koch, 1987).

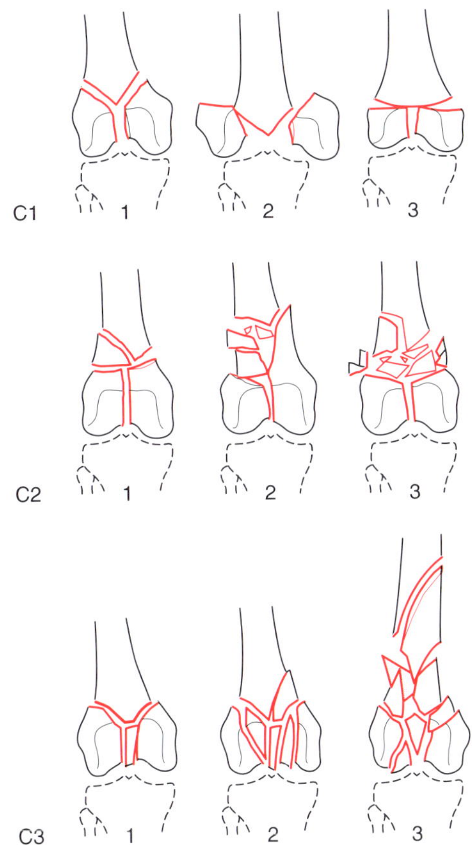

Figure 16.6. (*Continued*).

Type C2 Bicondylar, comminuted supracondylar fracture

Type C3 Bicondylar, comminuted supracondylar and intercondylar fracture

Chiron and Casey (1974) classified adult distal femoral fractures into three types which corresponded to the three main AO grades. Using the contemporary AO technique of angled blade plate fixation in 136 fractures, 85% of grade A fractures achieved a good or excellent outcome, compared with only 65% of grade B or C fractures. Similar results were obtained by Schatzker, Horne and Waddell (1975) who graded distal femoral fractures into three, which also corresponded to the three AO grades. Results in 71 fractures revealed that internal fixation using AO techniques were superior to conservative management, and outcome after surgery improved from AO grade C (60% good/excellent) to grade A (70% good/excellent). A good or excellent result after internal fixation was obtained in less than one-third of consevatively managed patients. Compared with his first publication, Schatzker's second paper classified fractures into three completely different types: either simple extra-articular, or comminuted extra-articular (equivalent to AO grades A/B), or intra-articular (equivalent to AO grade C). Seventeen distal femoral fractures were fixed by AO methods. Three-quarters of simple or comminuted extra-articular fractures obtained good or excellent results, compared with only 50% who obtained this outcome in the intra-articular group (Schatzker and Lambert, 1979).

The historical classification of bicondylar intra-articular distal femoral fracture types 'T' and 'Y' correspond to AO grade C fractures. These are some of the most difficult fractures of all to treat, and non-AO techniques have been extensively researched. The Zickel supracondylar device, for example, has been recommended for 'T' and 'Y' fractures (Zickel et al., 1977). In a cohort of 67 distal femoral fractures fixed with this device, however, only 48% of 'T' and 'Y' fractures obtained over 90° of knee flexion, compared with 80% who obtained this in simpler fractures (Zickel, Hobeika and Robbins, 1986). In contrast, Siliski, Mahring and Hofer (1989), used solely AO techniques to fix 52 fractures. Types C1/C2 achieved full weight bearing in 12 weeks, and 92% obtained good or excellent function. This compared with 16 weeks to full weight bearing, and 77% with good or excellent function in C3 fractures.

Comminuted fractures are AO grade A3 and C2/C3. Sanders, Regazzoni and Reudi (1989) used the AO dynamic condylar screw device in 49 fractures. They found that outcome was related to comminution, so that poor results were obtained in 45% of type A3 and 30% of C2/C3 fractures, compared with no poor results in the simple fracture types A1 and C1. The authors considered that tethering of quadriceps muscle or tendon might have reduced function in comminuted fractures. Problems with fracture stability in AO types C2/C3 are evident due to loss of medial cortex. Sanders et al. (1991) observed that in nine such injuries, standard lateral plating with a 95° fixed angled device failed to achieve stable fixation, with subsequent intra-operative varus angulation. As a result, in addition to bone graft, the medial cortex was plated. All fractures healed in anatomical position within 3 years. Yet knee flexion in excess of 90° was only achieved in six patients. Other non-AO techniques have been applied to comminuted distal femoral fractures to obtain stable fixation. Blatter et al. (1994) accepted non-anatomical shortening of the femur averaging 1 inch to achieve stability. Fifteen of 16 fractures united in 3 months and average knee flexion was 115°. Importantly, only one-third of patients required a heel lift, and all achieved at least a satisfactory overall outcome.

Open Fractures

Open fractures of the distal femur can again be classified using the Gustilo system (Gustilo, Mendoza and Williams, 1984). The same principles should be applied as those used in open femoral shaft fracture management. In grade III fractures this entails débridement and irrigation with 9 to 12 litres of saline or Hartmann's solution, followed by reconstitution of articular condylar anatomy with minimal disturbance, then stable fixation of condyles to shaft. As an alternative, primary external fixation across the knee is safe in the presence of grossly disrupted bone and soft tissue. Grade IIIC fractures imply popliteal artery damage, and this limb-threatening injury should be repaired within 6 hours of initial insult, if possible with skeletal stabilization prior to vascular repair (Matter and Rittman, 1978). If débridement and skeletal stabilization require a delay in excess of 6 hours, a temporary intraluminal shunt has been advocated to restore distal perfusion (Johansen et al., 1982). Internal fixation of open fractures of the distal femur can produce results as good as those in closed fractures. Shahcheraghi and Doroodchi (1993) analysed their experience with 31 open fractures. A good or excellent result was achieved in 80% of patients treated surgically. Even complex comminuted AO grade C2/C3 fractures can be adequately managed using rigid fixation. Bone grafting and supplementary plating should be delayed until the wound is considered clean (Sanders et al., 1991).

Fractures in the Elderly

In distal femoral fractures, patient age and bone quality is even more fundamental to management strategy than in femoral diaphyseal fractures. This is because simple falls in the elderly often result in comminution of the distal femur, and standard fixation methods tend to hold bone too poorly to allow even early protected mobilization, given the propensity of elderly patients to stumble. A proportion suffer from dementia, and operative failures frequently occur in this group (Pritchett, 1984). Furthermore, articular involvement in the elderly is much less critical than in younger patients, since the onset of arthritis 20 years hence is unlikely to be of consequence. Although the decision whether or not to operate is more finely balanced in distal femoral fractures, most recent reports suggest that all but the simplest fractures in the elderly are better managed operatively (Healy and Brooker, 1983; Sanders, Regazzoni and Reudi, 1989).

The inadequacy of a fixed angled device and lateral cortical plate alone in osteoporotic bone has led some authors to advocate conservative treatment in such patients (Schatzker and Lambert, 1979). A fixed angled lag screw or dynamic condylar screw, however, is likely to provide better purchase on osteopaenic bone and greater interfragmentary compression than an angled blade plate (Giles et al., 1982; Sanders, Regazzoni and Reudi, 1989).

Non-AO techniques have been applied to more difficult osteoporotic fractures in an attempt to achieve stability, and therefore early mobility. Brown and D'Arcy (1971) reviewed 22 elderly osteoporotic patients with fractures internally fixed with both lateral and medial plates. Although average knee flexion was only 55°, mobility was generally maintained. Papagiannopoulos and Clement (1987) treated 26 fractures with an intramedullary nail with wings to control rotation, and allow compression to a more stable position. They recommended its use in the elderly after 92% obtained good or excellent results. Outcome in 33 elderly patients treated with the Zickel supracondylar nail has been reviewed by Marks, Isbister and Porter (1994). Most were AO grade A1 fractures. In survivors, fractures united in an average of 12 weeks after mobilization in a cast brace, and all obtained an excellent final result.

Pathological Fractures

Pathological fractures due to metastases are frequently located in the distal femur (Haberman, 1982). Operative fixation of these fractures relieves pain and allows mobilization. In a review of pathological metastatic fractures of the lower limb, Harrington (1995) recommended the use of an intramedullary nail supplemented with cement for distal femoral fractures. He argued that this theoretically provided greater stability than a fixed angled device with lateral plate, due to the high risk of non-union in this type of fracture.

As in fractures of the femoral diaphysis, associated local injuries and multiple trauma all affect the decision making process in distal femoral fractures. Indications for internal fixation, rather than conservative management include significant polytrauma, bilateral femoral fracture, ipsilateral knee ligament disruption, and femoral shaft or tibial fracture (Browner et al., 1992).

Periprosthetic Fractures

Periprosthetic fractures of the distal femur are a recognized complication following total knee arthroplasty. Rheumatoid patients dependent on steroid therapy are at particular risk. An early report of 26 periprosthetic fractures treated conservatively identified that although 17 healed without complication, nine required revision due to non-union, malunion or loosening. Traction or cast bracing was recommended for patients with osteopaenic bone and those unlikely to make heavy demands on their prosthesis (Merkel and Johnson, 1986).

Neer's classification system has been used for periprosthetic fractures. DiGioia and Rubash (1991) carried out a literature review from which a classification based on Neer typing was suggested. A treatment algorithm in which the degree of displacement was a critical factor was formulated (criteria for displacement: translation <5 mm, angulation <10°, shortening <1 cm). The authors recommended that Neer type I/II fractures should be managed with cast brace immobilization, although displaced fractures were reduced first. Neer type III fractures should be treated by traction and brace, but if the bone/cement interface was involved, fixation or delayed revision should be employed. Chen, Mont and Bachner (1994) subclassified periprosthetic fractures as either displaced (Neer grade I) or undisplaced (Neer grade II/III). In a literature review of 195 fractures, the authors found that undisplaced fractures managed conservatively produced a satisfactory outcome in 83% of cases, compared with only 64% of those treated by operation. Approximately 65% of displaced fractures achieved a satisfactory result whatever the treatment regime.

Recent reports suggest that some periprosthetic fractures can be fixed without disturbing the prosthesis. Three reports totalling 14 cases have been published, in which retrograde supracondylar intramedullary nail fixation was employed through a small patellar-splitting incision to stabilize the fracture, with interlocked percutaneous screws. All healed without major complication (McLaren, Dupont and Schroeber, 1994; Jabczenski and Crawford, 1995; Murrell and Nunley, 1995). Alternatively, Rush rod fixation has been advocated by Ritter et al. (1995) who published their experience with 22 displaced fractures. All proceeded to union without loss of knee function.

References

Anderson, J.T. and Gustilo, R.B. (1980) Immediate internal fixation in open fractures. *Orthop. Clin. N.A.*, **11**, 569.

Baker, S.P., O'Neill, B. and Haddon, W.W. Jr (1974) The injury severity score: a method for describing patients with multiple injuries and evaluation emergency care. *J. Trauma*, **14**, 187–196.

Behrman, S.W., Fabian, T.C., Kudsk, K.A. and Taylor, J.C. (1990) Improved outcome with femoral fractures: early vs. delayed fixation. *J. Trauma*, **30**, 792–798.

Bergman, M., Tornetta, P., Kerina, M. and Sandhu, H. et al. (1993) Femur fractures caused by gunshots: treatment by immediate reamed intramedullary nailing. *J. Trauma*, **34**, 783–785.

Bethea, J.S., DeAndrade, J.R., Fleming, L.L. et al (1982) Proximal femoral fractures following total hip arthroplasty. *Clin. Orthop.*, **170**, 95–106.

Blatter, G., Konig, H., Janssen, M. and Mageri, F. (1994) Primary femoral shortening osteosynthesis in the management of comminuted supracondylar femoral fractures. *Arch. Orthop. Trauma Surg.*, **113**, 134–137.

Bone, L.B., Johnson, K.D., Weigelt, J. and Scheinberg, R. (1989) Early versus delayed stabilisation of femoral fractures: a prospective randomised study. *J. Bone Joint Surg.*, **71A**, 336–340.

Branch, H.E. (1994) March fractures of the femur. *J. Bone Joint Surg.*, **26**, 387–391.

Breederveld, R.S., Patka, P. and Van-Mourik, J.C. (1985) Refractures of the femoral shaft. *Neth. J. Surg.*, **37**, 114–116.

Brien, W.W., Kuschner, S.H., Brien, E.W. and Wiss, D.A. (1995) The management of gunshot wounds to the femur. *Orthop. Clin.*, **26**, 133–138.

Brown, A. and D'Arcy, J.C. (1971) Internal fixation for supracondylar fractures of the femur in the elderly patient. *J. Bone Joint Surg.*, **53B**, 420–424.

Browner, B.D., Burgess, A.R., Robertson, R.J. et al. (1984) Immediate closed anterograde Ender nailing of femoral fractures in polytrauma patients. *J. Trauma*, **24**, 921–925.

Browner, B.D., Jupiter, J.B., Levine, A.M. and Trafton, P.G. (1992) *Skeletal Trauma*, Vol. 2. Philadelphia: Saunders, pp. 1643–1683.

Brumback, R.J., Handal, J.A., Poka, A. et al. (1989) Intramedullary nailing of open fractures of the femoral shaft. *J. Bone Joint Surg.*, **71A**, 1324–1330.

Brumback, R.J., Reilly, J.P., Poka, A. et al. (1988a) Intramedullary nailing of femoral shaft fractures. Part I: decision-making errors with interlocking fixation. *J. Bone Joint Surg.*, **70A**, 1441–1452.

Brumback, R.J., Uwagie-Ero, S., Lakatos, R.P. et al. (1988b) Intramedullary nailing of femoral shaft fractures. Part II: fracture healing with static interlocking fixation. *J. Bone Joint Surg.*, **70A**, 1453–1462.

Cargile, J.S. III, Hunt, J.L. and Purdue, G.F. (1992) Acute trauma of the femoral artery and vein. *J. Trauma*, **32**, 364–371.

Chalmers, J. (1970) Subtrochanteric fractures in osteomalacia. *J. Bone Joint Surg.*, **52B**, 509–513.

Chapman, M.W. and Mahoney, M. (1979) The role of early internal fixation in the management of open fractures. *Clin. Orthop.*, **138**, 120–131.

Chen, F., Mont, M.A. and Bachner, R.S. (1994)

Management of ipsilateral supracondylar femur fractures following total knee arthroplasty. *J. Arthroplasty*, **9**, 521–526.

Chiron, H.S. and Casey, P. (1974) Fractures of the distal third of the femur treated by internal fixation. *Clin. Orthop.*, **100**, 160–170.

Cooke, P.H. and Newman, J.H. (1988) Fractures of the femur in relation to cemented hip prostheses. *J. Bone Joint Surg.*, **70B**, 386–389.

Copes, W.S., Champion, H.R., Sacco, W.J. et al. (1988) The injury severity score revisited. *J. Trauma*, **28**, 69–77.

Dabezies, E.J., D'Ambrosia, R., Shoji, H. et al. (1984) Fractures of the femoral shaft treated by external fixation with the Wagner device. *J. Bone Joint Surg.*, **66A**, 360–364.

DiGioia, A.M. III and Rubash, H.E. (1991) Periprosthetic fractures of the femur after total knee arthroplasty: a literature review and treatment algorithm. *Clin. Orthop.*, **271**, 135–142.

Dijkstra, S., Wiggers, T., Van-Geel, B.N. and Boxma, H. (1994) Impending and actual pathological fractures in patients with bone metastases of the long bones. A retrospective study of 233 surgically treated fractures. *Eur. J. Surg. Acta Chir.*, **160**, 532–542.

Fakhry, S.M., Rutledge, R., Dahners, L.E. et al. (1994) Incidence, management and outcome of femoral shaft fractures: a statewide population-based analysis of 2805 adult patients in a rural state. *J. Trauma*, **37**, 255–261.

Garland, D.E., Rieser, T.V. and Singer, D.I. (1985) Treatment of femoral shaft fractures associated with acute spinal cord injuries. *Clin. Orthop.*, **197**, 191–195.

Giles, J.B., DeLee, J.C., Heckman, J.D. and Keever, J.E. (1982) Supracondylar-intercondylar fractures of the femur treated with a supracondylar plate and a lag screw. *J. Bone Joint Surg.*, **64A**, 864–870.

Grover, J. and Wiss, D.A. (1995) A prospective study of fractures of the femoral shaft treated with a static intramedullary interlocking nail, comparing one versus two distal screws. *Orthop. Clin.*, **26**, 139–146.

Gustilo, R. (1990) *The Classification Manual*. St Louis: Mosby Year Book.

Gustilo, R.B., Mendoza, R.M. and Williams, D.N. (1984) Problems in the management of type III (severe) open fractures: a new classification of type III open fractures. *J. Trauma*, **24**, 742–746.

Haberman, E.T. (1982) The pathology and treatment of metastatic disease of the femur. *Clin. Orthop.*, **169**, 70–82.

Harrington, K.D. (1995) Orthopaedic management of extremity and pelvic lesions. *Clin. Orthop.*, **312**, 136–147.

Healy, W.L. and Brooker, A.F. Jr (1983) Distal femoral fractures. Comparison of open and closed methods of treatment. *Clin. Orthop.*, **174**, 165–171.

Hirsch, C., Cavandias, A. and Nachemson, A. (1955) An attempt to explain fracture types. Experimental studies on rabbit bones. *Acta Orthop. Scand.*, **24**, 8–29.

Jabczenski, F.F. and Crawford, M. (1995) Retrograde intramedullary nailing of supracondylar femur fractures above total knee arthroplasty: a preliminary report of four cases. *J. Arthroplasty*, **10**, 95–102.

Johansen, K., Bandyk, D., Thiele, B. and Hansen, S.T. (1982) Temporary intraluminal shunts: resolution of a management dilemma in complex vascular injuries. *J. Trauma*, **22**, 395–402.

Johansson, J.E., McBroom, R., Barrington, T.W. and Hunter, G.A. (1981) Fracture of the ipsilateral femur in patients with total hip replacement. *J. Bone Joint Surg.*, **63A**, 1435–1442.

Johnson, K.D. (1988) Current techniques in the treatment of subtrochanteric fractures. *Tech. Ortho.*, **3**, 14–24.

Johnson, K.D., Cadambi, A. and Seibert, G.B. (1985) Incidence of adult respiratory distress syndrome in patients with multiple musculoskeletal injuries: effect of early operative stabilisation of fractures. *J. Trauma*, **25**, 375–384.

Johnson, K.D. Johnston, D.W.C. and Parker, B. (1984) Comminuted femoral shaft fractures; treatment by roller traction, circlage wires and an intramedullary nail, or an interlocking intramedullary nail. *J. Bone Joint Surg.*, **66A**, 1222–1235.

Kaltenecker, G., Wruhs, O. and Quaicoe, S. (1990) Lower infection rate after interlocking nailing in open fractures of femur and tibia. *J. Trauma*, **30**, 474–479.

Karlstrom, G. and Olerud, S. (1977) Ipsilateral fracture of the femur and tibia. *J. Bone Joint Surg.*, **59A**, 240.

Lhowe, D.W. and Hansen, S.T. (1988) Immediate nailing of open fractures of the femoral shaft. *J. Bone Joint Surg.*, **70A**, 812–820.

McAndrew, M.P. and Pontarelli, W. (1988) The long-term follow-up of ipsilateral tibial and femoral diaphyseal fractures. *Clin. Orthop.*, **232**, 190–196.

McLaren, A.C., Dupont, J.A. and Schroeber, D.C. (1994) Open reduction internal fixation of supracondylar fractures above total knee arthroplasties using the intramedullary supracondylar rod. *Clin. Orthop.*, **302**, 194–198.

Magerl, F., Wyss, A., Brunner, C. and Binder, W. (1979) Plate osteosynthesis of femoral shaft fractures in adults. *Clin. Orthop.*, **138**, 62–73.

Marks, D.S., Isbister, E.S. and Porter, K.M. (1994) Zickel supracondylar nailing for supracondylar femoral fractures in elderly or infirm patients. A review of 33 cases. *J. Bone Joint Surg.*, **76B**, 596–601.

Mast, J., Jakob, R. and Ganz, R. (1989) *Planning and Reduction Techniques in Fracture Surgery*. New York: Springer-Verlag.

Matter, P. and Rittman, W. (1978) *The Open Fracture*. Bern: Huber.

Merkel, K.D. and Johnson, E.W. Jr (1986) Supracondylar fracture of the femur after total knee arthroplasty. *J. Bone Joint Surg.*, **68A**, 29–43.

Milgrom, C., Giladi, M. Stein, M. et al. (1985) Stress fractures in military recruits. A prospective study showing an unusually high incidence. *J. Bone Joint Surg.*, **67B**, 732–735.

Mize, R.D. (1989) Surgical management of complex fractures of the distal femur. *Clin. Orthop.*, **240**, 77–86.

Mize, R.D., Bucholz, R.W. and Grogan, D.P. (1982) Surgical treatment of displaced comminuted fractures of the distal end of the femur. *J. Bone Joint Surg.*, **64A**, 871–879.

Mont, M.A. and Maar, D.C. (1994) Fractures of the ipsilateral femur after hip arthroplasty: a statistical analysis of outcome based on 487 patients. *J. Arthroplasty*, **9**, 511–519.

Moran, C.G., Gibson, M.J. and Cross, A.T. (1990) Intramedullary locking nails for femoral shaft fractures in elderly patients. *J. Bone Joint Surg.*, **72B**, 19–22.

Müller, M.E., Allgower, M., Schneider, R. and Willeneger, H. (1991) *Manual of Internal Fixation*, 3rd edn. New York: Springer-Verlag.

Müller, M.E., Nazarian, S. and Koch, P. (1987) *Classification AO des Fractures*. New York: Springer-Verlag.

Murrell, G.A.C. and Nunley, J.A. (1995) Interlocked supracondylar intramedullary nails for supracondylar fractures after total knee arthroplasty: a new treatment method. *J. Arthroplasty*, **10**, 37–42.

Neer, C.S. II, Grantham, S.A. and Shelton, M.L. (1967) Supracondylar fracture of the adult femur. *J. Bone Joint Surg.*, **49A**, 591–613.

Nicholas, R.M. and McCoy, G.F. (1995) Immediate intramedullary nailing of femoral shaft fractures due to gunshots. *Injury*, **26**, 257–259.

Nutz, V. and Katholnigg, D. (1994) Einfluss der femurstabilisierung auf den verlauf des polytraumas mit schadel-hirn-trauma. *Unfallchirurg*, **97**, 399–405.

O'Brien, P.J., Meek, R.N., Powell, J.N. and Blachut, P.A. (1991) Primary intramedullary nailing of open femoral shaft fractures. *J. Trauma*, **31**, 113–116.

Olerud, S. (1972) Operative treatment of supracondylar-condylar fractures of the femur. Technique and results in fifteen cases. *J. Bone Joint Surg.*, **54A**, 1015–1032.

Papagiannopoulos, G. and Clement, D.A. (1987) Treatment of fractures of the distal third of the femur. A prospective trial of the Derby intramedullary nail. *J. Bone Joint Surg.*, **69B**, 67–70.

Pape, H-C., Regel, C., Dwenger, A. et al. (1993) Influence of thoracic trauma and primary femoral intramedullary nailing on the incidence of ARDS in multiple trauma patients. *Injury*, **24** (Suppl. 3), S83–S103.

Patterson, B.M., Routt, M.L.C. Jr, Benirschke, S.K. and Hansen, S.T. Jr (1995) Retrograde nailing of femoral shaft fractures. *J. Trauma*, **38**, 38–43.

Payne, W.K. III, Gabriel, R.A. and Massoud, R.P., (1995) Gunshot wounds to the thigh: evaluation of vascular and subclinical vascular injuries. *Orthop. Clin.*, **26**, 147–154.

Pritchett, J.W. (1984) Supracondylar fractures of the femur. *Clin. Orthop.*, **184**, 173–177.

Richards, R.R., Waddell, J.P., Sullivan, T.R. et al. (1984) Infra-isthmal fractures of the femur: a review of 82 cases. *J. Trauma*, **24**, 735–741.

Ritter, M.A., Keating, E.M., Faris, P.M. and Meding, J.B. (1995) Rush rod fixation of supracondylar fractures above total knee arthroplasties. *J. Arthroplasty*, **10**, 213–216.

Rittmann, W.W., Schibli, M., Matter, P. and Allgower, M. (1979) Open fractures: long term results in 200 consecutive cases. *Clin. Orthop.*, **138**, 132–140.

Salter, R.B. and Harris, W.R. (1963) Injuries involving the epiphyseal plate. *J. Bone Joint Surg.*, **45A**, 587–622.

Sanders, R., Regazzoni, P. and Reudi, T. (1989) Treatment of supra-intercondylar fractures of the femur using the dynamic condylar screw. *J. Orthop. Trauma*, **3**, 214–222.

Sanders, R., Swiontkowski, M., Rosen, H. and Helfet, D. (1991) Double-plating of comminuted unstable fractures of the distal part of the femur. *J. Bone Joint Surg.*, **73A**, 341–346.

Schatzker, J., Horne, G. and Waddell, J. (1975) The Toronto experience with the supracondylar fracture of the femur, 1966–1972. *Injury*, **6**, 113–128.

Schatzker, J. and Lambert, D.C. (1979) Supracondylar fractures of the femur. *Clin. Orthop.*, **138**, 77–83.

Seinsheimer, F. (1980) Fractures of the distal femur. *Clin. Orthop.*, **153**, 169–179.

Shahcheraghi, G.H. and Dodoodchi, H.R. (1993) Supracondylar fracture of the femur: closed or open reduction? *J. Trauma*, **34**, 499–502.

Siebert, C.H., Arens, S., Rinke, F. et al. (1995) Sekundare plattenosteosynthese von offenen frakturen der unteren extremitatimmer noch eine therapeutische alternative!? *Zentralbl. Chir.*, **120**, 32–36.

Siliski, J.M., Mahring, M. and Hofer, H.P. (1989) Supracondylar-intercondylar fractures of the femur. Treatment by internal fixation. *J. Bone Joint Surg.*, **71A** 95–104.

Sojbjerg, J.O., Eiskjaer, S. and Møller-Larsen, F. (1990) Locked nailing of comminuted and unstable fractures of the femur *J. Bone Joint Surg.*, **72B**, 23–25.

Su, J.-Y., Huang, C.-H., Lai, J.-H. and Wang, J.-P. (1988) Fracture of the femur during and following total hip arthroplasty. *J. West Pac. Orthop. Assoc.*, **25**, 17–21.

Taylor, M.T., Bannerjee, B. and Alpar, E.K. (1994) The epidemiology of fractures femurs and the effect of these factors on outcome. *Injury*, 25; 641–644.

ten Duis, H.J., Nijsten, M.W., Klasen, H.J. and Binnendi, J.K.B. (1988) Fat embolism in patients with an isolated fracture of the femoral shaft. *J. Trauma*, **28**, 383–390.

Winquist, R.A. and Hansen, S.T. (1980) Comminuted fractures of the femoral shaft treated by intramedullary nailing. *Orthop. Clin.*, **11**, 633–648.

Winquist, R.A., Hansen, S.T. Jr and Clawson, D.K. (1984) Closed intramedullary nailing of femoral fractures: a report of five hundred and twenty cases. *J. Bone Joint Surg.*, **66A**, 529–539.

Woodburne, R.T. and Burkel, W.E. (1994) *Essentials of Human Anatomy*. New York: Oxford University Press, pp. 597–599.

Wu, C.-C. and Shih, C.-H. (1991) Ipsilateral femoral neck and shaft fractures. Retrospective study of 33 cases. *Acta Orthop. Scand.*, **62**, 346–351.

Wu, C.-C. and Shih C.-H. (1993) Femoral shaft fractures associated with unstable pelvic fractures. *J. Trauma*, **34**, 76–81.

Wu, C.-C., Shih, C.-H. and Chen, L.-H. (1993) Femoral

shaft fractures complicated by fracture-dislocations of the ipsilateral hip. *J. Trauma* **34**, 70–75.

Zickel, R.E., Fietti, V.G. Jr, Lawsing, J.F. III and Cochran, G.V. (1977) A new intramedullary fixation device for the distal third of the femur. *Clin. Orthop.*, **125**, 185–191.

Zickel, R.E., Hobeika, P. and Robbins, D.S. (1986) Zickel supracondylar nails for fractures of the distal end of the femur. *Clin. Orthop.*, **212**, 79–88.

17

The knee

J. M. Latham

Introduction

The knee is the largest joint in the body, consisting of two condylar joints and the patellofemoral joint within a single cavity. The capsule, collateral and cruciate ligaments provide stability to the joint, whilst the two menisci allow gliding and rotation to occur during flexion and extension, as well as contributing to the stability of the knee. Knee injuries are common and disabling and cause a considerable financial burden to health care purchasers. They occur mainly in young adults and the elderly. High energy and sports injuries are more common in the young, whilst falls account for most of the injuries in the elderly. Osteoporosis and medical disorders commonly increase the morbidity and mortality associated with knee injuries in the elderly.

Injuries to the knee joint may involve both bone and soft tissues. Consequently classification is complicated by the need to consider all of the tissues involved. The classification is made simpler by considering in turn each of the structures that make up the knee joint, namely the distal femur, the proximal tibia, the capsule and ligaments and finally the soft tissue envelope. This chapter reviews the classifications of injuries to these structures. Classifications of injuries are only useful if they are easy to apply and reproducible. They should also allow the surgeon to select appropriate treatment and indicate prognosis.

The Distal Femur (see also Chapter 16)

Introduction

This section will examine the classification of distal femoral fractures. These fractures comprise 4–7% of all femoral fractures, they are usually the result of high energy trauma in the young or falls in the elderly with associated osteoporosis. The anatomical area to be considered comprises the supra- and intracondylar areas only. Early studies of these fractures reported significant problems and poor outcome with both closed and open methods of treatment, although later studies show considerable improvement in outcome following internal fixation.

Neer Classification

Neer, Grantham and Shelton (1967) classified distal femoral fractures into four groups (Table 17.1) based on the radiographic anatomy of 110 fractures occurring through the lower 7.5 cm of the femur (see Figure 16.4). Follow-up ranged from 1 year to 24 years after injury in 77 fractures. Neer suggested that the degree of displacement of the fracture fragments was related to the severity of injury, and that the severe injuries had a poor outcome, particularly after open reduction and internal fixation. In fact, they concluded that no fracture was suitable for internal fixation because of the unacceptable complication rate and poor outcome. All fractures treated with internal fixation were immobilized until there was evidence of union.

Table 17.1 Neer classification of fractures of the distal femur (Neer, Grantham and Shelton, 1967)

I	Minimal displacement
IIA	Medial displacement of the condyles
IIB	Lateral displacement of the condyles
III	Concomitant shaft and supracondylar fractures

Whilst this classification is simple, it lacks refinement and provides little information regarding prognosis and outcome, reflecting the less sophisticated treament options available at the time of the study.

Seinsheimer Classification

Seinsheimer (1980) developed the classification of distal femoral fractures further emphasizing the importance on outcome of fractures occurring within the knee joint (Table 17.2). The classification was based on the analysis of 47 femoral fractures occurring within 9 cm of the knee joint (see Figure 16.5). The influence of different methods of treatment on outcome was also studied. The average follow-up was 24 months. What emerges from the study is the superior outcome achieved by internal fixation followed by early motion, in intra-articular fractures, compared to conservative treatment. More importantly, it was found that there were poor results in all fractures treated by traction followed by immobilization.

AO Classification

The classification system developed by the AO group (Müller et al., 1990) in the 1970s was based on the analysis of more than 150 000 operatively treated fractures. From this archive a systematic classification of long bone fractures was produced, with particular emphasis on the morphology of a fracture. Fractures are arranged in increasing order of severity, reflecting 'anticipated difficulties of treatment, the likely complications, and the prognosis' (Müller et al., 1990). Fractures of the distal femur encompass the area bounded by a square, the sides of which are the same length as the widest part of the epiphysis (see Figure 16.6). Type A fractures involve the distal femoral shaft only and are extra-articular, type B fractures involve one of the condyles (partial articular) whereas type C fractures are 'T'- or 'Y'-shaped bicondylar fractures (complete articular). Within each group there are subgroups which define the degree of fragmentation (Table 17.3). The AO classification has recently been adopted by the Orthopaedic Trauma Association (1996).

Table 17.2 Seinsheimer (1980) classification of fractures of the distal femur

I	Non-displaced	
II	Distal metaphysis	
	A:	two parts
	B:	multi-fragmentary
II	Fractures involving the intercondylar notch	
	A:	medial condyle separate fragment
	B:	lateral condyle separate fragment
	C:	complete separation of both condyles from each other and the femoral shaft
IV	Fractures through the articular surface of the condyles	
	A:	medial condyle
	B:	lateral condyle
	C:	complex bicondylar fracture

Table 17.3 AO classification of distal femoral fractures (Müller et al., 1990)

A	Extra-articular	
	A1	simple
	A2	metaphyseal wedge
	A3	metaphyseal complex
B	Partial articular	
	B1	lateral condyle, sagittal
	B2	medial condyle, sagittal
	B3	frontal
C	Complete articular	
	C1	articular simple, metaphyseal simple
	C2	articular simple, metaphyseal multifragmentary
	C3	multifragmentary

Tibial Plateau Fractures

Introduction

The tibial plateau comprises the medial and lateral tibial condyles which articulate with the femoral condyles. Fractures of the tibial plateau commonly occur as the result of a medially directed force and/or axial compression. The resulting fracture pattern is determined by the magnitude and direction of the applied force and the quality of the bone. Disruption of the articular surface is common and these fractures will cause major problems if not treated properly.

The goal of treatment is a functional, pain-free and stable knee. One of the difficulties in assessing outcome following these injuries is the difference in criteria used in different studies. There have been many attempts to classify these fractures so that the optimum treatment can be offered in each case. Recent classifications have taken into account more sophisticated methods of assessing and treating intra-articular fractures. Classification systems of tibial plateau fractures are mostly anatomical and based on the radiographic appearances of the fracture pattern.

Hohl and Luck Classification

Hohl and Luck (1956) and Hohl (1967) reviewed initially 227 and finally 805 patients with tibial plateau fractures and classified them into undisplaced and displaced patterns (Table 17.4). An undisplaced fracture was one in which there was 'minimum fracture impaction or displacement (less than 3 mm) and little disruption of the articular surface'. Displaced fractures were classified according to the severity of injury to the articular surface (Figure 17.1). Undisplaced fractures were treated by traction and early motion. Displaced fractures were treated by internal fixation if there was depression of the joint surface of more than 5 mm. The outcome following internal fixation was related to the success of restoring a smooth stable articular surface so that early motion could be attained.

Kennedy and Bailey Classification

In this study of the aetiology of lateral tibial plateau fractures, Kennedy and Bailey (1968) subjected 44 cadaveric knees to valgus and compressive forces at various degrees of knee flexion. The rate of application of the forces was not clearly stated. Abduction forces applied to a flexed knee tended to produce a split fracture with varying amounts of joint depression. Compression forces produced depressed crush fractures of the tibial plateau whilst mixed forces of high magnitude produced variable fracture patterns. They classified fractures into abduction, compression, mixed or explosive types according to the anatomical pattern following testing. Whilst this classification gives a useful insight into the cause of tibial plateau fractures, it provides no clinical or prognostic information.

Table 17.4 Hohl (1967) classification of tibial plateau fractures

Undisplaced		
Displaced		
	Local depression	
		Central depression
		Split depression
	Total depression	
	Split	
	Comminuted upper end of tibia	

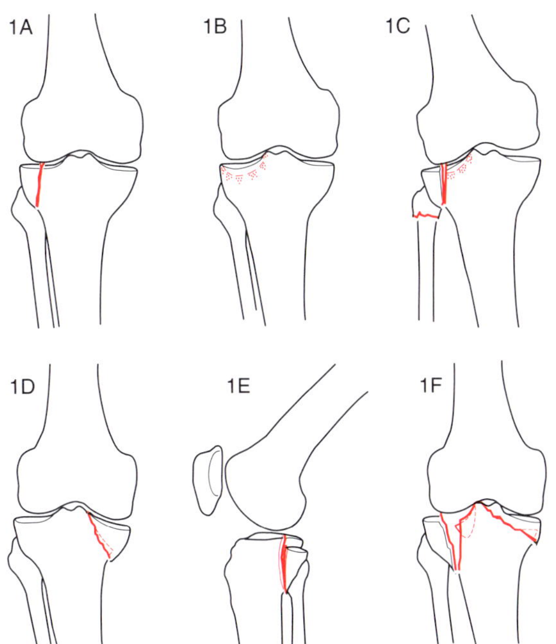

Figure 17.1. Hohl and Luck (1956) classification of tibial plateau fractures.

Schatzker Classification

Schatzker, McBroom and Bruce (1979) classified 94 tibial fractures into six groups based on the fracture pattern, and validated the system by correlating outcome with fracture type. In this system, the least severe injury was the simple cleavage fracture of the lateral tibial plateau, and the most severe injury was separation of the fractured tibial plateau from the tibial shaft (Table 17.5, Figure 17.2). The study clearly showed that displaced intra-articular fractures required internal fixation and early motion in order to achieve a satisfactory outcome. Immobilization resulted in joint stiffness in both the operated and non-operated groups and was not recommended. Also Schatzker makes the important observation that 'the results of a poor open reduction are far worse than the results of poor non-operative treatment'.

Schulak and Gunn Classification

This classification (Schulak and Gunn, 1975) was derived from an extensive literature review of classifications and outcome of tibial plateau fractures

Table 17.5 Schatzker classification of tibial plateau fractures (Schatzker, McBroom and Bruce, 1979)

I		Pure cleavage of the lateral plateau
II		Cleavage combined with depression of the lateral plateau
III		Pure central depression of the lateral plateau
IV		
	A	Cleavage of the medial plateau
	B	Depressed and/or comminuted fracture of the medial plateau
V		Bicondylar fractures
VI		Tibial plateau fractures with dissociation of the tibial metaphysis and diaphysis

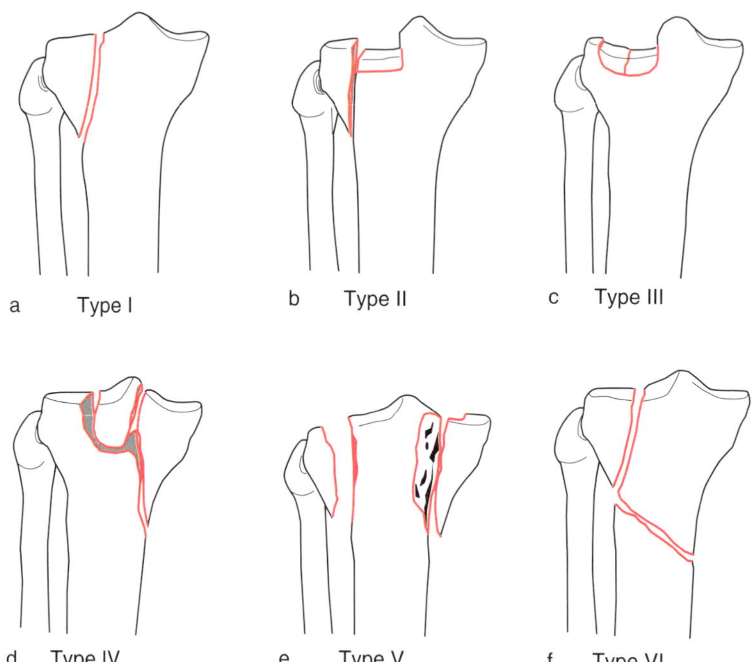

Figure 17.2. Schatzker, classification of tibial plateau fractures (Schatzker, MacBroom and Bruce, 1979). (a) Type I, pure cleavage fracture; (b) Type II, cleavage combined with depression; (c) Type III, pure central depression – there is no lateral wedge. The depression may also be anterior, posterior or involve the whole plateau; (d) Type IV, the medial condyle is either split off as a wedge (Type A, as illustrated) or it may be crumbled and depressed (Type B, not illustrated) in older, osteoporotic patients; (e) Type V – note continuity of the metaphysis and diaphysis; (f) Type VI, the fracture separates the metaphysis from the diaphysis.

(Figure 17.3). The classification was based on the fracture pattern reported in each series reviewed and there was a total of 599 fractures from 13 different studies. There are many similarities to the system used by Hohl (1967). Particular emphasis was placed on factors affecting outcome in each fracture type. Important factors included restoration of the articular surface and early motion.

Lansinger et al. Classification

In this 20-year follow-up study of 102 patients with tibial plateau fractures, Lansinger et al. (1986) produced a classification based on the fracture pattern

Table 17.6 Schulak and Gunn (1975) classification of tibial plateau fractures

I	Minimally displaced fractures			
II	Displaced lateral plateau fractures			
	A	Depression		
			(1)	Local
			(2)	Total
	B	Split		
	C	Mixed		
III	Displaced medial plateau fracture			
	A	Total depression		
	B	Split		
IV	Displaced bi-condylar fracture			

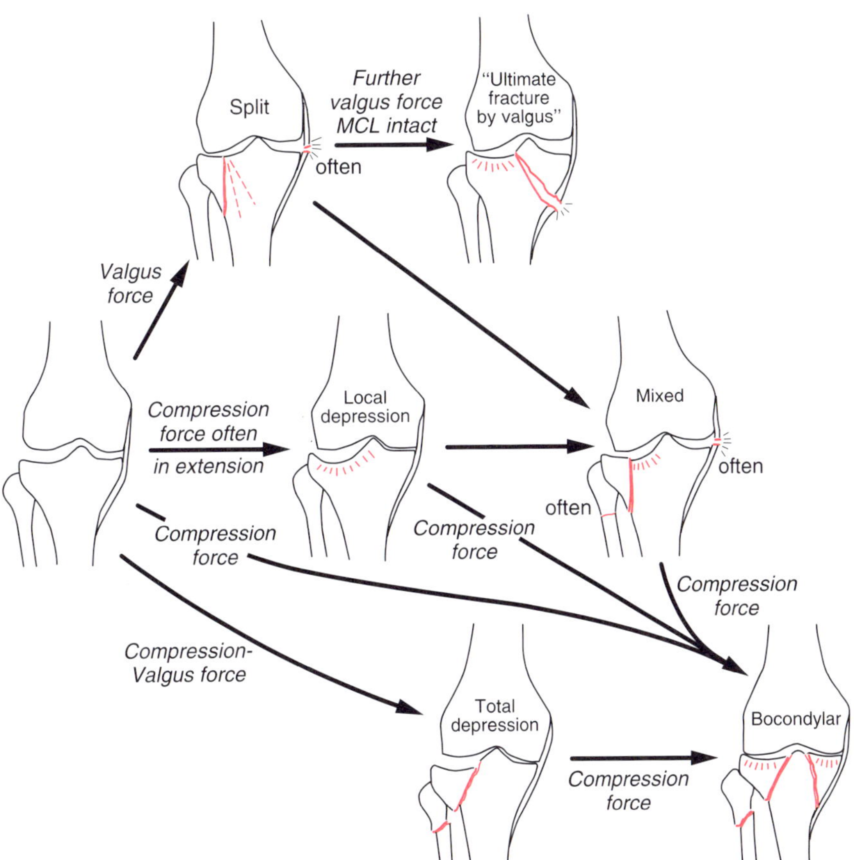

Figure 17.3. Schulak and Gunn (1975) classification of tibial plateau fractures. Medial collateral ligament (MCL) injuries commonly occur in split and mixed fracturers of the lateral plateau. In mixed injuries the fibula is often fractured. In total depression fractures a proximal fibula fracture and proximal tibio-fibula diastasis occurs.

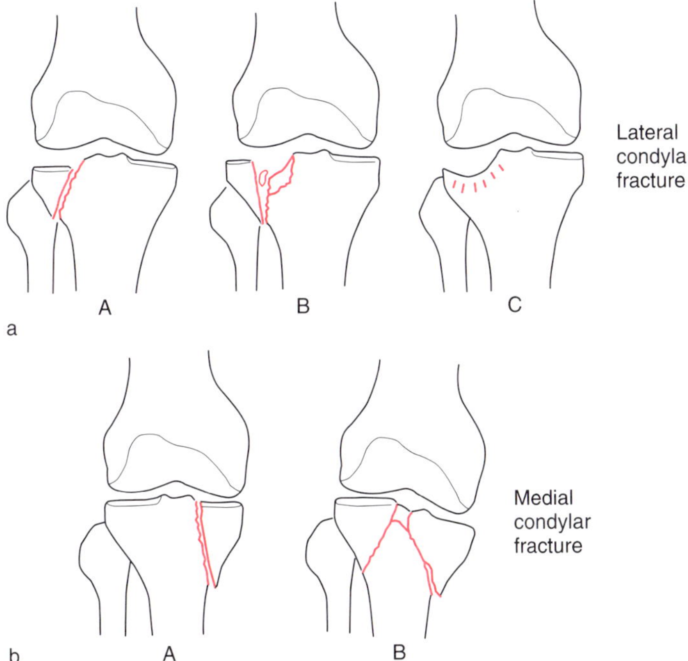

Figure 17.4. Lasinger et al. (1986) classification of tibial plateau fractures. (a) Lateral condylar fractures: A, split fracture; B, split-depressed fracture; C, depressed fracture. (b) Medial condylar fractures: A, split fracture; B, bicondylar fracture.

(Figure 17.4). The fractures were described as lateral condylar, medial condylar or bicondylar. The lateral condylar fractures were further classified as split, depressed or mixed. The indication for surgery was an increase of 10° varus or valgus in the injured knee compared to the normal knee, examined under anaesthesia. Surgery consisted of open reduction and internal fixation, whilst conservative treatment consisted of immobilization for 2 weeks, followed by flexion-extension exercises without weight bearing for a further 4 weeks. The outcome was generally good in most knees following this regime.

AO Classification

The segment of the proximal tibia that includes the tibial plateau is defined by the square method described above for distal femoral fractures (Müller et al., 1990). The description of the fracture follows the common principles of the AO method: type A fractures are extra-articular, type B partial articular and type C complete articular with detachment of the articular fragments from the diaphysis. Within each group there are subgroups defining the degree of fragmentation of the fracture (Figure 17.5). This classification has been adopted by the Orthopaedic Trauma Association (1996).

Tscherne and Lobenhoffer Classification

In this classification Tscherne and Lobenhoffer (1990) studied the fracture patterns of 244 tibial plateau fractures. They distinguished between plateau fractures, fracture dislocations and comminuted fractures (Figure 17.6). The rationale for this classification was the observation that certain fracture types were associated with significant soft-tissue injuries, usually as a result of high energy trauma (Moore, 1981). Plateau fractures were mainly stable whereas fracture-dislocations were unstable due to major ligamentous disruption. Comminuted fractures were of variable configuration and stability. Open reduction and internal fixation was the preferred tratment in all displaced and unstable fractures and the results compared favourably with most other published series.

Patellar Fractures

Introduction

The patella may be fractured by a direct blow or as a result of tensile forces acting through the extensor mechanism. Classification of patellar fractures is

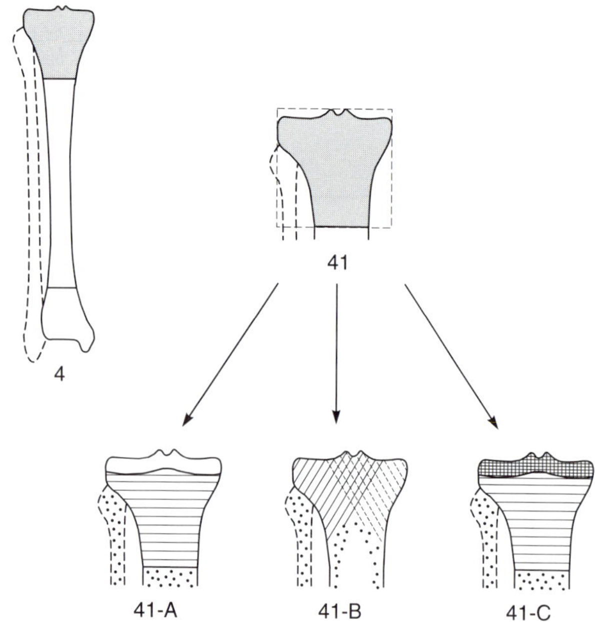

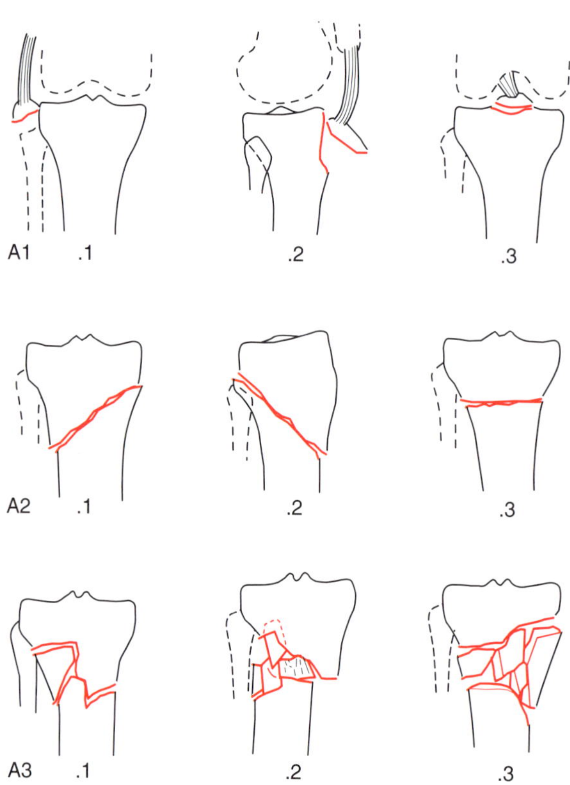

Figure 17.5. AO/ASIF classification of tibial plateau fractures (Müller et al., 1990). A, extra-articular.

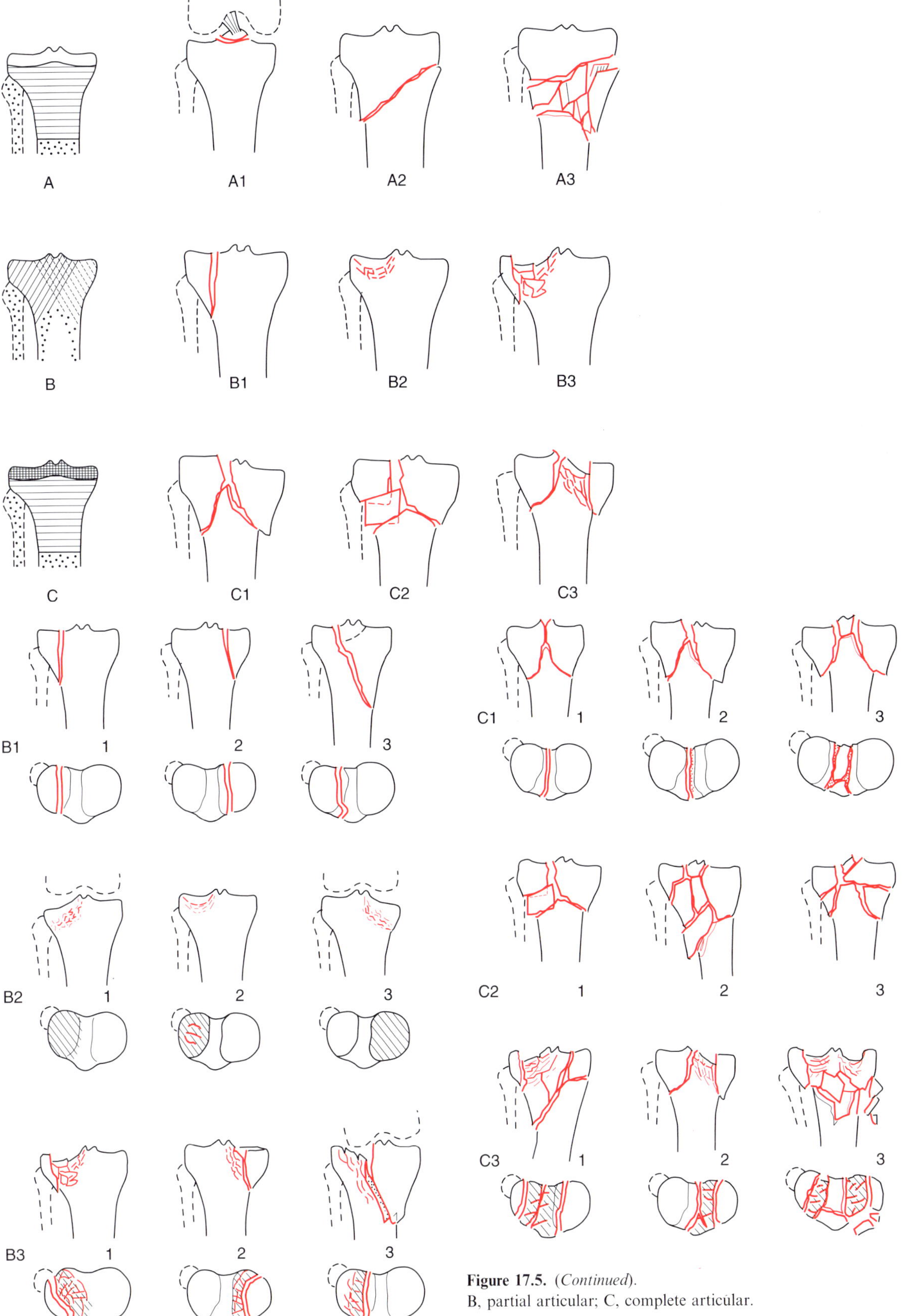

Figure 17.5. (*Continued*).
B, partial articular; C, complete articular.

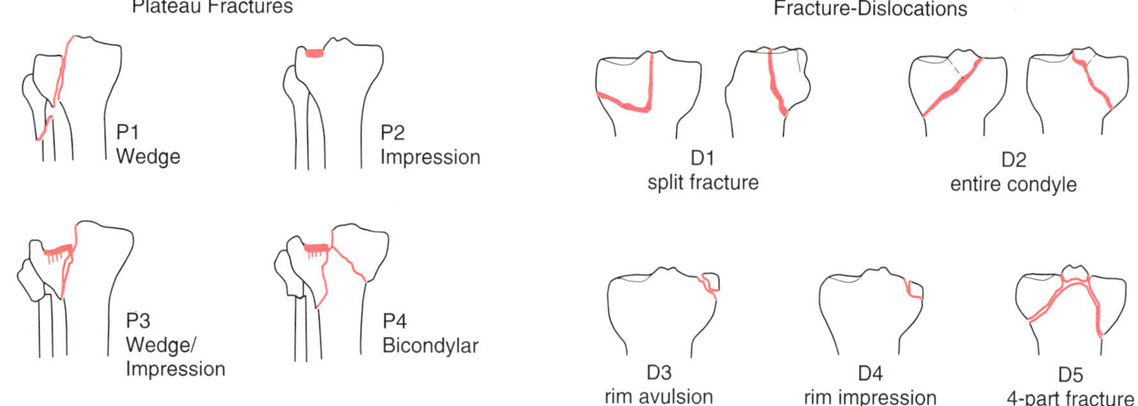

Figure 17.6. Tscherne and Lobenhoffer (1990) classification of tibial fractures.

Table 17.7 AO classification of tibial plateau fractures (Müller et al., 1990)

A:	Extra-articular fracture	
	A1	Avulsion
	A2	Metaphyseal simple
	A3	Metaphyseal multifragmentary
B:	Partial articular	
	B1	Pure split
	B2	Pure depression
	B3	Split depression
C:	Complete articular	
	C1	Articular simple, metaphyseal simple
	C2	Articular simple, metaphyseal complex
	C3	Multifragmentary

complicated by the variability in the pattern of fractures although broad categories can be identified, particularly the location and type of fracture (Bostrom, 1972; Lotke and Ecker, 1981). Treatment of patellar fractures includes conservative methods, open reduction and internal fixation, partial excision and total patellectomy. Total patellectomy produces results which are not as good as patellar reconstruction (Marya, Bhan and Dave, 1967).

Patellar Fracture Classification

Sanders (1992) analysed several outcome studies of patellar fractures and classified fractures into undisplaced and displaced patterns (Figure 17.7). Within the undisplaced group, stellate, transverse and vertical fracture patterns were identified. Displaced fractures were divided into non-comminuted and comminuted types. Non-comminuted fractures were either transverse (central) or polar (i.e. apical or basal). Comminuted fractures were stellate, transverse, polar or highly comminuted. Undisplaced fractures are normally treated conservatively with reasonable results. The treatment of displaced comminuted fractures is more controversial as both internal fixation and patellectomy are associated with considerable morbidity (Bostman, Kiviluoto and Nirhamo, 1981; Bunker, 1990). The Orthopaedic Trauma Association (1996) has recently classified patellar fractures into extra-articular, partial articular and complete articular, in the style of the AO classification of long bone fractures. These broad categories are further defined by the orientation and position of the major fracture line and the number of fragments.

Extensor Mechanism Injuries

Introduction

The extensor mechanism consists of the quadriceps tendon, the patellar tendon and the patella. Larsen and Lund (1986) in a series of 28 patients with extensor mechanism ruptures found that the quadriceps mechanism was more commonly injured in patients over 40 years old whereas the patellar tendon was more commonly injured in patients younger than 40. Midsubstance tendon injuries rarely occur in normal tendons, and most injuries occur at the musculotendinous junction or at the insertion of the tendon into bone (McMaster, 1933).

Classification

The Orthopaedic Trauma Association (1996) have classified these injuries into medial and lateral dislo-

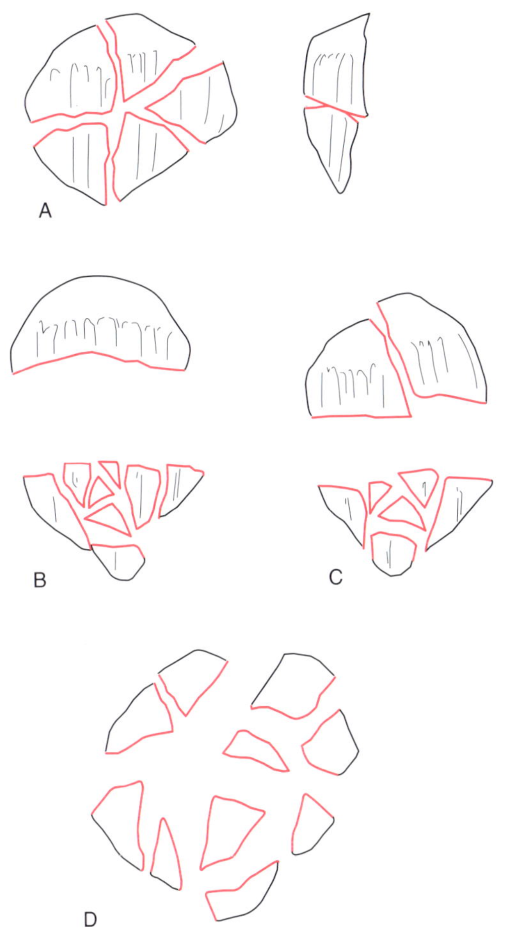

Figure 17.7. Sanders (1992) classification of patellar fractures. Selected examples: A, undisplaced fracture with any degree of comminution with <3 mm displacement. <2 mm of step of articular surface in lateral view; B, displaced transverse fracture with comminution of basilar poles; C, displaced transverse fracture with comminution of both apical and basilar poles; D, highly comminuted, highly displaced fracture.

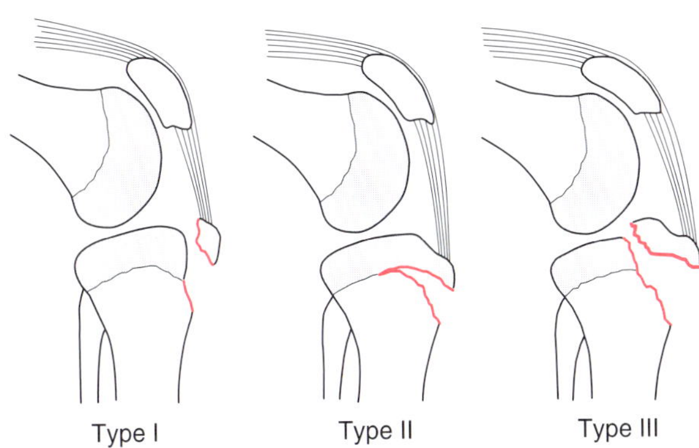

Figure 17.8. Watson-Jones (1976) classification of tibial tubercle fractures.

cations of the patellofemoral joint and disruption of the quadriceps or patellar tendon. Patellar tendon injuries are further defined by the presence of a bony avulsion or midsubstance tear. Operative treatment is associated with mainly good results (Larsen and Lund, 1986). In the adolescent, avulsion of the tibial tubercle can present following sudden contraction of the quadriceps. Watson-Jones (1976) classified these injuries into three types. In type I injuries there is a fracture involving the tibial tubercle alone; in type II, part of the proximal tibial epiphysis is involved and in type III, there is extension into the knee joint (Figure 17.8). In adults this injury is noted in the Orthopaedic Trauma Association (1996) classification.

Knee Dislocation

Introduction

Knee dislocations are uncommon but usually associated with significant soft-tissue injury and a high risk of injury to the popliteal artery. Many authors recommend arteriography in all cases of knee dislocation due to the risk of vascular injury and consequent high amputation rate.

Classification

The position of the tibia relative to the femur defines the direction of dislocation, i.e. anterior, posterior, medial, lateral and rotatory. The Orthopaedic Trauma Association (1996) has classified knee dislocation. In this classification, the injury is defined by the presence of specific ligament disruption, either mid-substance or bony avulsion. High energy injuries resulting from motor vehicle accidents and falls account for most knee dislocations, although rotational dislocations can occur as a result of lower energy sports injuries (Leffers, 1992). Rotational dislocations can be further classified as anteromedial, posteromedial, anterolateral and posterolateral. Most dislocations are in the anteroposterior plane. Treatment is controversial ranging from conservative to operative repair of all injured ligamentous structures.

Dislocation of the Proximal Tibiofibular Joint

This is an uncommon injury, classified by Ogden (1974). In this series of 43 patients the pattern of injury was subluxation, or anterolateral, posteromedial or superior dislocation. The mechanism of injury

was usually twisting of the flexed knee, and closed reduction was usually successful. The Orthopaedic Trauma Association (1996) has adopted a similar classification but has also included the presence or absence of an associated fibular or tibial fracture.

Ligament Injuries

Introduction

Injuries to the knee ligaments are common. Hughston et al. (1976a,b) divided the supporting structures of the knee into two groups—the static and dynamic stabilizers. The static stabilizers are the capsular and non-capsular ligaments, whilst the dynamic stabilizers are the muscles and tendons. Both the static and dynamic stabilizers have medial and lateral components. The anterior and posterior cruciate ligaments are referred to as the central ligaments. Injuries to the knee stabilizers produce degrees of instability of the knee joint, although the term instability is poorly defined. Ligamentous injuries are described as grade I when there is minimal tearing, grade II when there is a partial tear and grade III when there is a complete rupture (AAOS, 1984). One of the problems in the evaluation and documentation of knee injuries is a lack of conformity in assessment and terminology. The International Knee Documentation Committee (Hefti et al., 1993) have made valuable contributions in helping to resolve these problems.

Hughston (1976) Classification of Knee Instability

In Hughston et al.'s (1976a,b) studies, 157 patients with acute soft-tissue knee injuries were assessed clinically and arthroscopically. Clinical examination demonstrated the existence of ligamentous laxity of the knee, and this was classified accordingly as straight or rotatory instability. Instability was defined according to the degree of separation of the joint surfaces during stress testing, both clinically and under direct vision. Mild instability was present if the joint surfaces separated by 5 or less mm, moderate instability if the separation was between 5 and 10 mm, and severe if the separation was greater than 10 mm.

Straight Instability

Tears of the medial or lateral compartment ligaments in association with a posterior cruciate ligament tear produced a positive abduction stress test with the knee in full extension. This clinical test did not involve any rotation of the tibia with respect to the femur. Posterior instability was produced by a posterior cruciate tear in association with laxity or a tear of the posterior ligamentous structures, and was demonstrated by positive posterior drawer test. Anterior instability was caused by a torn posterior cruciate ligament and was demonstrated by a positive anterior drawer test with no rotation.

Rotatory Instability

Three types of rotatory instability were described. First, anteromedial instability in which there was a positive anterior drawer test with the tibia externally rotated. This was caused by a tear of the medial ligaments and was accentuated by a tear of the anterior cruciate ligament. Secondly, anterolateral instability was caused by a tear of the middle third of the lateral capsular ligament and was also accentuated by a tear of the anterior cruciate ligament. This was demonstrated by a positive anterior drawer test with the tibia internally rotated. Finally, in posterolateral instability there was a tear of the posterior musculoligamentous structures of the lateral compartment and this was demonstrated by a positive posterior drawer test.

Medial Collateral Ligament

The medial collateral ligament (MCL) is the most commonly injured knee ligament. Injuries usually occur as the result of a direct valgus force applied to the knee, and are graded according to the American Academy of Orthopaedic Surgeons (1984) classification. A grade I injury implies minor tearing of the ligament, grade II an incomplete tear and grade III a complete tear of the ligament. The grade of an MCL injury is determined by examining the knee under anaesthetic for ligamentous laxity, and treatment of isolated injuries is usually conservative. Where there is an associated anterior or posterior cruciate ligament rupture, operative treament of the cruciate ligament is usually recommended (Swenson and Harner, 1995). Thus, classification of injuries to the MCL should include the presence or absence of injuries to other ligamentous structures.

Lateral Collateral Ligament

The lateral collateral ligament (LCL) is injured by the application of excessive varus stress to the knee. Like the MCL, injuries are graded from I to III according to the AAOS (1984) classification. Grade III injuries represent a complete disruption and can be accompanied by anterior and posterior cruciate ligament injuries, as well as injuries to the posterolateral corner of the knee. Consequently, subluxation of the tibia on the femur occurs and this is referred to as rotatory instability. The direction of subluxation is determined by the structures injured.

Anterior Cruciate Ligament

The anterior cruciate ligament (ACL) is not only the primary restraint of anterior translation of the tibia, but also contributes to knee stability in other planes. Injuries to the ACL may be acute or chronic and can be simply classified into mid-substance tears or bony avulsions of the origin or insertion. This classification is useful because it is widely accepted that avulsions should be reattached (Schatzker, 1995). Tibial spine avulsions have been classified by Meyers and McKeever (1970). In this system, type I fractures represent a minimally displaced avulsion, type II an elevated fragment that is still attached, type III complete displacement, and type IV in which there is comminution of the tibial spine. Mid-substance tears of the ACL may be classified further into partial or complete, but in order to make this distinction arthroscopy is required. However, Kennedy et al. (1976) demonstrated that significant damage to the ultrastructure of the ligament could occur without macroscopic evidence of injury. Clinical examination, particularly the pivot shift and Lachman's test, and magnetic resonance imaging (MRI) can detect tears of the ACL but are not sensitive or reliable enough to differentiate partial from complete tears (Lintner et al., 1995; Umans et al., 1995). Measurement of tibial translation using the KT-1000 arthrometer quantifies the amount of ligamentous laxity but does not help to classify the ACL injury (Nogalski and Bach, 1994). A descriptive classification of mid-substance ACL injuries does not influence the management of the injured knee for several reasons. For example, the decision to reconstruct the ACL depends on many factors other than the presence of an ACL injury, such as the patient's age and willingness to undergo intensive postoperative rehabilitation. Also, many patients with an ACL-deficient knee are able to continue to participate in sports, or modify their lifestyles to place fewer demands on the injured knee (Buss et al., 1995). Furthermore knee function in the ACL-deficient knee is to a large extent determined by the integrity of the remaining ligaments and capsule. Thus a more useful classification system for ACL injuries would include an assessment of both the anatomical injury, knee function and the presence of injuries to other structures contributing to knee stability.

Posterior Cruciate Ligament

Swenson and Harner (1995) stated that recent improvements in the ability to diagnose posterior cruciate ligament (PCL) injuries have led to an apparent increase in the number of injuries that are recognized. The PCL prevents the tibia from displacing posteriorly and also acts as a secondary restraint of varus angulation and tibial external rotation (Grood, Stowers and Noyes, 1988). Injuries to the PCL may be classified as acute or chronic, avulsions or mid-substance tears, and occurring in isolation or in association with injuries to other structures in the knee. Like ACL avulsions, it is generally agreed that PCL avulsions should be fixed. Miller, Warner and Harner (1994) classified PCL injuries into six groups (Table 17.8) based on the amount of laxity found on the posterior draw test and the presence of injuries to other structures, although this classification does not appear to have been subjected to tests of inter and intra-observer reliability. Gross et al. (1992) suggested a classification of PCL injuries based on the appearance seen on MRI. In this system the presence of an increase in signal intensity indicated an injury to the PCL, but no distinction could be made between partial and complete injuries. Controversy surrounds the treatment of PCL ruptures, both isolated and complicated by other injuries (Dandy and Pusey, 1982).

Meniscal Injuries

Meniscal injuries commonly occur in association with other knee injuries. Menisci subjected to abnormal loads may tear or split and as a result lose the ability to function normally. There is no universally accepted classification of meniscal injuries, but descriptive classifications of the type, site, size and stability are commonly used (Swenson and Harner, 1995). The major types of meniscal tears are as follows:

a. Vertical longitudinal
b. Oblique
c. Degenerative
d. Radial
e. Horizontal

Müller et al. (1990) classified meniscal tears in association with ACL deficiency. In addition to the injuries noted above, he added secondary tears, crush

Table 17.8 Miller, Warner and Harner (1994) classification of posterior cruciate ligament injuries

Type	*Definition*	*Laxity*
I	PCL stretched	≤5 mm
II	PCL torn: MFL intact	5–9 mm
III	PCL and MFL torn	>10 mm
IVA	PCL and LCL	>12 mm
IVB	PCL and MCL	>12 mm
IVC	PCL and ACL	>15 mm

PCL, Posterior cruciate ligament; MFL, Meniscofemoral ligament; LCL, Lateral collateral ligament; MCL, Medial collateral ligament; ACL Anterior cruciate ligament

lesions and posterior detachment. A meniscal tear may be 'unstable' or 'stable', although these terms are not clearly defined. The propensity for a tear to displace on probing is often called instability, and is used as an indication for mensical resection or repair. The ability of a meniscal tear to undergo healing depends on an adequate blood supply. Anatomical studies have demonstrated that only the peripheral part of the meniscus has a blood supply and that the avascular central portion does not heal (King, 1936; Arnoczky and Warren, 1982). This has led to the classification of meniscal tears into zones reflecting their blood supply (Figure 17.9) (Miller, Warner and Harner, 1994). In the peripheral red zone (20–30% in the medial meniscus, 10–25% in the lateral meniscus) there is good potential for healing. In the inner white zone the potential for healing is poor, and in the transition zone (red/white) there is some potential for healing.

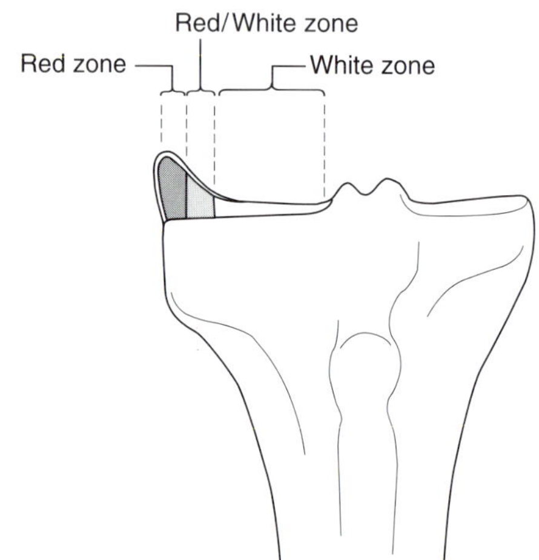

Figure 17.9. Miller classification of meniscal tears (Miller, Warner and Harner, 1994).

Conclusions

A synopsis of available classification systems for several of the major injuries about the knee has been given. Most of these classifications have not been subject to an analysis of intra- or interobsever error. Whilst it is useful to study the classification of an isolated fracture or ligamentous injury in order to determine the best method of treatment, it is important to remember that the knee is perhaps the most complex of the joints and injuries commonly affect more than one structure. Consequently, the study of the treatment and outcome of knee injuries is complicated by the lack of a uniform system of measuring outcome. Common methods of outcome measurement following treatment of injuries include factors such as pain, range of movement and time to union of a fracture. Unfortunately, in the knee these simplistic measures are inadequate in many cases and many studies are difficult to compare due to the adoption of different outcome criteria. This is not to say that classification of knee injuries is a futile exercise but rather to emphasize the need for uniformity of classification and measurement of outcome.

Group Discussion

The group concluded the following:

- The AO classification (Müller et al., 1990) was recommended for distal femoral fractures.
- The Schatzker, MacBroom and Bruce (1979) classification for tibial plateau fractures was considered the best option.
- For extensor mechanism injuries the Orthopaedic Trauma Association (1996) classification is probably the most useful but the subdivisions of patellar fractures are difficult to apply in many situations.

- The AAOS (1984) classification of ligamentous injuries is most useful. Knee dislocations are reasonably well classified on an anatomical basis.
- Meniscal injuries based on an anatomical classification with regard to site and blood supply are useful.

References

American Academy of Orthopaedic Surgeons (1984) *Athletic Training and Sports Medicine*. Chicago: American Academy of Orthopaedic Surgeons.

Arnoczky, S. and Warren, R. (1982) Microvasculature of the human meniscus. *Am. J. Sports Med.*, **10**, 90.

Bostman, O., Kiviluoto, O. and Nirhamo, J. (1981) Comminuted displaced fractures of the patella. *Injury*, **13**, 196–202.

Bostrom, A. (1972) Fracture of the patella. *Acta Orthop. Scand.*, **143** (Suppl.), 1–80.

Bunker T.D. (1990) The knee. In *Medicolegal Reporting in Orthopaedic Trauma* (M.A. Foy and P.S. Fagg, eds) Edinburgh: Churchill Livingstone.

Buss, D.D., Min, R., Skyhar, M. et al. (1995) Non-operative treatment of acute anterior cruciate ligament injuries in a selected group of patients. *Am. J. Sports Med.*, **23**, 160–165.

Dandy, D.J. and Pusey, R.J. (1982) The long-term results of unrepaired tears of the posterior cruciate ligament. *J. Bone Joint Surg.*, **64B**, 92–94.

Grood, E., Stowers, S. and Noyes, F. (1988) Limits of movement in the human knee: effect of sectioning the posterior cruciate ligament and the posterolateral structures. *J. Bone Joint Surg.*, **70A**, 88.

Gross, M.L., Grover, J.S., Bassett, L.W. et al. (1992) Magnetic resonance imaging of the posterior cruciate ligament. Clinical use to improve diagnostic accuracy. *Am. J. Sports Med.*, **20**, 732–737.

Hefti, F., Müller, W., Jakob, R.P. and Stäubli, H-U. (1993) Evaluation of knee ligament injuries with the IKDC form. *Knee Surg. Sports Traumatol. Arthroscopy*, **1**, 226–234.

Hohl, M. (1967) Tibial condylar fractures. *J. Bone Joint Surg.*, **49A**, 1455–1467.

Hohl, M. and Luck, J.V. (1956) Fractures of the tibial condyles. A clinical and experimental study. *J. Bone Joint Surg.*, **38A**, 1001–1018.

Hughston, J.C., Andrews, J.R., Cross, M.J. and Moschi, A. (1976a) Classification of the knee ligament instabilities. Part I. The medial compartment and cruciate ligaments. *J. Bone Joint Surg.*, **58A**, 159–172.

Hughston, J.C., Andrews, J.R., Cross, M.J. and Moschi, A. (1976b) Classification of the knee ligament instabilities. Part II. The lateral compartment. *J. Bone Joint Surg.*, **58A**, 173–179.

Kennedy, J.C. and Bailey, W.H. (1968) Experimental tibial plateau fractures. Studies of the mechanism and a classification. *J. Bone Joint Surg.*, **50A**, 1522–1534.

Kennedy, J.C., Hawkins, R.J., Willis, R.B. and Danylchuk, K.D. (1976) Tension studies of human knee ligaments. *J. Bone Joint Surg.*, **58A**, 350–355.

King, D. (1936) The healing of the semilunar cartilages. *J. Bone Joint Surg*, **18A**, 333.

Lansinger, O., Bergman, B., Korner, L. and Andersson, G.B.J. (1986) Tibial condylar fractures. A twenty year follow up. *J. Bone Joint Surg.*, **68A**, 13–19.

Larsen, E. and Lund, P.M. (1986) Ruptures of the extensor mechanism of the knee joint. *Clin. Orthop.*, **213**, 150–153.

Leffers, D. (1992) Dislocations and soft tissue injuries of the knee. In *Skeletal Trauma* (B.D. Browner, J.B. Jupiter, A.M. Levine P.G. and Trafton, eds). Philadelphia: Saunders.

Lintner, D.M., Kamaric, E., Moseley, J.B. and Noble, P.C. (1995) Partial tears of the anterior cruciate ligament. Are they clinically detectable? *Am. J. Sports Med.*, **23**, 111–118.

Lotke, P.A. and Ecker, M.L. (1981) Transverse fractures of the patella. *Clin. Orthop.*, **158**, 180–184.

McMaster, P.E. (1933) Tendon and muscle ruptures: clinical and experimental studies on the causes and location of subcutaneous reuptures. *J. Bone Joint Surg.*, **15**, 705–722.

Marya, S.K.S., Bhan, S. and Dave, P.K. (1987) Comparative study of knee function after patellectomy and osteosynthesis with a tension band wire following patellar fractures. *Int. Surg.*, **72**, 211.

Meyers, M. and McKeever, F. (1970) Fractures of the intercondylar eminence of the tibia. *J. Bone Joint Surg.*, **52A**, 1677.

Miller, M.D., Warner, J.J.P and Harner, C.D. (1994) Meniscal repair. In *Knee Surgery* (F.H. Fu, C.D. Harner and K.G. Vince, eds). Baltimore: Williams and Wilkins.

Moore, T.M. (1981) Fracture dislocation of the knee. *Clin. Orthop.*, **156**, 128–140.

Müller, M.E., Koch, P., Nazarian, S. and Schatzker, J. (1990) *The Comprehensive Classification of Fractures of Long Bones*. Berlin: Springer-Verlag.

Neer, C.S., Grantham, S.A. and Shelton, M.L. (1967) Supracondylar fracture of the adult femur. *J. Bone Joint Surg.*, **49A**, 591–613.

Nogalski, M.P. and Bach, B.R. (1994) Acute anterior cruciate ligament injuries. In *Knee Surgery* (F.H. Fu, C.D. Harner and K.G. Vince, eds). Baltimore: Williams and Wilkins.

Ogden, J.A. (1974) Subluxation and dislocation of the proximal tibiofibular joint. *J. Bone Joint Surg.*, **56A**, 145–154.

Orthopaedic Trauma Association Committee for Coding and Classification (1996) *J. Orthop. Trauma*, **10**, 1–158.

Sanders, R. (1992) Patellar fractures and extensor mechanism injuries. In *Skeletal Trauma* (B.D. Browner, J.B. Jupiter, A.M. Levine and P.G. Trafton eds). Philadelphia: Saunders.

Schatzker, J. (1995) Fractures of the tibial plateau. In *The Rationale of Operative Fracture Care* (J. Schatzker and M. Tile, eds). Berlin: Springer-Verlag, pp. 419–438.

Schatzker, J., McBroom, R. and Bruce, D. (1979) The tibial plateau fracture: The Toronto experience 1968–1975. *Clin. Orthop.*, **138**, 94–104.

Schulak, D.J. and Gunn, D.R. (1975) Fractures of the tibial plateaus. A review of the literature. *Clin. Orthop.*, **109**, 166–177.

Seinsheimer, F. (1980) Fractures of the distal femur. *Clin. Orthop.*, **153**, 169–179.

Swenson, T.M. and Harner, C.D. (1995) Knee ligament and meniscal injuries. Current concepts. *Orthop. Clin. North Am.*, **26**, 529–546.

Tscherne, H. and Lobenhoffer, P. (1990) Tibial plateau fractures. Management and expected results. *Clin. Orthop.*, **292**, 87–100.

Umans, H., Wimpfheimer, O., Harmati, N. et al. (1995) Diagnosis of partial tears of the anterior cruciate ligament of the knee: value of MR imaging. *Am. J. Roentgenol.*, **165**, 893–897.

Watson-Jones, R. (1976) Injuries of the knee. In *Fractures and Joint Injuries* (J.N. Wilson, ed.), 5th edn. Edinburgh: Churchill-Livingstone.

18

Tibial shaft fractures

C. L. M. H. Gibbons

Introduction

The tibia is the most commonly fractured long bone and the optimum treatment of tibial shaft fractures is bedevilled with controversy and debate. Nicoll (1964) in reviewing 705 such fractures stated; 'Fractures of the tibia are important for two reasons. The first is that they are common; the second is that they are controversial and anything that is both common and controversial must be important.' The spectrum of injury with shaft fractures is so varied that to fully accommodate all fracture types numerous classifications have been proposed and used. Fractures of the tibial shaft are notoriously difficult to classify and literature is well-endowed with descriptive and plagiarized classification systems.

The inherent difficulties in using a single classification framework is illustrated by Court-Brown and McBirnie (1995). They reviewed the epidemiology of 523 fractures using modern improved classifications and traditional data. They proposed that the fracture is usefully analysed using the AO system (Müller et al., 1990) for morphology with the Tscherne classification (Oesterne and Tscherne, 1984) for closed and the Gustilo model (Gustilo and Anderson, 1976) for open fractures. In addition, comminution is usefully assessed using the Winquist classification (Winquist and Hansen, 1980) and fractures were also described according to their level (proximal, middle and lower thirds) recording those which extended over more than one-third of the shaft—a total of six paradigms used.

In children fracture union is rapid and complications few. In adults the outcome is less certain, although most can expect a satisfactory result with either open or closed treatment. Significant complications of treatment include infection, non-union, delayed union and deformity which are commonly used as physical outcome measures. Infection, displacement and open fractures are associated with a worse outcome. Consequently, separate systems have been proposed for open and closed injuries. Some bones lend themselves readily to classification, e.g. femoral neck fractures, whereas the tibial shaft does not. This is an inherent difficulty in classifying these common and varied fractures.

Classifications of tibial shaft fractures have been proposed so that appropriate treatment can be selected and that useful information about prognosis and the results of that treatment can be made. A good classification should be simple and reproducible.

Anatomy

In most classifications the tibia shaft is not clearly defined. The AO group (Müller et al., 1990) defined the shaft (diaphysis) as that part of the bone excluding extra-articular fractures in the proximal and distal 5 cm of the tibia.

The adult tibia measures >30 to <50 cm in length. It is triangular in cross-section with an intramedullary diameter of 7–15 mm; the shaft or diaphysis is comprised of hard cortical bone and is felt as a palpable crest from the tubercle to just above the ankle. Approximately one-third of the bone is subcutaneous over the anteromedial aspect. This is significant in that the tibia is at risk from direct injury; an important factor in classification. Ten to twenty per cent of tibial fractures are associated with an open injury and significant damage to the soft-tissue envelope is common.

The blood supply of the tibial shaft is derived from the nutrient vessel (three ascending and one main descending branches) of the posterior tibial artery and periosteal vessels principally from the anterior tibial artery.

The leg comprises four compartments: anterior, lateral, posterior and deep posterior. The anterior contains the tibialis anterior, extensor digitorum longus, extensor hallucis longus, and peroneus tertius muscles with anterior tibial vessels and deep peroneal nerve. The lateral compartment comprises the peroneus longus and brevis and superficial peroneal nerve. The superficial deep compartment contains gastrocnemius, soleus, popliteus and plantaris with the sural nerve, long and short saphenous veins. Deep posterior compartment has tibialis posterior, flexor digitorum longus and flexor hallucis longus with the posterior tibial nerve and artery and peroneal artery. The tibialis posterior muscle has been described as a separate 5th compartment (Matsen and Clawson, 1975).

Outcome Measurements

In the classification of tibial fractures outcome has been evaluated in terms of physical measurements such as average time to bony union, risk of delayed union, risk of infection, and residual deformity. These are the commonly used clinical measurements of outcome. Although this is undoubtedly important to the surgeon, functional criteria and patient satisfaction may be more valid measures in the view of the patient.

Burnstein (1993) debated whether fracture classification systems work and asked are they really useful? For a classification to work it should have intra-observer and inter-observer reliability. If the classification is unreliable, then it is not useful in comparing outcomes.

There are no universally agreed or validated measures of outcome in tibial shaft fractures just as there are a plethora of classification systems.

Brumback and Jones (1993) demonstrated this problem of interpretation of a widely used classification in a survey of 245 orthopaedic surgeons. The surgeons each reviewed a video tape of 12 open tibial fractures and an accompanying history. They were then asked to classify the fracture according to Gustilo's classification. The average inter-observer agreement for all of these case presentations was 60% with agreement in the classification of each fracture ranging from 42% to 93%. They concluded that inter-observer agreement in the classification of open fractures was moderate to poor and that the classification system of Gustilo et al. (Gustilo, Mendoza and Williams, 1967) may be an inadequate model for decisions regarding treatment or comparing outcomes in open fractures.

Bridgman and Baird (1993) assessed deformity as a valid and repeatable measure of clinical outcome from 10 different monographs. They found that the best standard for assessing deformity as a measure of outcome in 52 of their series of tibial fractures was different in each paper (Table 18.1). Depending on whose outcome measures they used, they could show that between 4% and 42% of their patients could be said to have had a suboptimal outcome.

The Basis of Classification

Many classifications have been proposed for tibial shaft fractures. Each has its merits and drawbacks,

Table 18.1 Trafton's classification (from Trafton, 1992)

Fracture characteristic	*Minor*	*Moderate*	*Major*
Displacement	0–50% diameter	> 50%	Tibiofibular diastasis
Comminution	0 – minimal	0 or 1 butterfly fragments	≥ two free fragments or segmental
Wound	Open grade I; closed grade 0	Open grade II; closed grade I	Open grades III–V; closed grades II–III
Available energy (history)	Low	Moderate	High; crushing
Mechanism (fracture pattern)	Spiral	Oblique/transverse	Transverse/fragmented

After Ellis and Edwards, with Leach's modification. This system incorporates soft-tissue wound grading, after Gustilo and Tscherne, and is proposed as a general means for clinical grading of tibial shaft fractures. Use the factor of greatest severity to grade the fracture but in fact represents a snapshot of the treatment of tibial shaft fracture at that particular time.

Simple classifications exist for describing shaft fractures such as open and closed, stable and unstable (displaced), proximal and distal. However, more complex models are needed to draw in the factors which are considered to usefully guide treatment and determine which factors affect prognosis and outcome.

Two broad groups of classifications of tibial shaft fracture can be considered. The classification may be descriptive or dynamic. Descriptive classifications such as those proposed by Ellis (1958a,b), Nicoll (1964) and Chapman (1988) use clinical and radiographic evidence to describe the morphology of the fracture in terms of initial fracture displacement, comminution and soft-tissue injury.

More modern dynamic classifications have been proposed by Johner and Wruhs (1983) and the AO group (Müller et al., 1990) which take into account the dynamics of the injury and its relationship to the morphology of the fracture. This change in the philosophy of the assessment of tibial shaft fractures and the use of the dynamic model in classifying injury was prompted by Nicoll's observations on the 'personality' that each fracture exhibits.

Tibial shaft fractures can be treated by open and closed surgical methods which have been the subject of vigorous debate over the last two decades. An acceptable functional and cosmetic outcome can be expected with different treatments.

Sarmiento et al. (1989, 1995) demonstrated that closed treatment with cast bracing can prevent serious complications and open surgical procedures such as plating (Rüedi, Webb and Allgöwer, 1976), intramedullary nailing (Bone and Johnson, 1986; Puno et al., 1986; Court-Brown, Christie and McQueen, 1990) and the use of external fixators (Behrens and Searls, 1986) all have their adherents.

A classification should be able to allow meaningful comparison of results from these different treatments.

The most feared fractures are open tibial shaft fractures with extensive soft-tissue damage. As a result these in more recent times have been separately classified so that treatment and prognosis can be compared. Various classifications have been established for open tibial fractures, including those by Cauchoix (1978), Gustilo (Gustilo and Anderson, 1976) and Tscherne (Oesterne and Tscherne, 1984) which have gained wide acceptance in guiding treatment and comparing results. Doubts about the reliability of these systems have been expressed and research testing the usefulness of a classification is sparse. Brumback and Jones (1993) tested inter-observer reliability of the widely used Gustilo system for open fractures showing that inter-observer correlations were poor even among senior clinicians.

The classifications devised for closed fractures—both the earlier descriptive and more modern mechanistic models—have been more difficult to apply. This is partly due to treatment selection which influences the choice of factors perceived as important in influencing outcome. Johner and Wruhs' (1983) classification, on which the now widely accepted AO classification was based, excluded both fracture displacement and fracture level from their paradigm. They described their results from fractures treated by internal fixation and deemed that both these factors have little effect on outcome. A different conclusion may be drawn if these same fractures had been treated by closed methods where displacement has been shown to affect outcome. Ellis (1958a,b) classified 535 fractures managed conservatively and concluded that injury severity is determined by displacement as well as comminution and compounding. Similarly, Weissman, Herold and Engelberg (1966) and Leach (1984) based their classifications on displacement in fractures treated by conservative methods. Weissman et al. felt that initial displacement of the fragments was indicative of the severity of the trauma and so would ultimately affect outcome. Herein lies a problem with classification systems of tibial shaft fractures. Certain fractures where one factor, e.g. displacement, is viewed as the main determinant of outcome with one form of treatment may be of less significance in another form of treatment, yet the morphology of the fracture may be exactly the same. This will influence classification design.

Classification Design

Classifications of tibial shaft fractures have been based on:

1. Clinical appearance
2. Radiographic appearance
3. Mechanism of injury
4. Combination of the above.

From clinical appearance: deformity, extent of soft-tissue damage and the presence of an open wound can be assessed. Radiographs allow: fracture level, fracture pattern (morphology), comminution and displacement to be observed and measured. The accident mechanism that produces the fracture is the third main factor that influences the design of classification in tibial shaft fractures.

1. Open or Closed

Clinical examination of an injured limb will immediately classify the injury into an open or closed injury. This simple classification is of great importance in measuring outcome. Bauer, Edwards and Widmark (1962) proposed that the extent of soft-tissue injury rather than bone injury should guide the choice of treatment. They showed that the incidence of deep infection and poor results are directly correlated to the cause of fracture. Open injuries represent 10–20% of the total tibial fractures, yet have the poorest outcome. This was recognized in the earlier descriptive classifications such as Ellis (1958a,b) and it was Gustilo and Anderson (1976) and Oesterne and Tscherne (1984) who offered a classification of open fractures which attempted to predict accurately functional outcome. This system has been used to compare treatment. As a result, the importance of an open wound and extensive soft-tissue injury with the associated increased risk .of infection and ultimately poorer functional and cosmetic outcome is now better understood.

2. Fracture Level

The level of the fracture was an important feature of earlier classifications with a 3 or 4 fracture level classification included into most descriptive systems. Lower third fractures traditionally had the reputation of poorer outcome, i.e. longer healing times but several large series have showed clearly that speed of healing of shaft fractures is the same regardless of the level (Ellis, 1958a,b; Nicoll, 1964; Edwards, 1965; Hoagland and States, 1967). Modern classifications often do not include this feature but consider it more relevant to distinguish between diaphyseal and metaphyseal fractures, (Johner and Wruhs, 1983; Müller et al., 1990).

3. Comminution

Most classifications have recognized the importance of comminution and included this variable in the framework. Leach (1984) emphasized that comminution is a feature of a high velocity injury and comminuted bone will take longer to heal. Nicoll (1964) found an 80% incidence of delayed union in patients with comminution, compared with a 30% incidence where comminution was not present. With increasing comminution, non-union and delayed union increases. The AO system classifies fractures into types according to comminution which is a feature of injury severity. In historical descriptive classifications subcategories based on number of fragments, size of fragments, separate levels (double or segmental fractures) were included indicating fracture severity. Comminution in femoral shaft fractures has been graded by Winquist and Hansen (1980). They emphasized the importance of the amount of contact between fragments in the stability of the fracture. This classification although described in femoral shaft fractures is widely applied in classifying comminution in tibial shaft fractures (Court-Brown and McBirnie, 1995).

4. Fracture Pattern (Morphology)

This is an important and constant variable of fracture classification. Simple descriptions of spiral, transverse, longitudinal and double, e.g. Ekeland et al., (1988) have been replaced by mechanistic nomenclature such as torsion, bending or crush in modern dynamic classifications.

5. Displacement

Stability assessed from the radiograph has also been used in classification. A displaced fracture is viewed as unstable and attempts have been made to define and subclassify displacement in the earlier classifications. The radiograph may give a false impression as a severely displaced fracture may be reduced to a stable configuration before radiographic assessment.

Leach (1984) recognized the severity of injury as having a direct bearing on ultimate outcome. He noted that an open injury with severe initial displacement, comminution and soft-tissue injury with a subsequent infection as a result of high energy injury will have the worst prognosis. The prognostic significance of initial displacement is emphasized strongly in some of the descriptive classifications. Weissman, Herold and Engelberg (1966), in a series of 140 patients found that increasing displacement caused delayed union. Tibial shaft fractures with no displacement, or displacement of less than one-fifth of the shaft diameter, united in an average of 3 months and those with complete displacement of shaft took an average of 6 months to unite. Some of the later classification systems do not consider initial displacement as an important factor. Leach (1984) felt that the amount of initial displacement was the best prediction for bony healing, yet Johner and Wruhs (1983) believed that displacement was an unreliable factor.

The AO group have modified Johner and Wruhs' classification and included displacement as one of the indicators of the severity of the injury. Displacement is directly related to fracture union, but there is no evidence that ultimately it will affect the long-term functional result.

6. Accident Mechanism

This may be a direct or indirect injury and of either high or low energy. High energy injuries cause more open injuries, more soft-tissue damage and more skin loss as well as bone displacement and comminution. These injuries have a worse prognosis than low energy fractures (Hoaglund and States, 1967). Unlike the femur the tibia is not protected by a cushion of soft tissue over its subcutaneous anterior medial border, so is at risk from direct injuries and open wounds. In the AO classification, fracture morphology and mechanism of injury have been used to classify tibial shaft fractures. In contrast to direct injuries, indirect injuries of the tibia are usually twisting in nature and result in spiral or oblique fractures rather than comminuted or segmental patterns.

Fracture Classifications

Descriptive

1. Ellis classification

Ellis (1958a,b) used a simple classification of minor, moderate and major tibial shaft injury using displacement, comminution and compounding to determine the severity of injury and outcome. Of 535 fractures 86% had excellent clinical and anatomical results with conservative treatment. In addition to such outcome measures as average time to union and infection, he considered shortening, joint stiffness and return-to-work as important functional and outcome measures.

1. Minor fracture; an undisplaced non-angulated fracture with a minor degree of comminution or a minor open wound.
2. Moderate fracture; total displacement or angulation with a small degree of comminution or a minor wound.
3. Major fracture; complete displacement with major comminution or a major open wound.

In the three groups union took 10, 15 and 23 weeks and delayed union was 2%, 11% and 60% respectively.

2. Alms' classification

Allum and Mowbray (1980) reviewed 500 fractures using Alms' (1962) classification.

1. Transverse fracture (direct blow).
2. Spiral fracture (rotational force).
3. Oblique fracture (combination of these two forces).

Fracture level, comminution and accident mechanism were included in the classification. Allum recognized the importance of injury mechanism and confirmed the finding of Nicoll (1964) that major initial displacement was the most important factor in bone healing.

3. Nicoll classification

Nicoll (1964) critically looked at 705 tibial fractures and described the *personality* of the fracture. He identified four important factors in fracture classification:

1. Initial displacement.
2. Comminution.
3. Soft-tissue damage.
4. Infection.

All but infection are determined by the mechanism of injury which may be influenced by wound care and surgical technique. He described eight types based on the above personality rating where non-union and delayed union varied from 9% in the most favourable to 39% in thc least favourable type.

1. ND NC NW
2. ND NC SW
3. ND SC NW
4. SD NC NW
5. ND SC SW
6. SD NC SW
7. SD SC NW
8. SD SC SW

N = nil or slight; S = moderate or severe; D = displacement; C = comminution; W = wound.

4. Edwards classification

A controlled prospective study of 161 displaced tibial fractures which showed that the incidence of infection and poor results are correlated to the cause of the fracture. Edwards (1965) classification was based on:

1. Displacement
2. Comminution
3. Fracture level
4. Closed or open (2–5, 6–10, >10 cm)
5. Longitudinal or transverse morphology

Fracture types:

1. Displaced longitudinal closed

2. Displaced longitudinal open
3. Non-displaced longitudinal open
4. Displaced transverse closed
5. Displaced tranverse open
6. Non-displaced transverse open
7. Non-displaced transverse closed
8. Non-displaced longitudinal closed

This design was widely used by the Swedish school (Kristensen, 1979; Jensen et al., 1977) and later modified by Önnerfält (1978).

4. Weissman classification

Weissman, Herold and Engelberg (1966) based his classification on initial displacement of fragments as he felt that this was indicative of major trauma.

1. Minimal; with less than one-fifth overlap and 10° of angulation.
2. Mild; one to two-fifths overlap with angulation of 10–20°.
3. Marked; greater than one-half overlap.
4. Severe; total loss of contact.

5. Ekeland classification

Ekeland et al. (1988) classified 45 tibial fractures into four types. This was based on fracture pattern (morphology).

1. Transverse short oblique
2. Long oblique or spiral
3. Comminuted
4. Segmental

6. Chapman classification

Chapman's (1988) classification is based on fracture level and fracture pattern (morphology).

1. Type A. Transverse/short/oblique
2. Type B. Small butterfly
3. Type C. Large butterfly
4. Type D. Segmental
5. Type E. Spiral
6. Type F. Proximal one-quarter
7. Type G. Distal one-quarter

Most descriptive classifications of tibial shaft fractures are based upon fracture level (that is proximal, middle or distal thirds, or in Chapman's (1988) case quarters), morphology (transverse, spiral, oblique, comminuted, segmental) alignment (displacement and angulation) and associated factors, such as open wound or associated soft-tissue injury. The modern classifications have put greater emphasis on the importance of the condition of the soft-tissues and mechanism of injury (e.g. Müller et al., 1990).

Dynamic Systems

The methodology of classifications of tibial shaft fractures was changed by Johner and Wruhs (1983) who introduced the mechanism of injury into their classification as a predictor of severity of injury and ultimately functional outcome. This system was later modified by Müller et al. (1990) and drawn into the AO classification. The AO classification is a dynamic and complex paradigm which is effective in comparing the results of treatment and, although computer-friendly, is unwieldy in guiding initial treatment.

1. Johner and Wruhs' classification (Figure 18.1)

This classification recognized the important relationship between fracture pattern and injury mechanism.

A spiral pattern of injury is caused by torsion. An oblique or transverse pattern is caused by bending, often with a direct injury. Segmental or transverse highly comminuted patterns are caused by crushing.

In addition to mechanism, comminution was considered as an indicator of severity of injury. The degree of shattering of the bone will correlate with absorbed energy at the time of impact. There are three major categories:

1. Type A. Simple non-comminuted patterns.
2. Type B. Butterfly or wedge patterns.
3. Type C. Comminuted and segmental fractures.

This classification was used to assess the results of fractures treated with AO surgical techniques.

The three main types are further divided into three sub-types to make nine separate categories (Figure 18.1).

The fracture level, fracture displacement and extensive soft-tissue injury were not considered in this classification. This contrasts with the non-mechanistic classifications which rely on fracture level and displacement as important factors in determining severity of injury and outcome.

They showed that spiral and oblique fractures have the best prognosis after internal fixation. Crush injuries with comminution had significantly worse results. An A1, B1, C1 fracture can expect a 90% plus good or excellent outcome but a C3 fracture a 50% good or excellent outcome.

Johner and Wruhs (1983) used this classification to compare the treatment of fractures by different surgical methods and found a more rapid recovery with

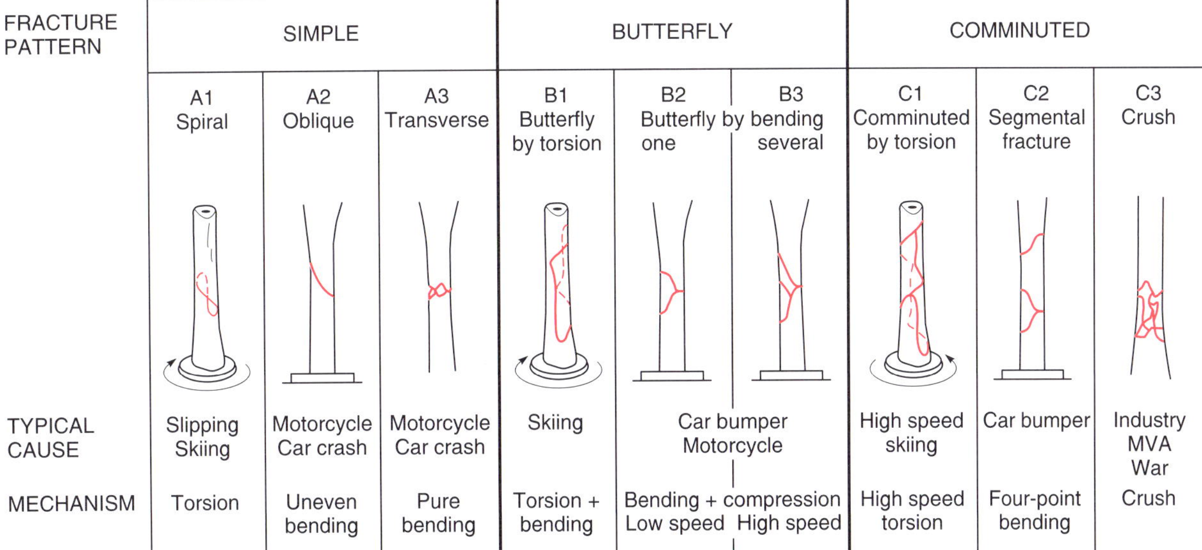

Figure 18.1. Johner and Wruhs' (1983) classification (from Trafton, 1992).

transverse high energy fractures treated with intramedullary nailing. In contrast they found more infection (17%) and implant failure (5%) after plate fixation of type B3 injuries.

2. AO classification (Figure 18.2)

The Johner and Wruhs (1983) classification was modified by the AO group (Müller et al., 1990) in their comprehensive classification of long bone fractures.

Four main factors were seen to influence the treatment, prognosis and outcome of tibial shaft fractures.

1. Accident mechanism (fractures produced by indirect trauma have a better prognosis than those produced by direct trauma).
2. Comminution (the risk of delayed union and non-union is increased by comminution).
3. Soft-tissue injury (open fractures have a higher infection rate than closed fractures).
4. Initial and final displacement (non-displaced fractures allow more simple treatment than displaced fractures).

There are nine morphological groups in this classification (Figure 18.2). There are three types: 42A (simple fracture), 42B (wedge) and 42C (complex). Forty-two is the code for the tibia shaft (4 = tibia and 2 = diaphysis). The code and classification have been designed so that they can be directly entered into a computer for analysis. This is a distinct advantage of this system in research and documentation.

The letters A, B, C (types) are used to represent an increasing degree of comminution and the numbers 1, 2, 3 represent criteria of the accident mechanism, i.e. direct and indirect impact and the amount of absorbed energy. These are divided into three groups:

Group 1 fractures include all spiral fractures produced mainly by indirect impact (torsion), i.e. A1, B1 and C1.
Groups 2 and 3 include fractures produced by direct impact (bending), causing oblique, transverse and segmental fractures namely B1, B2, B3 and C1, C2, C3.

Factors common to many classification systems which influence outcome relate to the presence or absence of cortical contact between the main, proximal and distal fragments (Ellis, 1958a,b; Nicoll, 1964; Johner and Wruhs, 1983). The AO system appreciates that if the main proximal distal fragments can be brought into contact with each other, rigid internal fixation can be achieved, the resulting stability is superior and final results and outcome are better than fractures without cortical contact.

It is a comprehensive but unwieldy system with 27 types (84 separate subtypes in total 42A(27), 42B(27), 42C(30)) of tibial shaft fractures. It is a mechanistic and complete classification in that it includes the accident mechanism and degree of comminution (energy absorbed by the bone) and takes into account the importance of the soft-tissue envelope.

It has been accepted as a valid classification system that allows comparison of results and predicts out-

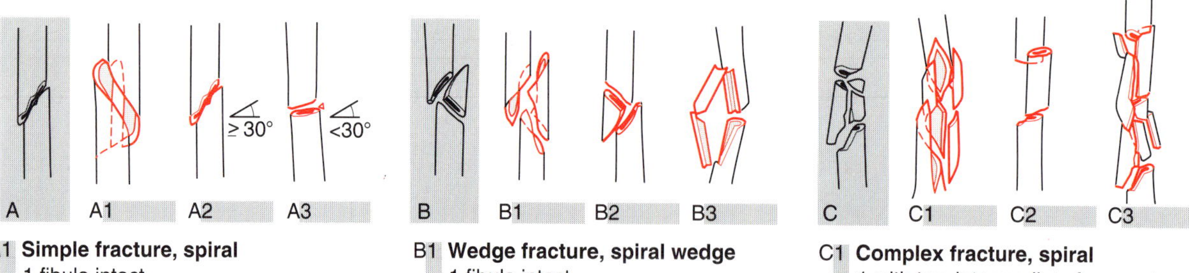

A1 Simple fracture, spiral
1 fibula intact
2 fibula fractured at another level
3 fibula fractured at the same level

A2 Simple fracture, oblique (≥ **30°**)
1 fibula intact
2 fibula fractured at another level
3 fibula fractured at the same level

A3 Simple fracture, transverse (< **30°**)
1 fibula intact
2 fibula fractured at another level
3 fibula fractured at the same level

B1 Wedge fracture, spiral wedge
1 fibula intact
2 fibula fractured at another level
3 fibula fractured at the same level

B2 Wedge fracture, bending wedge
1 fibula intact
2 fibula fractured at another level
3 fibula fractured at the same level

B3 Wedge fracture, fragmented wedge
1 fibula intact
2 fibula fractured at another level
3 fibula fractured at the same level

C1 Complex fracture, spiral
1 with two intermediate fragments
2 with three intermediate fragments
3 with more than three intermediate fragments

C2 Complex fracture, segmental
1 with one intermediate segmental fragment
2 with one intermediate segmental and additional wedge fragments(s)
3 with two intermediate segmental fragments

C3 Complex fracture, irregular
1 with two or three intermediate fragments
2 with limited shattering (< 4 cm)
3 with extensive shattering (≥ 4 cm)

Figure 18.2. AO/ASIF classification of tibial fractures (from Müller et al., 1990). The subgroups represent a variation within the group and the one illustrated is denoted within the shaded area of the text.

come. It gives a precise definition of fracture morphology. Fractures are arranged in increasing order of severity which anticipates difficulties in treatment so that likely complications and prognosis can be predicted.

3. *Tscherne classification*

Tscherne and his colleagues proposed four grades of severity for soft-tissue injury in closed fractures (Oesterne and Tscherne, 1984; Tscherne and Gotzen, 1984). With the emphasis on the soft-tissue envelope and compounding it is a complete design and all fractures are classifiable.

Grade 0. Injuries resulting from indirect forces with negligible soft-tissue damage.
Grade 1. Closed fractures of low or moderate energy mechanism with superficial abrasions or contusions of the soft-tissues overlying the fracture.
Grade 2. Significant muscle contusion with overlying contaminated skin abrasions. Fractures caused by direct violence, e.g. bumper fracture, and risk of compartment syndrome.
Grade 3. Closed fractures with extensive crushing, subcutaneous degloving or avulsion with a risk of arterial disruption or established compartment syndrome.

4. *Trafton classification*

Trafton (1992) proposed a dynamic classification incorporating soft-tissue injury.

The fractures are clinically graded using Ellis's (1958a,b) classification of minor, moderate and major categories.

The important fracture characteristics are considered to be displacement, comminution, wound, high or low energy, and mechanism of injury (see Table 18.1).

This classification was based on Ellis (1958a,b) and Leach's (1984) work and incorporated soft-tissue wound grading of Gustilo and Anderson (1976) and Tscherne and Gotzen (1984).

In grading fractures, according to severity, he recognized the pre-eminence of soft-tissue damage as Tscherne has emphasized, whether or not an open wound is present. He noted that when wound severity cannot be judged as in closed treatment or prior to adequate surgical exploration, the mechanism of injury in fact provides the best guide to the extent of soft-tissue damage. This may be indicated by accident history or the fracture pattern (morphology).

Although earlier descriptive classifications such as Ellis (1958a,b) and Edwards (1965) considered the importance of an open wound and soft-tissue damage, it is the recent dynamic classifications which have attempted to emphasize the pre-eminence of the soft-tissue in affecting outcome with or without compounding.

Classification of Open Fractures

1. Ellis classification

As noted above Ellis (1958a,b) recognized the importance of wound in fracture outcome.

Type 1 Minor. An undisplaced fracture with minor angular deformity and comminution with a minor wound.
Type 2 Moderate. A fracture with complete displacement and minor comminution with no more than a minor wound.
Type 3 Major. Comminuted fractures or a major open wound.

2. Gustilo classification

Gustilo and Anderson (1976) recognized three principle types of open fractures:

Type 1 An open fracture with a wound less than 1 cm long and clean.
Type 2 An open fracture with a laceration more than 1 cm long without extensive soft-tissue damage, flaps or avulsion.
Type 3 Either an open segmental fracture with extensive soft-tissue damage or a traumatic amputation.

Gustilo, Mendoza and Williams (1967) felt that the designation of Type 3 was too broad with too wide a range of severity of injury and predicted outcome. They therefore recommended a Type 3 open fracture should be divided into three subtypes in order of severity and the predicted poorer prognosis.

Type 3A Adequate soft-tissue coverage of a fractured bone, despite soft-tissue laceration of flaps of high energy trauma, irrespective of the size of the wound.
Type 3B Extensive soft-tissue injury loss with periosteal stripping and bone exposure, usually with massive contamination.
Type 3C Open fracture associated with arterial injury requiring repair.

Gustilo et al. found an increasing incidence of wound sepsis and amputation in Type 3 open fractures of different localization. Types 3A, 3B, 3C exhibited wound sepsis rates of 4%, 52% and 42%, and amputation rates of 0%, 16% and 42% respectively. This system is highly predictive of wound infection risk which with adequate management ranges up to 2% for Type 1, up to 7% for Types 2 and 3A, from 10% to 50% for Type 3B and from 25% to 50% for Type 3C. They specified that open fractures resulting from high energy trauma with segmental and comminuted fractures should be classified as Type 3 injuries, regardless of the size of wound. Caudle and Stern (1987) confirmed that these three subtypes were of prognostic value when they looked at 62 severe open tibial fractures with regard to the complication rate and outcome.

Scepticism is justified when treatment procedures and predicted outcomes are discussed with both closed and open tibial shaft fracture classifications. In the closed tibial classifications damage to the soft-tissue envelope is not fully appreciated nor considered in affecting outcome. With open classifications systems, such as that of Gustilo et al., fracture pattern (morphology) is over-emphasized in treatment considerations and prognosis.

3. Müller classification

The AO classification (Müller et al., 1990) graded three types of open fractures.

1. Type 1 first degree open fracture: the skin is pierced from within by bone fragments.
2. Type 2 second degree open fracture: the skin has been broken and crushed externally, causing moderate damage to the skin, subcutaneous tissue and muscle. The severity of the fracture is variable.
3. Type 3 third degree open fracture: the result of high energy injuries with extensive damage to the skin, subcutaneous tissues, muscles and neurovascular structures. This is often associated with a neurovascular injury and is usually severely infected. High velocity gunshot wounds are included in this category.

The new soft-tissue grading of Müller et al. (1990) allows further differentiation of skin injuries (IO integument open), the underlying muscle and tendon injury (MT) and the neurovascular injury (NV), thus allowing a much better grading of open fractures than with the old AO classification. This new AO classification, however, needs confirmation based on clinical trials.

4. Tscherne classification

Oesterne and Tscherne (1984) classification uses wound size, contamination and fracture pattern to grade open fractures.

Grade 1 First degree open (frOI): small puncture wound without skin contusion, negligible bacterial contamination and a low energy fracture pattern.

Grade 2 (frOII): small skin and soft-tissue contusion, moderate contamination and variable fracture patterns.

Grade 3 (frOIII): heavy contamination, extensive soft-tissue damage, often associated arterial and nerve injuries.

Grade 4 (frOIV): incomplete or complete amputations.

Unlike Gustilo's classification, Tscherne emphasizes grading of the severity of soft-tissue damage in closed as well as open injuries. He proposed four grades of severity for soft-tissue injury in closed fractures (Oesterne and Tscherne, 1984).

Grade 0 Injuries resulting from indirect forces with negligible soft-tissue damage.

Grade 1 Closed fractures: low or moderate energy mechanism with superficial abrasions or contusions of the soft-tissues overlying the fracture.

Grade 2 Significant muscle contusion which may have deep contaminated skin abrasions. Fractures caused by direct violence, e.g. bumper fracture, risk of compartment syndrome.

Grade 3 Closed fracture with extensive crushing, subcutaneous degloving or avulsion with a risk of arterial disruption or established compartment syndrome.

5. *Mangled extremity score (see also Chapter 1)*

Gregory et al. (1985) devised a severity grading system for multi-system injury of the extremity. With Gustilo Grade 3B and 3C fractures the mangled extremity scoring system (MESS) gives an accurate prediction of survivability and the lesser need for primary amputation. However with this system as well as Tscherne's and Gustilo's classifications, none of these grading systems gives an accurate prediction of function of the retained limb. Bondurant et al. (1988) assessed function after extensive reconstruction in Gustilo Grade 3B and C injuries and suggested that an amputation in the first few days may, in some cases, provide a better outcome than salvage surgery (see also Chapter 1).

Validation

1. Classification.
Are our classification tools valid?
2. Outcome measures (Tables 18.2 and 18.3).
What are the appropriate outcome measures in tibial shaft fractures?

The classic measures are physical, i.e. non-union, delayed union, and infection. There are no universally agreed standards and it has been conclusively shown that you can have a good or bad result according to the criteria that are chosen (Bridgman and Baird, 1993; Williams et al., 1995).

Watson (1994) reviewed the treatment of unstable fractures of the tibia shaft and concluded that few authors have classified closed fractures adequately in terms of severity of injury. It was evident that the use of non-operative treatment of tibial fractures that are displaced as a result of a high energy injury are associated with a high prevalence of mal-union, joint stiffness and poor functional outcome.

In addition to measures of physical outcome there should be an expanded view of endpoint to include functional and patient-scored criteria. Function, health status, pain, cost and patient expectation should be included. In the young active population who suffer tibial shaft fractures, functional outcome is important and can be characterized by work status and other measures such as sporting achievement following treatment.

Other physical outcome measures have been neglected and have an important effect on functional outcome such as ankle stiffness (Waddell and Reardon, 1983; Triffit et al., 1992); neurological complications (Koval et al., 1991; Williams et al., 1995), deformity such as shortening and rotation, and joint function. Bridgman and Baird (1993) showed in an audit of closed tibial fractures that agreed standards are based on little hard evidence.

Nicoll (1964) and Charnley (1961) recognized that a varus deformity is worse than valgus after tibial shaft fractures and there are no good studies to assess outcome which correlate the perception of cosmetic deformity. Waddell and Reardon (1983) and Bridgman and Baird (1993) considered these to be unacceptable outcomes with tibial shaft fracture treatment. Austin (1977) audited several large studies and demonstrated the difficulties in comparing outcome because of a lack of an agreed classification and of outcome measures and suggested a need for a large prospective study.

Table 18.2 Oxford versus conventional criteria for measuring malunion in tibial fractures. A good or unsatisfactory outcome is dependent on the standard used (Williams et al., 1995)

(a) Malunion results previously published

Author	*Method*	*Shortening*	*Sagittal*	*Coronal*	*Total (%)*
Nicoll, 1964	POP	2.5%(>2 cm)	3%(>10°)	3%(>10°)	9
Sarmiento, 1974	Functional bracing	21%(>4.1 mm)	9%(total >6°)	9% total	30
Oni et al., 1988	POP	5.3%(>1 cm)	13% (>5°)	20%(>5°)	38.3
Hooper et al., 1991	POP	21%	33% total		54
Ruedi et al., 1976	AO plating	0%	0%	0%	0 (3 plate failures)
Batten et al., 1978	AO plating	0%	0%	0%	0 (2 plate failures)
Fisher et al., 1978	AO plating	0%	0%	0%	0 (4 plate failures)
Bilat et al., 1994	AO plating	0%	0%	2%	4
Bone and Johnson, 1986	IMN	5% (>1 cm)	3% (total >5°)	3% total	8
Alho et al., 1990	IMN	14%(>1 cm)	6%(5°)	11%(5°)	31
Court-Brown et al., 1990	IMN	3% (>1 cm)	0%	0%	2.5
Hooper et al., 1991	IMN	7%	0%	0%	7
Moed, et al., 1994	IMN	0%	0%	0%	0
O'Dwyer et al., 1994	IMN	0%	0%	2% Rotation	2
Sargeant et al., 1994	IMN	0%	0%	0%	0

POP, plaster of Paris; IMN, intramedullary nailing

(b) Malunion of tibial shaft fractures after intramedullary nailing ('Oxford criteria')

Malalignment	*No. of episodes*
Shortening >0.5 cm	7
Distraction >0.5 cm	10
Sagittal (A.P.) >5°	7
Coronal (V/V) >5°	16
Malrotation >5°	6
Total episodes	46
Total no. of patients (one or more episodes of malalignment)	35 (34%)

A.P. anterior-posterior; V/V varus/valgus

(c) Malunion of tibial shaft fracture after intramedullary nailing ('conventional criteria')

Malalignment	*No. of episodes*
Shortening >1 cm	4
Distraction >1 cm	7
Sagittal (A.P.) >10°	1
Coronal (V/V) >10°	4
Malrotation >10°	4
Total episodes	20
Total number of patients (one or more episodes of malalignment in 94 patients)	12 (13%)

Table 18.3 Range of standards proposed in assessing functional outcome of tibial fractures (Bridgman and Baird, 1993)

Study	*Varus*	*Valgus*	*Ant/pst*	*Rot*	*Short*
Nicoll (1964)	10	10	10	10	20
Weissman et al. (1966)	5	5	5	–	–
Johner and Wruhs (1983)	5	5	10	10	5
Leach (1984)	5	5	10	–	8
Puno et al. (1985)	10	10	20	–	20
Trafton (1988)	5	5	10	10	15
Sarmiento et al. (1989)	5	5	5	–	10
	5	5	10	10	20

Ant/pst = anterior or posterior angulation; Rot = rotatory deformity; Short = shortening (mm)

Conclusions

Classification only serves a useful purpose if it is simple and accurate to apply, facilitates the choice of treatment and helps in predicting outcome. In classification design there has been a shift from emphasizing the importance of bone morphology to soft-tissue envelope damage in ultimately influencing outcome. Descriptive anatomical classifications, based on radiographic or clinical data, e.g. upper-middle-lower third or open-closed fractures have been superseded by dynamic models which stress accident mechanism and soft-tissue status in determining which are the important factors affecting functional outcome. However, simple rather than complex classifications often prove to be more durable and applicable to clinical practice. Few systems have been clinically validated and those in common use require testing.

The more complex the classification system, the more potential there is for intra- and inter-observer error and this should be borne in mind when clinically applying elaborate comprehensive systems such as that designed by the AO group.

A common, simple and accurate classification for both open and closed fracture (Batten, Donaldson and Aldridge, 1978) and an agreed set of validated end points is required if the comparison of treatment and results with these fractures is to be meaningful.

References

Allum, R.L. and Mowbray, M.A.S. (1980) A retrospective review of the healing of fractures of the tibial shaft with special reference to the mechanism of injury. *Injury*, **11**, 304–308.

Alms, M. (1962) Medullary nailing for fractures of the shaft of the tibia. *J. Bone Joint Surg.*, **44B**(2), 328–339.

Austin, R.T. (1977) Fractures of the tibial shaft: Is medical audit possible? *Injury*, **9**, 93–101.

Batten, R.L., Donaldson, L.J. and Aldridge, M.J. (1978) Experience with the AO method in the treatment of 142 cases of fresh fracture of the tibial shaft treated in the U.K. *Injury*, **10**, 108–114.

Bauer, G.L., Edwards, P. and Widmark, P.J. (1962) Shaft fractures of the tibia. Aetiology of poor results in a consecutive series of 173 fractures. *Acta Chir. Scand.*, **124**, 386.

Behrens, F. and Searls, K. (1986) External fixation of the tibia. Basic concepts and prospective evaluations. *J. Bone Joint Surg.*, **68B**, 246–254.

Bondurant, F.J., Cotler, H.B., Buckler, R. et al. (1988) The medical and economic impact of severely injured lower extremities. *J. Trauma*, **28**, 1270–1273.

Bone, L.B. and Johnson, K.D. (1986) Treatment of tibial fractures by reaming and intramedullary nailing. *J. Bone Joint Surg.*, **68A**, 877–887.

Bridgman, S.A. and Baird, K. (1993) Audit of closed tibial fractures: What is a satisfactory outcome? *Injury*, **24**(2), 85–89.

Brumback, R.J. and Jones, A.L. (1993) Inter-observer agreement in the classification of open fractures of the tibia. *J. Bone Joint Surg.* **76A**, 1162-1165.

Burnstein, A.H. (1993) Fracture classifications: do they work and are they useful? *J. Bone Joint Surg.*, **75A**, 1743–1744.

Cauchoix, J. (1978) In A. Grösse, I. Kempf, and D. Lafforgue, Le traitement des fracas, pertes de substance osseuse et pseudarthroses de fémur et du tibia par l'enclouage verrouillé. *Rev. Chir. Orthop.*, **64**, Suppl 2.

Caudle, R.J. and Stern, P.J. (1987) Severe open fractures of the tibia. *J. Bone Joint Surg.*, **69**, 801–807.

Chapman, M.W. (1988) Fractures of the tibia and fibula. In *Operative Orthopaedics* (M.W. Chapman, ed.). Philadelphia: Lippincott.

Charnley, J. (1961) *The Closed Treatment of Common Fractures*, 3rd edn. Edinburgh and London: Livingstone.

Court-Brown, C.M., Christie, J. and McQueen, M.M. (1990) Closed intramedullary nailing. Its use in closed and type one open fractures. *J. Bone Joint Surg.*, **72B**, 605–611.

Court-Brown, C.M. and McBirnie, J. (1995) The epidemiology of tibial fractures. *J. Bone Joint Surg.*, **77B**, 417–421.

Edwards, P. (1965) Fracture of the shaft of the tibia: 492 consecutive cases in adults: Importance of soft tissue injury. *Acta Orthop. Scand.*, **76** Suppl 2, 7–83.

Ekeland, A., Thoresen, B.O. Alho, A. et al. (1988) Interlocking intramedullary nailing in the treatment of tibial fractures. A report of 45 cases. *Clin. Orthop.*, **231**, 205–215.

Ellis, H. (1958a) The speed of healing after fracture of the tibial shaft. *J. Bone Joint Surg.*, **40B**, 42–46.

Ellis, H. (1958b) Disabilities after tibial shaft fractures: With special references to Volkmann's ischaemic contracture. *J. Bone Joint Surg.*, **40B**, 190–197.

Gregory, R.T., Gould, R.J., Peclet, M. et al. (1985) The Mangled Extremity Syndrome (M.E.S.): a severity grading system for multisystem injury of the extremity. *J. Trauma*, **25**, 1147–1150.

Gustilo, R.B. and Anderson, J.T. (1976) Prevention of infection in the treatment of 1025 open fractures of long bone. *J. Bone Joint Surg.*, **58A**, 453–458.

Gustilo, R.B., Mendoza, R.M. and Williams, D.N. (1967) Problems in management of types III (severe) open fractures: a new classification of type III open fractures. *Trauma*, **24**, 71–76.

Gustilo, R.B., and Merkow, R.L. (1990) Current concepts review. The management of open fractures. *J. Bone Joint Surg.*, **72A**, 299–303.

Hoaglund, F.T. and States, J.D. (1967) Factors influencing the rate of healing in tibial shaft fractures. *Surg. Gynecol. Obstet.*, **124**, 71–76.

Hooper, G.J., Kedell, R.G. and Penny, I.D. (1991) Conservative management or closed mailing for tibial shaft fractures. *J. Bone Surg.*, **73B**, 83–85.

Jensen, J.S., Hansen, F.W. and Johansen, J. (1977) Tibial shaft fractures. *Acta Orthop. Scand.*, **48**, 204–208.

Johner, R. and Wruhs, O. (1983) Classification of tibial shaft fractures and correlation with results after rigid internal fixation. *Clin. Orthop.*, **178**, 7–25.

Koval, K.J., Clapper, M.F., Brumback, R.J. et al. (1991) Complications of reamed intramedullary nailing of the tibia. *J. Orthop. Trauma*, **5**, 184–189.

Kristensen, K. (1979) Tibial shaft fractures. *Acta. Orthop. Scand.* **50**, 593–598.

Leach, R.E. (1984) Fractures of the tibia and fibula. In *Fractures in Adults* (C.A. Rockwood, Jr and D.P. Green, eds), Vol. 2. Philadelphia: Lippincott, pp. 1593–1664.

Matson, F.A. and Clawson, D.K. (1975) The deep posterior compartmental syndrome of the leg. *J. Bone Joint Surg.*, **57A**, 34–39.

Müller, M.E., Nazarian, S., Koch, P. and Schatzker, J. (1990) *The Comprehensive Classification of Fractures of Long Bones.* New York: Springer-Verlag.

Nicoll, E.A. (1964) Fractures of the tibial shaft: A survey of 705 cases. *J. Bone Joint Surg.*, **46B**, 373–387.

Oesterne, H.J. and Tscherne, H. (1984) Pathophysiology and classification of soft tissue injuries associated with fractures. In *Fractures with Soft Tissue Injuries* (H. Tscherne and L. Gotzen, eds). New York: Springer-Verlag, pp. 1–9.

Önnerfält, R. (1978) Fracture of the tibial shaft treated by primary operation and early weight-bearing. *Acta Orthop. Scand.*, **171** Suppl. 171, 1–63.

Puno, R.M., Teynor, J.T., Nagano, J. and Gustilo, R.B. (1986) Critical analysis of results of treatment of 201 tibial shaft fractures. *Clin. Orthop.*, **212**, 113–121.

Rüedi, T.H., Webb, J.K. and Allgöwer, M. (1976) Experience with dynamic compression plate (DCP) in 418 recent fractures of the tibial shaft. *Injury*, **7**, 252–257.

Sarmiento, A., Gersten, L.M., Sobol, P.A. et al. (1989) Tibial shaft fractures treated with functional braces. Experience with 780 fractures. *J. Bone Joint Surg.*, **71B**, 602–609.

Sarmiento, A., Sharpe, F.E., Ebramzadeh, E. et al. (1995) Factors influencing the outcome of closed tibial fractures treated with functional bracing. *Clin. Orthop.*, **315**, 8–24.

Trafton, P.G. (1992) Tibial shaft fractures. In *Skeletal Trauma* (B.D. Browner, J.B. Jupiter, A.M. Levine and P.G. Trafton, eds), Ch. 51. Philadelphia: Saunders, pp. 1771–1869.

Triffit, P.D., Konig, D., Harper, W.M. et al. (1992) Compartment pressures after intramedullary nailing closed tibial shaft fracture: their relation to functional outcome. *J. Bone Joint Surg.*, **74B**, 195–198.

Tscherne, H. and Gotzen, L. (1984) *Fractures with Soft Tissue Injuries.* New York: Springer-Verlag.

Waddell, J.P. and Reardon, G.P. (1983) Complications of tibial shaft fractures. *Clin. Orthop.*, **178**, 173–178.

Watson, J.T. (1994) Treatment of unstable fractures of the shaft of the tibia. *J. Bone Joint Surg.*, **76A**, 1575–1584.

Weissman, S.L., Herold, H.Z. and Engelberg, M. (1966) Fractures of the middle two thirds of the tibial shaft: results of treatment without internal fixation in 140 consecutive cases. *J. Bone Joint Surg.*, **48A**, 257–267.

Williams, J., Gibbons, M., Trundle, H. et al. (1995) Complication of closed tibial fractures. *J. Orthop. Trauma*, **9**, 476–481.

Winquist, R.A. and Hansen, S.T. (1980) Comminuted fractures of the femoral shaft treated by intramedullary nailing. *Orthop. Clin. North Am.*, **11**, 633–648.

19

The ankle

M. A. Green

Introduction

Pott, Dupytren, Cooper, Maisonneuve, Tillaux, Honigschmied, Stimson, Destot, Chaput, Quenu – and shall I not add Scudde, Cotton, Roberts and Speed? What can anyone say more at this late day? And yet the fact remains that there is no entirely satisfactory classification of ankle fractures in existence and many points of the mechanism of their production still are in dispute

Ashhurst and Bromer, 1922

Though we are perhaps far more knowledgeable about the mechanisms of injury than Ashhurst and Bromer were in 1922 we still find ourselves debating the best classification for ankle fractures. Attempts at classification were hampered prior to the discovery of X-rays, hence the number of eponyms ascribed to single fracture types, many quoted in the first paragraph. Historically, ankle fractures were first truly classified by Ashhurst and Bromer in 1922 who summarized the possible movements that were responsible for the production of ankle fractures. Following Ashhurst, Lauge-Hansen in 1942 (see Lauge 1948) produced a more detailed classification which was subsequently expanded (Lauge-Hansen, 1949) to include the combinations of ligament rupture and fractures that can occur. Danis (1949) later produced a classification modified by Weber and eventually adopted for use by the AO group. Denham (1964) used a similar classification to Ashhurst and Bromer but graded the fractures in each of the four major groups into three degrees (first, second and third) dependant on the quantity of movement of the talus relative to its proper position. These classifications (Ashhurst and Bromer, Denham) failed to progress in the literature. Burwell and Charnley (1965) produced a classification system that is loosely based on the Lauge-Hansen classification dividing the fractures into four groups. Generally all these classifications cover the vast majority of ankle fractures that are seen in clinical practice but more unusual fractures are described with an eponymous name rather than classified into differing severity grade or by mechanism of injury.

Lauge-Hansen Classification

Lauge-Hansen described in detail the pathogenesis of ankle fractures and based experimental work on cadavaric studies producing fractures he broadly classified into four types, each with a distinct radiographic appearance. It should be noted that the cadavaric fractures were all produced by immobilization of the heel and foot with nails and then producing force at the ankle by hand. As few as four and at maximum eight experiments were conducted in each of the four groups he described. These he classified as to how the foot was positioned at the time of the injury (i.e. pronated or supinated) and the direction of the injuring force (i.e. abduction, adduction, eversion and inversion). Each of the four injury patterns is then further divided into stages that represent the degree of damage/dissipation of force from rupture of ligament complexes to fractures (see Appendix 1). Substitution of the words external rotation for the word eversion clarifies the Lauge-Hansen classification considerably. The information gained from classifying the ankle fracture in the Lauge-Hansen classification was to be of use when seeking to perform a closed manipulation and reduction of an ankle fracture prior to the advent of internal fixation. It is suggested that the classification is complete enough to include all (about 96.3% (Burwell and Charnley, 1965), 99% (Yde, 1980) but a minority of fractures.

Danis Weber AO Classification

Danis' (1949) classification was modified by Weber (1972) and more recently modified by the AO group (see Appendix 1). It is based on the location of the fibula fracture relative to the syndesmosis and the horizontal portion of the tibiotalar joint. This classification was introduced with the advent of rigid anatomical internal fixation where the emphasis is on accurate open reduction rather than a closed reversal of the injury. While this system is undoubtedly simpler than the Lauge-Hansen classification, initially it ignored the medial malleolar injuries that were thought to be biomechanically unimportant. The classification has three types of fracture A, B and C. Essentially the higher the fibular fracture the more extensive the damage to the tibiofibular ligaments and fibula and the greater the chance of instability of the ankle mortise (see Appendix 1). Some authors have advocated the expansion of this classification to incorporate the medial injury and this would almost certainly increase the complexity of the classification to be similar to that of Lauge-Hansen (Harper, 1992). In essence, this is exactly what the recent AO modification has done. The main weakness is the lack of ability for the Danis Weber classification system to distinguish between the possible fracture types which may be contained within its group B type fracture. The different way in which they may behave leads to difficulty in interpreting the outcome of certain studies where this system has been used (Bauer et al., 1985).

Burwell and Charnley

Burwell and Charnley (1965) produced a system loosely based on the Lauge-Hansen classification by dividing the fractures into four groups:

1. Eversion (lateral rotation) fractures, including Lauge-Hansen supination, eversion and pronation eversion type fractures.
2. Adduction fractures.
3. Abduction fractures
4. Compression fractures.

This system failed to gain acceptance by other authors and appears to have only been used in one study, by workers from another unit, to classify fractures (Daly et al., 1987). The main criticism of the classification by these authors is that, although fractures could be classified, within this classification the severity and mechanism of the fractures could not be identified.

Similarities of the Lauge-Hansen and the Danis Weber systems

The two major systems are not completely interchangeable and have important differences. Although both systems give much information based purely on radiographs neither should be divorced from clinical examination. Type A (Weber, 1972) fractures can include supination/adduction injuries, pronation/abduction injuries (type 1) and pronation/eversion injuries (type 1). Type B injuries include supination/eversion and, possibly, pronation/abduction (type 2) and pronation/eversion (types 2 and 4). Type C consists of pronation eversion (type 3) and pronation/abduction (type 3). Types A, B and C also include patterns of other groups depending on the stage of the injury.

Prognosis Outcome and Treatment by Classification

Type B injuries (AO and Weber) as highlighted above include both supination/eversion (stage 2) injuries and supination/eversion (stage 4) injuries. This has led to particular problems in attempting to compare outcome, treatment and prognosis based on the two classifications. Currently (though it is acknowledged that the latest AO modification of the Weber system is the most precise), all the literature to date has used the simple A, B, C classification (see Appendix 1) in comparison with the Lauge-Hansen classification. Bauer, Jonsson and Nilsson (1985) showed that, of 103 patients with type B injuries, 18 had arthrosis of varying degrees on radiography at follow-up. When classified according to Lauge-Hansen principles, then two had minimal changes in the stage 2 group; one in the stage 3 group; and 14, with varying degrees of degeneration, were present in the stage 4 group. This study suggests that the accuracy of results of such studies as those by Cedell (1967) and Yde and Kristensen (1980 a,b) are doubtful particularly if the supination/eversion (stage 2) erroneously also include some stage 4 injuries.

Seligson and Frewin (1986) suggest that using the Lauge-Hansen classification as a guide to treatment of those patients who have a supination/eversion (type 2) injury the majority should be treated with closed conservative methods, and of those with supination/eversion (type 4) the majority should be treated with open reduction and internal fixation. This treatment algorithm is lost when using the Weber classification. They also found that the supination/eversion (type 2) injuries did well whatever the form of the treatment while the outcome of supination/eversion (type 4) injuries was poor. Again these are results that may be lost by the Weber classification. Operative treatment of these injuries did marginally better than non-operative treatment. Overall, the Weber B group appeared to do well, but not if subdivided into the Lauge-Hansen groups. This would seem to suggest a predominance of the supination/eversion (type 2) within the Weber B class in this work.

Current literature suggests that the outcome of certain fracture groups, such as pronation/eversion injuries (Seligson and Frewin, 1986) may be worse compared with other groups. Neither classification scheme has any features, independent of the method of treatment, that will predict the outcome of a fracture (Wilson, 1990; Broos and Bisschop, 1991; Lindsjo, 1985). They may provide a guide to the method of treatment (Yde, 1980a,b; Bauer et al., 1985a; Cedell, 1985; Kristensen and Hansen, 1985). Bauer et al. (1985b) certainly suggested that those patients whose fractures were within the more severe stages of the Lauge-Hansen classification (e.g. PE stage 4 and SE stage 4) were those who developed moderate to severe arthrosis. This should be expected as they have more severe fractures anyway. It should also be noted that some of Bauer's work was also based on a highly selected group of patients (e.g. patients with Weber type C fractures were omitted from his study, as were patients with open fractures, other limb trauma, undisplaced fractures and those with complicating disease). Overall, it is generally recognized that the AO (Danis Weber) system is simpler to use in everyday usage but it does lack the detail that may be necessary to produce accurate clinical trials.

Both systems contain groups that can be treated by either closed and open methods. They also contain injuries that may be open or closed. These factors may affect eventual outcome possibly more than classification.

Reproducibility of Classifications

Since radiographic interpretation is an important feature of both classifications both inter- and intra-observer during this step may occur and be of clinical importance. Certainly appropriate instruction when using the Lauge-Hansen classification can reduce the inter-observer variation (Rasmussen, Madsen and Bennicke, 1993). Nielsen, Dons-Jensen and Sorensen (1990) demonstrated that inter-observer variation is large. They concluded that the Lauge-Hansen classification was difficult to apply in a reproducible fashion. Despite training with the classification system, only 43% of 118 fractures were classified identically for class and stage by the four observers and 68% by class only. Intra-observer variation ranged from 64 to 82% for class and stage and 81 to 95% for class only.

Only one study compares the Lauge-Hansen and Danis Weber classifications for inter- and intra-observer variability (Thomsen et al., 1991). On the first interpretation of the test radiographs, only the Danis Webber classification system showed acceptably high levels of precision when tested against the possibility of obtaining results by chance alone. Both classifications showed acceptable levels of precision after the second attempt but neither reached a level that could be called excellent. Thomsen also found that the category 'pronation/abduction' injuries were particularly difficult to distinguish by all observers (reaching statistical significance compared to the ability of observers to distinguish between other categories in the classification). This indicates that it has a particularly poor radiographic definition. What cannot be said is whether this variation makes any difference to the eventual clinical outcome.

In theory, ankle fractures should occur in equal frequency in similar populations. This should allow comparison of outcome from differing units. Lindsjo (1985), combining five studies of ankle fractures (including the work by Baek Kristensen, 1949; Yde, 1980; Lindsjo, 1981), reviewed the incidence of each fracture group between the various series published. Of all the different series published the only common finding was that the majority of the ankle fractures were the supination/eversion type. All the other fracture patterns varied significantly in their incidence. Baek Kristensen suggested that the difference in distribution between his work and that of Lauge-Hansen was related to the recruitment from an urban as opposed to a rural area, but the use of different selection criteria may also be of importance. It is not clear from the literature as to the proportions of the various Weber classes contained within the various series published. Difference in observer interpretation may also play a role in determining the relative frequency particularly of the Lauge-Hansen groups. Long-term reviews may be confounded by population changes. In Malmo, the Lauge-Hansen supination/eversion (type 4) fractures have been shown to have shifted from a predominantly young group to occuring in a predominantly elderly female group. The ratio between similar type fractures class and grade, e.g. supination/eversion (type 2) and supination/eversion (type 4) may also change (Bauer et al., 1987a,b).

Epiphyseal Injuries of the Ankle

Distal tibial physis injuries have been reported as the most common physeal injury in children (Mann and Rajmaira, 1990). There have been many attempts to classify ankle growth plate injuries: Carothers (1955) classified them according to adult type classifications such as Ashhurst and Bromer (1922) and the other now popular adult classification systems. However, the additional complication of how the physis behaves may result in a fracture classified into a Lauge-Hansen supination but being either a Salter Harris type 2, 3 or 4 depending on how the physis behaves (Salter and Harris, 1963). Other unique injuries exist in children such as the perichondral ring type injury (Tisa, Brandreth and Reinherz, 1988). Outcome may be related to the Salter Harris grading. The most frequent and serious complication of these fractures is the premature and asymmetric closure of the growth plate. This is commonest in the medial malleolar fractures of Salter Harris type 3 or 4 (Crawford A, 1995). Speigal, Cooperman and Laros (1978) grade outcome into low risk, high risk and unpredictable outcome by the part of the ankle involved (fibula or tibia) and the Salter Harris type of fracture. However, other factors such as the fracture comminution, the fracture displacement, the adequacy of reduction and the skeletal maturity of the patient also influence the outcome. In Speigal's work, the Salter Harris type 2 fractures were found to have a higher risk of epiphyseal complications than the minimally displaced type 3 or 4.

Tillaux and triplane fractures (both two and three part) occur in older children in whom the asymmetric closure of the physis has started to occur. These are really descriptive terms for certain unique pattens of fracture in children's ankles rather than a form of classification.

Distal Tibial Fractures including Pilon Fractures

These fractures result from axial and rotational forces and cause variable degrees of metaphyseal disruption, articular damage and malleolar displacement. The common denominator is a vertical compression force. The force may act singly on the tibia, the area of destruction being determined by the position of the foot at the time of impact, or may act in combination with shear and rotational forces. They may overlap with fractures of the distal tibia. It is not uncommon for some of these fractures to have a pilon component. Many may be open (Gustilo and Anderson, 1976). There have been several attempts to classify these fractures: the Ashurst and Bromer (1922) classification of ankle fractures included the distal tibial fracture, as did the work of Lauge-Hansen (Lauge, 1948; Lauge-Hansen, 1950, 1952). Lauge-Hansen believed that forced dorsiflexion was involved in all of this type of fractures and, therefore, added a fifth classification to his group of four—the pronation/dorsiflexion fracture (Lauge-Hansen, 1953).

deSouza (1993) suggested a classification based on severity of the injury from A to C. Group A included the low energy injuries resulting from rotational or shear stresses (the so-called spiral extension fractures of Maale and Seligson (1980)). These, it is suggested by the authors, should have a better outcome than those pilon fractures associated with central compression. Type B includes injuries where there has been high energy compressive and rotational injuries including Lauge-Hansen's stage 4 injuries in the pronation/dorsiflexion, supination/external rotation and pronation/external rotation groups. Group C fractures are high energy axial compressive injuries. This last group has been further classified by Rüedi and Allgower (1969) into three groups 1–3, depending on the displacement and comminution of the joint surface (Figure 19.1):

- Group 1 — Cleavage fractures of the articular surface without major dislocation of the articular surface.
- Group 2 — Significant fracture and dislocation but without comminution.
- Group 3 — The same fracture as group 2 but with significant comminution and impaction of the distal tibia.

Ovadia and Beals (1986) in a retrospective analysis extended this three category division of Rüedi into five types (Figure 19.2):

- Grade I — Non-displaced articular fractures.
- Grade II — Minimally displaced articular fracture.
- Grade III — Displaced articular fracture with several large fragments.
- Grade IV — Displaced articular fracture with multiple fragments and a large metaphyseal defect.

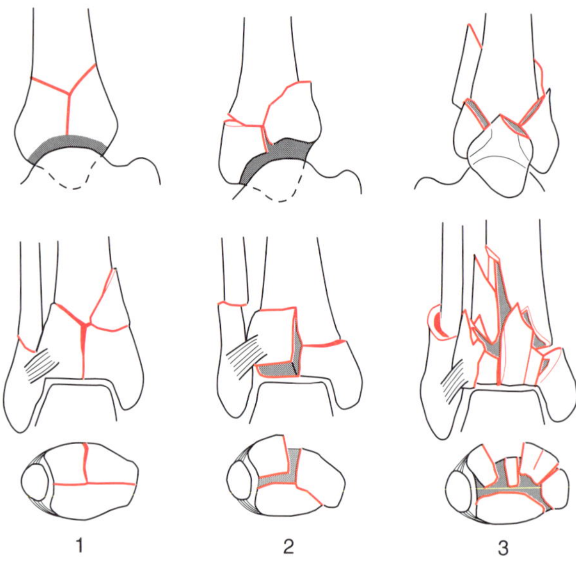

Figure 19.1. Rüedi and Allgower's (1969) classification of distal tibial fractures.

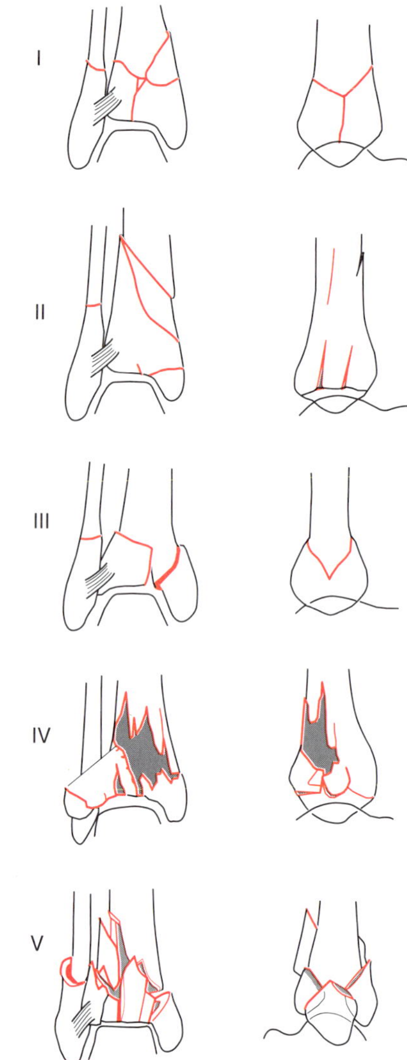

Figure 19.2. Ovadia and Beals' (1986) extended classification of distal tibial fractures.

Grade V   Displaced articular fracture with severe comminution.

It should be noted that Rüedi's first series includes injuries predominantly from skiing. Ovadia's patients were a mixed group of high-velocity motorcycle or automobile injuries with a few skiing injuries. However, later work of Rüedi and Allgower (1979) was slightly more balanced with 47% of the series having suffered injuries produced by skiing. Overall, the results were slightly poorer than his initial work.

Mast, Speigel and Pappas (1986) amalgamated systems in an effort to reach a form of classification that would give a guide to prognosis and outcome. They suggested a three part classification: Type 1 (essentially malleolar fractures that had a significant axial load at the time and, therefore, had large posterior plafond fragments); Type 2 (fractures with spiral extension (e.g. those of Maale and Seligson); and Type 3 (those fractures with central compression injuries by impaction of the talus onto the distal tibia with or without a fibula fracture and subdivided by the same method as Rüedi into groups A–C). According to Mast, the outcome of types 1 and 2 in this grouping will be better than that of group 3 as they are not pure compressive injuries and are less comminuted. However, he submitted no clinical evidence to support this.

Kellam and Waddell (1979) proposed a more prognostic classification. Type A (or rotational fractures) were the result of lower energy with minimal anterior tibial cortical comminution and large articular fragments usually associated with a short oblique or transverse fibular fracture above the level of the plafond. Type B (or compressive fractures) had severe anterior tibial cortical comminution multiple articular fragments and metaphyseal impaction often not associated with a fibular fracture. Kellam and Waddell demonstrated that the outcome of Type A fractures was 84% good compared with 53% good in Type B injuries, at 18 months follow-up.

The latest and most comprehensive classification is that of Müller and the AO group (Müller, Nazarian and Koch, 1987). They encorporate the pilon fractures into fractures of the distal tibia, as three types: A, extra-articular; B, partial articular; and C, complete articular. Each type is subdivided into three groups and then subdivided as for AO fractures into three further subgroups (see Appendix 2). In this classification all fracture types can be adequately identified. The prognosis for a poor result is said to increase from group 1 to group 3 and through the subgroups. There are no direct studies to prove this and any such assumptions will have been made based on previous work using different classification systems.

Outcome

McFerran et al. (1992) retrospectively reviewing 52 plafond fractures, classified according to the Rüedi classification. They showed that the incidence of complications was very similar between the three groups with respect to wound breakdown and infection. Sadly, no data were presented on the long-term follow-up of the patients.

Bone, McNamara and Seibel (1993), in 20 patients, showed a zero rate of infection (0%), even within the severe open fractures, compared with a 30% rate found by Ovadia and Beals (1986), suggesting that the use of an external fixator rather than primary fixation is beneficial in these fractures that may be associated with severe soft-tissue damage. Thus, method of treatment and techniques may be as important to prognosis as classifying the fracture correctly. Ovadia and Beals (1986) claimed to be evaluating the difference between alternative methods of treatment (including the use of external fixation). Of 145 fractures, 80 were treated with internal fixation and 65 by other methods. Of this latter group, only five were treated with external fixation while the rest were treated with a variety of internal stabilizing techniques and only 32 being treated purely with closed reduction in a cast. They demonstrate that functional results were much worse as severity of fracture progressed from grade 1 to grade 5 (percentage of good/excellent results: 100% for grade 1, 90+ % grade 2, 89% grade 3, 50% grade 4 down to 40% for grade 5). Open reduction and internal fixation were shown to be better treatment compared with other methods, in all but the least displaced injuries. The quality of the reduction was also found to be a factor in the outcome in this study. In all fracture groups, outcome is better in those in which anatomical reduction is achieved when compared with similar grades of fracture in which there has been a failure to achieve an anatomical reduction (Bourne, Rorabeck and McNab, 1983). Teeny and Wiss (1993) found that the incidence of infection correlated with the Rüedi grade of fracture (0% grades 1 and 2, and 37% grade 3), an increasing number of patients required ankle fusion as the Rüedi grade increased (10% for grades 1 and 2 and 26% for grades 3). Lower levels of other complications such as skin sloughing with lower grade fractures and results when classified into outcome (poor, fair to good/excellent) outcome correlated well with the initial radiograph/Rüedi grading. Again, good results were found after Rüedi grading the plafond fractures of Etter and Ganz with satisfactory results in 86% of grades 1–2 and 80% of grade 3. Helfet (1994) also classified his tibial plafond fractures according to Rüedi, showing that a greater percentage of good results and a lower percentage of poor or adequate fractures were in the grade 2 compared with grade 3.

Grade 3 pilon fractures may contain quite a heterogeneous group of fractures varying from relatively low velocity with minimal soft-tissue damage to high velocity injuries with severe soft-tissue damage. Comparison of the reported results of operative intervention on grade 3 fractures give 62% good results (Rüedi and Allgower, 1979), while Kellam and Waddell (1979) and Bourne, Rorabeck and McNab (1983) achieve only 25–37%. This may highlight an advantage for the Ovadia classification which grades the more complex fractures slightly more carefully and may relate more closely to the amount of energy involved in creating the injury.

Appendix 1

Lauge-Hansen system of fracture classification (Figure 19.3)

Supination/Adduction

Stage 1 Transverse fracture of the lateral malleolus at or below the tibial plafond or a rupture of the lateral collateral ligament.

Stage 2 A vertical fracture of the medial malleolus.

Supination/Eversion

Stage 1 Rupture of the anterior tibiofibular ligament (occasionally with avulsion of the tubercle of Chaput).

Stage 2 A spiral oblique (helical) fracture of the fibula begining at the joint line and running vertically.

Stage 3 Rupture of the posterior tibiofibular ligament with or without a fracture of the posterior malleolus.

Stage 4 An avulsion fracture of the medial malleolus or rupture of the deltoid ligament.

Pronation/Abduction

Stage 1 Avulsion fracture of the medial malleolus or rupture of the deltoid ligament.

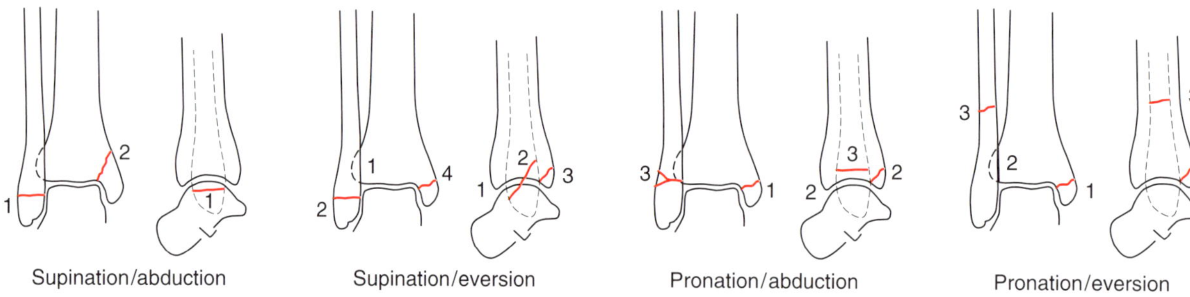

Figure 19.3. Lauge–Hansen system of distal tibial fracture classification.

Stage 2 Rupture of the anterior and posterior tibiofibular ligaments, often with a fracture of the posterior malleolus.

Stage 3 Short oblique fracture of the fibula beginning at the joint line, there may be lateral comminution.

Pronation/Eversion

Stage 1 An avulsion fracture of the medial malleolus or rupture of the deltoid ligament.

Stage 2 Rupture of the anterior tibiofibular ligament and interosseous membrane.

Stage 3 Oblique fracture of the fibula at some distance proximal to the lateral malleolus (usually 7–8 cm).

Stage 4 A fracture of the posterior malleolus secondary to avulsion by the posterior inferior tibiofibular ligaments.

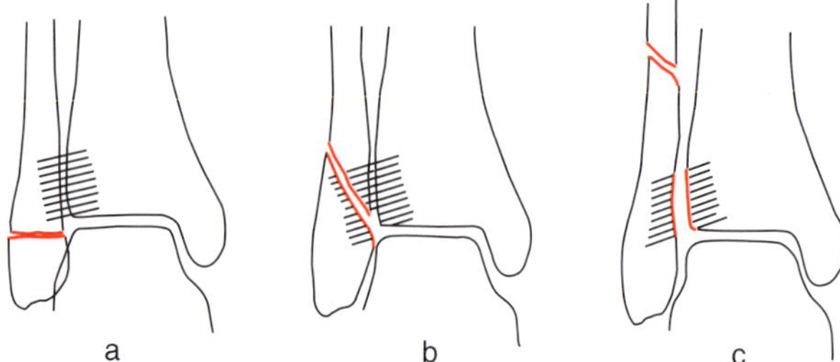

Figure 19.4. Classification of luxation fractures of the ankle by Weber (1972).

Danis Weber and (AO) classifications (Figure 19.4)

Type A

Primarily caused by forced supination and result in fibular fracture below the level of the tibiotalar joint. Fibular fracture shows a transverse avulsion fracture at or below the level of the ankle joint or rupture of the lateral collateral ligament (medial malleolar fractures if present are almost vertical in their orientation). Posterior edge of the tibia is nearly always intact but when a posterior malleolar fracture occurs it is on the medial side and is an extension of the medial malleolar fracture. Tibiofibular ligamentous complex is always intact.

Type B

Fractures occur at the level of the syndesmosis and involve a 50% chance of injury to the syndesmosis. Hypothesized mechanism is forced outward rotation of the talus. Fibular fracture shows a helical fracture at the level of the syndesmosis. The medial malleolus is intact or has a horizontal avulsion fracture or rupture of the deltoid ligament. In the tibiofibular ligament complex the interosseous membrane remains intact, the posterior tibiofibular ligament may be disrupted or intact as a result of the avulsion of its attatchment to the tibia. The anterior syndesmosis remains intact if the spiral fracture starts below the level of the ankle joint. If the fracture begins at the level of the ankle joint the anterior syndesmosis may be partially or completely torn, the syndesmotic ligament may also be avulsed from either its tibial or fibular attachment. If a posterior malleolar fracture is present then it is on the lateral side and is an avulsion off the Volkmann triangle.

Type C

Fibular fracture above the syndesmosis and as far as the fibular head with subsequent syndesmotic injury. This is primarily an external rotation injury. The fibula may be fractured anywhere from the syndesmosis to the fibular head. Rarely a proximal tibiofibular joint dislocation results. The medial malleolus has a horizontal avulsion or a rupture of the deltoid. At the posterior edge of the tibia there is always disruption of the posterior tibiofibular ligament. Posterior malleolar fractures would be as type B.

AO modification of the Danis Weber classification (Figure 19.5)

The more recent modification of this classification by the AO group is into a more precise subdivision of each three groups into three further sets 1–3 and each of these into three subgroups.

A = Infra syndesmotic lesion
  A1 Infra syndesmotic lesion, isolated
    .1 Rupture of the lateral collateral ligament
    .2 Avulsion of the tip of the lateral malleolus
    .3 Transverse fracture of the lateral malleolus
  A2 Infra syndesmotic lesion, with fracture of the medial malleolus
    .1 Rupture of the lateral collateral ligament
    .2 Avulsion of the tip of the lateral malleolus
    .3 transverse fracture of the lateral malleolus
  A3 Infra syndesmotic lesion, with posteromedial fracture
    .1 Rupture of the lateral collateral ligament
    .2 Avulsion of the tip of the lateral malleolus
    .3 Transverse fracture of the lateral malleolus
B = Trans-syndesmotic fibular fracture
  B1 Trans-syndesmotic fibular fracture, isolated
    .1 Simple
    .2 Simple with rupture of the anterior syndesmosis
    .3 Multifragmentary
  B2 Trans-syndesmotic fibular fracture, with medial lesion
    .1 Simple with rupture of the medial collateral ligament and rupture of the anterior syndesmosis
    .2 Simple with fracture of the medial malleolus and rupture of the anterior syndesmosis
    .3 Multifragmentary
  B3 Trans-syndesmotic fibular fracture, with medial lesion and a Volkmann (fracture of the posterolateral rim)
    .1 Fibula simple with rupture of the medial collateral ligament
    .2 Fibula simple with fracture of the medial malleolus
    .3 Fibula multifragmentary with fracture of the medial malleolus
C= Suprasyndesmotic lesion
  C1 Suprasyndesmotic lesion, diaphyseal fracture of the fibula, simple
    .1 With rupture of the medial collateral ligament
    .2 With fracture of the medial malleolus
    .3 With fracture of the medial malleolus and a Volkmann (= Dupytren)
  C2 Suprasyndesmotic lesion, diaphyseal fracture of the fibula, multifragmentary
    .1 With rupture of the medial collateral ligament
    .2 With fracture of the medial malleolus
    .3 With fracture of the medial malleolus and a Volkmann (= Dupytren)
  C3 Suprasyndesmotic lesion, proximal fibular lesion
    .1 Without shortening, without Volkmann
    .2 With shortening, without Volkmann
    .3 Medial lesion and a Volkmann

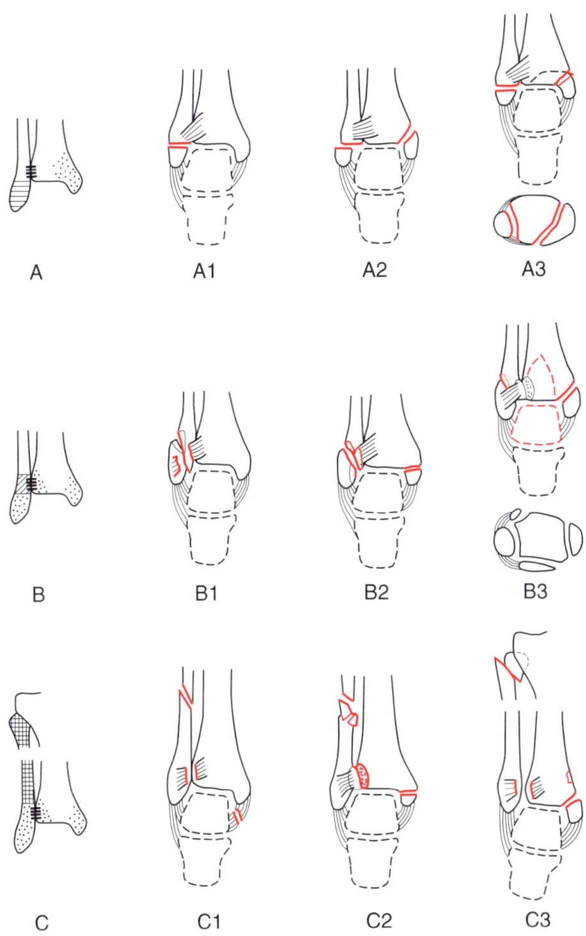

Figure 19.5. AO/ASIF classification of ankle fractures (Dawes and Weber 1990).

Appendix 2

AO classification of distal tibial fractures including pilon fractures (Figure 19.6)

A = Extra-articular fracture
  A1 Extra-articular fracture, metaphyseal simple
    .1 Spiral
    .2 Oblique
    .3 Transverse

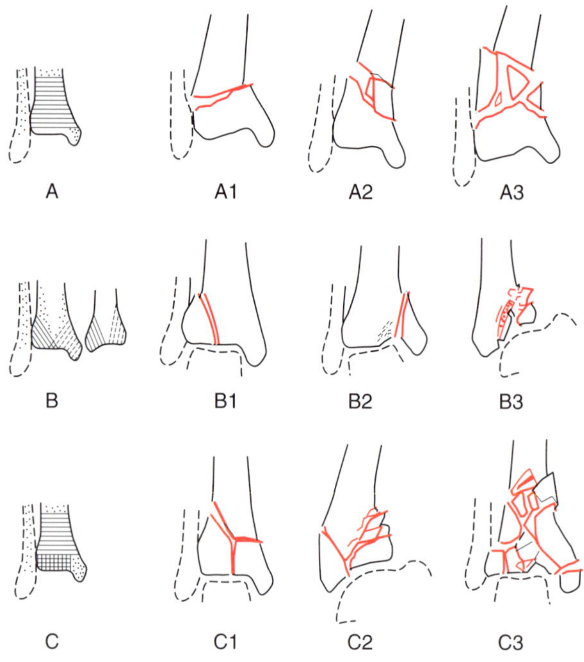

Figure 19.6. AO/ASIF classification of distal tibial fractures (Weber 1990).

A2 Extra-articular fracture, metaphyseal wedge
.1 Posterolateral impaction
.2 Anteromedial wedge
.3 Extending into the diaphysis
A3 Extra articular fracture, metaphyseal complex
.1 Three intermediate fragments
.2 More than three intermediate fragments
.3 Extending into the diaphysis
B = Partial articular fracture
B1 Partial articular fracture, pure split
.1 Frontal
.2 Sagittal
.3 Metaphyseal multifragmentary
B2 Partial articular fracture, split depression
.1 Frontal
.2 Sagittal
.3 Of the central fragment
B3 Partial articular fracture
.1 Frontal
.2 Sagittal
.3 Metaphyseal multifragmentary
C = Complete articular fracture
C1 Complete articular fracture, articular simple, metaphyseal simple
.1 Without depression
.2 With depression
.3 Extending into the diaphysis
C2 Complete articular fracture, articular simple, metaphyseal multifragmentary
.1 With assymetric impaction
.2 Without asymmetric impaction
.3 Extending into the diaphysis
C3 Complete articular fracture, multifragmentary
.1 Epiphyseal
.2 Epiphyseo-metaphyseal
.3 Epiphyseo-metaphyseo-diaphyseal

References

Ashurst, A.P.C. and Bromer, R.S. (1922) Classification and mechanism of fractures of the leg bones involving the ankle. *Arch. Surg.*, **4**, 51–129.

Baek Kristensen, T. (1949) Treatment of malleolar fractures according to Lauge-Hansen's method. Preliminary results. *Acta Chir. Scand.*, **97**, 362–379.

Bauer, M., Bergstrom, B., Hemborg, A. and Sandegard, J. (1985) Malleolar fractures; non-operative versus operative treatment. A controlled study. *Clin. Orthop.*, **199**, 17–27.

Bauer, M., Jonsson, K. and Nilsson, B. (1985) Thirty year follow up of ankle fractures. *Acta Orthop. Scand.*, **56**, 103–106.

Bauer, M., Bengner, U., Johnell, O. and Redlund-Johnell, I. (1987a) Supination – eversion fractures of the ankle joint; changes in incidence over 30 years. *Foot and Ankle*, **8**, 26–28.

Bauer, M., Johnell, O., Redlund-Johnell, I. and Johnsson, K. (1987b) Ankle fractures. *Foot and Ankle*, **8**, 23–25.

Bone, L., McNamara, K. and Seibel, R. (1993) External fixation of severely comminuted and open tibial pilon fractures. *Clin. Orthop. Rel. Res.*, **292**, 101–107.

Bourne, R.B., Rorabeck, C.H. and McNab, J. (1983) Intra-articular fractures of the distal tibia: The pilon fracture. *J. Trauma*, **23**, 591–596.

Broos, P.L.O. and Bisschop, A.P.G. (1991) Operative treatment of ankle fractures in adults; correlation between types of fractures and final results Injury. *Br. J. Accident Surg.*, **22**, 403–406.

Burwell H.N. and Charnley A.D. (1965) The treatment of displaced fractures of the ankle by rigid internal fixation and early joint movement. *J. Bone Joint Surg.*, **47B**, 634–659.

Carothers, C.O. (1955) Clinical significance of a classification of epiphyseal injuries at the ankle. *Am. J. Surg.*, **89**, 879–889.

Cedell, C.A. (1967) Supination outward rotation injuries of ankle. *Acta Orthop. Scand.*, Suppl. 110.

Cedell, C.A. (1985) Is closed treatment of ankle fractures advisable? *Acta Orthop. Scand.*, **56**, 101–102.

Crawford, A.H. (1995) Ankle fractures in children. *Int. Course Lect.*, **44**, 317–324.

Daly, P.J., Fitzgerald, R.H., Melton, J. and Ilstrup, D.M. (1987). Epidemiology of ankle fractures in Rochester, Minnesota. *Acta Orthop. Scand.*, **58**, 539–544.

Danis, R. (1949) Les Fractures malleolaires. In *Theorie et*

Pratique de L'osteosynthese (R. Danis, ed.). Paris: Masson et Cie, pp. 133–165.

Denham, R.A. (1964) Internal fixation for unstable ankle fracture. *J. Bone Joint Surg.*, **46B**, 206–211.

Gustilo, R.B. and Anderson, J.T. (1976) Prevention of infection in the treatment of one thousand and twenty-five open fractures of long bones. Retrospective and prospective analyses. *J. Bone Joint Surg.*, **58A**, 453–458.

Harper, M.C. (1992) Ankle fracture classification systems; a case for integration of the Lauge-Hansen and AO-Danis-Weber schemes. *Foot and Ankle*, **13**, 404–407.

Helfet, D.L., Koval, K., Pappas, J., Sanders, R.W. and Di Pasquale, T. (1994) Intra-articular 'pilon' fractures of the tibia. *Clin. Orthop.*, **298**, 221–228.

Kellam, J.F. and Waddell, J.P. (1979) Fractures of the distal tibial metaphysis with intra articular extension—the distal tibial explosion fracture. *J Trauma*, **19**, 593–595.

Kristensen, K.D. and Hansen, T. (1985) Closed treatment of ankle fractures. Stage 2 supination eversion fractures followed for 20 years. *Acta Orthop. Scand.*, **56**, 107–109.

Lauge, N. (1948) Fractures of the ankle; 1 Analytical historical survey as the basis of new experimental, roentgenological and clinical investigations. *Arch. Surg.*, **56**, 259–317.

Lauge-Hansen, N. (1949) 'Ligamentous' ankle fractures: Diagnosis and treatment. *Acta Chir. Scand.*, **97**, 544–550.

Lauge-Hansen, N. (1950) Fractures of the ankle. Combined experimental–surgical and experimental–roentgenologic investigations. *Arch. Surg.*, **60**, 957–985.

Lauge-Hansen, N. (1952) Fractures of the ankle. Clinical use of genetic roentgen diagnosis and genetic reduction. *Arch. Surg.*, **64**, 488–500.

Lauge-Hansen, N. (1953) Fractures of the ankles V: Pronation–dorsiflexion fracture. *Arch. Surg.*, **67**, 813–820.

Lindsjo, U. (1981) Operative treatment of ankle fracture. *Acta Orthop. Scand*, **52** (Suppl.), 189.

Lindsjo, U. (1985) Classification of ankle fractures: the Lauge-Hansen or the AO system? *Clin. Orthop.*, **199**, 12–16.

Maale, G. and Seligson, D. (1980) Fractures through the distal weight bearing surface of the tibia. *Orthopaedics*, **3**, 517–521.

Mann, D.C. and Rajmaira, S. (1990) Distribution of physeal and nonphyseal fractures in 2650 long bone fractures in children aged 0–16 years. *J. Paediatr. Orthop.*, **10**, 713–716.

Mast, J.W., Speigel, P.G. and Pappas, J.N. (1986) Fractures of the tibial pilon. *Clin. Orthop. Rel. Res.*, **230**, 68–82.

McFerran, M.A., Smith, S.W., Boulas, H.J. and Schwartz, H.S. (1992) Complications encountered in the treatment of Pilon fractures. *J. Orthop. Trauma*, **6**, 195–200.

Müller, M.E., Nazarian, S. and Koch, P. (1987) *AO Classification of Fractures*. Berlin: Springer-Verlag, pp. 170–179.

Nielsen, J.O., Dons-Jensen, H. and Sorensen, H.T. (1990) Lauge-Hansen classification of malleolar fractures; an assessment of the reproducibility in 118 cases. *Acta Orthop. Scand.*, **61**, 385–387.

Ovadia, D.N. and Beals, R.K. (1986) Fractures of the tibial plafond. *J. Bone Joint Surg.*, **68A**, 543–551.

Rasmussen, S., Madsen, P.V. and Bennicke, K. (1993) Observer variation in the Lauge-Hansen classification of ankle fractures. *Acta Orthop. Scand.*, **64**, 693–694.

Rüedi, Th. and Allgower, M. (1969) Fracture of the lower end of the tibia into the ankle joint. *Injury*, **1**, 92–97.

Rüedi, Th. and Allgower, M. (1979) The operative treatment of intra-articular fractures of the lower end of the tibia. *Clin. Orthop.*, **138**, 105–110.

Salter, R.B. and Harris, W.R. (1963) Injuries involving the epiphyseal plate. *J. Bone Joint Surg.*, **45A**, 587–622.

Seligson, D. and Frewin, P. (1986) Ankle fractures classification as a guide to treatment. *Unfallchirurg.*, **89**, 1–7.

deSouza, L.J. (1993) Fractures and dislocations about the ankle. In *Fractures and Dislocations* (R.B. Gustilo, R. F. Kyle and D. Templeman, eds), Ch. 30. St Louis: Mosby.

Speigel, P.G., Cooperman, D.R. and Laros, G.S. (1978) Epiphyseal fractures at the distal ends of the tibia and fibula: A retrospective study of two hundred and thirty-seven cases in children. *J. Bone Joint Surg.*, **60A**, 1046–1050.

Teeny, S.M. and Wiss, D.A. (1993) Open reduction and internal fixation of tibial plafond fractures variables contributing to poor results and complications. *Clin. Orthop. Rel. Res.*, **292**, 108–117.

Thomsen, N.O.B., Overgaard, S., Olsen, L.H. et al. (1991) Observer variation in the radiographic classification of ankle fractures. *J. Bone Joint Surg.*, **73B**, 676–678.

Tisa, L.M., Brandreth, D.L. and Reinherz, R.P. (1988) Identification of epiphyseal ankle injuries. *J. Foot Surg.*, **27**, 345–349.

Weber, B.G. (1972) Die Verletzungen des oberen Sprunggelenkes In *Aktuelle Probleme i der Chirurgie No. 3,* 2nd edn. Bern: Hans Huber.

Wilson, F.C. (1990) The pathogenesis and treatment of ankle fractures; Historical studies. *Inst. Course. Lect.*, **39**, 73–77.

Yde, J. (1980) The Lauge-Hansen classification of malleolar fractures. *Acta Orthop. Scand.*, **51**(1), 181–192.

Yde, J. and Kristensen, K.D. (1980a) Ankle fractures supination eversion fractures of stage 2. Primary and late results of operative and non-operative treatment. *Acta Orthop. Scand.*, **51**, 695–702.

Yde, J. and Kristensen, K.D. (1980b) Ankle fractures supination - eversion fractures of stage 4. Primary and late results of operative and non-operative treatment. *Acta Orthop. Scand.*, **51**, 981–990.

20

The foot

A. W. B. Hanna

Introduction

The foot is the Cinderella of orthopaedic surgery. Many consider the foot as a simple appendage used in locomotion; in fact it is well adapted to perform a variety of functions that make it a marvel. It has to carry the whole body weight and accommodate through the different circumstances of surface irregularity and changing speed of motion during running, jumping and walking.

The real function of the foot is to adjust to the differing ground conditions to obtain a smooth gait. Thus abnormality of foot structures such as joint stiffness, destruction of normal bony architecture or disruption of soft-tissue will have an effect on the function. The foot is a structure that is particularly vulnerable to problems because of its many articulations and its thin soft-tissue coverage, facts one must be aware of during management of injuries. Correct classification of injuries of the foot will give a clear view and help to plan the treatment of choice and predict the outcome of that treatment.

Many classifications are listed in the literature describing injuries of many parts of the foot. None of the studies have assessed either the reliability of these classifications, or recommended specific ones for use on the basis of being good, reliable tools for planning treatment and predicting outcomes of these injuries.

Anatomical Considerations

The bony morphology of the foot comprises three segments, the tarsus, the metatarsus and the phalanges which develop from proximal to distal in the embryo. From the anatomical stand point the foot has three divisions:

1. Hind foot: consisting of calcaneus and talus.
2. Midfoot: consisting of navicular, cuboid and three cuneiform bones.
3. Forefoot: consisting of five metatarsals and fourteen phalanges

The hind foot articulates with the midfoot through Chopart's joint (midtarsal joint), while the midfoot articulates with the forefoot through Lisfranc's joint. There are certain types of injuries which may involve any anatomical part of the foot. These include:

- Stress fractures: which are due to repeated minor trauma, e.g. march fractures of the second metatarsal bone.
- Neuropathic fractures and joint injuries.
- Multiple high energy injuries (crush injuries): these are often open injuries.

Injuries of the Hind Foot

The Talus

Fractures of the talus have generated great interest among surgeons because the unique anatomy of this bone makes it difficult to treat. It has no muscle attachments, only ligaments that stabilize it to all the surrounding bones; 70% of its surface is covered by articular cartilage and it has a special blood supply. The anatomical blood supply and its susceptibility to damage in certain injuries determines the fracture complications of arthritis and avascular necrosis (Figure 20.1)

Classifications dealing with fractures involving the whole talus

There are several general classifications in the literature describing talar injuries. Bohler (1936) described one of the early classifications which was descriptive without correlation between types and their prognosis (Table 20.1). Bonnin (1950) classified injuries into single, double and triple dislocation according to which joints were affected but he did not differentiate between fractures and dislocations (Mindell et al., 1963). Coltart (1952) published a schematic classification (Table 20.2) for talar injuries relating the choice of treatment to the type of fracture and grading each type according to the likelihood of late complications. Coltart's classification was the basis for describing talar fractures in many studies as well as the modifications made by other authors such as Mindell et al. (1963) (Table 20.3). Kenwright and Taylor (1970) modified Coltart's classification by adding a fourth type which is dislocation and fracture-dislocation of the midtarsal joint. There was a tendency to include all injuries of the talus in a single classification system.

Grob et al. (1985) used a combination of the classification systems of Weber (1974), Marti (1978) and Lehner (1968), the 'Marti-Weber' classification, in their study (Table 20.4; Figure 20.2) . They included injury of either the subtalar or the ankle joints in a single group and correlated the outcome to the type of fracture. This was logical in relation to the severity of the injury but making dislocation of the talus from the ankle joint equivalent to dislocation at the subtalar joint does not provide a useful measure of comparison.

Szyszkowitz, Reschauer and Seggl (1985) used another general classification for describing talar injuries. This resembles the Marti-Weber method of

Table 20.1 Bohler classification (1936)

Type 1	Simple fracture
Type 2	Fracture of the neck of the talus with angulation of the fragment with an angle opened upwards
Type 3	Fracture of the neck of the talus with foot displaced forward and to medial side underneath body of talus
Type 4	Compression fracture of the body

Table 20.2 Coltart classification (1952)

Type 1 Fractures	Chip or avulsion fractures Compression fracture of the head Simple fracture of the neck Simple fracture of the body
Type 2 Fracture-dislocations	Fracture of the body with subtalar dislocation Fracture of the neck with subtalar dislocation Fracture of the neck with posterior dislocation of the talar body, i.e. dislocation of the body from the ankle joint
Type 3 Dislocation	From the subtalar, ankle and midtarsal joint
Type 4 Miscellaneous	For example: Fracture dislocation of the ankle with subtalar dislocation Fracture neck of talus + fracture dislocation of the midtarsal joint
Type 5	Dislocation of the tarsus at the subtalar joint without fracture of the talus

Table 20.3 Mindell et al. classification (1963)

Type A Fractures	Of the talar neck, body, posterior lip
Type B Dislocation	Single, double, triple
Type C Fracture-dislocation	Fracture of the body with subtalar or total talar dislocation Fracture of the neck with subtalar or total talar dislocation

Table 20.4 The Marti-Weber classification used by Grob et al. (1985)

Type I	Minor fracture in the distal neck, flakes or avulsion fractures or fractures of the talar head
Type II	Undisplaced body or neck fractures
Type III	Fracture of the body or the neck with dislocation or subluxation of the subtalar or tibiotalar joint
Type IV	Fracture of the body or the neck with dislocation of subtalar and tibiotalar joints Extremely comminuted fractures of the body

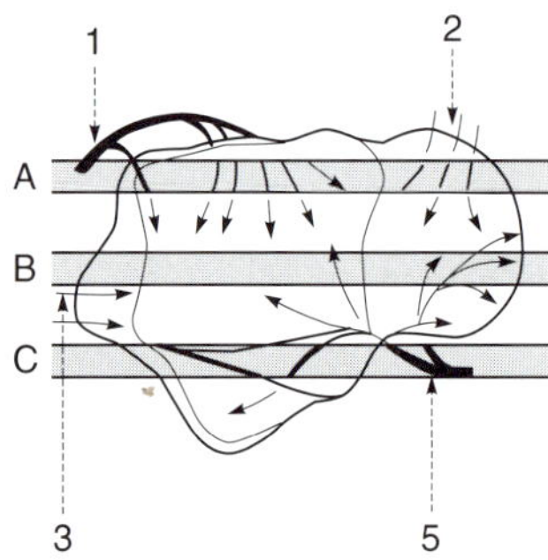

Dorsal aspect of the talus with sagittal displayed, (A) medially, (B) centrally (C) laterally

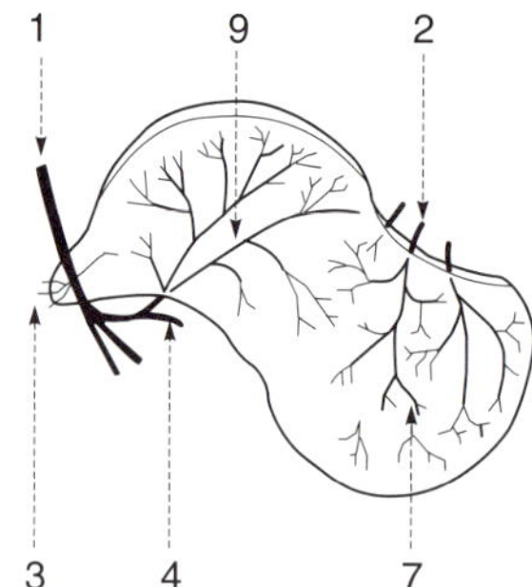

A) Blood supply of the medial one-third of the talus

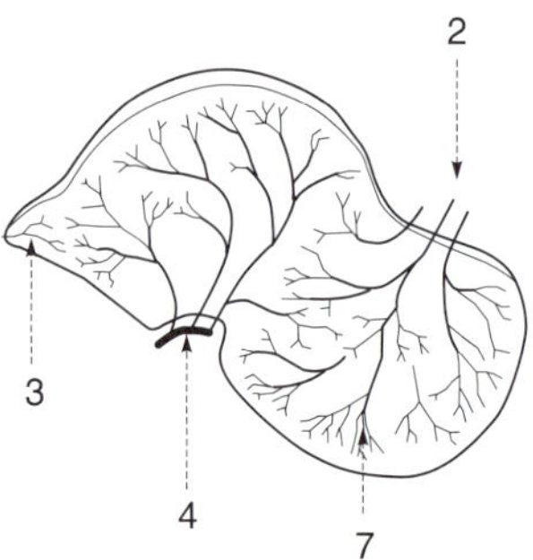

B) Blood supply of the central one-third of the talus

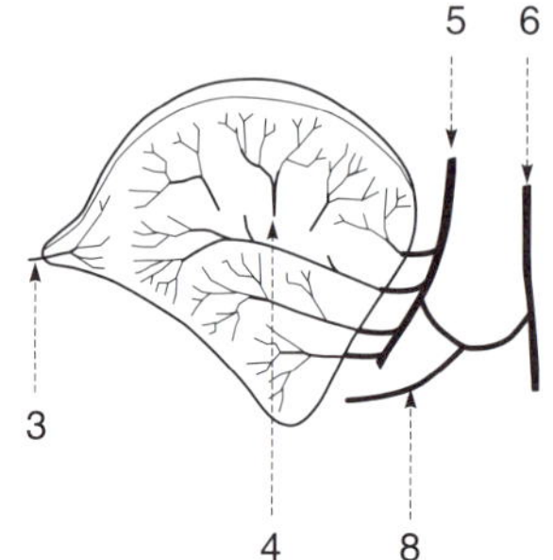

C) Blood supply of the lateral one-third of the talus

1 - Posterior tibial artery
3 - Posterior tubercle branches
5 - Perforating peroneal artery
7 - Tarsal sinus branches
9 - Deltoid branches

2 - Dorsalis pedis branches
4 - Artery to tarsal canal
6 - Lateral tarsal artery
8 - Artery of the tarsal sinus

Figure 20.1. Blood supply of the talus (Grob et al., 1985).

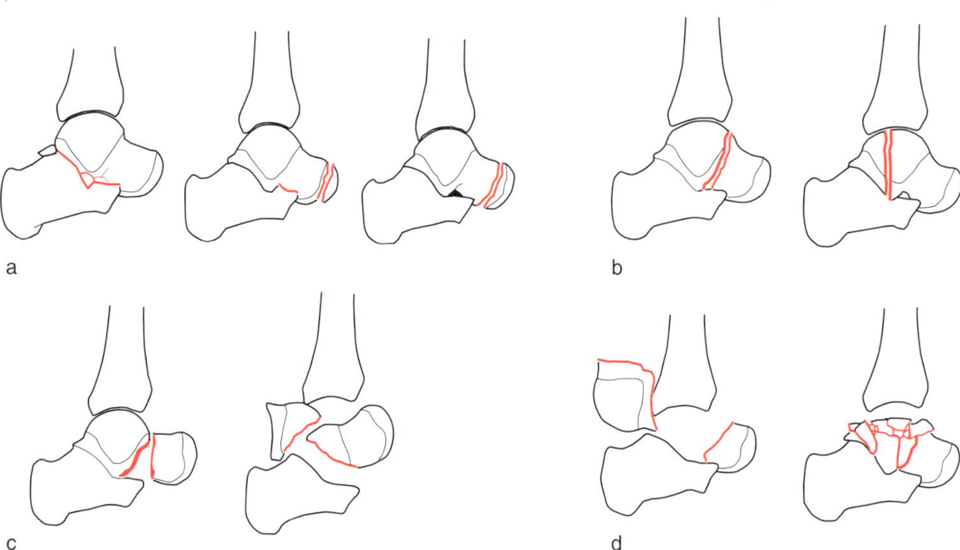

Figure 20.2. Marti–Weber classification of talar fractures used by Grob et al. (1985).

placing patterns of fractures that seem to behave in the same way for the same grade. Proximal neck and body fractures are placed in a single grade called 'central' fractures; they separate ankle dislocations from subtalar dislocations. Their classification was based on fracture pattern and the final outcome as measured by avascular necrosis and post-traumatic arthritis (Table 20.5; Figure 20.3). Their study of 85 talar fractures with a long term follow up of 5 to 8 years, gives an excellent review. It includes the incidence of neck fractures relative to those of the body; joint involvement with both talar neck and body fractures; and the incidence of open injury in each type (Table 20.6).

Frawley, Hart and Young (1995) used the same classification in a retrospective assessment of 26 cases with a follow-up of 2–9 years. They did not assess its observer reliability. They concluded that the Szyszkowitz classification is a good predictor for the outcome of these injuries.

Table 20.5 Szyszkowitz et al. classification (1985)

Type I	Peripheral fractures: • Including lateral process, posterior process, distal neck and head fractures • Circulation is intact • No avascular necrosis
Type II	Central fracture without displacement: • Proximal neck and body • Circulation mainly intact • Seldom avascular necrosis occurs
Type III	Central fracture with displacement: • The proximal neck or body fragments are displaced with subluxation of the subtalar joint and normal ankle joint • Interrupted intraosseous circulation • Avascular necrosis usually occurs

Classifications dealing with injuries involving different parts of the talus

a. Talar neck injuries

Hawkins (1970) used Coltart's (1952) landmark between body and neck to define talar neck fractures. Body fractures pass through the trochlea involving both the ankle joint and the posterior articular facet

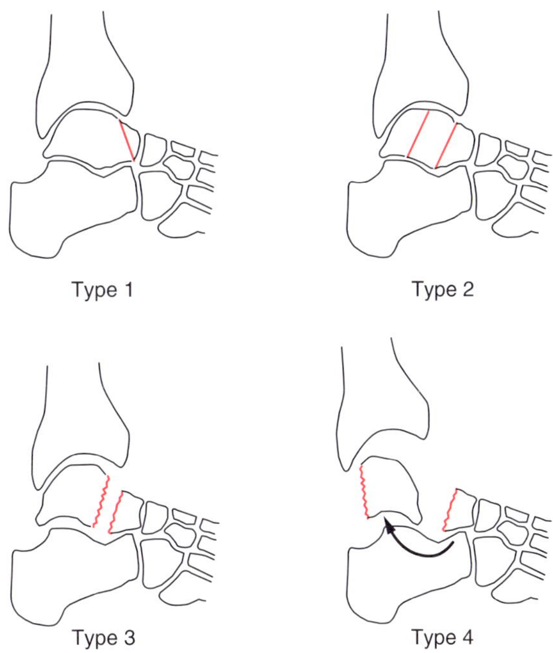

Figure 20.3. Szyszkowitz et al. classification (1985).

Table 20.6 Incidence types of talar fractures in study by Szyszkowitz et al. (1985)

	Type of fracture	*Total*	*Closed*	*Open GI*	*Open GII*	*Open GIII*
Type I	Peripheral fractures	14	12	–	1	1
Type II	Central fractures. No displacement	22	18	–	1	3
	Proximal neck	8	6	–	–	2
	Body	14	12	–	1	1
Type III	Central fracture + displacement	15	14	–	1	–
	Proximal neck	9	8	–	1	–
	Body	6	6	–	–	–
Type IV	Dislocation fractures	34	20	1	8	5
	Neck	23	16	1	5	1
	Body	9	3	–	2	4

of the posterior subtalar joint. Neck fractures lie between the posterior and middle articular facets of the talus and do not reach the ankle joint. Adelaar's (1989) illustration of the demarcation between the two parts makes the difference clear (Figure 20.4).

Hawkins devised a classification based on the degree of the injury and the blood supply of each part. He recognized avascular necrosis as the most important complication to affect the outcome of these injuries. The classification was used to define treatment (Table 20.7; Figure 20.5).

This classification has not been tested for observer reliability or evaluated for its ability to predict outcome. However, it is widely used to define and manage talar neck fractures (Peterson, Goldie and Irstam, 1977; Penny and Davis, 1980; Freund, 1988; Adelaar, 1989).

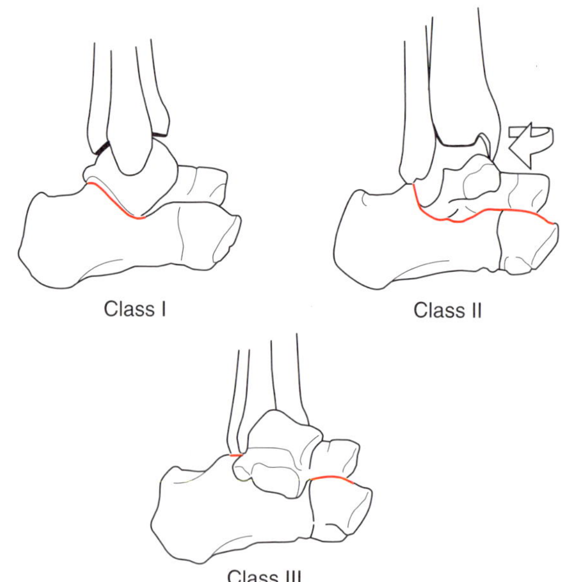

Figure 20.5. Hawkins (1970) classification of talar neck fractures.

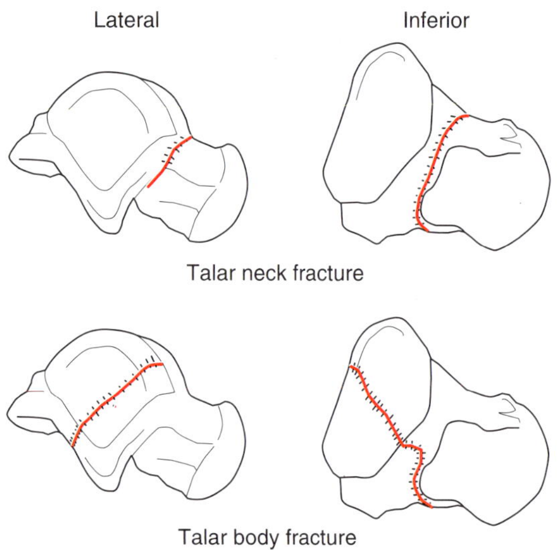

Figure 20.4. Fractures of the talar neck do *not* involve the posterior articular facet. Talar body fractures *do* involve the articular facet (Adelaar, 1989).

Modifications of Hawkins' classification

Hawkins' classification has been modified by some authors. Canale and Kelly (1978) added another grade where the fracture is accompanied by a dislocation of the head of the talus from the talonavicular joint (type IV). They state that it could accompany type II or III injuries and they expected 100% avascular necrosis and consequently poor results. Since then some studies have used this classification as a guide for description and management of this injury (Comfort et al., 1985; Daniels and Smith, 1993; Baumhauer and Alvarez, 1995).

Bodamer et al. (1987) reported four cases of type IV injury with avascular necrosis of the head without body involvement. Daniels and Smith (1993) commented that avascular necrosis may occur in the head as well as the body. Khazim and Salo (1995) reported a case of avascular necrosis of the head alone after fracture neck of talus with only talonavi-

Table 20.7 Hawkins classification (1970)

Group I	Undisplaced vertical fracture of the talar neck. Fracture line enters subtalar joint between middle and posterior facets. No injury to blood vessels. No incidence of avascular necrosis
Group II	Displaced fracture of talar neck with subtalar joint subluxed or dislocated but with normal ankle. The fracture line frequently enters a portion of the body and posterior facet of the talus. Two of the main sources of blood supply are interrupted. Usually there is avascular necrosis
Group III	Displaced neck fracture with the talus dislocated from both subtalar and ankle joint. Talar head is in normal relation with the navicular. Very high incidence of avascular necrosis

cular dislocation but no disruption of the ankle and subtalar joint. Schatzker and Tile (1993) did not agree about the importance of type IV injuries. They believed that the main essence of the injury is the talar body dislocation.

Comfort et al. (1985) put emphasis on the displacement of the talus from its joints to grade the severity of injury rather than on the fracture displacement (Table 20.8). They noticed that some of Hawkins' type I cases (judged by the clinical photographs published in the original paper) showed mild displacement although Hawkins' description was undisplaced; in other words, they believed that Hawkins' description was inaccurate since all displaced fractures must have subluxation or dislocation from subtalar or tibiotalar joints. They also recommended open reduction and internal fixation of displaced type I fractures in contrast to others who advised conservative treatment. In their study they had 100% good or excellent final results of type I fracture in comparison with Peterson, Goldie and Irstam (1977) and Penny and Davis (1980) who had 75% and 40% final good and excellent results for type I fractures respectively. This modification is now generally understood as the Hawkins' classification.

Although Hawkins' classification is the most commonly used, there remains debate about its validity in determining prognosis. Baumhauer and Alvarez (1995) believed that the degree of comminution (which is not covered by Hawkins) plays an important role in defining the outcome of these fractures, as it predisposes to malunion either in malalignment or malrotation which in turn leads to post-traumatic arthritis. Canale and Kelly (1978) considered less than 5 mm residual displacement to be acceptable. Delee (1986) and Canale (1990) demonstrated a strong correlation between varus malunion and subsequent arthritis with more limitation of whole foot function.

Daniels and Smith (1993) believed that avascular necrosis of the talus does not necessarily give poor results. Peterson, Goldie and Irstam (1977) reported 75% excellent results in the presence of avascular necrosis. Similar findings have been reported by Grob et al. (1985).

Table 20.8 Comfort et al. classification (1985)

Type I	Fracture without displacement of subtalar or tibiotalar joints
Type II	Fracture + displaced subtalar + undisplaced tibiotalar joints
Type III	Fracture + displaced subtalar and tibiotalar joints
Type IV	Fracture + displaced talonavicular joint

b. Talar body fractures

Sneppen et al. (1977) devised a classification for describing injuries of the talar body recognising five types (Table 20.9; Figure 20.6) and suggesting a plan of treatment for each group. They discussed the variants for each type and tried to relate the prognosis to the type of fracture (Table 20.10). Types B and E had the worst prognosis. They also stressed the importance of types C and D and advised more care in treating these fractures as they are often complicated by arthritis of the subtalar joint (5% in type C and 11% in type D).

This classification has been used and modified by others. The original classification distinguished group A (compression fractures) from transchondral fractures. Baumhauer and Alvarez (1995) and Adelaar (1989) considered group A as transchondral fractures.

Type A transchondral fractures were previously called osteochondritis dissecans (Kappis, 1922). The name implies an inflammatory condition of cartilage and bone. Berndt and Harty (1959) found a high proportion of these cases (80–92%) are due to trauma to the ankle and changed the name to transchondral fractures. Their classification has four types, according to the site of the lesion and mechanism of injury (Table 20.11; Figure 20.7). The classification has been

Table 20.9 Sneppen's et al. classification (1977)

a. Compression fractures	Involves only the medial or lateral part of trochlea in the ankle joint
b. Shear fractures of the body	(i) Coronal or (ii) sagittal types. Coronal type, may be mistaken with neck fractures but fracture line is more posterior involving the trochlea
c. Fracture of the lateral process	
d. Fracture of the posterior process	
e. Crush fractures	Entails fragmentation of entire trochlea with massive incongruence in both talocrural and subtalar joints

Table 20.10 Incidence of complications with each body fracture type in the study by Sneppen et al. (1977)

Type of fracture	*Union with post-traumatic arthritis*		*Union with avascular necrosis*		*Incidence of severe disability*	
	Complicated cases	Total no. of group	Complicated cases	Total no. of group	Complicated cases	Total no. of group
Type A	5	10	–	–	2	12
Type B	11	17	5 Dislocated cases 13	17 Dislocated cases 17	11	17
Type C	8	11	–	–	0	11
Type D	1	9	–	–	0	9
Type E	3	4	3	4	4	4

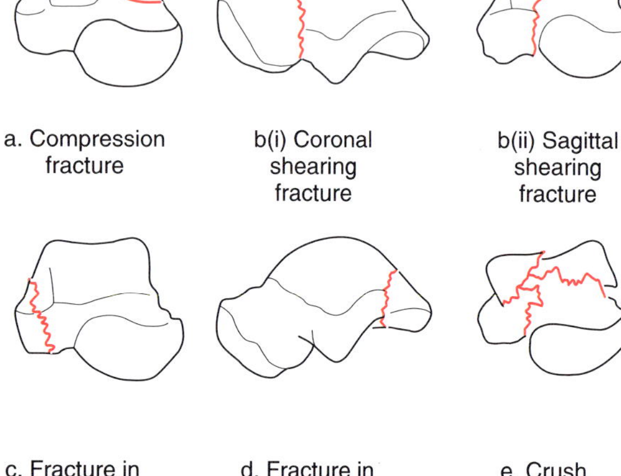

Figure 20.6. Sneppen et al. (1977) classification of talar body fracture.

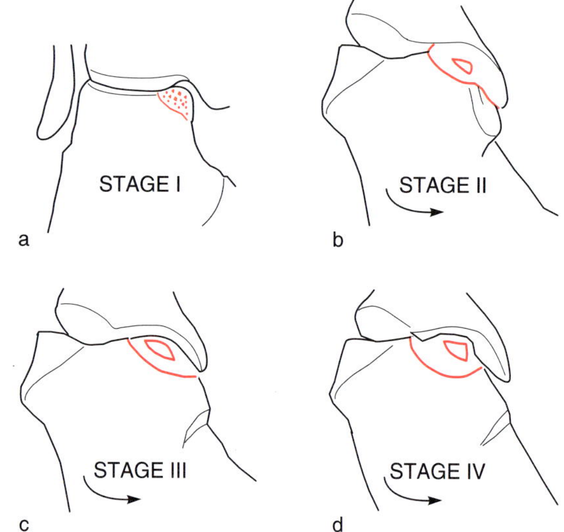

Figure 20.7. Stages of transchondral medial lesion of talar body (Cavale and Gelding, 1980).

Table 20.11 Berndt and Harty classification (1959)

Grade	*Description*
Type I	Talar dome compression, without break in the cartilage, plus sprain of the ligament
Type II	Partially detached osteochondral fracture plus rupture of the ligament
Type III	Complete non-displaced osteochondral fracture plus rupture of the ligament
Type IV	As type 3 but with displacement of the fragment

used in more recent literature to describe and plan treatment for this type of fracture (Canale and Belding, 1980).

c. Posterior process fractures

This fracture has been classified into posterolateral tubercle fracture (Shepard, 1882) and posteromedial tubercle fracture (Cedell, 1974). Sneppen et al. (1977) grouped it as C type. Recently, Ebraheim, Skie and Podeszwa (1994), in a case report, identified another type in which the whole of the posterior process is fractured. Posterior process fractures can be best classified into:

a. Posterolateral fracture (Shepard): commonest.
b. Posteromedial fracture (Cedell): rare (usually due to avulsion of the deep component of the deltoid ligament).
c. Whole posterior process fracture: rarest.

d. Lateral process fractures

Schrock (1952) found that these injuries constitute 24% of all talar body fractures. Dimon (1961)

Table 20.12 Hawkins' classification of lateral process of talus fractures (1965)

Type 1	Simple fracture of the lateral process extending from the talofibular articulation down to the posterior talocalcaneal articulation
Type 2	Comminuted fracture involving both talofibular and posterior calcaneal articular surface of the entire lateral process
Type 3	Chip fracture of the inferior and anterior portion of posterior articular process of the talus. No extension to talofibular articulation

described the pathological anatomy of lateral process fractures as involving both the talofibular and the posterior aspect of subtalar articulation. Hawkins (1965) divided this fracture into three types (Table 20.12). This has been used in various published studies, including Mukherjee, Prigle and Baxter, 1974; Shelton and Pedowitz, 1982; Heckman and Mclean, 1985). Fehnel et al. (1994) emphasized the size and the degree of displacement of the fragment in planning treatment. They recommended open reduction and internal fixation in cases with fragments greater than 1 cm or if displaced by more than 1 mm. Sneppen et al. (1977) described lateral process fractures as one entity (type D).

e. Talar head fractures

Adelaar (1989) described fractures of the head as undisplaced or displaced and advised the following criteria for open reduction in displaced fractures:

- An articular step off (displacement undefined).
- Involvement of 50% or more of the articular surface.
- Talonavicular joint instability.

He otherwise advised conservative treatment. Most talar head fractures are not displaced because of the strong soft-tissue attachment (Baumhauer and Alvarez, 1995).

Talar Fractures in Children

No special classification has been suggested for managing paediatric talar fractures. Many authors have included cases of young children with adult cases in their series (Coltart, 1952; Hawkins, 1970; Kenwright and Taylor, 1970).

Conclusions

Two methods of classifying talar injuries are described. One is to deal with all the injuries in a general classification and the other is to deal only with specific anatomical sites of the talus, assuming that each site behaves in a certain way different from the others, e.g. neck and body fractures.

When comparing the outcome of talar neck and body fractures, no great difference can be found except for the incidence of each (neck fractures are more common). In both groups, avascular necrosis is common (48–58%). Similar results are also found for the incidence of osteoarthritis.

The two recent studies have combined proximal neck and body fractures into one group, calling them central fractures (Szyszkowitz, Reschauer and Seggl, 1985; Frawley, Hart and Young, 1995), without evidence to the contrary these simpler systems are a more favourable choice.

Dislocation of the Talus

Although dislocation or subluxation may occur at the same time as talar fractures, it may also occur without a fracture. Some classification systems have included dislocation of the talus, e.g. Coltart (1952) and Kenwright and Taylor (1970), while others preferred to look at them separately. Generally, two types of dislocation are identified:

A Subtalar dislocation

Dislocations of the distal articulation of the talus, i.e. the subtalar and the talonavicular joints with an intact ankle joint.

This lesion was also called peritalar dislocation by Barber, Briker and Haliburton (1961), it is a simultaneous dislocation of the distal articulations of the talus. Most authors divide this injury into two types:

Type I: medial dislocations

In which the foot is displaced medially with the calcaneus lying medially and the head of the talus prominent dorsolaterally. The navicular is medial or sometimes even dorsal. Grantham (1964) called this type 'basketball player's foot'.

Type II: lateral dislocations

Lateral dislocation with the deformity occurring in the opposite direction.

Type I was reported to be more common and to have a better prognosis (Delee and Curtis, 1982). Other variants have been reported in the literature. Grantham (1964) reported rare cases of anterior and posterior dislocation, accompanied by a degree of medial or lateral dislocation. Main and Jowett (1975) described a further variant and called it a 'medial swivel dislocation'. This involved dislocation of the talonavicular joint and a medial subluxation of the subtalar joint with an intact calcaneocuboid joint.

B Total talar dislocation

When the ankle joint is also involved.

This may be a medial or lateral talar dislocation. Detenbeck and Kelly (1969) reported the largest series of this type of injury (nine cases). Pinzur and Meyer (1978) reported an exceedingly rare case of posterior talar dislocation without fracture. Continuing the forces causing medial subtalar dislocation will lead to total lateral talar dislocation and vice versa (Leitner, 1955). This injury has a poor prognosis in all the published reports in the literature.

Fractures and Dislocation of the Calcaneum

Calcaneal fractures

Fractures of the calcaneus constitute 60% of all major foot injuries and 2% of all fractures. Historically, it is a testament to the complexity of the injury that well over 150 papers addressing calcaneal fractures were published in the English literature between 1720–1930 (Lindsay and Dewar, 1958). The debate continues.

Böhler (1936) was one of the early authors who published a classification on calcaneal fractures (Table 20.13). He showed the importance of the tuber joint angle (Figure 20.8) on the lateral plain

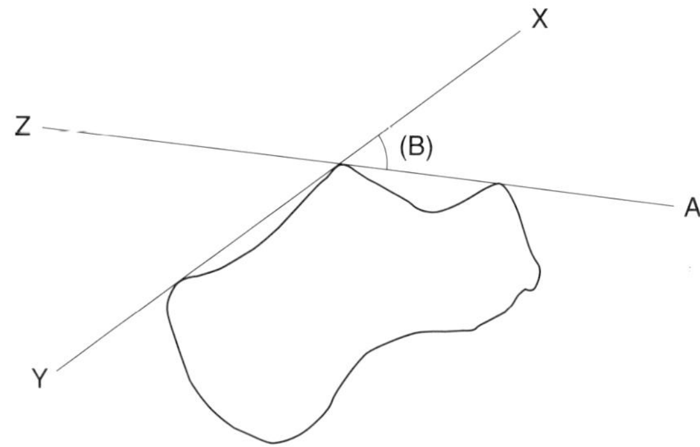

Figure 20.8. Illustration of Böhler's angle (1936). Böhlers angle (B) is the angle lying at the bisection of the two lines: XY, a line passing on the posterior–superior aspect of the calcaneus; and AZ, a line extending back from the superior aspect of the anterior part of the calcaneus to the upper midpoint.

radiograph in planning and assessing treatment. Using this classification he showed that groups 1–3 are easier to treat with a better outcome, while groups 4–8 are more difficult to treat and have a worse prognosis.

Schofield (1936) modified Böhler's classification, dividing the fractures into five groups (Table 20.14). He arranged the grades in ascending order of severity according to the degree of functional loss and used Bohler's angle to define each type and in turn to predict the final outcome.

The study of Palmer (1948) helped in understanding calcaneal fractures. He considered that a fracture is formed from simple linear fracture lines which develop from three different force types:

1. Avulsion
2. Compression
3. Shearing causing either medial or lateral shear fracture of the tuber calcanei

Table 20.13 Bohler's classification (1936)

Grade	*Description*
Group 1	Fracture of upper part of tuberosity (beak fracture)
Group 2	Fracture of medial part of tuberosity + displacement
Group 3	Fracture of sustentaculum tali alone
Group 4	Fracture of body. No displacement of joint surfaces articulating with the talus
Group 5	Fracture of the body + displacement of lateral part of joint surface of the posterior articular facet
Group 6	Fracture of the body + displacement of the whole posterior articular surface widespace seen in lateral view
Group 7	Fracture of the body with displacement of lateral portion of the posterior articular process plus subluxation of the mid tarsal joint
Group 8	Fracture body + (crushing and dislocation of anterior portion from cuboid)

Table 20.14 Schofield classification (1936)

Grade	*Description*
Type I	Including avulsion fracture and medially displaced sustentaculum tali
Type II	Fracture of body with no displacement and no articular involvement
Type III	Fracture of the medial process of tuberosity
Type IV	Fracture of the trochlear process, anterior portion of the body which may involve cuboid and talar articulation. No tuberosity injury and no disturbance of Bohler's angle
Type V	Comminuted body + displacement + compression impaction + disturbance of angle + subtalar and calcaneocuboid joint involvement

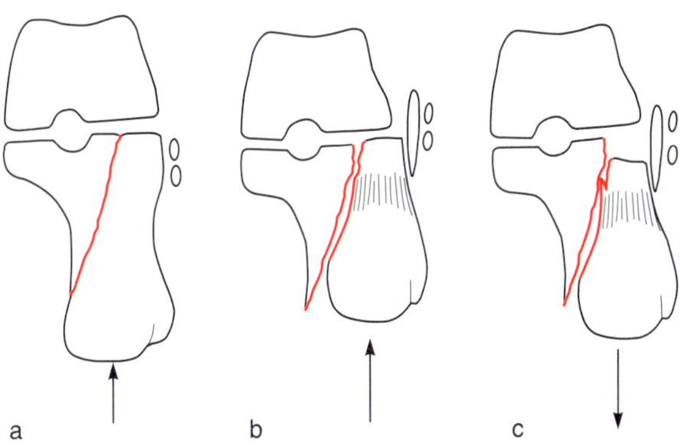

Figure 20.9. Mechanism of comminuted fractures of calcaneal fractures: (a) Primary shearing fracture; (b) compression and maximum dislocation at moment of accident; (c) the lateral block, with impacted fragment due to secondary degree compression, recoils downward, forming a ledge across the surface of the joint.

Palmer further classified fractures with extension to the posterior subtalar joint (lateral shear fractures) into two types according to the shape of the secondary fracture line and the resultant secondary compression fracture fragment:

Type I: The fracture line passes around the posterior articular facet creating a fragment formed of articular surface with a fragment of body which is dipped into the cancellous bone of the calcaneal body.

Type II: This has the secondary fracture lines passing posterior to the tuberosity. The main pathology of this injury is the impaction of the fractured fragment into the rest of calcaneal body (Figure 20.9). Palmer suggested surgical intervention which was formed of bone grafting to elevate these depressed segments without internal fixation. He recorded good results with only one failure of the 23 cases he treated.

Essex-Lopresti (1952) modified Böhler's system (Table 20.15; Figure 20.10). He recognized that the most important factor determining the outcome of these fractures is the involvement of the articular surface of the subtalar joint. He divided fractures into those involving the subtalar joint and those which do not. Classifying the fracture in this way helped to select treatment. This study recommended different methods in managing different fracture types. Tongue-type fractures were managed by semi-invasive methods (spiked reduction), while central depression types were managed operatively, elevating the depressed fragment.

Many authors still use this classification in evaluating their cases, believing that it still has the most clinical relevance (Parmer, Triffit and Cregg, 1993; Chan and Ip, 1995; Melcher et al., 1995). Others have added simple modifications but always follow the main principles of its originators.

Warrick and Bremner (1953) classified fractures into two groups: Group I are isolated (primary)

Table 20.15 Essex-Lopresti classification (1952)

I. Fractures not involving subtalar joint	*II. Fractures involving subtalar joint*
A. Tuberosity fractures:	A. Without displacement
Beak type	B. With displacement
Avulsion medial border	1. Tongue type
Vertical	2. Centrolateral depression type
Horizontal	3. Sustentaculum tali fracture alone
	4. With gross comminution from below Severe tongue and joint depression type
B. Calcaneocuboid joint only:	
Parrot nose	5. From behind forward with dislocation
Various	of subtaloid joint

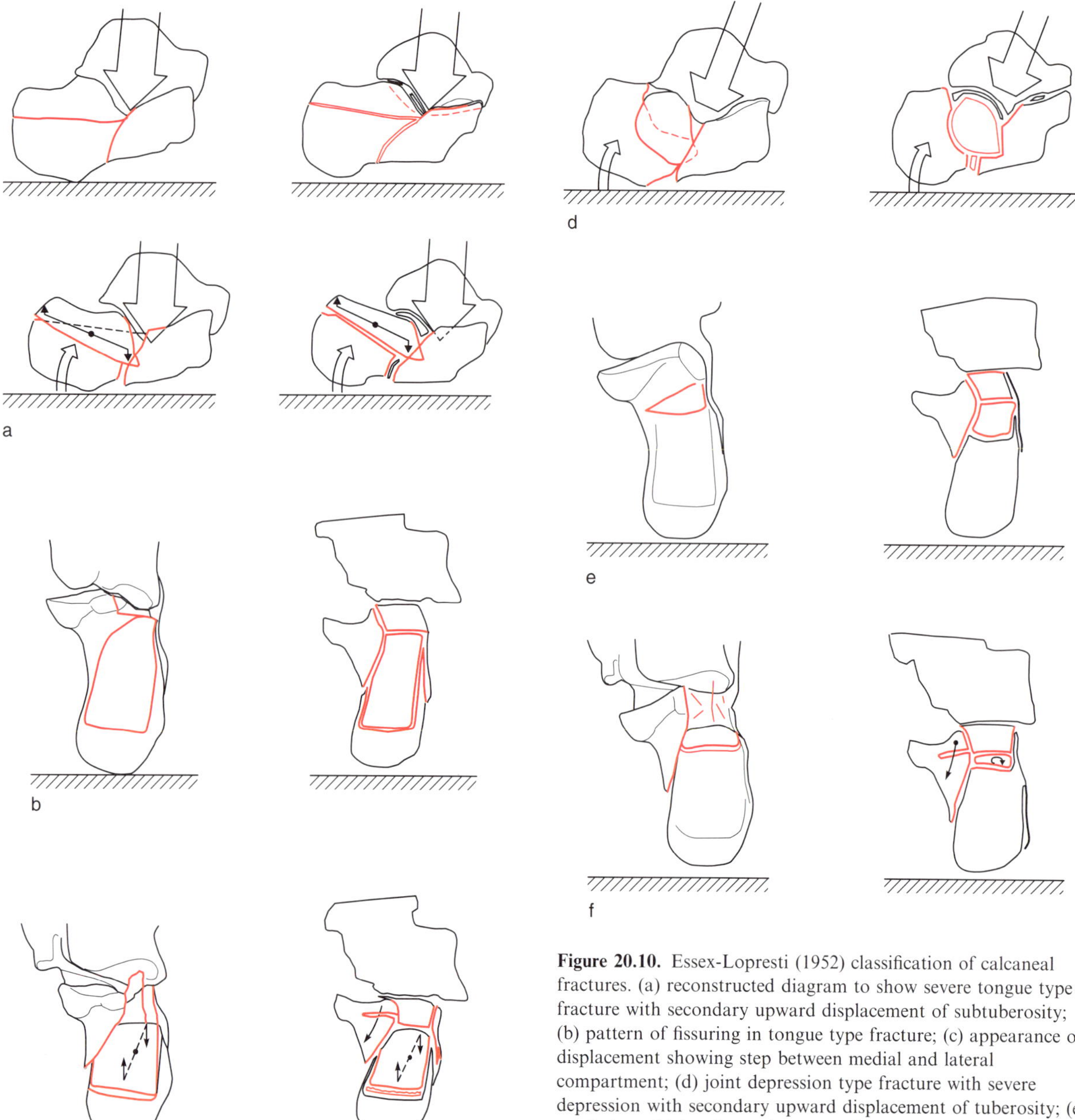

Figure 20.10. Essex-Lopresti (1952) classification of calcaneal fractures. (a) reconstructed diagram to show severe tongue type fracture with secondary upward displacement of subtuberosity; (b) pattern of fissuring in tongue type fracture; (c) appearance of displacement showing step between medial and lateral compartment; (d) joint depression type fracture with severe depression with secondary upward displacement of tuberosity; (e) axial view of joint depression type of fracture, showing pattern of fissuring; (f) displaced variety to show step between medial sustenacular component and depressed joint surface.

fractures while group II are compression shearing (secondary) fractures (Table 20.16; Figure 20.11).

They divided group II according to the number of fragments produced by the primary and secondary fracture lines. Diminution of tuber joint angle does not necessarily mean involvement of the subtalar joint. Stephenson (1987) adopted this classification.

Watson-Jones (1955) combined the work of Essex-Lopresti (1952) and Warrick and Bremner (1953) with Palmer (1948) and published another classification (Table 20.17; Figure 20.12).

Protheroe (1969) demonstrated that a beak fracture is actually a variant of avulsion fractures by showing the anatomical expanded insertion of the tendo achillis. This had been previously classified as a separate entity by Böhler (1936), Essex-Lopresti (1952), and Watson-Jones (1955). Protheroe advised surgery rather than conservative treatment.

Nade and Monahan (1973) described a classification based on that of Watson-Jones. They believed it was a simple classification adaptable to clinical problems (Table 20.18; Figure 20.13) and based on grad-

Table 20.16 Warrick and Bremner classification (1953)

Group I	*Group II*
Isolated fracture of anterior end of bone	With division of bone into two fragments: • Fracture involving posterior articular facet with no displacement • Fracture involving the facet + lateral displacement of the main lateral fragment • Not involving the articular facet but with diminution of the tuber joint angle (retrothalamic fracture of continental writer)
Isolated fracture of sustentaculum tali	With two main fragments + secondary fracture line in the main lateral fragment: • Fracture with displacement of lateral part of the posterior articular facet: *Type 1*: the articular fragment depressed into underlying bone *Type 2*: the depressed articular surface is continuous with upper part of the main lateral fragment • Fracture with displacement of the entire posterior articular facet: *Type 1*: the articular fragment depressed into underlying bone *Type 2*: the depressed articular surface is continuous with upper part of the main lateral fragment
Fracture of the tuberosity: a. Beak fracture b. Vertical fracture	

Table 20.17 Watson-Jones classification (1955)

Fractures not involving the subtalar joint	*Fractures involving the subtalar joint*
Vertical fracture of tuberosity	Adjacent but not inside the joint
Horizontal fracture (beak fracture)	With displacement of the lateral part of subtalar joint
Fracture of sustentaculum tali	With central crushing of the whole subtalar joint and may be accompanied by calcaneocuboid joint involvement
Fracture of anterior end of calcaneus	

Table 20.18 Nade and Monahan classification (1973)

Type	*Description*
Type A	Isolated fracture of calcaneus with no involvement of posterior subtalar joint (may be comminuted)
Type B	Fracture of calcaneus with minimal injury to posterior subtalar joint (may not be apparent on plain radiograph)
Type C	Comminuted fracture of calcaneus with significant displacement and involvement of posterior subtalar joint

ing the behaviour of each fracture in each group, in other words the final results of each fracture. They agreed with Thoren's (1964) conclusion that involvement of the posterior subtalar joint is the most important predictor of outcome. Their classification does not differentiate between patterns of intra-articular involvement.

Soeur and Remy (1975) stated that the distribution of cortical and cancellous bone in the calcaneus is a significant factor in the pattern of fracture and its management. They used Warrick and Bremner's work on pathological anatomy and correlated it with their own work on calcaneal structure. They stress the importance of the thalamic portion of the calcaneus (Table 20.19; Figure 20.14). Their classification described the fracture both by understanding the forces applied and by identifying the resulting fragments. They showed that, following impaction, the fragments need rotatory manipulation during reduction rather than simple elevation as previously

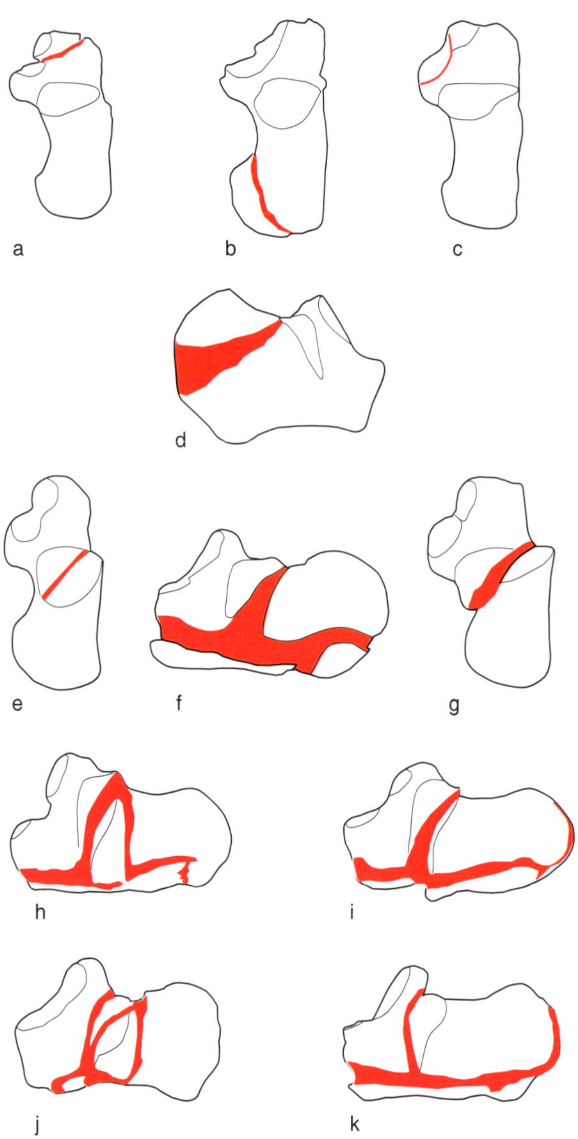

Figure 20.11. Warrick and Bremner (1953) classification of calcaneal fractures: (a) fracture of anterior end of calcaneum; involving calcaneo-cuboid joint; (b) vertical fracture of tuberosity; (c) isolated fracture of sustacular talio; (d) beak fracture of the tuberosity; (e) fracture involving, but not displacing, the posterior articular facet; (f) fracture with dimunition of the tuber-joint angle, but with no involvement of the posterior articulate facet; (g) fracture involving the posterior articular facet with lateral displacement of the main lateral fragment; (h) fracture with depression of the lateral part of the articular surface (Type 1); (i) fracture with depression of the lateral part of the articular surface (Type 2); (j) fracture with depression of the entire posterior articular process (Type 1); (k) fracture with depression of the entire posterior articular process (Type 2).

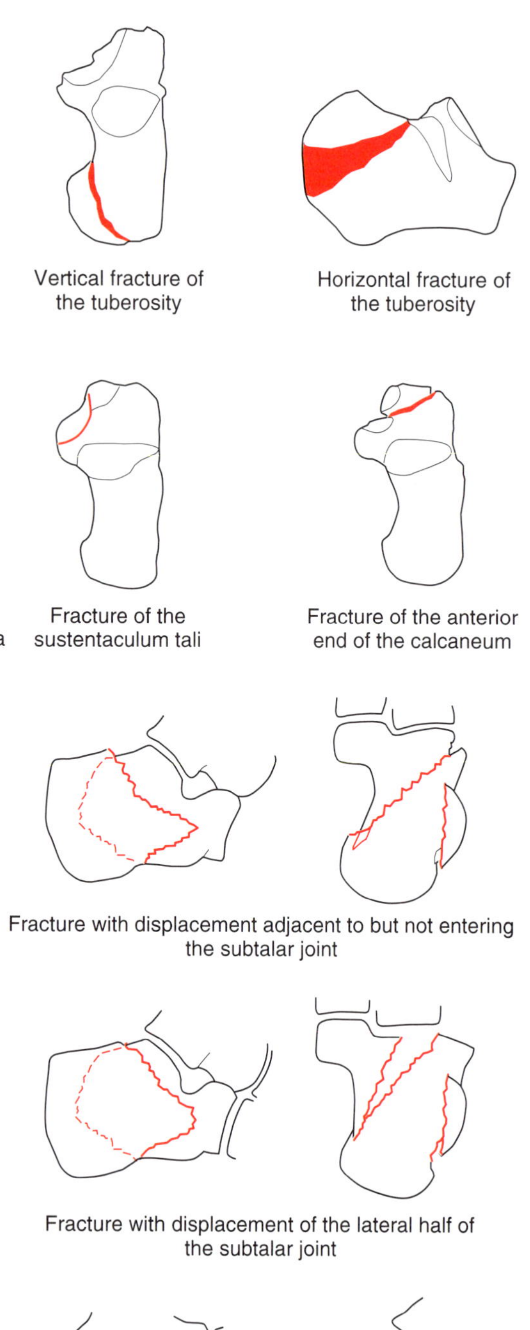

Figure 20.12. Watson-Jones (1955) classification.

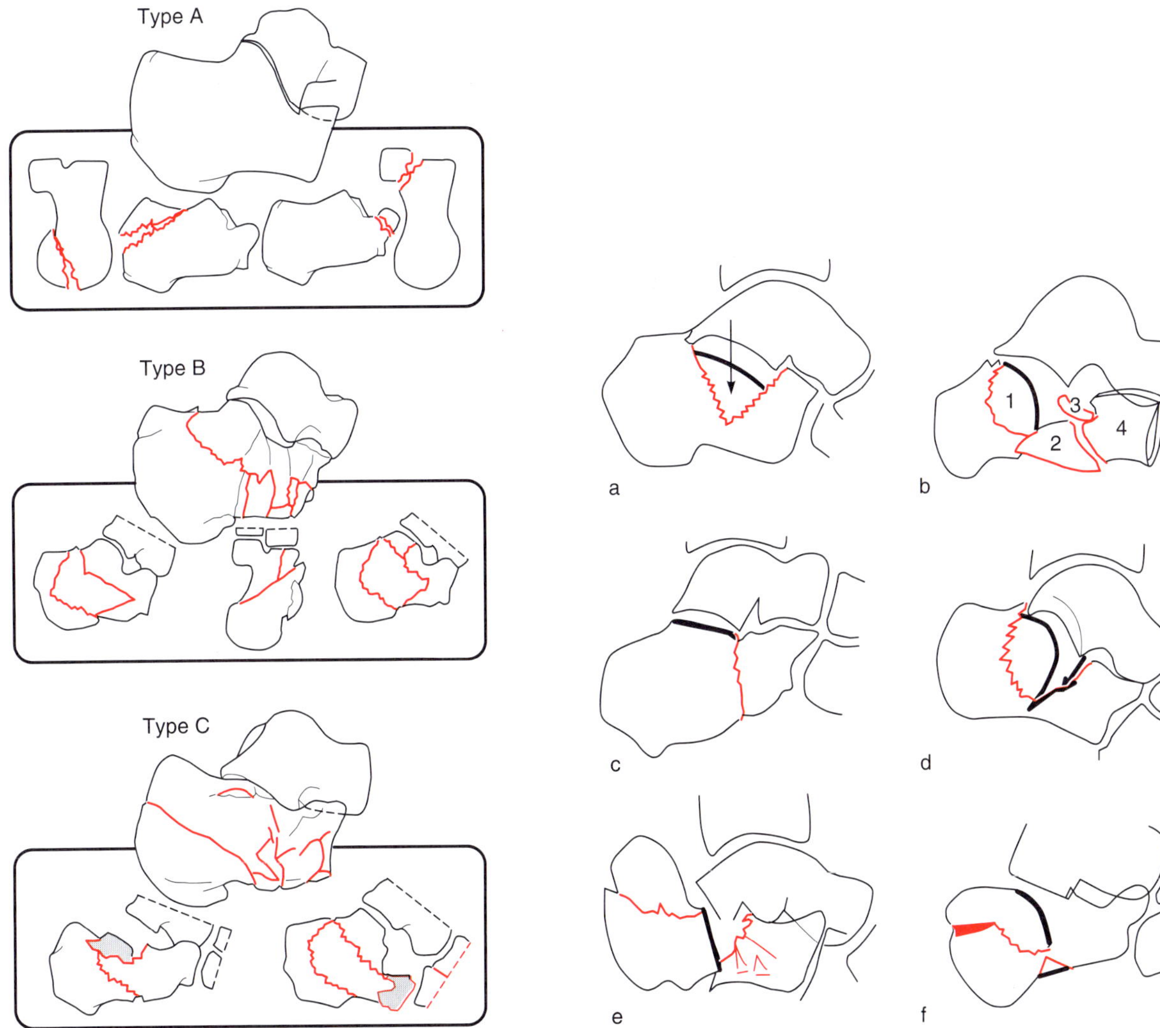

Figure 20.13. Nade and Monahan (1973) classification.

Figure 20.14. Soeur and Remy (1975) classification.

Table 20.19 Soeur and Remy classification (1975)

Non-thalamic	*Thalamic*
• Fracture of the anterior end	Vertical compression: Pure vertical depression of the posterior articular facet
• Fracture of the medial tubercle	Shearing and compression:
• Fracture of the lateral tubercle	*First degree*: Shearing alone, 2 main fragments: i. anteromedial, having the anterior part of the bone + sustentaculum tali ii. posterolateral, having the thalamic portion + remainder of bone No or minimal displacement
• Fracture of the body of calcaneus	
• Fracture of sustentaculum tali	*Second degree*: Shearing and compression secondary fracture line in both fragments; in the main, the posterior fragment runs around the thalamic portion separating a comet fragment
• Fracture of plantar spur	
• Posterior beak fracture	*Third degree*: Comminuted fractures. Extensive comminution, comet fragment is hardly identified and buried near the plantar surface between the comminuted fragments

believed (Palmer, 1948). They correlated satisfactory results with their treatment plan based on this classification. Soeur and Remy's classification was a modification of Essex-Lopresti and it has been adopted by other authors for describing and managing their cases. It is also the basis for the CT classification described by Sanders et al. (1993).

Burdeaux (1983) collated factors affecting outcome. The most important were:

1. Loss of calcaneal height.
2. Increase in calcaneal width and length.
3. Outward bulging of the lateral wall of the tuberosity fragment.
4. Damage of the posterior articular surface.

Burdeaux pointed out that changes in Böhler's angle do not have a direct effect on the final outcome. Many fractures of the joint depression types have no change in the angle. He concluded that most of the morbidity following calcaneal fractures comes from failure to reduce the width and lateral bulging.

Ross and Sowerby (1985) published a classification which is another modification of the Essex-Lopresti classification (Table 20.20). This is based on the site of the primary fracture line which is medial, central or lateral to the posterior articular facet. In this study fractures with the line passing through the facet behaved better (78% good results) than those with the line medial to the facet (50% good results). Tongue type fractures had a better outcome than central depression fractures. Their study aimed at achieving anatomical reduction of displaced intra-articular or extra-articular fracture by operative techniques. The amount of residual displacement was used to evaluate the results. They considered that displacement of less than 1 mm is an excellent reduction, 1–2 mm a good reduction and more than 2 mm displacement a poor reduction. They showed it was feasible to reduce anatomically each type of fracture and to expect a good outcome when there is an accurate reduction.

Paley and Hall (1989) published a classification for intra-articular fractures of the calcaneus which depended on the amount of comminution of the fractured fragments. They tried to identify the patterns of fracture lines in the fractured calcaneus from plain radiographs and recognized two groups: posterior secondary fracture lines, which are similar to those described by Essex-Lopresti (tongue and central depression fractures); and anterior secondary fracture lines extending either to the plantar surface just proximal to the calcaneocuboid joint (plantar type), or into the joint (calcaneocuboid type). They used the Degan, Morrey and Braun (1982) classification of the anterior process of the calcaneus:

Type I — Undisplaced.
Type II — Displaced fracture with no articular involvement.
Type III — Large displaced fragment with involvement of the calcaneocuboid joint.

Paley and Hall used the secondary fracture lines to identify the degree of comminution: Those with a single anterior/posterior line are non-comminuted fractures. More than one line anteriorly and posteriorly are defined as a moderately comminuted fractures. Comminuted types are those where it is impossible to differentiate between the many fracture lines.

Paley and Hall believed that this classification is simple, identifies the degree of comminution by plain radiograph and the fracture patterns are

Table 20.20 Ross and Sowerby classification (1985)

Type	*Description*
Type I Involving subtalar joint displacement	*Tongue type* • with primary fracture line medial to posterior facet • with primary fracture line central to the facet • with the line lateral to posterior facet *Joint depression type* • with the primary line medial to posterior facet • with the primary line central to the facet
Type II Avulsion fractures of the tuberosity	*Extra-articular* *Intra-articular*, i.e. extending to posterior facet

arranged in ascending order of severity. Their results had shown that tongue types gave better results than central depression types and moderately comminuted cases gave better results when compared with the severely comminuted ones. Paley and Hall published another study in 1993 using the same classification with minor modifications (Table 20.21; Figure 20.15) in a trial to demonstrate its reproducibility. They identified risk factors for bad outcome including age more than 50 years, heavy body weight, heavy labour type of work, decreased Bohler's angle, and disturbance of the calcaneal affecting the calcaneofibular space. They believed that heel height, fat pad height, arch angle, talocalcaneal angle and the length of tendoachilles fulcrum has no effect on the outcome.

Segal, Marsh and Leiter (1985) published one of the first papers reporting the use of CT scanning to evaluate calcaneal fractures. They demonstrated that CT could define intra-articular fracture lines, displacement of the articular facet, loss of integrity of the medial wall, as well as the size and position of the displaced articular fragment.

Lowrie et al. (1988) reported a classification system based on CT scanning of intra-articular fractures of the calcaneus. They compared the plain radiographic classification of the fractures described by Essex-Lopresti and the pattern of these fractures on the CT scans. This showed that CT helped to delineate the fracture morphology and plan the operative treatment (Table 20.22). This classification is a purely descriptive one without correlation between each class and its outcome.

Crosby and Fitzgibbons (1990) devised another classification for intra-articular fractures (Table 20.23; Figure 20.16) depending mainly on the extent of damage to the posterior articular surface of the calcaneus as seen on coronal CT scans. Grades were

Table 20.21 Paley and Hall classification (1993)

Grade	*Description*
Type A	Shear fracture two-part fragment
Type B	Tongue fracture: B1 no comminution B2 with moderate comminution
Type C	Central depression C1 no comminution C2 with moderate comminution
Type D	Severely comminuted can not be easily classified due to numerous fracture lines and major displacement

Table 20.22 Lowrie et al. (1988) CT classification of intra-articular fracture of the calcaneus

Type	*Description*
Type 1 Inverted 'Y' pattern	Posterior facet split into two approximately equal fragments diverging from each other and a large fragment of the tuberosity seen posteriorly
Type 2 Large fragment type	Fractured fragment has whole of the posterior facet + additional small medial or lateral fragment
Type 3 Longitudinal split type	Vertical fracture line producing nearly equal size fragments
Type 4 Comminuted	Multiple fracture fragments

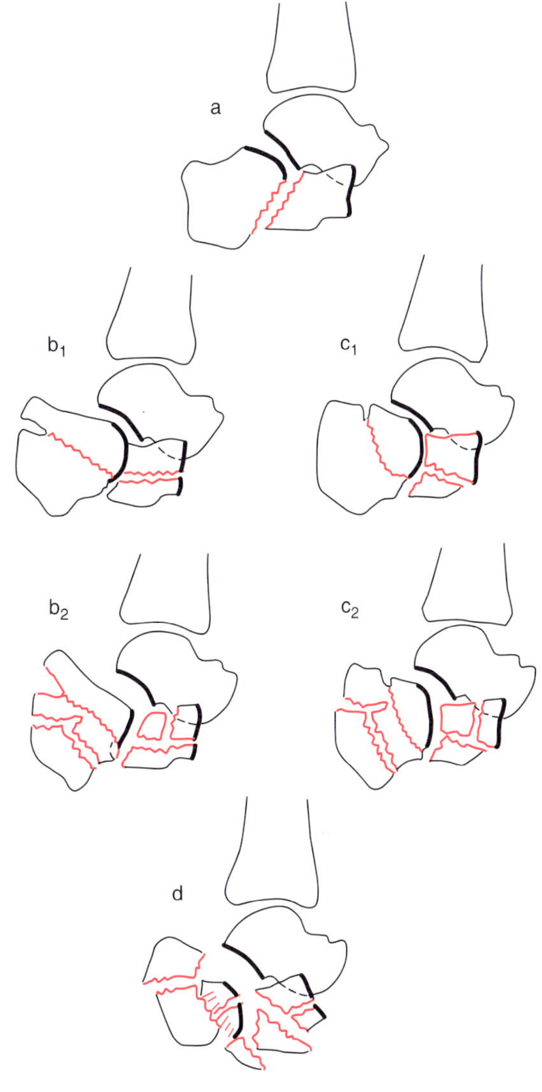

Figure 20.15. Paley and Hall (1989) classification.

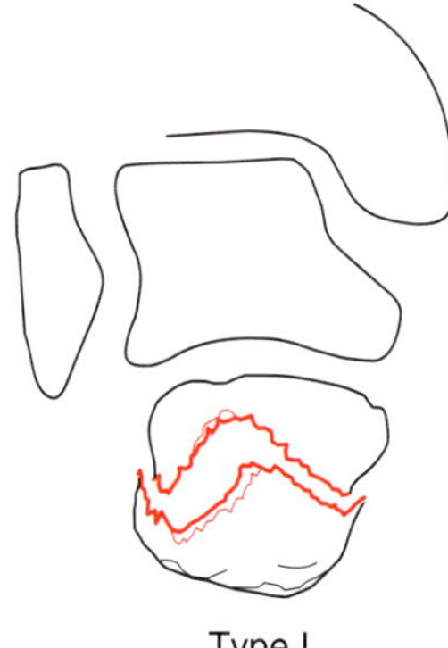

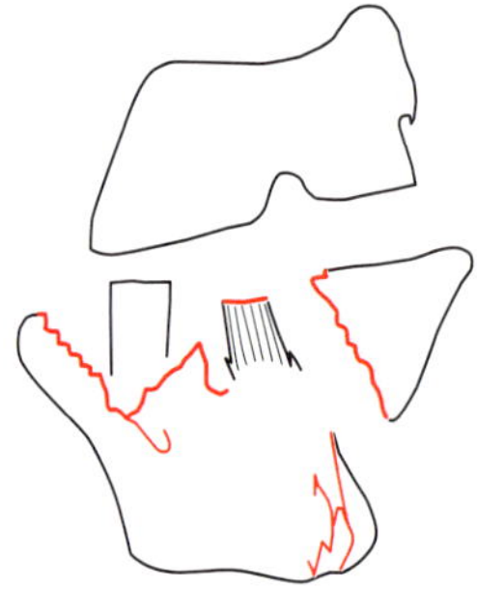

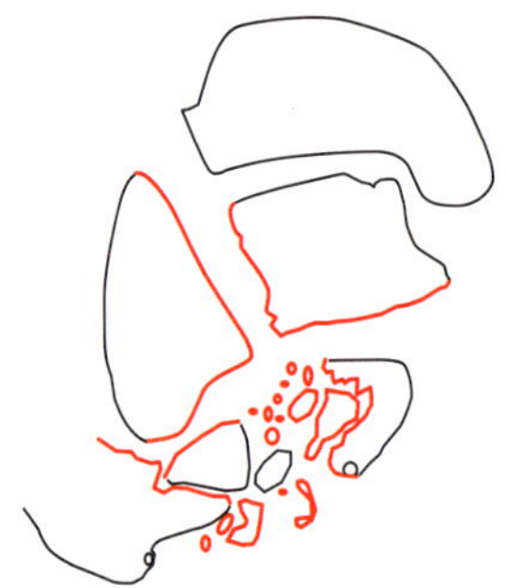

Figure 20.16. Crosby and Fitzgibbons (1990) CT classification of intra-articular calcaneal fractures.

Table 20.23 Crosby and Fitzgibbons (1990) CT classification of intra-articular calcaneal fracture

Type	*Description*
Type I	Non- or minimally displaced fracture < 2 mm diastasis or depression
Type II	Displaced disrupted fragments > 2 mm fragments are large. No major comminution
Type III	Severely comminuted fractures involving the posterior articular facet the fragments are multiple and small

arranged according to the amount of comminution and the degree of displacement of the fragments. They correlated the outcome with the fracture pattern in their study after analysing the results of treating 30 cases of intra-articular fractures of the calcaneus conservatively. They recommended conservative treatment for their type I fractures but offered no recommendation for treating type II fractures. They showed that type III cannot be treated conservatively. The reliability of their classification was not assessed.

Hutchinson and Huebner (1994) used the same classification to evaluate their cases which were treated operatively. They found that type II fractures had a better outcome than type III. They concluded that this was due to the quality of anatomical reduction. They also discussed the importance of restoration of the calcaneofibular space in obtaining a better outcome and recommend CT. They found no difference in the final outcome between the plain radiographic tongue and joint depression types.

Monsey et al. (1995) used the same classification to compare the results of operative and non-operative treatment. They agreed with Crosby and Fitzgibbons treatment plan for types I and III fractures. They recommended that further prospective studies with large number of patients to clarify the optimal treatment for type II fractures.

Sanders et al. (1993) published a classification based on the number of fragments and location of the fracture lines (Table 20.24; Figure 20.17). They used the work of Souer and Remy (1975) to determine the expected sites of these fracture lines. They divided the posterior articular facet of the calcaneus into three equal columns giving three zones, medial, central and lateral using two lines (A) and (B). Another line (C), separates the posterior facet of the calcaneus from the sustentaculum tali creating a fourth column. Thus, eight types of fractures could be identified.

They correlated the final outcome with the type of fracture and found that classifying the fracture this way gave a good indicator for each class. The final outcome deteriorates as the number of fragments increases; this appears to be directly related to the difficulty in obtaining good anatomical reduction. Sanders and Gregory (1995) demonstrated the reproducibility of this classification.

Eastwood, Gregg and Atkins (1993) devised another CT classification when they correlated findings seen on plain radiograph with CT scans of a consecutive series of 120 displaced intra-articular fractures treated surgically. They used the concept of primary and secondary fracture lines on plain radiography, as described by Palmer (1948), Essex-Lopresti (1952), Thoren (1964), and Burdeaux (1983). Three-part fractures, described by Essex-Lopresti (1952) as tongue and central depression types, were further classified by CT into three differ-

Table 20.24 Sanders et al. (1993) CT classification of intra-articular calcaneal fracture. A and B: Two fracture lines separating posterior articular facet from lateral to medial into three potential pieces; lateral, central and a medial columns. C: A third line separating posterior facet from sustentaculum fragment

Type	*Description*
Type I	Non-displaced irrespective of the number of fracture lines
Type II	Two parts fragment or split IIA, IIB, IIC
Type III	Three parts or split depression IIIAB, IIIAC, IIIBC
Type IV	Four parts or highly comminuted

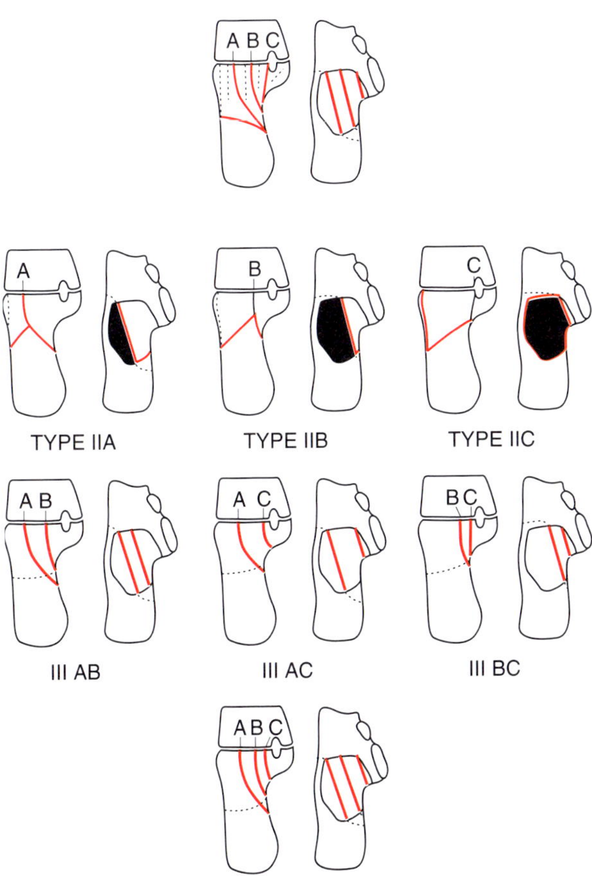

Figure 20.17. Sanders et al. (1993) CT classification of intra-articular calcaneal fractures.

ent types (Table 20.25). Eastwood and her colleagues defined the three main fragments as:

1. The body fragment.
2. The lateral joint fragment which is the part carrying a portion of the posterior articular facet.
3. The sustentacular fragment, which is the antero-medial fragment.

They studied the orientation and displacement of these fragments (Figure 20.18). They showed that the lateral joint fragment is depressed in only 7% of the cases (type III) and is more commonly rotated (a finding in agreement with Souer and Remy (1975)). This is in contrast to other authors (Palmer, 1948; Thoren, 1964; Burdeaux, 1983; Stephensen, 1987; Leung et al., 1989) who believed it is nearly always depressed. They looked in more detail at the lateral wall of the fractured calcaneus than previous authors, as this is the first part of the bone to be seen on a lateral exposure. Their classification helped them to decide what to do intra-operatively to approach and reduce all the fragments involved in this type of fracture

Eastwood et al. also showed that widening of the calcaneus (which is one of the determinant factors of calcaneal fractures outcome) is due to the rotatory and lateral displacement of the body fragment and not due to broadening of the body by the direct impact as it was believed before. This classification helped them to determine surgical treatment. Its prognostic value remains to be demonstrated.

Zwipp et al. (1993) devised a classification system that is based on a combination of that of Essex-Lopresti (1952) and Ross and Sowerby (1985). They included a soft-tissue component using Tscherne's (1984) classification for grading closed and open soft-tissue injuries. This is based on:

1. The number of fragments.
2. The number of joints affected.
3. The degree of soft-tissue injury (Figure 20.19).

They emphasized the importance of the anterior subtalar fragment in both treatment and outcome.

They used three radiographic (anteroposterior, lateral and axial), four Broden plain radiographs and axial and coronal CT scans to evaluate the fracture pattern. To calculate the fracture score, the number of fragments were added to the degree of soft-tissue involved. Other points were added according to degree of soft-tissue involvement (Table 20.26). They found that classifying the fracture this way helped them to judge the severity of the fracture, select the surgical approach, understand the best way of reducing the fracture, choose the

Two part fracture

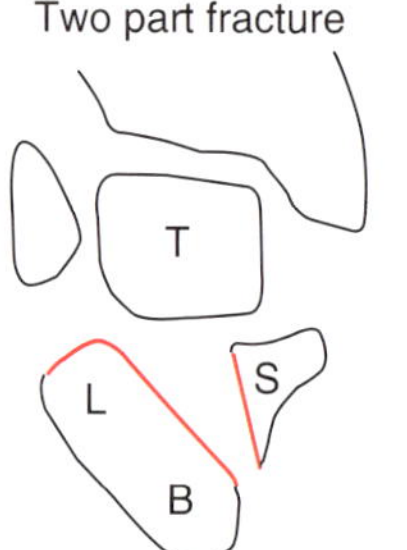

The combined body and lateral joint fragment is subluxed lateral to the talus

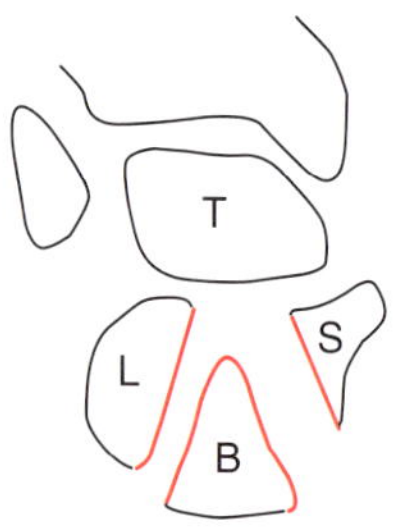

Type I
Lateral wall of fracture bone is formed by lateral joint fragment in valgus and away from subtalar by upward impact of wedge body fragment

Three part fracture

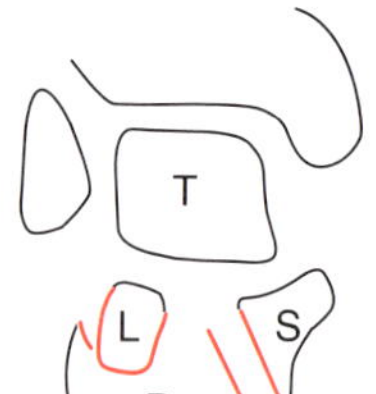

Type II
Lateral wall formed of lateral joint fragment above and body fragment below

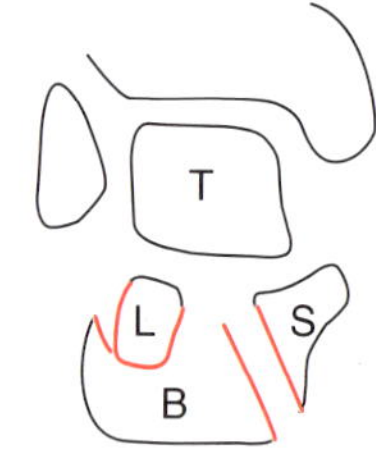

Type III
Lateral joint fragment impacted within body fragment lateral wall is formed by lateral wall of the body lateral joint fragment away from talus

Figure 20.18. Eastwood, Gregg and Atkins (1993) CT classification of intra-articular calcaneal fractures.

Table 20.25 Eastwood, Gregg and Atkins (1993) CT classification of intra-articular calcaneal fractures

Type	*Description*
Type I	Two part fragment: calcaneus is divided into sustentacular fragment and a lateral main fragment
Type II	Three part fragment: the apparent lateral wall is formed solely by the lateral joint fragment
Type III	Three part fragment: the lateral wall is formed by the lateral joint fragment superiorly and the body inferiorly
Type IV	Three part fragment: the lateral wall is formed solely by the body fragment

implant to be used, and predict the outcome (Figure 20.20).

Unfortunately, this complex classification does not give guidance in the English literature to judge displacement. It is dependant on the accuracy of the CT cuts and the reproducibility has not been tested. It omits some fracture patterns, such as a three-part fracture with involvement of the calcaneocuboid joint only.

Corbett et al. (1995) suggested another CT-based classification of calcaneal fractures. This depended on posterior facet disruption, the calcaneocuboid joint as well as lateral wall blowout and medial wall displacement (Table 20.27; Figure 20.21). Their criteria of significant displacement are:

- Articular disruption in either posterior facet or calcaneocuboid joints >2 mm.
- Angulation of the fractured fragment $>10°$.
- Medial wall overlap or displacement >1 cm.
- Lateral wall bulge beyond the line drawn perpendicular to the incisura fibularis.

Each case is given a plus or minus sign according to the involvement of the calcaneocuboid joint, applying the above criteria.

They concluded that this system is readily available, easy to apply, and is reproducible among independent observers. Unfortunately, there is no mention of an investigation into the reliability of this classification.

They compared their results with those of Crosby and Fitzgibbons (1990). They claimed to be able to subtype Crosby and Fitzgibbons' Type II.

Dislocation of the calcaneus

Calcaneal dislocation from cuboid and talar articulation is rare (Hamilton, 1949; Carey, Lance and Wade, 1965; Viswanath and Shephard, 1977), it is usually a lateral dislocation.

Discussion of Calcaneal Fracture Classification in Adults

There are numerous classifications of calcaneal fractures which look very similar. There has been a development in the understanding of these complex injuries which is reflected in these classifications. The recognition of the difference in behaviour between extra- and intra-articular fractures, the patterns of intra-articular fragments, the optimum surgi-

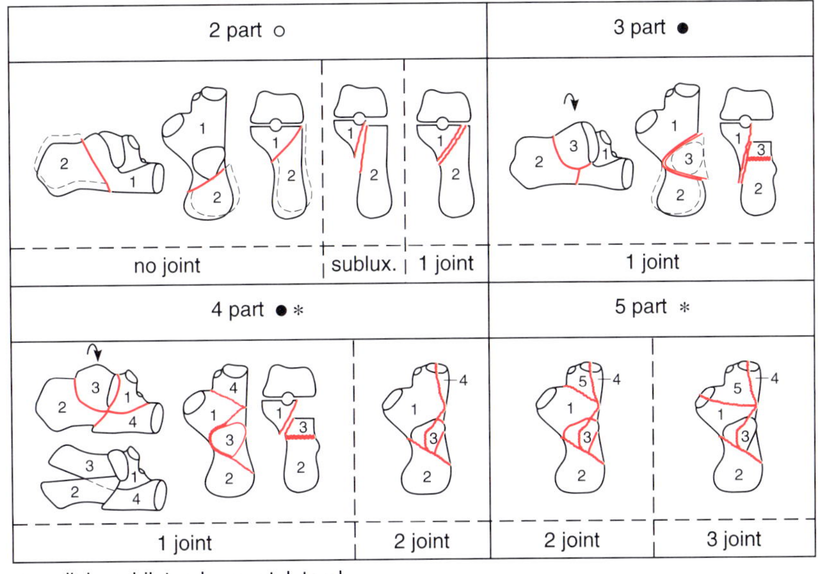

Figure 20.19. Zwipp et al. (1993) classification of calcaneal fractures.

cal approaches and the advent of CT instead of plain radiographs are all important landmarks.

In spite of this work, there is but a single attempt to validate the reliability of a classification (Sanders and Gregory, 1995). No clear winner has emerged which can define both treatment and prognosis. Formal prospective studies are needed to do this with larger numbers of patients.

Paediatric Injuries

Calcaneal injuries in children tend to be more benign than those in adults. In one study, 63% of childhood calcaneal injuries occur in an extra-articular pattern (Schmidt and Weiner, 1982). Many authors have included children in their studies on calcaneal injuries in adults. They have not elaborated specifically on the

Table 20.26 Zwipp et al. (1993) classification of intra-articular fracture of the calcaneus

Grade	*Description*
Type I	Two part fragment. Fracture line passes between tuberosity and sustentacular fragments: Type IA No joint involvement. Extra-articular Type IB Posterior subtalar joint subluxation Type IC Posterior subtalar joint involvement with minimal displacement
Type II	Three part fragment + posterior subtalar joint affection only. Fracture line passes between sustentacular fragment and tuberosity with another line separating the joint fragment from the tuberosity fragment.
Type III	Four part fragment in which an additional fracture line separates the anterior process from the sustentacular fragment. This may be: Type IIIA Affection of the posterior subtalar joint only Type IIIB Affection of the posterior subtalar + calcaneocuboid when the line passes through the joint
Type IV	Five part fragment in which two fracture lines (described in type IIIB) are both present and at the same time create another fragment, the anterior subtalar joint fragment. Type IVA The above plus two joint affection – posterior subtalar + calcaneocuboid Type IVB The above plus three joint affection – posterior subtalar + calcaneocuboid + anterior subtalar

N.B. If there is comminution of the main fragment or a fracture of talus, cuboid or navicular bones a (+) sign is added

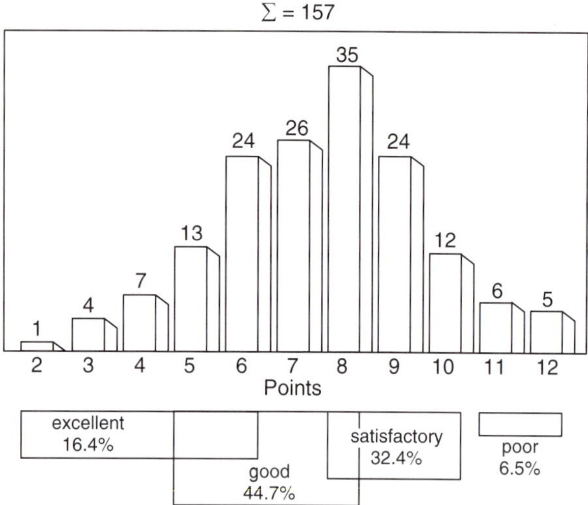

Figure 20.20. Diagram relating results to fracture score (Zwipp et al. (1993) classification).

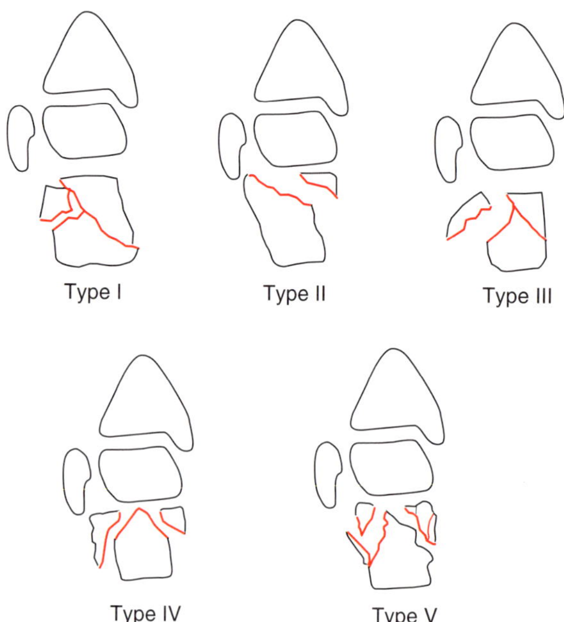

Figure 20.21. Diagram of Corbett et al. (1995) classification.

type of injury in children (Essex-Lopresti, 1952; Rowe et al., 1963). Since paediatric injuries tend to be benign, it is inappropriate to include them in an adult series.

Schmidt and Weiner (1982) studied this fracture pattern in children only. Their model was devised from the classifications of Essex-Lopresti (1952), Rowe et al. (1963) and Chapman (1977). This involved dividing the fracture pattern into six types (Table 20.28).

Types 1, 2, 3 and 6 were considered as extra-articular injuries, whilst types 4 and 5 as intra-articular (Figure 20.22).

Other authors have used the Schmidt and Weiner (1982) classification for calcaneal fractures in children. Drvaric and Schmitt (1988) described a case of anterior process fracture (type 1c) but used the Degan, Morrey and Braun (1982) classification because it included a large intra-articular fragment. Shantz and Rasmussen (1988) used the Schmidt and Weiner classification but excluded type 6.

Walling et al. (1990) studied calcaneal apophyseal fractures. They compared the pattern of injuries to injuries of the growth plate of long bones as described

Table 20.27 Corbet et al. (1995) classification

Type 1	Non- or minimally displaced intra-articular fracture. No significant displacement. No lateral wall blow-out
Type 2	Displaced fractures in which the fracture line misses or involves less than 10% of the posterior articular surface. Main part of the articular surface remains attached to the sustentacular fragment
Type 3	A secondary fracture line running vertical and in lateral orientation gives a third fragment which is significantly displaced but there is no significant medial wall overlap
Type 4	As type 3 but with medial wall overlap > 1 cm. Note that the posterior articular facet is not severely comminuted, still reconstructable
Type 5	As type 4 but with comminuted articular facet

Table 20.28 Schmidt and Weiner (1982) classification of calcaneal fractures in children

Type 1	a. Fracture of tuberosity or apophysis
	b. Fracture of the sustentaculum tali
	c. Fracture of the anterior process
	d. Distal inferolateral aspect, a benign linear fracture of the distal inferolateral surface of the calcaneus or calcaneocuboid joint
	e. Small avulsion off the body
Type 2	a. Beak fracture
	b. Avulsion fracture of the insertion of the tendo achillis
Type 3	Linear fracture not involving the subtalar joint
Type 4	Linear fracture involving the subtalar joint
Type 5	Compression fracture of the subtalar joint
	a. Tongue type
	b. Joint depression type
Type 6	An open fracture with significant bone loss of the posterior aspect of the calcaneus and loss of Achilles tendon insertion

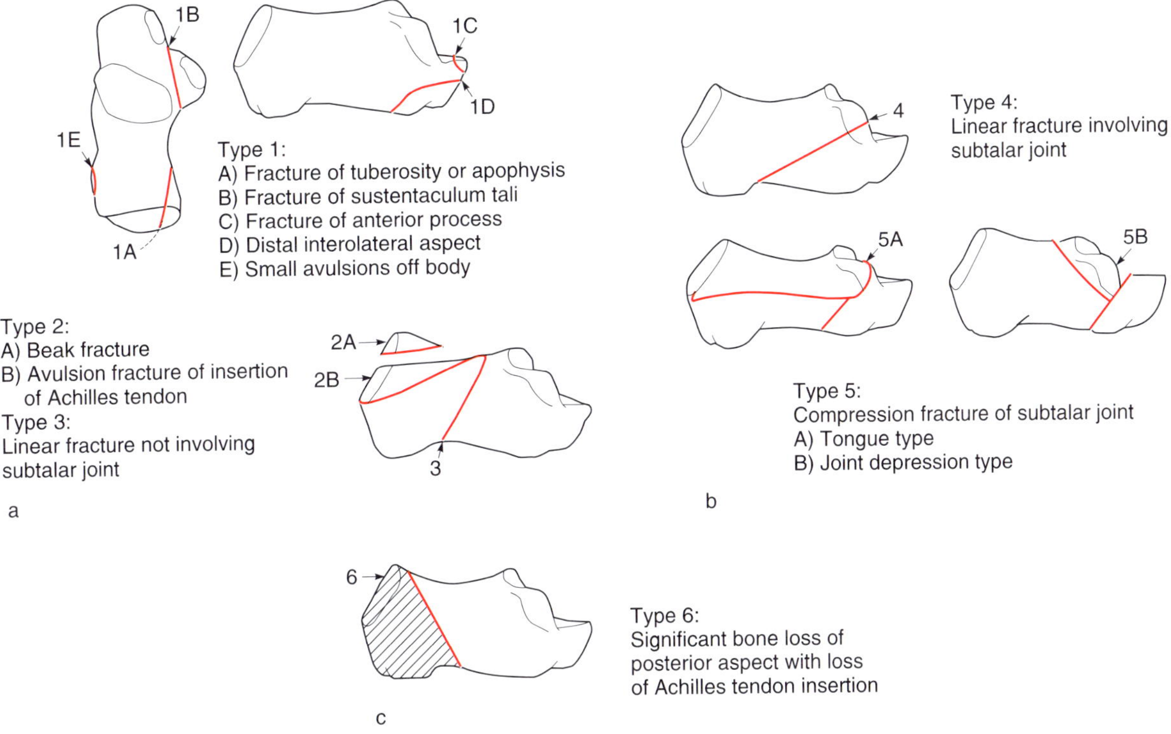

Figure 20.22. Schmidt and Weiner (1982) classification of calcaneal fractures in children.

by Ogden (1990). This described four types of calcaneal apophyseal injuries, using the same principles as the Salter Harris classification of injuries to the growth plate (1963) (see Chapter 19):

Type 1 Pure separation of the apophysis through the growth plate.
Type 2 Separation of the apophysis with a triangular metaphyseal fragment.
Type 3 Split fracture of the apophysis and growth plate without metaphyseal extension.
Type 4 Split fracture of the apophysis and growth plate with extension to the metaphysis of the calcaneal body (Figure 20.23).

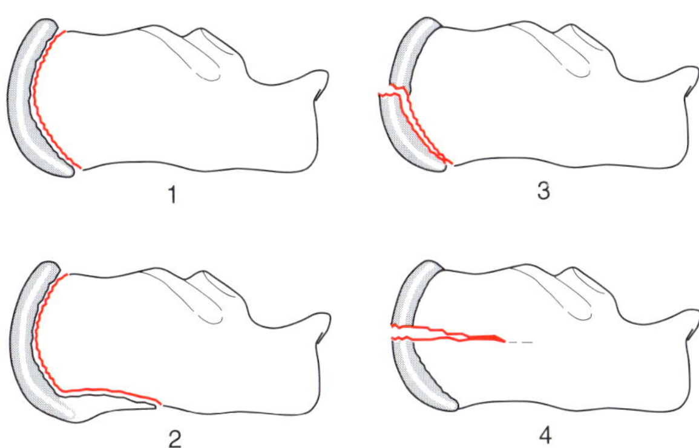

Figure 20.23. Walling et al. (1990) classification of calcaneal apophysis fracture.

Injuries of the Midfoot

Midtarsal Joint Injuries

The midtarsal joint or Chopart's joint is the articulation between the hind foot and the midfoot and is represented by the talonavicular and calcaneocuboid joint.

Injuries to this joint were classified by Main and Jowett (1975). They used the work of Hermel and Gershon-Cohen (1953), Dewar and Evans (1968), Stark (1973) and classified the injury according to the direction of forces applied to the foot at the time of impact which in turn produces deformities in the same direction as these forces. Five types were recognized:

1. Medial
2. Lateral
3. Longitudinal
4. Plantar
5. Crush injuries

There are listed in ascending severity and worsening prognosis. Each of these injuries may be either:

- a fracture sprain (a term used by Bradford and Larsen, 1951)
- a fracture subluxation or dislocation.

The medial type has a swivel dislocation subtype, in which a medially directed force dislocates the talonavicular joint leaving the calcaneocuboid joint intact and the subtalar joint subluxed. The calcaneus does not invert as in a medial subtalar dislocation but rotates on the interosseous talocalcaneal ligament without tearing it.

This classification has been used to evaluate the severity of injuries, planning of treatment and prediction of outcome (Howie, Hooper and Hughess, 1986; Rymaszewski and Robb, 1988; Heckman, Rockwood and Green, 1991).

Midtarsal occult subluxation was originally termed a 'nut cracker' fracture by Hermel and Gershon-Cohen (1953) but later classified as a fracture sprain by Main and Jowett (1975). Davis, Lubowit and Thordarson (1993) treated this injury by open reduction and internal fixation, rather than conservative methods, in order to prevent the development of a progressive valgus foot. They used the Degan, Morrey and Braun (1982) classification of fractures of the anterior process of the calcaneus to evaluate and manage these calcaneocuboid injuries.

Isolated Injuries to the Midfoot Bones

Navicular injuries

Eichenholtz and Levine (1964) classified navicular fractures into three types:

Type I Avulsion fracture of the tuberosity.
Type II A fracture involving the dorsal lip.
Type III A fracture through the body of the navicular.

Torg et al. (1982) used the same classification in their study and added another type (Type IV) which is a navicular stress fracture. Type III is rarely encountered in isolation and usually occurs in conjunction with midtarsal joint injuries (Main and Jowett, 1975).

Sangeorzan et al. (1989) classified navicular body fractures into three types:

Type 1 Transverse fracture line in the coronal plane with no angulation of the forefoot.
Type 2 The major fracture line is dorsolateral to plantar-medial and the forefoot is displaced laterally.
Type 3 Comminuted fracture (either central or lateral) of the body with the forefoot displaced laterally.

One fracture out of the 21 used in their series did not fit this classification. They identified a satisfactory reduction to be restoration of more than 60% of joint surface in both anteroposterior and lateral planes at which joint subluxation is reduced. They suggested open reduction and internal fixation for each type and were able to predict the outcome of each fracture type using this classification.

Isolated total navicular dislocation occurs very rarely.

Cuboid injuries

These are usually a part of midtarsal joint injuries. Isolated cuboid fractures may be a stress type or as a result from a direct blow.

Cuneiform injuries

These may occur as a result of a direct blow, There is no specific classification. Subluxation and fracture-dislocation of cuneiform bones may be involved in midtarsal or tarsometatarsal joint injuries.

Injuries of the Forefoot

Injuries to the Tarsometatarsal Joint (Lisfranc's Joint Complex Injury)

This injury was named after Lisfranc who was a surgeon in Napoleon's army. Although he amputated at the level of this joint, he is not known to be involved with injuries to the joint. Willpula (1967) reported the prevalence to be 1.6% of all foot injuries. Vuori and Aro (1993) reported the figure to be five times greater (7.9%).

A useful classification system for tarsometatarsal injuries is difficult to create because of the many possible fracture combinations (Rosenberg and Patterson, 1995). Quénu and Küss (1909) produced the earliest classification for this injury and divided it into three types:

Type 1 Homolateral dislocation with all five metatarsals dislocated together.
Type 2 Isolated dislocation with one or two metatarsals are displaced from the others.
Type 3 Divergent dislocation in which there is a separation between the first and the second

metatarsals both in sagittal and coronal planes.

Several, more modern, classification systems have been described (Francesconi, 1925; Jeffreys, 1963; Wilson, 1972; Bonnel and Barthelemy, 1976). All of these depend on the mechanism of injury and direction of forces to classify the fracture. These are of limited value since the mechanism of injury is always complex and variable. These classifications do not offer any advantage in defining treatment or prognosis.

Hardcastle et al. (1982) used the classification of Quénu and Küss (1909) as a basis for a new classification. They divided the injury into three types:

Type A — A total displacement in which there is incongruity of the entire tarsometatarsal joint and the displacement is in a single plane (sagittal, coronal or combined).

Type B — A partial displacement with part of the joint incongruent. The displaced segment is in one plane and of two subtypes:

1. Medial partial, where the displacement involves the first metatarsal in isolation or in combination with one or more of the second, third or fourth metatarsals.
2. Lateral partial, where the displacement affects one or more of the lateral four metatarsals but not the first ray.

Type C — Is a divergent type which may have partial or total incongruity. In the anteroposterior view, the first metatarsal is displaced medially while any combination of the lateral four metatarsals are displaced laterally (i.e. a sagittal plane displacement occurs in conjunction with the coronal displacement).

They devised a plan of treatment using this classification and concluded that the classification is simple to apply and gives an indication of the prognosis.

Myerson et al. (1986) used the same classification in evaluating their results and added a simple modification dividing type B into B1 and B2 according to whether the displacement is medial or lateral respectively. Type C was also subdivided into two groups. C1, where a divergent pattern is partially incongruent, with the first ray medially displaced and the lateral four rays in any other concomitant pattern. C2 is a totally incongruent joint (Figure 20.24). They concluded that this classification is helpful in planning the treatment but could not demonstrate any value in prognosis.

Brunet and Wiley (1987) similarly found some unsatisfactory functional results related to type A rather than type B or C. They suggested this was due to the disruption of the medial column of the foot. Vuori and Aro (1993) found some relationship between the type of fracture with the mechanism of injury but this was statistically insignificant.

Myerson (1989a) recognized a group of patients who had an apparently minor injury, with no fracture, with prolonged recovery. He called these a 'subtle luxation injury', in which there was a widening or diastasis of the first and second metatarsals or between the middle and medial cuneiform bones. Similar widening may occur between the second and third metatarsals, as well as between the middle and lateral cuneiforms. The latter positive finding may have a Fleck sign, where a small avulsed fragment

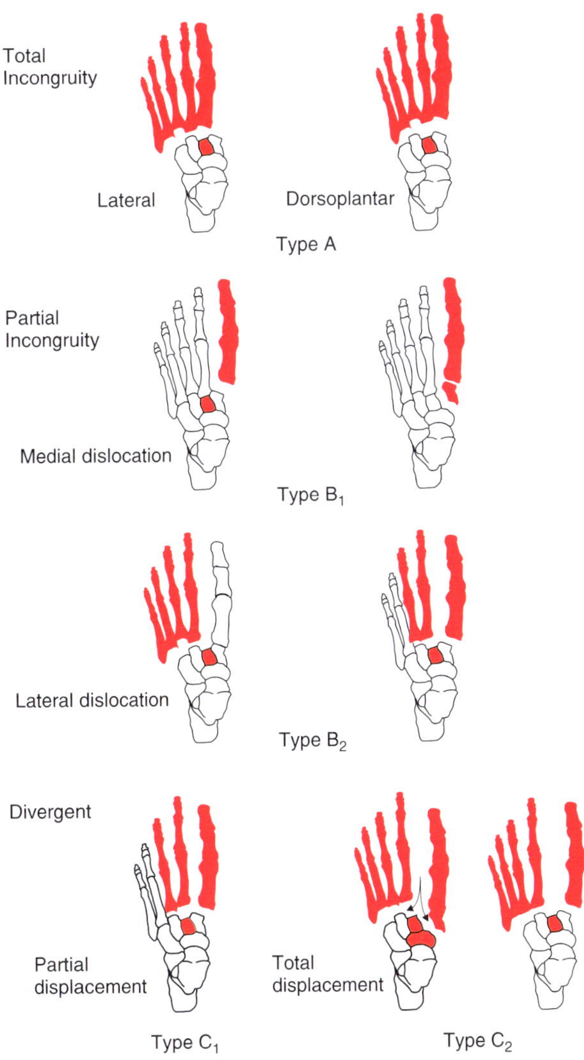

Figure 20.24. Myerson et al. (1986) classification of Lisfranc fracture dislocation.

of the medial aspect of the base of the second metatarsal bone is seen.

Faciszewski, Burks and Manaster (1990), in a study of 15 patients, recommended a plan of treatment for the subtle luxation injury group. This depends on the integrity of the longitudinal arch rather than the extent of the diastasis. They correlated a good outcome to the absence of flattening of the longitudinal arch of the foot and advised open reduction and internal fixation of this type of injury if this flattening is present. They used a weight-bearing lateral radiograph of the foot to assess the amount of flattening by measuring the distance between the plantar surfaces of the medial cuneiform and that of the base of fifth metatarsal bone.

Curtis, Myerson and Szura (1993) included low velocity injuries with the previously published classification by Myerson et al. (1986) and graded them as:

- First degree sprain.
- Second degree sprain: both of them with varying degrees of partial tear of the ligaments of the Lisfranc joint.
- Third degree sprain: with total rupture of the ligaments leading to diastasis at the level of the joint (subtle luxation group).

Injuries to the Metatarsals

Shaft fractures

No specific classification exists in the literature for metatarsal shaft fractures. They may be seen in crush injuries of the foot involving the metatarsal shafts. There may be severe soft-tissue injuries, as well as other indirect injuries (e.g. Lisfranc joint complex injuries). The location of the fracture depends on the metatarsal involved, the activity of the patient and the foot shape (Sammarco and Carrasquillo, 1995).

Most authors describe metatarsal shaft fractures as being displaced or undisplaced. Dorsal or plantar displacement of the second, third, and fourth metatarsals and medial displacement of the first or lateral displacement of the fifth may require extra care. Hardaker (1989) called the spiral fracture of the distal fifth metatarsal the 'ballet dancer's fracture'. Conwell and Reynolds (1961) recommended special attention to fractures of the first metatarsal as it carries much of the body weight.

Fractures of the base of the fifth metatarsal

Sir Robert Jones (1902) was the first to describe this fracture after he sustained this injury himself while dancing. Stewart (1960) and Dameron (1975) identified two patterns of fractures of the fifth metatarsal. The first is a fracture occurring at least 1.5 cm distal to the tuberosity near the diaphysis. The second is an avulsion fracture of the tuberosity involving the proximal tip of the bone and the articular surface of the joint. Although the original paper of Sir Robert Jones showed a plain radiograph of a fracture 1.5 cm distal to the tuberosity, some have thought that the Jones fracture was an avulsion fracture (Dameron, 1975; Anderson, 1977) while others (Kavanaugh, Brower and Mann, 1978) specified it to be the transverse diaphyseal fracture of the base.

Due to its variable prognosis and its relative high incidence, Jones fractures have been studied by many authors. Torg et al. (1984) classified the fracture according to the presence of an area of sclerosis into:

- Acute fracture type: sharp fracture line with no hypertrophy.
- Delayed union type: where the fracture has sclerosis with narrowing of the medullary canal.
- Non-union type: sclerosis obliterating the medullary canal.

They found that this classification helped both selecting treatment and estimating prognosis. Josefsson et al. (1994a) modified the classification by recognizing the delayed union type as a stress fracture and the non-union type to be a refracture.

Clapper, O'Brien and Lyons (1995) classified injuries occurring through the fifth metatarsal bone into:

Type 1 An avulsion fracture of the tuberosity.
Type 2 True Jones fracture.
Type 3 Fractures of the neck and shaft.

They found the Torg et al. (1984) classification reliable in describing, planning treatment and predicting the outcome of Jones fractures. They also agreed with Dameron (1975) who was the first to recommend a special plan of management for athletes.

Metatarsal head fractures

Hardaker (1989) identified this injury as a fracture occurring at the head of the metatarsal with the fracture line running obliquely to involve that part of the head which is intra-articular but without capsular attachment. He noted that it has a tendency to progress to avascular necrosis although this did not occur in the Heckman, Rockwood and Green (1991) series.

Injuries to the Metatarsophalangeal Joints

The First Metatarsophalangeal Joint

Bowers and Martin (1976) identified an injury to the first metatarsophalangeal joint in ballet dancers as a separate entity and called it 'turf toe'. Sammarco and Miller (1982) recognized two types of sprain which are either plantar sprain in soccer players, due to hyper-extension injury, or a dorsal sprain in ballet dancers, with hyper-flexion injuries.

Jahss (1980) classified the injuries of the first metatarsophalangeal joint into two types:

- Type I — Dorsal dislocation with intact sesamoidal mass.
- Type IIA — Dislocation with rupture of the intersesamoidal ligaments.
- Type IIB — Dislocation with one of the sesamoids (commonly the medial) being fractured transversely with the distal fragment intra-articular while the proximal one in its normal relation with the other sesamoid.

Jahss called irreducible dislocations 'complex', this is equivalent to Kaplan's (1957) complex dislocations of the second metacarpophalangeal joint. This classification has been frequently used since its publication with some modifications given in later reported cases.

Copeland and Kanat (1991) added an extra type. IIC, where there is both complete disruption of the intersesamoid ligaments and a transverse fracture of either sesamoid (i.e. IIA + IIB). Heckman, Rockwood and Green (1991) classified the injury into sprains, dislocation, complex dislocation and fracture of the sesamoid bones. Another variant was reported by Nabarro and Powell (1995) with dislocation and proximal migration of the sesamoid mass and avulsion of the medial and lateral short sesamophalangeal ligaments. Plantar dislocation of the first metatarsophalangeal joint was described for the first time by Mata, Ovejero and Grande (1995).

The Lesser Metatarsophalangeal Joints

Coughlin (1989) classified the dislocation of these joints as either acute or chronic. Stephenson, Beck and Richardson (1994) recorded the acute injury as either a simple or a complex dislocation in any direction. English (1964) described a special type of dislocation in association with dislocation of the adjacent metatarsal base.

Injuries to the Toes

Injuries to Hallux

Heckman, Rockwood and Green (1991) described fractures of the big toe to be either non-displaced, displaced or displaced with intra-articular extension, with differing treatment and prognosis for each type. Dislocation of the interphalangeal joint of the big toe was classified by Miki, Tamamuro and Kitai (1988) in a similar way to metatarsophalangeal joint dislocation classification but with two subtypes of the complex dislocation group:

1. Dislocation with displacement of the volar plate inside the joint.
2. Dislocation with displacement of the volar plate dorsally over the neck of the proximal phalanx of the toe which is locked in hyperextension.

Injuries to the Lesser Toes

There is no specific classification described for fractures of the phalanges. Katayama, Murakami and Takahashi (1988) studied 33 cases of interphalangeal dislocations of the lesser toes and classified them in the same way as those of the big toe.

Midfoot and Forefoot Injuries in Children

No specific classification for injuries to these sites in children is published, apart from growth plate injuries of the metatarsal heads. Johnson (1981) described a paediatric Lisfranc joint injury in a child at the level of the first tarsometatarsal joint and named it a 'bunk-bed fracture'. It may occur as a subtle injury that is overlooked and cause fracture-dislocation or fracture-subluxation of the first tarsometatarsal joint. It may also include fractures of the first and second metatarsals.

Orthopaedic Trauma Association (OTA) Classification of Fractures Around the Foot in Adults

With the increasing need for a uniform classification, describing all fractures, for easy communication between orthopaedic traumatologists and for standardization of research, the Orthopaedic Trauma Association Committee for Coding and Classification extended the AO/ASIF classification (Müller et al., 1990) to remaining non-classified

CALCANEUS

A. Calcaneus, extra-articular (73-A)

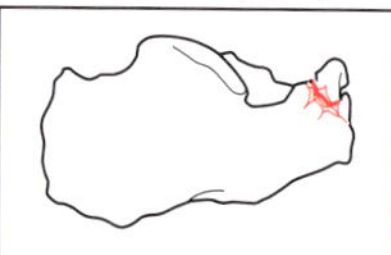

1. Calcaneus, extra-articular, avulsion (73-A1)

Subgroups and Qualifications: Calcaneus, extra-articular, avulsion (73-A1)
1. Anterior process (73-A1.1)

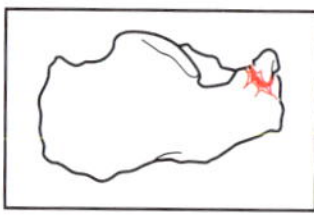

2. Medial process (sustentaculum) (73-A1.2)

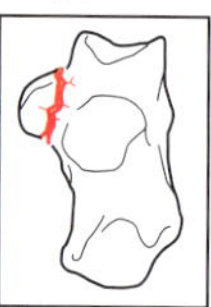

3. Posterior tuberosity (73-A1.3)

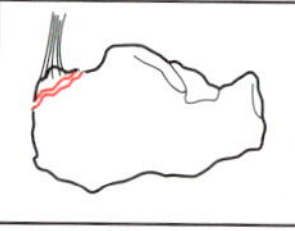

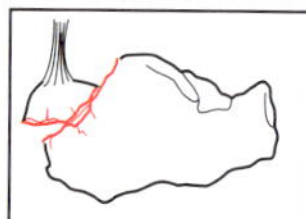

B. Calcaneus, isolated body (73-B)

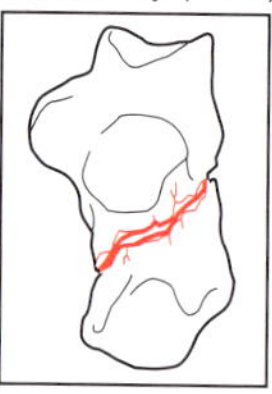

1. Calcaneus, isolated body, undisplaced (73-B1)

Simple (73-B1.1)

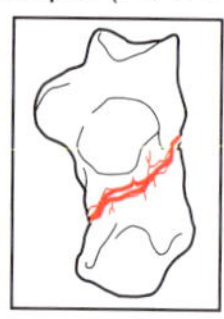

Multifragmentary (73-B1.2)

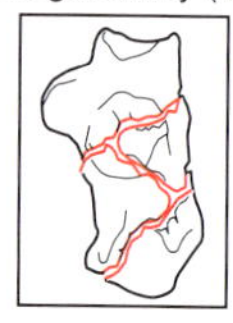

2. Calcaneus, isolated body, displaced (73-B2)

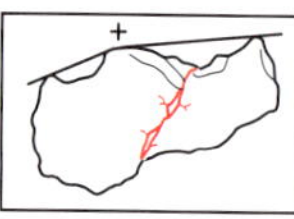

Calcaneus, isolated body, displaced (73-B2)
1. Decreased tuber angle but > 0° (73-B2.1)

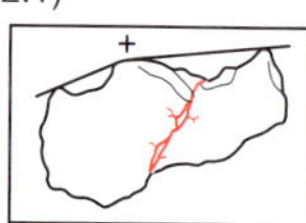

2. Flat tuber angle (0°)(73-B2.2)
3. Negative tuber angle (<0°) (73-B2.3)

3. Calcaneus, isolated body, fracture into calcaneal-cuboid joint (73-B3)
(1) undisplaced
(2) displaced

Calcaneus, isolated body, fracture into calcaneal-cuboid joint (73-B)

1. Split (73-B3.1)

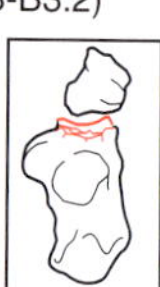

2. Depression (73-B3.2)

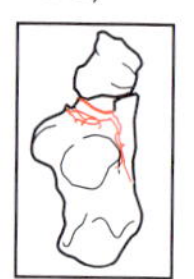

3. Split/depression (73-B3.3)

C. Calcaneus, articular (73-C)

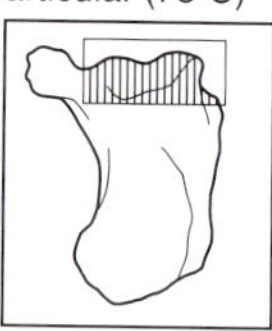

1. Calcaneus, articular, two part (73-C1)

1. Posterior facet fracture, lateral (73-C1.1)

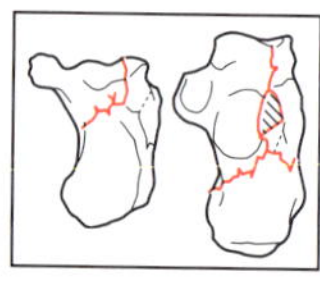

2. Posterior facet fracture, central (73-C1.2)

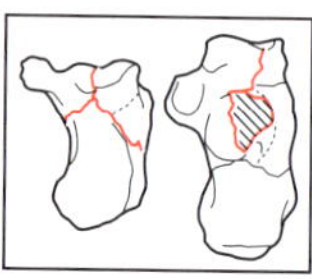

3. Posterior facet fracture, medial (73-C1.3)

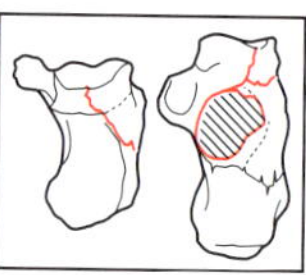

2. Calcaneus, articular, three part fracture (73-C2)

Calcaneus, articular, three part fracture (73-C2)
1. Posterior facet fracture, lateral-central (73-C2.1)

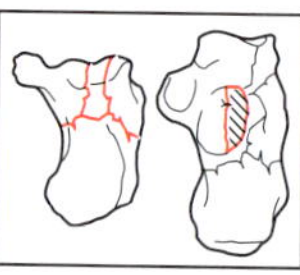

2. Posterior facet fracture, lateral-medial (73-C2.2)

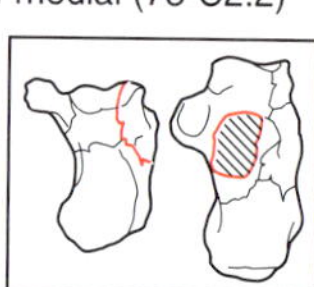

3. Posterior facet fracture, central-medial (73-C2.3)

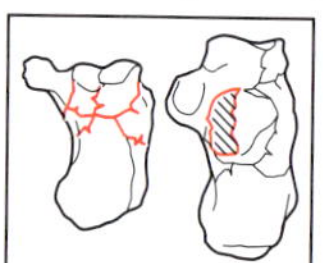

3. Calcaneus, articular, 4 plus parts (73-C3)

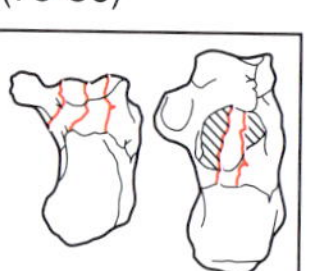

Figure 20.25. OTA classification of bones of the foot

TALUS

Types:
A. Talus, extra-articular (72-A)

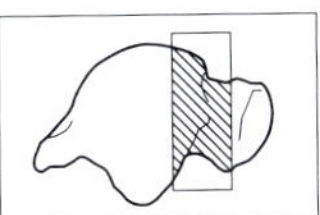

Groups:
Talus, extra-articular (72-A)
1. Talus, extra-articular, neck (72-A1)
(1) undisplaced
(2) displaced

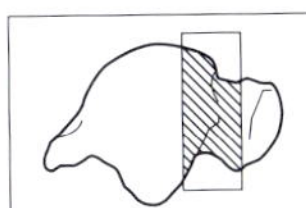

Subgroups and Qualifications:
Talus, extra-articular, neck (72-A1.1)
1. Simple (72-A1.1)

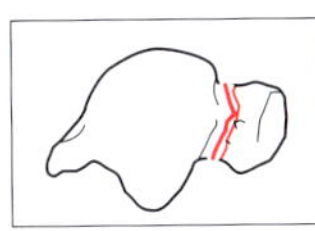

2. Multifragmentary (72-A1.2)

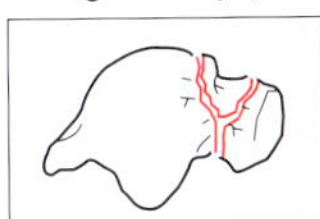

3. Associated with dislocation of talus (72-A1.3)
(1) body
(2) body and head

2. Talus, extra-articular, avulsions (72-A2)

1. Superior, from neck, ankle capsule (72-A2.1)

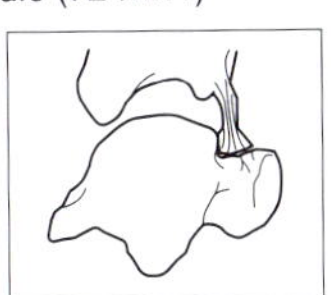

2. Lateral process (72-A2.2)

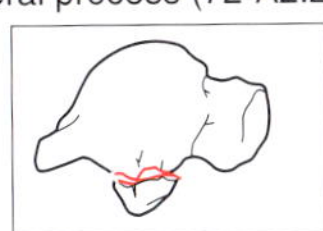

3. Posterior process (72-A2.3)

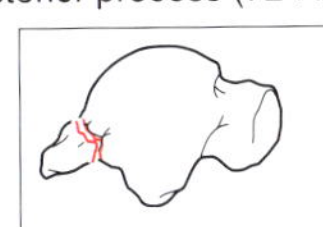

B. Talus, partial articular (72-B)

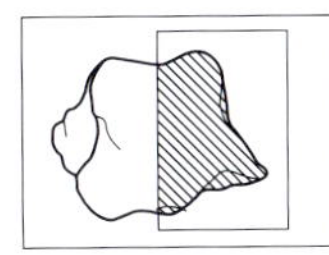

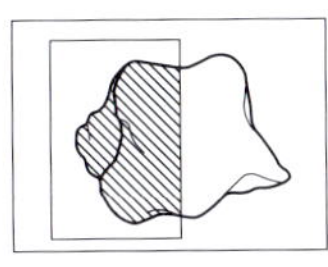

Talus, partial articular (72-B)
1. Talus, partial articular, lateral half body (72-B1)

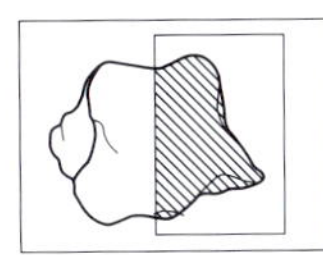

2. Talus, partial articular, medial half body (72-B2)

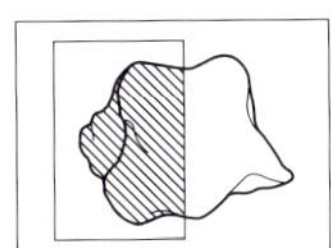

Talus, partial articular, lateral half body (72-B1)
1. Split (72-B1.1)

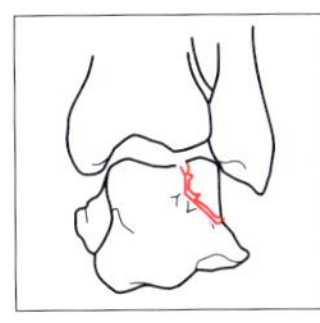

2. Depression (72-B1.2)

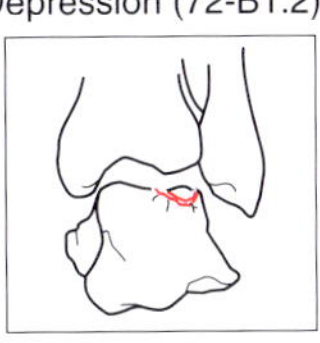

3. Split/depression (72-B1.3)

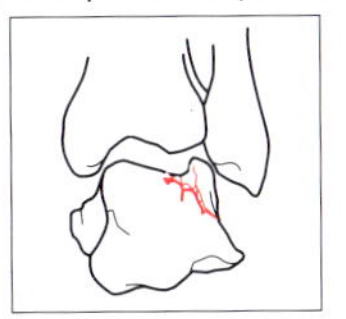

1. Split (72-B2.1)

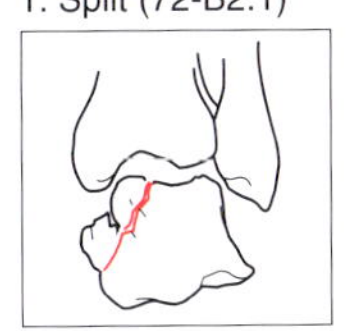

2. Depression (72-B2.2)

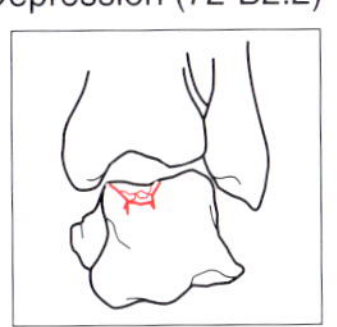

3. Split/depression (72-B2.3)

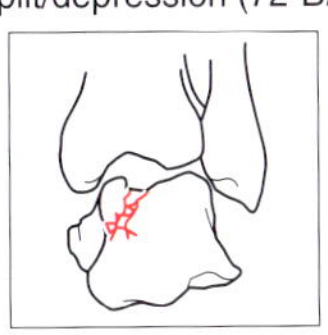

3. Talus, partial articular, coronal (72-B3)

1. Split (72-B3.1)

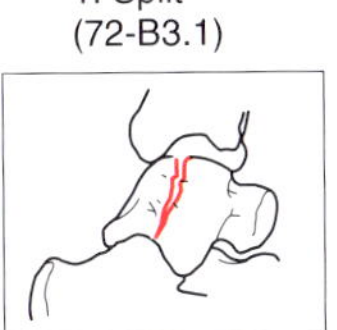

2. Depression (72-B3.2)

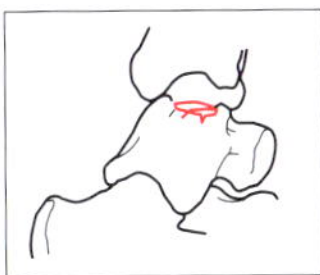

3. Split/depression (72-B3.3)

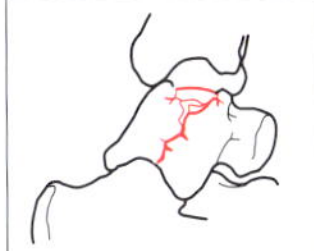

C. Talus, articular (72-C)

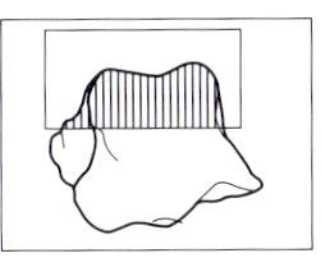

1. Talus, articular, simple (72-C1)

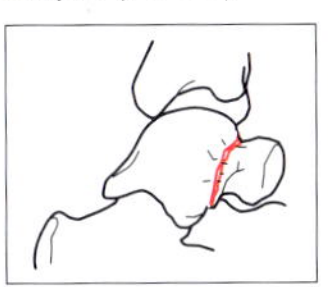

1. undisplaced (72-C1.1)
(2) displaced (72-C1.2)

2. Talus, articular, multifragmentary (72-C2)

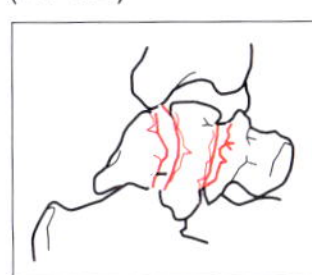

1. undisplaced (72-C2.1)
(2) displaced (72-C2.2)

Figure 20.25. (*Continued*)

NAVICULAR

Navicular, extra-articular avulsion (74-A1)
1. Anterior surface (74-A1.1)

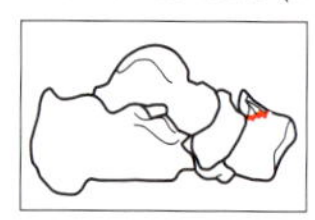

2. Medial surface (74-A1.2)

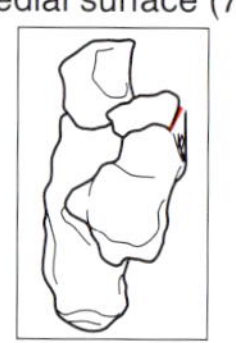

3. Miscellaneous (74-A1.3)

A2

Navicular, extra-articular, coronal body split (74-A2)

A3

Navicular, extra-articular, multifragmentary body (crush) (74-A3)

Navicular, partial-articular, sagittal lateral half (74-B1)
1. Split (74-B1.1)

2. Depression (74-B1.2)

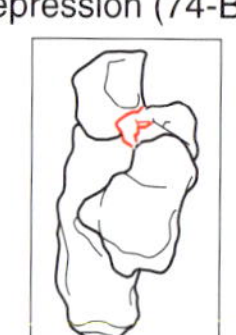

3. Split/depression (74-B1.3)

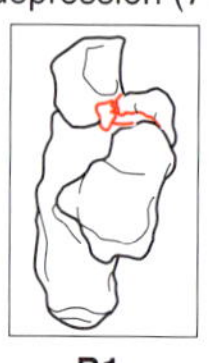

B1

Navicular, partial-articular, sagittal medial half (74-B2)
1. Split (74-B2.1)

2. Depression (74-B2.2)

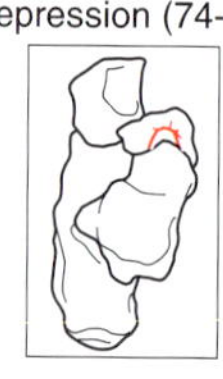

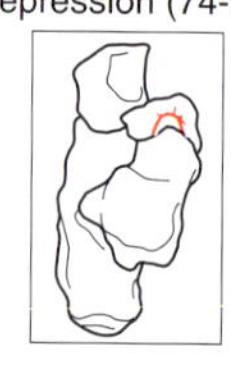

3. Split/depression (74-B2.3)

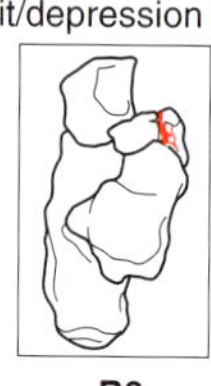

B3

Navicular, partial-articular, horizontal fracture (74-B3)
1. Split (74-B3.1)

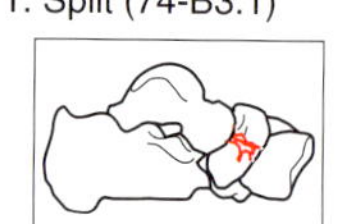

2. Depression (74-B3.2)

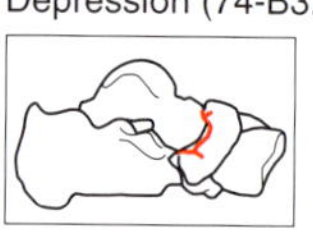

3. Split/depression (74-B3.3)

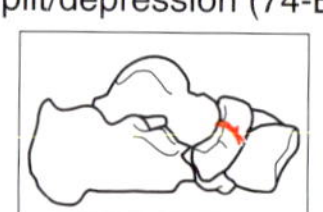

B3

Navicular, articular (both joints involved) (74-C)
1. Navicular, articular (both surfaces) multifragmentary (74-C1)
(1) undisplaced
(2) displaced

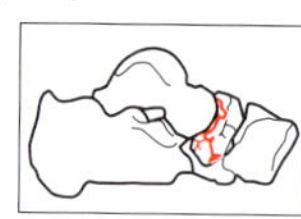

C1

CUBOID

Groups:
Cuboid, extra-articular (no joint involved)(76-A)
1. Cuboid, extra-articular, avulsion (76-A1)

Subgroups and Qualifications:
Cuboid, extra-articular, avulsion (76-A1)
1. Anterior (76-A1.1)

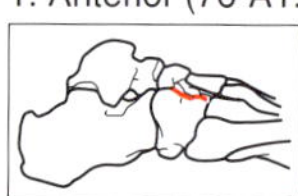

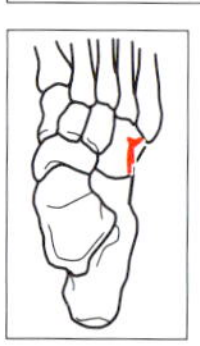

3. Plantar (76-A1.3)

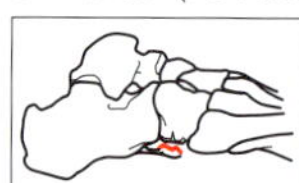

2. Cuboid, extra-articular, coronal (76-A2)

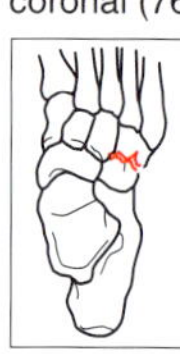

3. Cuboid, extra-articular, multifragmentary (crush)(76-A3)

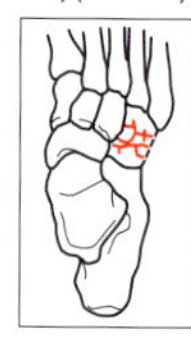

Cuboid, partial articular (one joint involved)(76-B)
1. Cuboid, partial articular, sagittal (76-B1)

1. Split (76-B1.1)

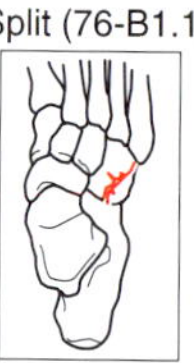

2. Depression (76-B1.2)

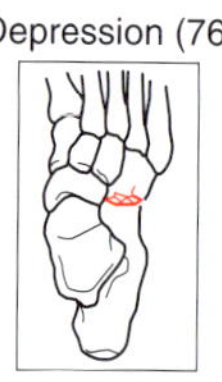

3. Split/depression (76-B1.3)

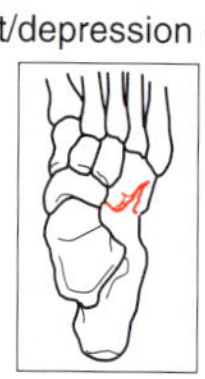

2. Cuboid, partial articular, horizontal (76-B2)

1. Split (76-B2.1)

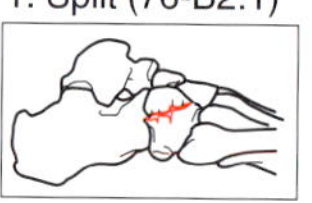

2. Depression (76-B2.2)

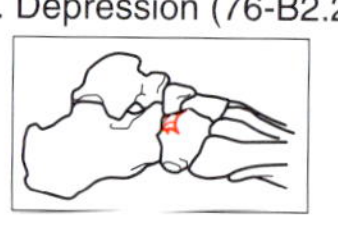

3. Split/depression (76-B2.3)

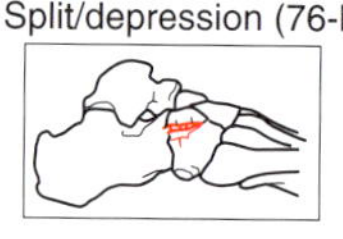

Cuboid, articular (both joints involved)(76-C1)
1. Cuboid, articular, multifragmentary (76-C1)
(1) undisplaced
(2) displaced

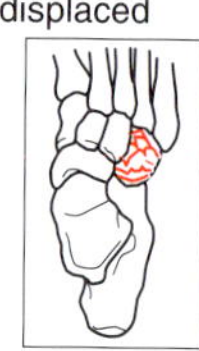

Figure 20.25. *(Continued)*

CUNEIFORM BONE

Groups:
Cuneiform, extra-articular (no joint involved)(75-A)

1. Cuneiform, extra-articular avulsion (75-A1)

1. Anterior (75-A1.1)

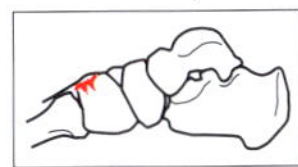

2. Medial (75-A1.2)

3. Plantar (75-A1.3)

2. Cuneiform, extra-articular coronal (75-A2)

1. Simple split (75-A2.1)

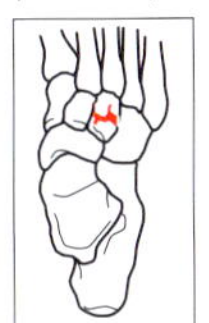

3. Cuneiform, extra-articular muitifragmentary (75-A3)

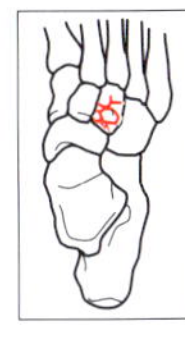

A

Cuneiform, partial articular (one joint involved)(75-B)

1. Cuneiform, partial articular, sagittal (75-B1)

1. Split (75-B1.1)

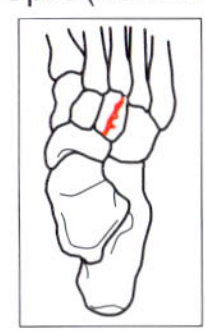

2. Depression (75-B1.2)

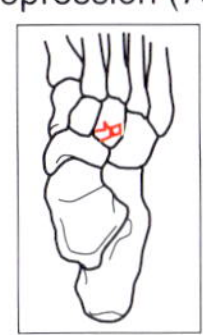

3. Split/depression (75-B1.3)

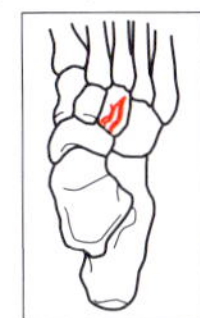

2. Cuneiform, partial articular, horizontal (75-B2)

1. Split (75-B2.1)

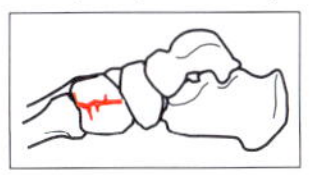

2. Depression (75-B2.2)

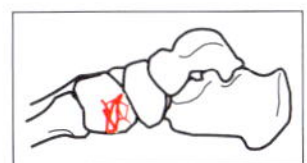

3. Split/depression (75-B2.3)

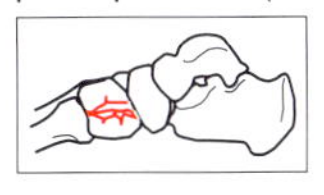

B

Cuneiform, articular (both joints involved) (75-C)

1. Cuneiform, articular, multifragmentary (75-C1)

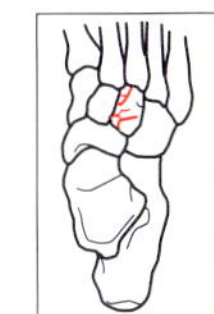

C

METATARSALS

Groups:

1. Metatarsals, proximal extra-articular (81-A1)

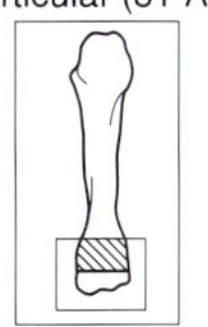

2. Metatarsal, diaphysis simple (81-A2)

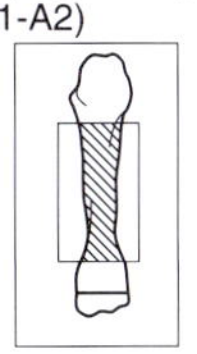

3. Metatarsal, distal extra-articular (82-A3)

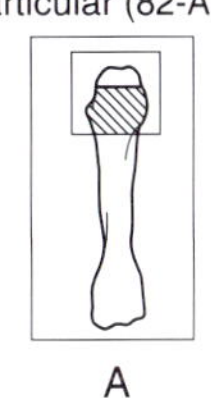

A

1. Metatarsals, proximal partial articular (81-B1)
(1) unicondyle medial
(2) unicondyle lateral

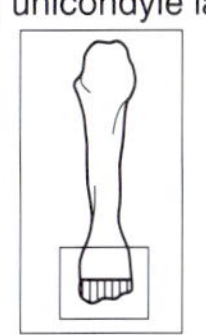

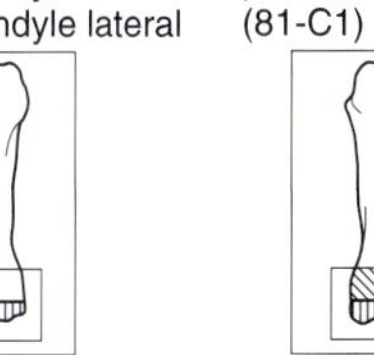

2. Metatarsal, diaphysis wedge (81-B2)

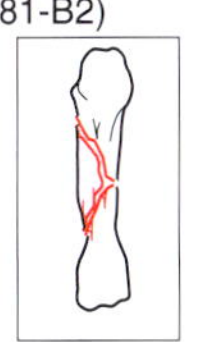

3. Metatarsal, distal partial articular (81-C3)
(1) unicondyle lateral
(2) unicondyle medial

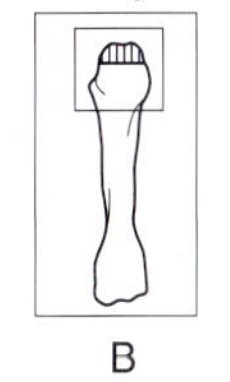

B

1. Metatarsals, proximal articular (81-C1)

2. Metatarsal, diaphysis complex (81-C2)

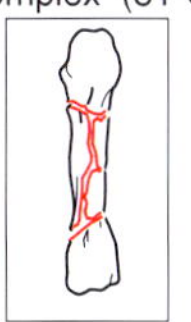

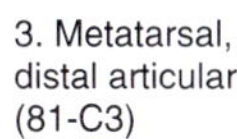

3. Metatarsal, distal articular (81-C3)

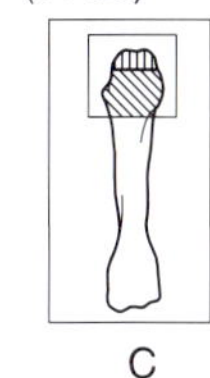

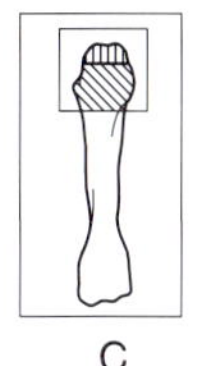

C

PHALANGES

Groups:

1. Phalanx, proximal extra-articular (82-A1)

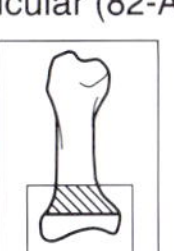

2. Phalanx, diaphysis simple (82-A2)

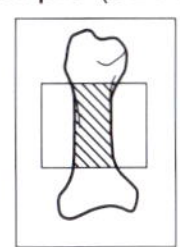

3. Phalanx, distal extra-articular (82-A3)

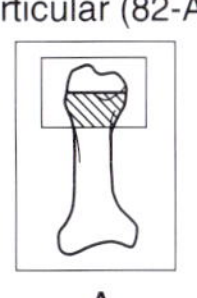

A

1. Phalanx, proximal partial articular (82-B1)
(1) split
(2) depression
(3) split depression

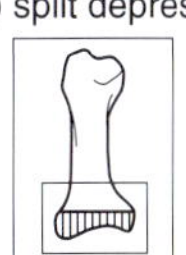

2. Phalanx, diaphysis wedge (82-B2)

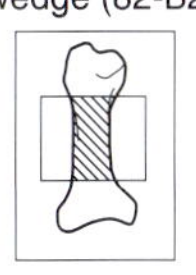

3. Phalanx, distal partial articular (82-B3)

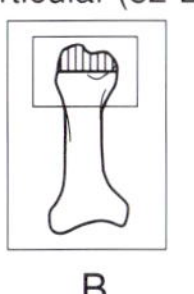

B

1. Phalanx, proximal articular (82-C1)
2. Phalanx, diaphysis complex (82-C2)

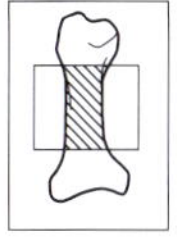

3. Phalanx, distal articular (82-C3)

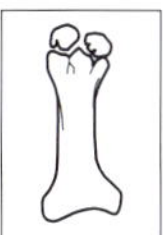

C

Figure 20.25. (*Continued*)

bones. This combination of the Comprehensive Long Bone Classification System with the classification of the other bones reordered into the alphanumeric system of the comprehensive classification forms the OTA classification (1996).

The same alpha-numerical coding system that was used for classifying long bone fractures was used again for classifying the remaining previously non-classified bones.

The OTA gave the bone location codes for the foot with (7) for the hind foot and midfoot and (8) for the forefoot. For each individual bone the coding system was as follows:

(72) for talus, (73) for calcaneus, (74) for navicular, (75) for cuneiform, (76) for cuboid. The Lisfranc joint was given the code number (80), metatarsals (81) and phalanges (82).

The triple coding (A, B, C) used for severity as well as the subgrouping system (1, 2, 3) was the same as the AO codes. The OTA classification of each bone in the foot is shown in Figure (20.25a–d) with the illustrations used for this purpose.

The classification includes codes for dislocations. They gave the letter (D) to express dislocation. Midtarsal joint is expressed by (77 D), in which they recognized two types. The Lisfranc joint is given the code (80-D) while metatarsophalangeal and interphalangeal are (81-D) joints.

Although this is an attempt to produce a standard classification system that can cover all fractures of the skeleton, the method of assessing some fractures into morphological types is not very clear. For example, in calcaneal fractures there is no recommendation as to which kind of imaging is to be used to differentiate between the different groups of this type of fracture.

It is still too early to judge if the OTA classification will thrive, particularly as there is little or no clinical evidence of its value in selecting treatment and predicting outcome.

Crush Injuries of the Foot

Despite the improvement in managing soft-tissue trauma in the foot (Myerson, 1989b; Papa and Myerson, 1991) and the increased attention being paid to diagnosing and treating compartment syndromes (Myerson, 1990; Myerson and Henderson, 1993), morbidity after crush injury remains high.

Reis and Michaelson (1986) recommended a standardized plan of treatment, incorporating principles of rigid skeletal internal fixation, early soft-tissue coverage and aggressive rehabilitation to ensure the best possible outcome. Hardin and Robinson (1954) tried to correlate the direction of force and the time of application of these forces to the pattern of injury produced. Myerson et al. (1994) classified crush injuries into three types according to the magnitude of trauma to soft-tissue and bone:

1. Compressive injuries: When a heavy object falls or comes into contact with the foot it may result in:
   a. Minor injury of subcutaneous tissue-skin only.
   b. Major injury with fractures, dislocations or fracture-dislocations of the bony skeleton of the foot, with or without open wounds.
2. Shear or degloving injuries: When a tangential force is applied to the surface of the foot, leading to avulsion of the soft tissues.
3. Mangling injuries: With marked disruption of bone and soft tissue.

A standardized protocol of treatment was used according to the type of injury. Although the overall outcome of these injuries is fair, following the protocol of treatment may lead to an improvement in results. No soft-tissue injury classification system seems to be helpful in predicting the outcome. A variety of inter-related factors are likely to be responsible for the morbidity associated with crush injuries of the foot. These include not only the magnitude and type of injury but also loss of viable soft-tissue envelope or neuroischaemia following contusion. The latter may occur because of either direct trauma to the peripheral nerves, or intraneural and extraneural fibrosis after oedema of the injured foot, resulting in chronic persistent pain.

References

Adelaar, R.C. (1989) Treatment of complex fractures of the talus. *Orthop. Clin. North Am.*, **20**, 691–707.

Anderson, L.D. (1977) Injuries of the forefoot. *Clin. Orthop.*, **122**, 18–27.

Barber, J.R., Briker, J.D. and Haliburton, R.A. (1961) Peritalar dislocation of the foot. *Can. J. Surg.*, **4**, 205–210.

Baumhauer, J.F. and Alvarez, R.G. (1995) Controversies in talar fractures. *Orthop. Clin. North Am.*, **26**, 335–351.

Berndt, A.L. and Harty, M. (1959) Transchondral fractures of the talus. *J. Bone Joint Surg.*, **41A**, 988–1020.

Bodamer, W.J., Torre, R.J., Cotch, M.T. and Goldman, F.D. (1987) Avascular necrosis of the head of the talus. A complication of group IV fractures of the talar neck. *J. Am. Podiatric Assoc.*, **77**, 217–222.

Böhler, L. (1936) *The Treatment of Fractures*, 4th edn. Bristol: John Wright, p. 466.

Bonnel, F. and Barthelemy, M. (1976) Traumatismes de

l'articulation de Lisfranc entroses graves, luxations, fractures: etude de 39 observations personnelle et classification biomecanique. *J. Chir. (Paris)*, **111**, 573–592.

Bonnin, J.G. (1950) *Injuries to the Ankle*. New York: Grune & Stratton.

Bowers, K.D. Jr and Martin, R.B. (1976) Turf-toe: A shoe surface related football injury. *Med. Sci. Sports*, **8**, 81–83.

Bradford, C.H. and Larsen, I. (1951) Sprain fractures of the anterior lip of the os calcis. *N. Engl. J. Med.*, **244**, 970.

Brunet, J.A. and Wiley, J.J. (1987) The late results of tarsometatarsal joint injuries. *J. Bone Joint Surg.*, **69B**, 437–440.

Burdeaux, B.D. (1983) Reduction of calcaneal fractures by the McReynolds medial approach. Technique and its experimental basis. *Clin. Orthop.*, **177**, 87–103.

Canale, S.T. (1990) Fractures of the neck of the talus. *Orthopedics*, **13**, 1105–1110.

Canale, S.T. and Belding, R.H. (1980) Osteochondral lesion of the talus. *J. Bone Joint Surg*, **62A**, 97–102.

Canale, S.T. and Kelly, F.B. Jr (1978) Fractures of the neck of the talus. Long term evaluation of 71 cases. *J. Bone Joint Surg.*, **60A**, 143–156.

Carey, E.J., Lance, E.M. and Wade, P.A. (1965) Extra-articular fracture of the os calcis. *J. Trauma*, **5**, 362–372.

Cedell, C.A. (1974) Rupture of the posterior talotibial ligament with the avulsion of bone fragment from the talus. *Acta Orthop. Scand.*, **45**, 454–461.

Chan, S. and Ip, F.K. (1995) Open reduction and internal fixation for displaced intra-articular fractures of the os calcis. *Injury*, **26**, 111–115.

Chapman, H.G. (1977) Os calcis fracture in childhood. *J. Bone Joint Surg.*, **59B**, 510.

Clapper, M.F., O'Brien, T.J. and Lyons, P.M. (1995) Fractures of the fifth metatarsal. Analysis of a fracture registry. *Clin. Orthop.*, **315**, 238–241.

Coltart, W.D. (1952) Aviator astragalus. *J. Bone Joint Surg.*, **34B**, 545.

Comfort, T.H., Behrens, F., Gaither, D.W. et al. (1985) Long term results of displaced talar neck fractures. *Clin. Orthop.*, **199**, 81–87.

Conwell, H.E. and Reynolds, F.C. (1961) Metatarsal fractures. In *The Management of Fracture Dislocations and Sprains* (J.A. Key and H.E. Conwell, eds), 7th edn. St Louis: Mosby, p. 1068.

Copeland, C.L. and Kanat, I.O. (1991) A new classification for traumatic dislocation of the first metatarsophalangeal joint. Type IIC. *J. Foot Surg.*, **30**, 234–237.

Corbett, M., Levy, A., Abramowitz, A.J. and Whitelaw, G.P. (1995) A computer tomographic classification system for the displaced intraarticular fracture of the os calcis. *Orthopaedics*, **18**, 705–710.

Coughlin, M.J. (1989) Subluxation and dislocation of the second metatarsophalangeal joint. *Orthop. Clin. North Am.*, **20**, 535–551.

Crosby, A.L. and Fitzgibbons, S.T. (1990) Computerised tomographic scanning of acute intra-articular fractures of the calcaneus. *J. Bone Joint Surg.*, **72A**, 852–859.

Curtis, M.J., Myerson, M. and Szura, B. (1993) Tarsometatarsal joint injuries in the athletes. *Am. J. Sports Med.*, **21**, 497–502.

Dameron, T.B. Jr (1975) Fractures and anatomical variations of the proximal portion of the fifth metatarsal. *J. Bone Joint Surg*, **57A**, 788–792.

Daniels, T.R. and Smith, J.W. (1993) Talar neck fractures. Foot Fellow Review. *Foot and Ankle*, **14**, 225–234.

Davis, C.A., Lubowitz, Z.J. and Thordarson, D.B. (1993) Midtarsal fracture subluxation. Case report and review of the literature. *Clin. Orthop.*, **292**, 264–268.

Degan, T.J., Morrey, B.F. and Braun, D.P. (1982) Surgical excision of the anterior process fracture of the calcaneus. *J. Bone Joint Surg.*, **64B**, 519–524.

Delee, J.C. (1986) Fractures and dislocations of the foot. In *Surgery of the Foot* (R.A. Mann, ed.), 5th edn. St Louis: Mosby, pp. 656–715.

Delee, J.C. and Curtis, R. (1982) Subtalar dislocation of the foot. *J. Bone Joint Surg.*, **64A**, 433–437.

Detenbeck, L.C. and Kelly, P.J. (1969) Total dislocation of the talus. *J. Bone Joint Surg.*, **51A**, 283–288.

Dewar, F.P. and Evans, D.C. (1968) Occult fracture subluxation of the midtarsal joint. *J. Bone Joint Surg.*, **50B**, 386–388.

Dimon, J.H. (1961) Isolated displaced fracture of the posterior process of the talus. A case report. *J. Bone Joint Surg.*, **43A**, 275–281.

Drvaric, D.M. and Schmitt, E.W. (1988) Irreducible fracture of the calcaneus in a child. Case report. *J. Orthop. Trauma*, **2**, 154–157.

Eastwood, D.M., Gregg, P.J. and Atkins, R.M. (1993) Intra-articular fractures of the calcaneum. Part 1 Pathological anatomy and classification. *J. Bone Joint Surg.*, **75B**, 183–188.

Ebraheim, N.A., Skie, M.C. and Podeszwa, D.A. (1994) Medial subtalar dislocation associated with fracture of the posterior process of the talus. A case report. *Clin. Orthop.*, **303**, 226–230.

Eichenholtz, S.N. and Levine, D.B. (1964) Fracture of the tarsal navicular bone. *Clin. Orthop.*, **34**, 142–157.

English, T.A. (1964) Dislocation of the metatarsal bone and adjacent toe. *J. Bone Joint Surg.*, **46B**, 700–704.

Essex-Lopresti, P. (1952) Mechanism, reduction techniques and results in fractures of the os calcis. *Br. J. Surg.*, **39**, 395–419.

Faciszewski, T., Burks, R.T. and Manaster, B.J. (1990) Subtle injuries of Lisfranc joint. *J. Bone Joint Surg.*, **72A**, 1519–1522.

Fehnel, D.J., Baumhauer, J.F., Huber, B. and Peterson, L.F. (1994) Fracture of the lateral processs of the talus. A functional outcome study. Presented Foot and Ankle summer meeting. Quoted by Baumhauer and Alvarez (1995).

Francesconi, F. (1925) Sopra un caso di lussazione di Lis Franc. *Chir. Organi Mov.*, **9**, 589–604.

Frawley, P.A., Hart, J.A. and Young, D.A. (1995) Treatment outcome of major fractures of the talus. *Foot and Ankle*, **16**, 339–345.

Freund, G.K. (1988) Complicated fractures of the neck of the talus. *Foot and Ankle*, **8**, 203–207.

Grantham, S.A. (1964) Medial subtalar dislocation. Five cases with a common aetiology. *J. Trauma*, **4**, 845–849.

Grob, D., Simpson, L.A., Weber, B.G. and Bray, T. (1985) Operative treatment of displaced talus fractures. *Clin. Orthop.*, **199**, 88–96.

Hamilton, A.R. (1949) An unusual dislocation. *Med. J. Aust.*, **1**, 271.

Hardaker, W.T. Jr (1989) Foot and ankle injuries in classical ballet dancers. *Orthop. Clin. North Am.*, **20**, 621–627.

Hardcastle, P.H., Reschauer, R., Kutscha-Lissber, G.E. and Schoffmann, W. (1982) Injuries to the tarsometatarsal joint. Incidence, classification and treatment. *J. Bone Joint Surg.*, **64B**, 349–356.

Hardin, C.A. and Robinson, D.W. (1954) Coverage problems in the treatment of wringer injuries. *J. Bone Joint Surg.*, **36A**, 292–298.

Hawkins, L.G. (1965) Fractures of the lateral process of the talus. A review of thirteen cases. *J. Bone Joint Surg.*, **47A**, 1170–1175.

Hawkins, L.G. (1970) Fractures of the neck of the talus. *J. Bone Joint Surg.*, **52A**, 991–1002.

Heckman, J.D. and Mclean, M.R. (1985) Fractures of the lateral process of the talus. *Clin. Orthop.*, **199**, 108–113.

Heckman, J.D., Rockwood, C.A. and Green, D.P. (1991) Fractures and dislocations of the foot. In *Fractures in Adults*, Vol. II, 3rd edn. Philadelphia: Lippincott, pp. 2151–2169.

Hermel, M.B. and Gershon-Cohen, J. (1953) Nutcracker fracture of the cuboid by indirect violence. *Radiol.*, **60**, 850–854.

Howie, C.R., Hooper, G. and Hughes, P.F. (1986) Occult midtarsal subluxation. *Clin. Orthop.*, **209**, 206–209.

Hutchinson, F. and Huebner, M.K. (1994) Treatment of os calcis fractures by open reduction and internal fixation. *Foot and Ankle*, **15**, 225–232.

Jahss, M.H. (1980) Traumatic dislocation of the first metatarsophalangeal joint. *Foot and Ankle*, **1**, 15–21.

Jeffreys, T.E. (1963) Lisfranc's fracture dislocation; a clinical and experimental study of tarsometatarsal dislocation and fracture dislocation. *J. Bone Joint Surg.*, **45B**, 546–551.

Johnson, F.G. (1981) Paediatric Lisfranc injury: bunk-bed fracture. *Am. J. Roentgenol.*, **137**, 1041–1044.

Jones, R. (1902) Fracture of the base of the fifth metatarsal bone by indirect violence. *Ann. Surg.*, **35**, 697–700.

Josefsson, P.O., Karlsson, M., Johnell, I.R. and Wendeberg, B. (1994a) Closed treatment of Jones fracture. Good results in 40 cases after 11–26 years. *Acta Orthop. Scand.*, **65**, 545–547.

Josefsson, P.O., Karlsson, M., Johnell, I.R. and Wendeberg, B. (1994b) Jones fracture. Surgical versus non surgical treatment. *Clin. Orthop.*, **299**, 252–255.

Kaplan, E.B. (1957) Dorsal dislocation of the metacarpophalangeal joint of the index finger. *J. Bone Joint Surg.*, **39A**, 1081–1086.

Kappis, M. (1922) Weilere zur tramatisch mechanischen entestebung der spontanen knorpe labloosungen (sogen. osteochondritis, dissecans). *Dtsch Z. Chir.*, **171**, 13.

Katayama, M., Murakami, Y. and Takahashi, H. (1988) Irreducible dorsal dislocation of the toe. Report of three cases. *J. Bone Joint Surg.*, **70A**, 769–770.

Kavanaugh, J.H., Brower, T.D. and Mann, R.V. (1978) The Jones fracture revisited. *J. Bone Joint Surg.*, **60A**, 776–782.

Kenwright, J. and Taylor, R.G. (1970) Major injuries of the talus. *J. Bone Joint Surg.*, **52B**, 36–48.

Khazim, R. and Salo, P.T. (1995) Talar neck fractures with talar head dislocation and intact ankle and subtalar joint. A case report. *Foot and Ankle*, **16**, 44–48.

Lehner, M. (1968) *Talusfraktur und Talusnekrose*. Inauguraldissertation, Medizinische Fakultät Universität Bern. Luzern, Maihof.

Leitner, B. (1955) The mechanism of total dislocation of the talus. *J. Bone Joint Surg.*, **37A**, 89–95.

Leung, K.S., Chan, W.S., Shen, Wy. et al. (1989) Operative treatment of intra-articular fractures of the os calcis. The role of rigid internal fixation and primary grafting. Preliminary results. *J. Orthop. Trauma*, **3**, 232–240.

Lindsay, R.N. and Dewar, F.P. (1958) Fractures of the os calcis. *Am. J. Surg.*, **95**, 555.

Lowrie, I.G., Finlay, D.B., Brenkel, I.J. and Gregg, P.J. (1988) Computed tomographic assessment of the subtalar joint in calcaneal fractures. *J. Bone Joint Surg.*, **70B**, 247–250.

Main, B.J. and Jowett, R.L. (1975) Injuries of the midtarsal joint. *J. Bone Joint Surg.*, **57B**, 89.

Marti, R. (1978) Talus und calcaneusfrakturen. In *Die Frakturenbahandlung bei Kindern und Jugendlichen* (B.G. Weber, C. Brunner and F. Freuler, eds). Berlin: Springer-Verlag.

Mata, S.G., Ovejero, A.H. and Grande, M.M. (1995) Plantar dislocation of the first metatarsophalangeal joint during lactation. *Int. Orthop.*, **19**, 65–66.

Melcher, G., Degonda, F., Leulenegger, A. and Ruedi, T. (1995) Ten years follow up after operative treatment for intra-articular fractures of the calcaneus. *J. Trauma*, **38**, 713–716.

Miki, T., Tamamuro, T. and Kitai, T. (1988) An irreducible dislocation of the great toe. Report of two cases and review of the literature. *Clin. Orthop.*, **230**, 200–206.

Mindell, E.B., Cisek, E.E., Kartalian, G. and Dziab, J.M. (1963) Late results of injuries to the talus. *J. Bone Joint Surg.*, **45A**, 221.

Monsey, R.D., Levine, B.P. B.A., Trevino, S.G. and Kristiansen, T.K. (1995) Operative treatment of acute displaced intraarticular calcaneal fractures. *Foot and Ankle*, **16**, 57–63.

Mukherjee, S.K., Prigle, R.M. and Baxter, A.D. (1974) Fractures of the lateral process of the talus: report of thirteen cases. *J. Bone Joint Surg.*, **56B**, 263–267.

Müller, M.E., Nazarian, S., Koch, P. and Schatzker, J. (1990) *The Comprehensive Classification of Fractures of Long Bones*. New York: Springer-Verlag.

Myerson, M.S. (1989a) The diagnosis and treatment of

injuries to the Lisfranc joint complex. *Orthop. Clin. North Am.*, **20**, 655–664.

Myerson, M.S. (1989b) Split thickness skin excision: Its use for immediate wound care in crush injuries of the foot. *Foot and Ankle*, **10**, 54–60.

Myerson, M.S. (1990) Diagnosis and treatment of compartment syndrome of the foot. *Orthopaedics*, **13**, 711–717.

Myerson, M.S., Fisher, R.T., Burgess, A.R. and Kenzora, I.E. (1986) Fracture dislocation of the tarsometatarsal joints: End results correlated with pathology and treatment. *Foot and Ankle*, **6**, 225–242.

Myerson, M.S. and Henderson, M.R. (1993) Clinical applications of a pneumatic intermittent impulse compression device after trauma and major surgery to the foot and ankle. *Foot and Ankle*, **14**, 198–203.

Myerson, M.S., McGarvey, W.C., Henderson, M.R. and Hakim, J. (1994) Morbidity after crush injuries to the foot. *J. Orthop. Trauma*, **8**, 343–349.

Nabarro, M.N. and Powell, J. (1995) Dorsal dislocation of the metatarsophalangeal joint of the great toe: a case report. *Foot and Ankle*, **16**, 75–78.

Nade, S. and Monahan, P.R.W. (1973) Fractures of the calcaneum: a study of the long term prognosis. *Injury*, **4**, 201–207.

Ogden, J.A. (1990) *Skeletal Injuries in the Child*, 2nd edn. Philadelphia: Saunders, pp. 876–883.

OTA (1996) The Orthopaedic Trauma Association Committee for coding and classification (1996): Fractures and dislocation compendium. *J. Orthop. Trauma*; **10** (Suppl. 1), 98–120.

Paley, D. and Hall, H. (1989) Calcaneal fractures controversies. Can we put humpty dumpty together again? *Orthop. Clin. North Am.*, **20**, 665–677.

Paley, D. and Hall, H. (1993) Intra-articular fractures of the calcaneus. A critical analysis of results and prognostic factors. *J. Bone Joint Surg.*, **75A**, 342–353.

Palmer, I. (1948) The mechanism and treatment of fractures of the calcaneus. Open reduction with the use of cancellus graft. *J. Bone Joint Surg.*, **30A**, 2–8.

Papa, J. and Myerson, M.S. (1991) Soft tissue coverage in the management of foot and ankle trauma. Part 1. *Contemp. Orthop.*, **22**, 509–520.

Parmer, H.V., Triffit, P.D. and Cregg, P.J. (1993) Os calcis fractures. *Curr. Orthop.*, **7**, 184–192.

Penny, T.N. and Davis, L.A. (1980) Fractures and fracture dislocations of the neck of the talus. *J. Trauma*, **20**, 1029–1037.

Peterson, L., Goldie, I.F. and Irstam, L. (1977) Fractures of the neck of the talus. A clinical study. *Acta Orthop. Scand.*, **48**, 696–706.

Pinzur, M.S. and Meyer, P.R. Jr (1978) Complete posterior dislocation of the talus. Case report and discussion. *Orthop*, **131**, 205–209.

Protheroe, K. (1969) Avulsion fractures of the calcaneus. *J. Bone Joint Surg.*, **51B**, 118–122.

Quénu, E. and Küss, G. (1909) Etude sur les luxations du metatarse (luxations metatarsotarsiennes) du diastasis entre le ler et le 2e metatarsien. *Rev. Chir.*, **39**, 281–336, 720–791, 1039–1134.

Reis, N.D. and Michaelson, M. (1986) Crush injury to the lower limbs. Treatment of the local injury. *J. Bone Joint Surg.*, **68A**, 414–418.

Rosenberg, G.A. and Patterson, B.M. (1995) Tarsometatarsal (Lisfranc) fracture dislocation. *Am. J. Orthop.*, **24**(2) Suppl. 7–16.

Ross, S.D.K. and Sowerby, M.R. (1985) The operative treatment of fractures of the os calcis. *Clin. Orthop.*, **199**, 132–143.

Rowe, C.R., Sakellarides, H.T., Freeman, P.A. and Sorbie, C. (1963) Fractures of the os calcis: A long term follow-up study of one hundred and forty-six patients. *JAMA*, **184**, 920–923.

Rymaszewski, L.A. and Robb, J.E. (1988) Mechanism of fracture dislocation of the navicular. Brief report. *J. Bone Joint Surg.*, **70B**, 492.

Sammarco, G.J. and Carrasquillo, H.A. (1995) Intramedullary fixation of metatarsal fractures and non union. Two methods of treatment. *Orthop. Clin. North Am.*, **26**, 265–272.

Sammarco, G.J. and Miller, E.H. (1982) Forefoot conditions in dancers, II. *Foot and Ankle*, **3**, 93–98.

Sanders, R., Fortin, P., Di Pasquale, T. and Walling, A. (1993) Operative treatment in 120 displaced intra-articular calcaneal fractures. Results using a prognostic computed tomography scan classification. *Clin. Orthop.*, **290**, 87–95.

Sanders, R. and Gregory, P. (1995) Operative treatment of intra-articular fractures of the calcaneus. *Orthop. Clin. North Am.*, **26**, 203–214.

Sangeorzan, B.J., Benirschke, S.K., Mosca, V. et al. (1989) Displaced intra-articular fracture of the tarsal navicular. *J. Bone Joint Surg.*, **71A**, 1504–1510.

Schatzker, J. and Tile, M. (1993) The management of fractures and dislocations of the talus. In *Major Fractures of the Pilon, the Talus and the Calcaneus: Current Concept of Treatment* (H. Tscherne and J. Schatzker, eds). Berlin, Heidelberg: Springer-Verlag, pp. 87–104.

Schmidt, T.L. and Weiner, D.S. (1982) Calcaneal fractures in children. An evaluation of the nature of the injury in 56 children. *Clin. Orthop.*, **171**, 150–155.

Schofield, R.O. (1936) Fractures of the os calcis. *J. Bone Joint Surg.*, **18A**, 566–580.

Schrock, R.D. (1952) Fractures of the foot: fractures and dislocation of the astragalus. *Inst. Course Lect.*, **9**, 361–365 (quoted from Baumhauer and Alvarez, 1995).

Segal, D., Marsh, J.L. and Leiter, B. (1985) Clinical application of computerised axial tomography scanning of calcaneal fractures. *Clin. Orthop.*, **199**, 114–123.

Shantz, K. and Rasmussen, F. (1988) Good prognosis after calcaneal fractures in childhood. *Acta Orthop. Scand.*, **59**, 560–563.

Shelton, M.L. and Pedowitz, W.J. (1982) Injuries to the talus and mid foot. In *Disorders of the Foot* (M.H. Jahss, ed.), Vol. 2. Philadelphia: Saunders.

Shepard, F.J. (1882) A hitherto undescribed fracture of astragalus. *J. Anat. Physiol.*, **18**, 79–81.

Sneppen, O., Christensen, S.B., Krosoe, O. and Lorentzen, J. (1977) Fractures of the body of the talus. *Acta. Orthop. Scand.*, **48**, 317–324.

Soeur, R. and Remy, R. (1975) Fractures of the calcaneus with displacement of the thalamic portion. *J. Bone Joint Surg.*, **57B**, 413–421.

Stark, W.A. (1973) Occult fracture subluxation of the mid-tarsal joint. *Clin. Orthop.*, **93**, 291–292.

Stephenson, J.R. (1987) Treatment of displaced intraarticular fractures of the calcaneus using medial and lateral approaches, internal fixation and early motion. *J. Bone Joint Surg.*, **69A**, 115–130.

Stephenson, K.A., Beck, T.L. and Richardson, E.G. (1994) Plantar dislocation of the metatarsophalangeal joints: Case report. *Foot and Ankle*, **15**, 446–449.

Stewart, I.M. (1960) Jones fracture of the base of fifth metatarsal bone. *Clin. Orthop.*, **16**, 190–198.

Szyszkowitz, R., Reschauer, R. and Seggl, W. (1985) Eighty-five talus fractures treated by open reduction and internal fixation with five to eight years follow up study of sixty-nine patients. *Clin. Orthop.*, **199**, 88–96.

Thoren, O. (1964) Os calcis fracture. *Acta Orthop. Scand.*, **70** Suppl. 1–116.

Torg, J.S., Pavlov, H., Cooley, L.H. et al. (1982) Stress fracture of the tarsal navicular. A retrospective review of 21 cases. *J. Bone Joint Surg.*, **64A**, 700.

Torg, J.S., Balduini, F.C., Zelko, R.P. et al. (1984) Fracture of the base of the fifth metatarsal distal to the tuberosity. *J. Bone Joint Surg.*, **66A**, 209.

Tscherne, H. (1984) *Fractures with Soft Tissue Injuries* (H. Tscherne and L. Gotzen, eds). Berlin: Springer-Verlag.

Viswanath, S.S. and Shephard, E. (1977) Dislocation of the os calcis. *Injury*, **9**, 50–52.

Vuori, J.P. and Aro, H.T. (1993) Lisfranc joint injuries: trauma mechanisms and associated injuries. *J. Trauma*, **35**, 40–45.

Walling, A.K., Grogan, D.P., Carty, C.T. and Ogden, J.A. (1990) Fractures of the calcaneal apophysis. *J. Orthop. Trauma*, **4**, 349–355.

Warrick, C.K. and Bremner, A.E. (1953) Fractures of the calcaneum. *J. Bone Joint Surg.*, **35B**, 33–49.

Watson-Jones, R. (1955) *Fractures and Joint Injuries*, 4th edn, Vol. 2. Edinburgh and London: Livingstone, p. 867.

Weber, B.G. (1974) Knöchel, fusswerzel und mittelfuss. In *Chirurgie der Gegenwart*, Bd. 4a: Unfallchirurgie. München: Urban und Schwarzenberg.

Willpula, E. (1967) Metatarsal fractures and Lisfranc joint dislocation. Thesis, University of Helsinki.

Wilson, D.W. (1972) Injuries of the tarso-metatarsal joints. *J. Bone Joint Surg.*, **54B**, 677–686.

Zwipp, H., Tscherne, H., Thermann, H. and Weber, T. (1993) Osteosynthesis of displaced intra-articular fractures of the calcaneus. Results in 123 cases. *Clin. Orthop.*, **290**, 76–86.

Index